Security
Second Edition

Security
Second Edition

A Guide to Security System Design and Equipment Selection and Installation

EUR ING Neil Cumming, C. Eng.
Dodd Cumming and Love Chartered Consulting Engineers
Plymouth, Devon, England

Butterworth–Heinemann
An Imprint of Elsevier

Boston Oxford London Singapore Sydney Toronto Wellington

Library of Congress Cataloging-in-Publication Data

Cumming, Neil
   Security : a guide to security system design,
equipment selection, and installation / Neil Cumming.
— 2nd ed.
         p.   cm.
   Includes bibliographical references and index.
   ISBN-13: 978-0-7506-9624-1   ISBN-10: 0-7506-9624-9
   1.   Electronic security systems — Equipment and
supplies. I. Title.
TH9737.C83   1992
621.389'28 — dc20   91–37966
                                                  CIP

British Library Cataloguing-in-Publication Data

   Cumming, Neil
      Security, -2nd ed
      I. Title
   690
   ISBN-13: 978-0-7506-9624-1
   ISBN-10: 0-7506-9624-9

Butterworth-Heinemann
An Imprint of Elsevier
313 Washington Street
Newton, MA 02158-1626

Transferred to Digital Printing 2011

Dedication

To my wife, Catherine, who is still my wife despite this book; to my children, Thomas, Jennifer, and Peter, who spent the winter in the garden while I wrote it; and to Claire Kemp, who translated my handwriting into her native French and then typed it into perfect English!

Contents

Appendices

Acknowledgments

I am indebted to the following companies for the advice, assistance, and photographs provided to me in the preparation of this book. Thank you!

Abloy Security Group Ltd., 7–6
Adams Rite (Europe) Ltd., 6–10, 6–11, 6–14, 6–23, 6–24
Ademco, 2–8, 2–13, 2–14, 4–1
American Society for Industrial Security (ASIS)
ASSA Locks (UK), 6–9, 6–12, 6–13, 6–25
Burle Industries, Inc., 5–8, 5–10
C & K Systems, Inc., 4–65
Casi-Rusco, 7–4
Chubb Alarms, Ltd., Great Britain, 1–2, 2–9, 2–11, 2–12, 3–1, 3–3, 4–7, 4–8, 4–13, 4–55, 4–64, 4–66, 5–3, 5–6, 5–10, 7–3, 7–7, 7–8
Chubb Electronics, North America
Corkey
D/A Technology
Dedicated Micros, (GB), 5–17
Diebold, Inc.
ECCO Industries
Eyedentify, Inc., 7–2
Focus CCTV, (GB)
Galaxy Control Systems, 7–9
Indala Corporation, 7–1
Math Associates, Inc., 2–25, 2–26
Medeco
Microwave Sensors, Inc., 4–60, 4–61
Morse Security Group, Inc., 2–10
Mosler, Inc., 5–14, 5–15, 5–16
National Institute of Technology and Standards
Newmark Technology, (GB) WEY
NSCIA, (GB), 1–1
Phase Security
Phillips Lighting, Ltd., 8–1, 8–2, 8–5, 8–7, 8–19, 8–20
Pulnix America, Inc., 3–13 through 3–16, 4–33, 4–34, 4–36, 4–37
Racal-Guardall, Inc., 3–12, UK Ltd., 3–9, 4–14, 4–15, 4–25, 4–30, 4–39
Radionics, 7–10
Robot UK Ltd., 5–12, 5–13
Seaboard Electronics, 2–27
Security Door Controls
Security Industry Association
Sentrol, 4–3, 4–4, 4–14, 4–15
Siedle, Inc., 7–5
Stellar Systems, 3–4 through 3–6, 3–26
Thorn Lighting (GB) Ltd.
Underwriters Laboratories, 1–3
Weyrad Electronics (GB) Ltd. (Viper), 4–22 through 4–24
Yale Ltd., 6–19

Preface

During the preparation of this book, I was very fortunate to have been introduced to James W. Owen, Jr., and I would like to extend a very special "thank you" to him. Jim has been my consulting editor and his guidance and advice were invaluable. Jim was involved in many of the developments and applications of devices covered in this book. If they didn't work, wouldn't work, or could be defeated easily, he told me and I am grateful for his expertise. Thank you, Jim.

I wrote the first edition of this book on the design selection and installation of security systems because I needed to know more than salesmen could tell me and couldn't find a suitable source of information.

Early in my career in security systems consultancy I was enlisted to help select the most appropriate video door entry system for 20 14-story apartment blocks in London. The area was what is now called an *urban ghetto* and the systems were to be used to give control of the circulation, space, escalators, etc., back to the tenants of the blocks and to deny access to local hoodlums. The system had to work well under a pretty rough environment, so I was anxious to ensure the product was up to the job.

I interviewed ten company representatives and asked each to give a simple description of their product, how it worked, how it could be made not to work, etc. I might as well have talked to little green men from Mars! All the "salesmen" but one knew virtually nothing technical about their products, had never spent time actually installing their products, and hadn't considered the possible products' failings. The one salesman who *did* know his company's product won the contract *and* his product was the best. It has proved itself robust and still works well.

The other "salesmen" had successfully duped thousands of unsuspecting, ill-prepared (or uninterested) architects, engineers, and building managers for years. Their sales "technique" was to take up the entire presentation by describing the system features and either avoid direct pertinent questions or counter them with jargon and gobbledygook. This tactic often worked, because interviewers were too embarrassed at their lack of knowledge of this "language" to request that terms be explained in simple English. The interviewers' attitudes seemed to be, "If I cannot understand their language or terms, they *must* know what they are talking about."

As an engineer, this lack of very basic knowledge on both sides of the buy/sell setup was extremely annoying, so I took the obvious step of looking for a book, which I could give to my architect, engineer, and building manager friends, that they could use as a subject primer and reference source. The book simply needed to explain clearly what the risks were, what the systems or sensors could do to help, how to choose a system, select a company, and make sure the system was installed properly.

To my horror, I found that there weren't any books that covered all the major elements of security hardware in an easy-to-read, semitechnical manner. There were books that covered closed circuit television (CCTV) in great technical detail and "alarms" in two pages, but there was nothing comprehensive for the busy layperson who invariably has to decide what type of system should be installed.

My colleagues and friends got fed up with waiting for me to find them this book. That was 10 years ago. Now, here is the second edition. I hope you find it valuable. My friends have, but they they would or they would no longer be my friends!

This book aims to

provide much of the essential information needed for the intelligent, careful design, selection, purchase, installation, commissioning, and maintenance of common security systems

demystify the "black art" and "black box" approach to security systems and their components by explaining in plain English (without the sales pitch) what these systems do and more importantly what they don't do

arm the potential designer, purchaser, and system user with enough product knowledge to be able to quickly identify available options and to ask sensible questions of those persons trying to sell systems or services

raise the level of awareness of the thought, time, and

effort that should be put into security system specification, design, and purchase; these systems should be placed on a par with fire safety systems in this respect

reduce the numbers of poorly specified, designed, and installed systems by showing all parties involved the points that should be considered

Security system purchasers and installers are often amazed by the number of false alarms, disputes, etc., that exist within the security industry. I am not. I hope that after reading this book, you will agree that the information you gained was worth the time spent reading it—and was worth the purchase price! If this is so, I will be happier, and you and your company will be safer.

To this end, great emphasis is placed on basic presentation of information. These presentations take the form of tabular data, diagrams, lists of dos and don'ts, illustrations, etc. I will make no apologies for the absence of pages of flowing prose, as in my experience, these pages are seldom read by those individuals requiring *answers*. This lack of prose is not because of illiteracy or any hatred of reading, but because, invariably, your time is someone else's money. I hope that by using this approach, you will be able to quickly identify particular areas of interest and, with minimal effort, avail yourself of basic information that is of practical use.

The chapters of this book are laid out in the manner in which security problems should be approached—as an integral part of the other functions of the building. Each chapter stands alone, giving all the relevant information with as little reference to other chapters as possible, to facilitate your search for specific answers. However, I cannot stress strongly enough that the effectiveness of any security solution depends on careful consideration of the materials in all chapters.

The information in this book is divided into a number of chapters and the chapters are generally divided into these sections:

Introduction—gives a quick reference to each device

Principles of operation—describes the basic physical theory and operation of the device and its constituent parts, and lists the basic types of devices within a particular group

Equipment types and selection—expands upon the differences between devices in the same group, lists the principal features for which to look in any good quality device, and provides questions that should be answered by the specifier or supplier before you select any particular model

Uses—outlines the primary uses to which the device is put

Application—gives advice on the suitability of a particular application, what to avoid, where and how to install it, what it can and cannot do, etc.

Limitations, advantages, and disadvantages—itemizes those properties that will help you make the choice beween different devices for a particular application

1
Stages of System Design, Selection, and Installation

The first stage in considering security systems is to perform a *security overview*. The order of the security overview considerations should be

1. crime prevention
2. risk control and management
3. physical security countermeasures
4. electronic security countermeasures
5. maintenance and vigilance

THE DESIGN OF INTRUDER DETECTION AND ALARM SYSTEMS: SECURITY OVERVIEW

Crime Prevention

The goal of true crime prevention is to reduce the temptation of committing a criminal act. Any business would do well to remember that crime prevention is infinitely preferable to crime detection and detention. If a person commits a crime against you, you not only lose capital, you also pay to keep the criminal in prison!

Crime is all-encompassing in its scope and occurs due to poor moral attitude (e.g., what values you instill in your family and work force), poor social conditions (e.g., a warm blanket may be worth killing for if you are freezing to death), and poor education (e.g., if poorly educated, a person may lack the reasoning power to avoid peer pressure and may really believe statements such as "everybody steals").

These primary crime prevention considerations often seem too daunting for individuals or companies to combat. However, they are not and, unless you want to buy every device discussed in the latter chapters of this book, you should consider these items before all else. Ask yourself (or, if you are designing a system for a client, ask the client) to consider,

What do *you* "steal" from your company? Time, money, goods, services, expenses? This may cause jealousy and resentment among employees and pilfering will proliferate as the work rate drops.

Conditions. Are you *really* a paternalistic employer? If you pay the going rate, have a nursery for mothers who work, and have good social and recreational facilities, are your employees going to steal from you or give information to those who might? Paternalism encourages loyalty.

Education. If you educate and enhance the skills of those who work for you, they gain self-esteem, have greater loyalty, and are more re-employable if your company goes bankrupt or you make them redundant. Unemployed, unemployable, unskilled people are ten times more likely to commit crimes than their better trained and educated neighbors.

You should form *long-term primary* philosophies for your company.

Secondary crime prevention techniques are more tangible and deserve even more consideration in the *short term*. These include

neighborhood watch schemes
business and trade intelligence and information sharing
group communication schemes
police/community joint projects

Until quite recently crime was seen as something an *individual* suffered and individuals enlisted the help of the police to detect and detain.

Thankfully, crime is becoming recognized as a thing that affects the whole community in which the individual who suffered lives and works. Use of the social crime prevention techniques listed above has the following effects:

Reduces the attractiveness of crime. To take on an individual building or guard is easy. To attempt to enter an area of buildings where everybody is watching out for everybody's interests is downright foolish!

Enhances genuine community feelings, allows for many ideas on crime prevention to be aired and assessed, increases awareness, and allows focusing on the potential risks.

Brings the police closer to the community they serve. In Britain (which has an unarmed police force) it is the maxim that you can only *effectively* police with the consent of the majority of people in the community. It is, therefore, extremely important that everyone knows the local policemen, their crime problems, and how they can be helped. It is equally important that the police recognize you and your employees, know what you do and how you do it, and know what risks you face. How can they really help if they do not know these basic facts? If policemen in your business area don't know anyone, they have two alternatives—treat everybody with suspicion and possibly detain the wrong people, thereby alienating innocent citizens and hampering police efforts, or ignore everybody except the obvious criminal (i.e., the one with the gun). Unfortunately, in this case, the man leaving your premises in broad daylight with a large suitcase seems to pose no threat. The fact that he does not work for you, does not live in the area, has not been seen before, and has a suitcase full of your dearest possessions is missed. In small towns, people know and trust their local sheriff. City dwellers should learn from this and "adopt" their patrol officer!

It cannot be overemphasized that crime prevention should be considered before asset protection, crime detention, and all the associated technology and money they involve. Crime prevention is cheaper than the alternatives by a long way.

Risk Control Management

Risk control management techniques support crime prevention. Having done your best to reduce the causes of crime, you must accept that risks still exist. The amount of time and capital you now spend on risk control and management depends on the value of the product you wish to protect. Many large companies and government agencies spend vast sums of money employing full-time security officers to protect their products. Risk control and management are part of their everyday business. However, what if your company does not wish to employ security officers? What should you look for when carrying out your own risk control management exercise?

Depth of Study.   The first thing to decide is the type and depth of consideration you are going to give the problem. There are three main levels from which to choose.

1. *In-depth security analysis*—this covers every aspect of an operation/business, from changing processes that use expensive materials to ones that use cheap material, to retrospectively checking employee references, records, and so on.
2. *Overall security plan*—this usually accepts current processes and employees, and considers security guard usage, action plans, and contingency plans to cope with intrusion or loss. It includes a fabric and alarm study.

3. *Fabric and alarm study*—this is by far the most common type of study. The physical strength of building, barriers, alarm systems, etc., are considered.

The next step in risk control management is to identify those risks you need and want to control. This is called *risk identification and anticipation (RIA)*.

You must ascertain the type of risk you want to protect against and the probability of it occurring. This may be to protect people *and* property, to protect people only, and to protect your organization from criminals outside of or within your employ. You can do this by listing all known assets of every type along with their value to you and value to others—*asset appreciation*. For instance, certain common chemicals may be very useful to a bomb maker who cannot purchase them elsewhere without being identified. You should also list all likely, probable, possible, and highly unlikely methods of attack on your premises. Then decide which methods to protect against and which methods economics and sense dictate you have to accept will succeed. Opportunist petty criminals can be more easily protected against than suicidal terrorists or religious fanatics.

Remember that yesterday's unusual risk or form of attack may be commonplace tomorrow. Ten years ago, drunks often tried to steal the odd bottle from liquor stores; today crack cocaine addicts may steal the entire stock, burn down the premises, and attack you if you object. In order for your security system to work, you must anticipate changes and regularly review your system.

Risk Control.   Having established your at-risk assets or targets and the risks they face, you now have the fundamental problem of what to do about them. The first step in risk control is to significantly reduce the attractiveness of the assets to potential intruders. This policy is the direct opposite of the *Fort Knox Policy,* which gathers the valuables together (advertising their presence) and then provides them with the ultimate in protection. The Fort Knox Policy is a political policy, not a security policy. A *country* must be able to demonstrate its wealth of capital reserves; very few *businesses* need to do this. Antique galleries and bullion dealers are a few of the notable exceptions.

Reduction in asset attractiveness can be achieved using a variety of techniques:

Asset removal—moneyless transactions, barter deals, and staged assembly of valuable composite goods in many different sites

Rapid asset turnover—stockless production and manufacture to demand time. An absence of finished goods or valuable raw materials on the premises transfers the asset risk to shipping and freight companies, suppliers, and distributors. It is no longer your problem.

Asset dissipation policy—the antithesis of the Fort Knox

Policy. Spread out your assets in small amounts and in many geographically distant, well-protected caches.

Asset insurance — the *rainy day dollar principle.* Determine the true implications of potential loss and hold separate reserves for others, including reserve capital for replacement (self-insurance), insurance policies to replace cost of goods within a certain number of days, etc.

Asset transfer — to bonded warehouses, security vaults, etc. Make security somebody else's business.

Having done all you can to control the risk to assets by either major surgery or minor pruning, you are now left with the *core risk.* This is the risk you *must* take to carry out your work profitably.

The core risk can be categorized in two ways — protectable asset core risk and unprotectable asset core risk.

The former is usually easier to identify than the latter. If you add value to gold by using it in small quantities to produce jewelry, the asset risk might be employee pilferage of very small quantities of gold over long periods of time. You will probably decide that carefully monitored stock control, accounting, security, and access control will make this a protectable asset core risk.

However, what if your business develops computer software algorithms? How can you protect your intellectual property if your best computer programmer becomes dissatisfied with aspects of her job and does not tell you? If another company suddenly develops a very similar program, did she tell them or do great minds think alike? This is an unprotectable asset core risk and you had better consult your line management and personnel department to ensure you are one big happy family!

Physical Security Countermeasures

Physical security countermeasures are not the subject of the remainder of this book. They are, however, preeminent to all other security measures. Even the three little pigs of nursery rhyme fame knew the value of physical security countermeasures and had two prototypes built and two near fatalities before they found the best solution to their highly unusual wolf problem!

Physical security techniques include

managing your environment by landscaping to reduce cover, managing traffic flow, etc.

hardening your target by including ever-decreasing volumes of physically protected space with ever-increasing strength (e.g., fence, walls of building, vault walls, safe)

Some excellent sources of information on physical security measures (walls, fences, etc.) are given in Appendix 1 of this book.

Seven Essentials for Security.

Security is established by the following measures. From this list, you can see that the actual detection of intruders comes low on the list. If the preceding measures are not given full consideration, true security will never be established and the installed intruder detection system will probably prove worthless and troublesome.

Deter the potential attacker/intruder — design the building to appear strong and solid. Make it self-evident that the structure is guarded, occupied, and equipped with a security system. This need not detract from the company image or architectural aesthetics.

Demarcate — establish defined boundaries. Do not allow the site to become an open house, shortcut, play area, or attraction to vandals. Create a defendable space with fences, hedges, gates, etc.

Prohibit — only allow entry or exit to the premises through a limited number of doors, gates, and barriers. Always make people establish their right to be in that area. Never leave any area open during times of risk (e.g., night, shift changes).

Delay — create a sound, solid structure that will delay unauthorized entry for sufficient time (minutes rather than seconds) to significantly increase the risk of intruder detection.

Detect — if, despite the strength of your physical barriers, the intruder gains entry, ensure that entry is to areas patrolled by the guards or monitored by the intruder detection system and that his route to his objective will result in his being monitored by that detection system for the longest possible period.

Communicate alarm — make sure that somebody *who is prepared to react* is extremely likely to hear the alarm that has been raised. Send the signal to the security company or police, ring multiple local bells, etc.

Deny — deny easy access to information, keys, ladders, plans, computers, switch rooms, etc., all of which might prove to be of use to a potential intruder.

The principles of these essentials are explained with clarity and in depth in companion books *Security of Premises,* Stanley L. Lyons, and *Design Against Crime: Beyond Defensible Space,* Barry Payner. I strongly advise anyone attempting to provide or improve security to read or refer to these books. No amount of intruder detection equipment or other hardware can compensate for a building or site that is fundamentally physically insecure.

Mapping Out the Response.

Having considered all the preceding material, the wise building owner will now have an asset/risk register and plan that shows the protectable assets and where they are located. It should also identify, for each area, the highest probable core risk (i.e., from highly skilled intruder to semiskilled employee).

The common security use of this asset/risk register and plan is to draw *zones of protection* around each area to be considered. These are geographic areas or lines that may be literally plotted on scale drawings of the building complex. They include

- building site boundary
- building site grounds and approaches
- building outline (exterior walls, doors, etc.)
- building interior and routes to location of asset (i.e., direct via corridor or indirect through weaker walls)
- asset enclosure or final protection boundary (e.g., safe, strong-room wall)

Having done this, and constantly bearing in mind the probable *overall* threat, consider the essentials for security in relation to these zones to find an appropriate solution. The appropriate solution is normally thought to be one that will sufficiently delay the ingress of the intruder to allow the alarm to be raised and the response force to arrive to detain the intruder. *Ingress* time is important, as you want the element of surprise to reside with the response force. If the response time is so long that the intruder is already making an escape, your response force will have to search a much wider area.

Security designers have special knowledge of intrusion/response times. Have them plot position and skill level for each asset type. For example,

exterior fence	1 min. to overcome
interior grounds	2 mins. to cover
building perimeter	3 mins. to cut through wall
corridor/interior	2 mins. to asset area
asset time protection	3 mins. to overcome
total ingress time	*11 mins.*

Use the total ingress time for each asset type to enhance existing physical security measures and provide electronic interior or exterior security systems that will delay the intruder long enough for the response force to arrive.

Electronic Security Countermeasures

Electronic security countermeasures form the major part of the remainder of this book. I believe and hope to show that security systems

are good value for the money

aid the police in detaining 1 out of 15 criminals committing crimes of *all* types

allow police and guards to carry out deterrence by patrolling premises, neighborhoods, etc., thereby enhancing cooperation, public visibility, and feelings of safety

increase the detection rate of crime and reduce court time (e.g., CCTV identification of bank robbers has had dramatic successes)

actively and continually deter all but the very professional criminal

What is an intruder alarm system? This might seem to be a simple question to answer, but it is not. It is easier to turn the question around and explain what an intruder alarm system *is not:*

It *is not* a means of physical protection. No alarm system ever actively prevented a guard from being assaulted.

It *is not* a means of restricting access to only those with authority. This is the role of the building materials, fences, barriers, locks, restricted access systems, etc.

It *is not* a means of identifying a person who claims to have a right to be within the protected area.

It *is not* a means of apprehending intruders. Only the response force, the police, or a private security guard can physically restrain and arrest an intruder.

Intruder alarm systems are not the ultimate weapon in the war against crime; they are simply tools that can be used or misused. An intruder alarm system remotely senses the presence of a person within the space covered by the system sensors. The system then notifies someone who might react in the system owner's interest. This could prevent a crime from taking place or help catch an intruder. This is the most that can be hoped for. If the system achieves this goal only once, it will probably have paid for its total installation and maintenance costs!

Before considering any intruder detection and alarm system, the specifier should ensure that it will not be compromised or upset as a result of lack of integration within the overall planning and layout of the building, and its position on the site.

Having given due consideration to the site, structure, and physical security, what should an intruder detection and alarm system be? A *good* intruder detection and alarm system should be

selected as the result of a full, exhaustive client brief that defines the system's functions and the method of alarm (covert or overt)

provided only as a secondary, backup system to supplement adequate, physical protection measures

simple to operate, understand, install, and maintain, but very effective

very difficult to overcome, tamper with, or attack from both within and outside of the building

acceptable to users, insurers, fire officers, police, and all other persons with whom it interfaces

reliable in use

A *good* intruder alarm system must work the first time and every time, and only when a real attack is taking place. It should not give rise to false alarms caused by environmental conditions, birds or mammals, or electrical interference. The system should *always* justifiably reassure the user.

Unfortunately, *good* intruder alarm systems are as rare as good cars, and the parts that go into making them are just as complex. Good cars are well designed, engineered, and assembled; good intruder alarm systems are well *specified*, designed, engineered, and installed. The first of these requirements (i.e., specification) is the most important part of the whole sequence. An intruder alarm system will be built to a standard, be assembled to a particular specification on your premises, and only detect intruders attacking the building in specific ways, using specific techniques. Just as a small family car will not achieve 200 miles per hour (322 kilometers per hour), an intruder system or sensor designed for residential use will not necessarily be satisfactory for a commercial building. The *only* persons who can assess the actual risks faced and values of goods to be protected and, hence, the system required, are the clients or their trusted and fully informed representatives. The police, security staff, and alarm or insurance company representatives may assist in forming the specification, but only the building owners and designers are fully cognizant of every aspect of the building and its use. They must act as the final arbiters of what is required for their building.

Maintenance and Vigilance

No security system is maintenance-free and all system cost budgets should allow for regular, specialized testing, challenging, and maintenance.

Vigilance is the responsibility of those long-term employees who are responsible for security within a company. Crime types and risks change at an increasingly rapid rate and vigilance and awareness of these changes is the key to long-term security success.

THE DESIGN PROCEDURE

The Steps to Success

The design of a security system should be a very logical progression of events leading to the installation of a system that provides adequate protection, is easy to operate and maintain, is reliable, and, above all, is effective. To this end, in new construction projects, the professional security manager, the engineer, or the architect should be the central figure and take the project through the following stages. Where the building exists or is being converted, the stages are still relevant, although

other members of the client's staff or design team may play the leading role.

So you, the owner, or you, the security adviser, have decided that a security system may be needed. What are the logical steps to success?

1. Initial meeting and survey
2. System budget consideration
3. Formation of outline brief
4. Design team formation
5. Preliminary survey
6. Design team review
7. Quality of advice (threat analysis, asset review)
8. Feasibility study
9. Report and approval
10. Design development
11. Presentation, approval, and specification
12. Selection and use of security companies
13. Competitive tender and contracts
14. Installation
15. Testing, commissioning, setting to work, and handing over
16. Operation
17. Maintenance
18. Review
19. Standards

In their haste to concentrate on more visible elements of building design, architects, engineers, and building owners often forget that really effective security is the result of an integrated design process. Just because security systems can be installed retrospectively does not mean that they should be!

The Engineer's Joint Contracts Documentation Committee lists the phases of any design as study, preliminary design, final design, bidding (tendering) and registration, construction, and operation. This system, used to develop building designs, corresponds well with the previously listed security design steps. It is essential that at every step of a new building design architects, engineers, and security system designers coordinate and integrate their actions. There is no point in telling architects 6 months after they have agreed to the structural loads that a 32-ton safe must be located on the 12th floor and will they design a secure stairway to the room in which it will stand! Clients or their representatives should consider the following.

Initial Meeting and Survey. The purpose of this meeting should be to write down the client's accumulated knowledge of how the business works, the perceived risks, where the risks are located, and the possible operational restrictions on the system. Also, walk the site with the client and rely on the abilities of your eyes and ears. Familiarize yourself with the area. Then establish a budget for the works.

At the end of the meeting, compile a list of objectives and basic information for meeting attendees to take away and consider and review carefully. After due consideration, all people involved should write additional comments *on this list* to help in the preparation of an outline brief.

Remember, no matter how good you feel your information-gathering skills are, you are likely to collect only a fraction of the information actually needed to achieve an optimum solution. The design team will include professionals who will ask many more searching questions. Do not take this as a criticism of your ability. Cooperation is the key to success.

System Budget Consideration.   Clients need to assess how much they can afford to spend on security. Many clients often consider security as a luxury or of relatively low priority and, hence, allow far too little in their budgets for the installation of appropriate security systems. In order to obtain a realistic cost estimate, clients must determine how much they can afford to spend *after* they consider increased insurance premiums due to lack of adequate security (insurance companies are often the prime instigators of system installation) and the consequences, in human and financial terms, of

> physical damage to building fabric
> physical damage to building contents, materials, etc., caused by vandalism, fire, etc.
> lost production due to stolen raw materials
> damage to production equipment
> damage to building services
> failure to meet delivery deadlines or other business goals
> loss of privacy due to espionage, computer crime, etc.
> cost of replacement of goods and materials

If only a fraction of 1% of the cumulative cost of these is spent effectively, then clients and insurers will be satisfied!

Once a confidential budget is agreed to by the client, the security manager, engineer, or architect will be in a position to ensure that the proposals of the design team do not stray toward the absurd and are likely to stay within the predefined budget.

Formation of Outline Brief.   The purpose of the outline brief is to gather together, in a structured way, all of the in-house wisdom of the owner and employees, and to summarize this knowledge in writing. This written brief will then become the working document toward which members of the design team work to design a good, workable, affordable solution.

A good outline brief will provide all relevant information without needing to ask every employee what they think should be done and receiving a thousand different answers. Your outline brief should establish the following so that it may be conveyed to the design team:

I. Building
   A. Up-to-date plans and future expansion
   B. Materials of construction and openable windows and doors
   C. Fire escape routes
   D. Current entry and exit doors
   E. Uses of adjacent buildings and hours of use
II. Operation
   A. What does the business do?
   B. What is its turnover?
   C. What goods are stored, imported, exported, and used on-site? Where are they stored?
   D. How many people work in each area?
   E. How do these people enter and leave an area?
   F. How many employees have been employed for more than 3 years?
   G. How many contract or temporary staff members are employed? What is the turnover rate of the temporary staff?
   H. Who visits the site regularly?
   I. Who maintains electrical and other services?
   J. Which areas have unrestricted access? Which are restricted? Restricted to whom? Why? When? How?
   K. What shifts are worked?
   L. When do cleaners work?
III. Existing security provisions
   A. Guards, systems, etc.
   B. Perceived gaps in security
   C. Locking regimes

Remember, time is money and good designers do not come cheap. Even if a security company is designing a system for you, you pay in one way or another. So, the more time you spend getting a good outline brief together, the quicker and better the design proposal devised and presented.

After the brief is compiled, you need to select your design team. Remember to delegate, listen, and learn. You or your client will be able to contribute and eventually decide on the system proposed, but do not try and impose your thinking on the design team.

Design Team Formation.   Assemble the design team. This advisory team should include, at initial meetings,

> the clients—specifically, the actual operational managers (i.e., the people who can identify assets that require protection, risks, and how the building and its operation affects security or vice versa). Many company owners or directors/managers like to think they know what is going on, that they are in touch with everyday practices, etc. This is rarely the truth. If you want to know how to get into a building when the manager thinks it's locked, ask the cleaner or lowest office clerk!

the building services engineer or security manager

a security installation company representative or specialist independent security consultant

the clients' insurance representative (security surveyor)

a local crime prevention police officer

a local fire officer (if conflicts in access and emergency escape are likely to arise)

Clients should be aware that these individuals will approach their problems from different angles and with different objectives, many of which may conflict. In my view,

the objectives of the fire officer in providing safe means of escape are preeminent. No security system should prevent them. No escape/security conflict is insurmountable.

the advice of your *local* police crime prevention officers is *always* valuable. They will know of local crime trends, local restrictions or laws, etc.

the role of your insurance company is often poorly defined and of limited use — *it should not be*. Most insurance companies do very little to promote or improve really effective security. Why should they? Provided that losses paid do not exceed income, a few claims each year will justify increased premiums and will increase profit. Encourage your insurance company to specify what they feel is needed *and* make them justify their claims. If they cannot, change your insurer!

Remember that you are actually insured by insurance underwriters. They will ask an insurance company to survey the risk on their behalf. If the insurance company is doing its job poorly, they will ask a local, *nonspecialized* insurance representative to visit you and survey the risk. If the local specialist security representative does a better job, the underwriters should appoint this independent security/insurance surveyor. This person will consider all security aspects, compare these to the risks, and (following discussions with the building owner) prepare a report and recommendation to the underwriters via the company. Far too often, this report simply lists "Requirements for Upgrading for Acceptance." This list is often ill-considered isolated and incomplete without the input from others in the design team. Insurers could and should do more to establish national codes and standards that would be freely and universally accepted as *minimum* standards.

Preliminary Survey. Appoint a lead design consultant who has full responsibility for eventual proposals. Ask the design team to consider the site and the risks, outline the building proposals showing room layouts and building orientation, and carry out preliminary surveys.

Design Team Review. Hold a design team review to identify

- any special site restrictions (state laws)
- any particularly high risks (adjoining roofs, etc.)
- level of appropriate security
- police and fire brigade recommendations
- resolution of any fundamental conflicts

Quality of Advice. You have selected your design team. You are paying them directly or indirectly. How do you judge the quality of what they are producing? Well, if you selected your team members based on personal recommendations, references, etc., their basic ideas should be sound. If your team members are very experienced, they should provide high-quality advice (although a mix of young blood and enthusiasm with experience is often better).

Security consultants operate in many specialized fields, including physical and electrical hardware and system design. "Good" consultants are normally listed in the Security Letter Source Book, which gives references and qualifications. But this is not enough. Further down the line, your team members will produce and present a study. At that stage, you must make an assessment — how good is it?

You may judge quality by assessing what they know, what they have told you, and how clearly they have presented information. Lack of real understanding is often hidden behind early specification of devices and instant solutions (i.e., "You need three XXXS32s and a shotgun!").

Look for their knowledge of

your adversaries

you, your business, and your site

the actual end user and system operator

the value of the protected product

the device or system

Knowledge of the Adversary: "Know Thy Enemy." To know the adversary, you must know the

scale of threat — opportunist thief to terrorist group; which type and why

type of adversary — external staff, internal staff, ex-employee, "telephone repairman"

type of attack an adversary might use — most use brute force, but if value is high, skill level will be high

social pressures — poor neighborhood or highly contentious operation in politically aware neighborhood

political pressures — this year's savior is next year's demon (i.e., nuclear power plant, fur importer)

local awareness — corrupt security organizations; police or politicians with vested interests; gang warfare/drug target. Does your intruder have a shopping list?

Knowledge of You, Your Business, and Your Site. Let's assume you have something really worth stealing and a skilled intruder will attack. The design team must be able to think like this intruder. The stages of the intruder's planning (and, hence, your design team's planning) might be

I. Research your target
    A. Company activity
    B. Potential assets
    C. Motive
    D. How will you sell goods or dispose of them?
II. Study the site from a distance
    A. Position of building (e.g., alone, among others?)
    B. Overlooked by other buildings?
    C. Overhung or undermined by other structures?
    D. Multitenanted building?
    E. How many staff? Staff turnover? Do they all recognize one another at the gate or reception area?
    F. Roads, paths, fences, means of escape
    G. How do people enter and leave?
    H. Who sets the alarm? Where?
    I. Do guards patrol? How, where, and why?
III. Study building in detail
    A. Are plans readily available? Can you obtain keys or access cards?
    B. Do you know people who will tell you how things are done in the building, where the main electricity point is, etc.?
    C. Security devices, bells, entrances, exits
    D. Shortcuts, secret entrances
    E. Can you con your way inside the building?

Having put themselves in the shoes of the intruder, the design team should give advice that demonstrates

> how you, your trusted friends, etc., gain entry
> how the business functions, the management structure, the opportunities for crime, whether you may have underpaid or undervalued employees
> how your operations take place, where you take delivery and how often, and where you store goods
> practical and impractical solutions (There have been many ideal security solutions that cannot work in real life—card access entry, of 100,000 fans, to a football stadium through four gates.)
> how to upset you or your preconceived ideas. If you believe your pet dog is a good guard dog, the *good* adviser will be prepared to show you how easy it is to drug it, shoot it, or frighten it. If your cousin installs security systems, the *good* adviser will risk your anger by properly and tactfully demonstrating how your cousin is unfit to plug in an electric appliance, let alone do security work properly.

how value is added. For instance, denim cloth is relatively cheap and a clothes store in Russia storing denim cloth would not be a high-value target. If the next store is after the Levi labels that have been sewn on, half the organized black market in Russia will have desires upon them. Remember, if your product had no value, you would not be in business. their persistence and inquisitiveness. If, during your survey walk, you say, "Oh, that's just a service duct behind that locked door," do they insist on seeing it for themselves?

Knowledge of the Actual End User. Who will run the system? For example, there is little point in installing a sophisticated CCTV-monitoring station with split screens, motion detection, and tracking if the end user has an IQ of 20 or end users change so frequently that continuous training is needed.

Knowledge of the Value of the Protected Product. There are many types of value. The true value, when considering the potential risk/expenditure quotient, may be a sum or a multiple of

- absolute, capital sales value
- intrinsic value
- sentimental value (e.g., an heirloom)
- availability, security, replacement value
- production continuity value
- embarrassment value
- value to a competitor
- pricelessness
- insurance value

Knowledge of the Device or System. Have your advisers used the proposed device or system elsewhere? Did it work? Do they tell you why it fails? How often? How it can be *made* to work? The quality range? Why your neighbors or competitors do not use them?

Feasibility Study. Ask the security manager or lead consultant to provide a full feasibility study with order-of-cost estimates based on information gathered during the preceding stages. This study should include sections on

> types of risk envisioned
> methods of protection appropriate for each risk area
> how systems operate and interact with fire systems, police, and private security company personnel
> budget costs broken down into capital for system and annual maintenance, replacement, and operating costs
> block and schematic diagrams showing outline proposals

It is usually advisable to have this feasibility work done for a set fee and not as a free survey service, as the latter is often not of sufficient depth. Free surveys are commonly offered by security system installers as a means of attracting customers. If the system is simple or for a low-risk structure (e.g., small building, house, shop), this type of survey will suffice. If the system is any larger, then a detailed feasibility study is required. Established companies have security surveyors with years of practical experience; smaller companies tend to use inexperienced general salespeople. If the system is to include external security, security lighting, CCTV, or access systems installation, then a comprehensive survey and feasibility study is essential.

Report and Approval.   Discuss the feasibility study with the authors and clients, and modify it as necessary. Get client approval to proceed with a named specialized designer or installer (security company) or to prepare design drawings and specifications for competitive bidding (security consultant).

Design Development.   Allow your designer to proceed with the design. Act as a focal point. Ensure security system designers are notified of every change to building design, landscaping, materials of construction, changes in room use, etc. Ensure systems designers have consulted with local authorities regarding bylaws (noise), kept police involved in design, and taken full account of fire escape procedures and agreed signaling methods.

The security designer is the one person on the design team who has a true need to know everything concerned with the building—its contents, management structure, operations, etc. This person should be provided with

room data sheets detailing number of people, room size, finishes and materials of construction, window type and glazing, heating and ventilation methods

room activity sheets detailing functions of rooms, materials stored and equipment used, interconnection with other rooms, deliveries to rooms, and levels of stock

personnel sheets detailing roles of persons within a defined space, security clearance to other spaces, key holders, whether there are shifts and late hours, early starts envisioned, and opening and closing times

design drawings, including plans 1:100/1:50 for all floors, basements, and elevations; roof plans; site layout and landscaping; perimeter protection; mechanical and electrical services; window and door joinery details (if needed); and ironmongery schedules for inspection, comment, or specification

At the conclusion of this stage of the design, the design team should prepare a presentation document that includes

system descriptions

operational use explanations

system outline drawings showing equipment, positions, coverage, etc.

a listing of the headings for a specification to be constructed as a *Bill of Quantities* (i.e., a list of major parts with approximate plus-profit costs against each item and totaled to a grand sum; at this stage an accuracy of ± 20% on the expected tender bids should be the target)

Presentation, Appraisal, and Specification.   The scheme should be explained to the client (using the presentation document) and, when necessary, insurers, police, etc., in a single meeting. Any modification resulting from the presentation should be made in a final design, and either the scheme should be adopted if the price estimate or quote is right or competitive tenders should be invited. Do not bully the designer into unwise, cost-cutting reductions!

It is essential at this stage that, before clients approve adoption of the design scheme or release of tender documents, they

know what the system *cannot* do

understand that the system will only operate successfully with full cooperation in its management, maintenance, etc.

know the full cost

have the approval of their insurers

understand the duties they must carry out (e.g., managing system operation; ensuring the building is secured properly, windows closed, etc.; checking that nothing fouls system sensors or is moved that could give rise to false alarms)

understand the role of the security system installer, and the limit of their responsibility and liability. *Codes of Practice* give the standard format and extent of responsibility to be written into an agreement (see Figures 1–1 and 1–2 for typical examples).

have seen and appreciate the content of any agreement into which they enter

At the presentation, it is also essential that confidence in the capabilities of the system results from a true exchange of questions and answers. The designer must explain every proposed item to the satisfaction of those assembled, who are prepared to ask, How? What if? etc., until they are truly satisfied with the answers.

Designers are advised to prepare for a presentation properly. Record the questions and answers, if possible, to note any design changes that result from the presentation.

After the presentation has been approved and design variations have been resolved, the design team should proceed with the final design.

QUOTATION

OUTRIGHT PURCHASE AGREEMENT

INSTALLATION/EXTENSION AND SERVICE

INTRUDER FIRE ALARM CCTV ACCESS CONTROL EMERGENCY LIGHTING

Name of Customer:

Date

We are pleased to submit this Quotation in respect of the System at:

('the Premises')

details of a new installation or extension to an existing installation are given in the enclosed Specification reference:

dated:

1 Installation Charge*	£	Service Charge* first year	£ per
2 Extension Charge*	£	Additional Service Charge*	£ per

*All Charges are plus VAT at the rate prevailing at the date of invoice.

1. INSTALLATION CHARGE — covers supply and installation of the System and is payable on or before Installation Date. Service Charge covers servicing of the System and is payable on or before Installation Date and annually thereafter except for the fire alarm Standard Service Charge which is payable against invoice after each service visit.

2. EXTENSION CHARGE — covers supply and installation of an extension of the System and is payable on or before Extension Date. The proportionate part of the Additional Service Charge for the period from Extension Date to the next anniversary of the Installation Date of the original System will be invoiced separately and is payable on receipt of invoice. Each subsequent Service Charge is payable annually thereafter except for the fire alarm Standard Service Charge which is payable against invoice after each service visit.

The Quotation is subject to the enclosed Specification if provided and to the Terms and Conditions following. Please read them carefully.

Unless otherwise agreed in writing, this Quotation may be accepted up to 30 days from the date of quotation after which the Company reserves the right to amend it or to withdraw it.

The Installation, Extension or Service Charge is based on installation, extension or service work being carried out during Normal Working Hours, and is conditional on our having continuous and uninterrupted working and assumes unhindered access to the areas where the work is to be carried out. It does not cover extraneous work, carpet lifting or refitting, building work or decoration and should the Company agree to carry out such work at the request of the Customer during installation or service, then the Company shall not be liable for any damage caused to such items. All additional costs arising from any alteration to the enclosed Specification required by you and interruption or delays by you, your employees, agents or customers or other trades during the course of installation or service may result in additional charges.

The System is not designed or adapted for use in adverse industrial atmosphere or extremes of weather or abnormal operating conditions of any kind.

Signed for and on behalf of Chubb Alarms Limited.

BSIA
THE BRITISH SECURITY
INDUSTRY ASSOCIATION

BFPSA

**Chubb
Alarms
Limited**

TERMS AND CONDITIONS

DEFINITIONS

In this Agreement the following expressions have the meanings given to them below:—

(a) the Quotation, the enclosed Specification if provided and the Acceptance together with these Terms and Conditions.

Agreement

(b) Chubb Alarms Limited.

Company

(c) the intruder alarm, fire alarm, emergency lighting, closed circuit television or access control system installed in the Premises or the Chubb System.

System

(d) the installation more fully described in the Specification.

Chubb System

(e) the date on which the installation of the original System is completed by the Company.

Installation Date

(f) the date on which the extension to the System is completed by the Company.

Extension Date

(g) Comprehensive Service — for a Chubb System a minimum period of three years from the Installation Date. For a System not installed by Chubb a minimum period of three years from the date of this Agreement.

Standard Service — for a Chubb System a minimum period of one year from the Installation Date. For a System not installed by Chubb a minimum period of one year from the date of this Agreement.

Extension Agreement — a minimum period of one year from the Extension Date.

continuing thereafter unless terminated by not less than three months notice in writing given by either party to the other expiring at any time after the minimum period.

Service Agreement Period

(h) a period of 12 months from Installation Date or Extension Date for an extension of the System (response during Normal Working Hours only).

Warranty Period

(i) 0830 to 1700 hours, subject to alteration by the Company, Monday to Friday except for Statutory and Common Law, Public and National Holidays.

Normal Working Hours

THE CUSTOMER

(a) shall obtain and pay for all necessary consents including listed building consent for the erection of an alarm bell for the installation of the System and shall give to the Company access to the Premises at all reasonable times for the purpose of doing anything which the Company is required or entitled to do under this Agreement. The Customer acknowledges that it may be necessary for some work to be carried out outside Normal Working Hours.

Consent and Access

(b) shall make available at no cost to the Company all necessary ladders and scaffolding or other items required for access to the equipment for fire alarm service which shall be safe to use and comply with relevant legislation.

Access Equipment

(c) shall assume all risks in the Chubb System and its constituent components upon and from delivery of the components to it or to its premises or as otherwise directed by it or collection by it or its agents or servants but title to the Chubb System shall not pass to the Customer until the Company shall have received payment in full of the Installation or Extension Charge indicated on the front of the Quotation. In the event of termination of the Agreement prior to such payment the Company may enter the Premises and may repossess the System or any part thereof.

Title and Risk

(d) shall operate the System only in accordance with the written information and instructions which may from time to time be supplied by the Company to the Customer.

System Operation

(e) shall notify the Company forthwith (confirming such notice in writing) of any defect appearing in the System and shall permit the Company to take such reasonable steps as it shall consider necessary to remedy such defect.

Non-Interference with the System

(f) shall reimburse the Company any charge made by the Police, Fire or other authority to the Company from time to time in connection with the System.

Authority Charges

(g) shall pay for the cost of any work required to be carried out to the System and materials therefor or any attendance by the Company at the Premises which may be required by the Police, Fire or other authority or any other circumstances arising outside the control of the Company, which shall include attendance for the purpose of system reset following accidental operation or abortive attendance.

**Work to System/
Attendance at Premises**

(h) shall notify the Company of any proposed structural alterations to the Premises affecting or which may affect the System; any reasonable extension or alteration to the System which may thereby become necessary shall be carried out by the Company at the expense of the Customer.

Structural Alterations

FIGURE 1–1 Typical security quotation and terms of agreement

Damage to the System

(i) shall pay for the reasonable cost of all work to be carried out to the System and materials therefor due to damage to the System unless caused by the negligence of the Company or its servants or agents.

Indemnity

(j) (i) shall indemnify and keep indemnified the Company against any claims whatsoever for damage penalties costs and expenses and against all liabilities in respect of any patent, registered design or any industrial copyright of any third party where the System is made or procured for the Customer and supplied by the Company to other than the Company's design at the Customer's request or is used by the Customer in conjunction with other equipment not supplied by the Company

(ii) shall indemnify the Company against all liability loss damage penalties costs and expenses whatsoever caused and howsoever arising and by whomsoever made including but not limited to any claim made against the Company by the Police or Fire Authority due to a false alarm signal from the System unless such false alarm signal is solely attributable to the Company's defective equipment in the System.

Services

(k) shall advise the Company of the existence of concealed pipes, wires and cables for water, gas, electricity, telephone or other services affecting the Premises and shall confirm the location of such services to the Company's technician before work commences. In the absence of such notice the Company accepts no liability for damage to such services or any loss damage or injury whatsoever incurred or sustained in consequence thereof as the Customer hereby acknowledges and the Customer shall indemnify the Company against any claim whatsoever for loss damage or injury resulting from damage to such services as aforesaid.

THE COMPANY

Installation Date or Extension Date

(a) shall install or extend the System within a reasonable time of date of acceptance by the Customer of this quotation or as otherwise agreed by the Company in writing Provided Always that time shall not be of the essence

Contract Maintenance

(b) shall, in the event the Customer confirms a requirement for the System to be serviced during the Service Agreement Period, carry out routine servicing and emergency service more fully set out in the System Service Schedule/Service Agreement* dated . . . *delete whichever does not apply

Warranty

(c) shall, in the event the Customer does not require contract maintenance during the Warranty Period make good by repair or at the Company's option by the supply of a replacement, defects which under proper use appear in such part or parts of the Chubb System during the Warranty Period which arise solely from faulty materials or workmanship. The warranty is conditional on the system or equipment being serviced and maintained throughout the Warranty Period in accordance with the relevant British Standards or BSIA Codes of Practice

THE COMPANY'S LIABILITY

The Company has no special knowledge of the nature and value of the contents of the Premises for which the System has been specified and in which it is to be installed or serviced or of the nature of the risks to which the Premises and their contents will be or may be from time to time exposed. The potential loss or damage which the Customer might suffer is likely to be disproportionate to the sums that can reasonably be charged by the Company under agreements of this nature. As the Customer knows or should know the extent of such potential loss or damage and is therefore in the best position to do so it should insure against all likely risks Accordingly, the Company limits its liability to the Customer as set out in this Part 4, which specifies the entire liability of the Company including liability for negligence.

Acceptance of certain liability

(a) The Company accepts liability

(i) for death or personal injury resulting from negligence of the Company, its servants or agents acting in the course of their employment;

(ii) arising out of any breach of the obligations as to title implied by statute;

(iii) where the Customer deals as consumer for any breach of any condition or warranty implied by statute as to the correspondence of the System with description or sample or as to its quality or fitness for purpose or particular purpose,

(iv) up to the sum of £250,000 for direct physical damage to the Premises or their contents to the extent to which such damage or loss is caused by the negligence of the Company or its servants or agents whilst working on the Premises in the course of their employment.

The provisions of this Part 4 do not affect the statutory rights of the Customer as a consumer.

Submission of claims

(b) The Customer is required to notify the Company of any claim in (a) above as soon as reasonably possible and in any event within three months of the act, omission or occurrence giving rise to the alleged damage or loss except that any claim under clause (iv) above shall be notified to the Company within thirty days of the Customer suffering any alleged damage or loss.

Exclusion of Certain losses

(c) (i) Except as provided in clause (a) above the Company shall have no liability whether in contract tort (including negligence) or otherwise for any loss damage or injury where such loss damage or injury arises indirectly or unforeseeably from or is consequential or contingent upon a wrongful act or omission on the part of the Company or of its servants or agents acting in the course of their employment.

(ii) The Company shall have no liability in any circumstances whatsoever whether in contract tort (including negligence) or otherwise and whether caused directly or indirectly for financial loss or loss of profits contracts business anticipated savings use or goodwill

Overall financial limits

(d) Except in the circumstances described in (a) (i), (a) (ii) or (a) (iii) above when no limit will apply and (a) (iv) above the Company's liability whether in contract tort or otherwise will in no circumstances whatsoever exceed for each claim the aggregate of the initial installation charge together with subsequent extension charges and the current annual service charge subject to an overall financial limit of £50,000.

Extension of protection to employees

(e) Save as provided in clause (f) below, for the purposes of any exclusion or limitation of liability contained in this Part 4 of this Agreement the Company is or shall be deemed to be contracting both on its own behalf and also as agent for and/or trustee of any servant or agent employed by the Company and such servant or agent shall to this extent be or be deemed to be in contractual relationship with the Customer and to be entitled to the benefit of any exclusion or limitation of liability as aforesaid.

Deliberate wrongful acts of employees

(f) Under no circumstances shall the Company have any liability for any deliberately wrongful act, default or omission by any employee of the Company acting in the course of his employment unless such act, default or omission could have been avoided by the exercise of due care and diligence on the part of the Company as employer.

GENERAL

Termination after breach of contract

(a) If any payment shall be more than one month in arrear the Company shall have the right to withhold further deliveries of constituent components of the Chubb System and to withdraw immediately the service provided for the System. Time for payment shall be of the essence of this Agreement. Written notice of withdrawal of the service will be given to the Customer.

(b) If either party shall commit any breach of this Agreement then the other may by seven days' notice in writing terminate this Agreement. This right of termination shall be in addition to the Company's rights under (a) above and shall be without prejudice to its right to recover any sum due from the Customer.

(c) the Company reserves the right to charge interest on any sum due to the Company and not paid on the due date at the rate of 4% per annum above the base rate of Barclays Bank plc from time to time compounded monthly on all amounts overdue until payment thereof such interest to run from day to day and to accrue after as well as before any judgement.

Terminated by the Customer

(d) the Customer shall be entitled subject to the payment to the Company of any arrears of any Service Charge and any other payments due under the Agreement to the expiry of the Service Agreement Period to terminate this Agreement by not less than three months' notice in writing to the Company expiring at any time after twelve months from the Installation Date or Extension Date.

Notification of Increased Charges

(e) the Company may increase the Service Charge at any time after 12 months from the Agreement date by giving notice in writing to the Customer stating the new Service Charge and the date (not being earlier than the date of the notice) on and after which the new Service Charge shall become effective. The Customer may within 14 days after the service of any notice of the new Service Charge give three months notice in writing to the Company terminating this Agreement. If the Customer shall give such notice of termination the new Service Charge shall not be effective

Certificates

(f) following completion of the installation/extension of the Chubb System the Company will issue to the Customer its Handover Certificate stating the Installation/Extension Date and such certificate shall be conclusive evidence of such completion.

Right of Assignment

(g) the Company may assign all or any of its rights under this Agreement and perform any of its obligations through sub-contractors. The Company may not assign its obligations save to a subsidiary Company of Racal Electronics Plc, without the prior written consent of the Customer not to be unreasonably withheld.

Antecedent Agreements

(h) this Agreement supersedes and terminates any antecedent agreement relating to the System at the Premises without prejudice to any liabilities or obligations of either party to the other outstanding upon such termination

ACCEPTANCE BY THE CUSTOMER

I/We the Customer accept this Quotation in respect of a Chubb System at the Premises subject to the enclosed Specification and to the Terms and Conditions above which I/we have read and which I/we understand.

I/We confirm I/we require Contract Maintenance (delete if inapplicable).

Invoices and statements to

WITNESS — SCOTLAND ONLY

Signature of witness

Address of witness

Signature of witness

Address of witness

Signed by
duly authorised to sign on behalf of the Customer

(Print name)

Position held

Date

Chubb Alarms Limited
Branch Address:

SPECIAL NOTES FOR THE CUSTOMER'S ATTENTION AND CONSIDERATION

The System you have selected is not designed to perform its function unaided. No system can do that.

To ensure you obtain maximum benefit from the system, you are advised to:

(a) follow the procedures laid out in the operating instruction for setting and resetting an intruder alarm system;

(b) carry out the periodic checks and tests recommended by BS 5839 for a fire alarm system;

(c) keep keys and security cards in a safe place;

(d) make sure changes to furniture layout etc. do not obstruct infra red beams and space detectors;

(e) retain the information supplied concerning the method of operation, design, installation and maintenance of the system;

(f) contact our local Office if a change in your circumstances warrants a review of your system;

(g) INSURE YOUR PROPERTY AND VALUABLES AGAINST THEFT OR DESTRUCTION.

An alarm system is not a substitute for insurance. It is a device designed to give early warning of a fire in the case of a fire alarm, whereas an intruder alarm is a protective measure to deter intrusion. No intruder alarm can detect all forms of unauthorised entry or be completely tamper-proof. With your full co-operation the system you have chosen will achieve its maximum level of efficiency. If you have any doubts or questions on any aspect of your alarm system, contact our local Office immediately for free advice.

Transfer of System

(i) this Agreement is personal to the Customer and may not be assigned or otherwise transferred by the Customer

The Company on written request by the Customer and subject to payment of all sums due to the Company by the Customer, may be prepared to enter into a new agreement (at current rates and on its standard terms and conditions then in force) with a new occupier of the Premises in respect of the System

Improvements to Specification

(j) the Company's policy is one of constant improvement and the Company reserves the right to alter the specification of any component part or parts of the System at its discretion at any time without notice. the Company also reserves the right because of difficulties in obtaining supplies, to use at its discretion equipment and materials other than those specified provided this does not materially affect the performance of the System

Force Majeure

(k) any failure by the Company to perform any of its obligations hereunder by reason of strikes, lock-outs, labour disputes, weather conditions, traffic congestion, mechanical breakdown, obstruction of any public or private road or highway, or any cause beyond the control of the Company shall not be deemed to be a breach of this Agreement

Notices

(l) any notice required to be given hereunder shall be sufficiently given if properly addressed and sent by post to, in the case of a Company, its registered office and, in the case of the Customer, its last known address and shall be deemed to have been properly served at the time when in the ordinary course of transmission it would reach its destination

Complete Agreement

(m) no terms or representations express or implied other than those expressly embodied in this Agreement shall be binding upon the Company unless accepted by the Company under the hand of a Director, the Secretary or Authorised Officer in writing

---

SYSTEM SERVICE SCHEDULE

The Company will provide Comprehensive/Standard* Service
*delete whichever does not apply.

Comprehensive Service

The Company shall during the Agreement Period carry out   routine service visits in every 12 months during Normal Working Hours and provide emergency service facilities in accordance with the relevant British Standard Any necessary repairs or replacements to the System caused by inherent defect or fair wear and tear will be carried out by the Company at its own expense

NOTES: The cost of access control cards, closed circuit television vacuum tube devices, video cassette recorder record and playback heads, consumable items, replacement fire alarm cable, fire alarm batteries, extinguishing media or propellant cartridges discharged for whatever reason, testing/replacement of pressure vessels/cartridges in accordance with relevant British Standards or other legislation, full installation tests of fire alarm wiring to BS5839 and any other specified exclusions, is chargeable to the Customer.
The cost of materials and labour incurred for work carried out to repair accidental or malicious damage to the System or to reset the System after misoperation by the Customer or his agents or servants is chargeable to the Customer.

Standard Service

The Company shall during the Agreement Period carry out   routine service visits in every 12 months during Normal Working Hours and provide emergency service facilities in accordance with the relevant British Standard

NOTES: The cost of materials only used in repair work carried out during routine service visits is chargeable to the Customer.
The cost of materials and labour incurred for work carried out during any emergency service visit is chargeable to the Customer.
During the Warranty Period defective equipment will be replaced free of charge except for goods not manufactured by the Company where the Company will pass on to the Customer the extent that it is so able the benefit of any guarantee received by it.

Emergency Service

The emergency service is available 24 hours a day under normal circumstances for Intruder Alarms and Fire Alarms but only during Normal Working Hours for Closed Circuit Television and Access Control Systems.

Equipment Schedule

To be completed if no Specification is to be provided

Control Equipment:

Signalling.

Devices

Registered Office: Western Road, Bracknell, Berkshire, England.
Registered in England No. 524469

The Electronics Group RACAL

RCA/ird
July 1990

GUIDELINES FOR INTRUDER ALARM CONTRACT FOR OUTRIGHT SALE AND MAINTENANCE

These guidelines indicate the minimum requirements for an intruder alarm contract for outright sale and maintenance only. They do not apply to contracts for rental/maintenance. They do not preclude the addition of other appropriate sections or paragraphs but it is important that the layout is followed so that customers shall be in no doubt regarding the scope of the contract.

The guidelines do not in themselves have any legal status. The actual words and content in individual alarm company's contracts shall remain their responsibility and they are strongly recommended to continue to seek legal advice in this respect.

A contract consists of:

> Offer documents, which should include:
> Quotation
> Specification
> Terms and Conditions
> and an acceptance Document

It should also be remembered that supporting literature may be taken to be part of the offer documents.

Quotation

The Quotation should make the following items clear

1. Address of Installation
2. The sale price of the Installation (including alarm equipment) to be sold to the customer
3. The annual maintenance charge

V.A.T. liability must now be stated by law and in the cases of fixed price quotations of this nature it is permissible to quote the tax rate rather than the actual amount.

e.g. Sale price of the installation (including alarm equipment £ . . . . . . (no V.A.T. payable)
Annual Maintenance £ . . . . . . plus 15% V.A.T.

The Quotation should also state:

"The installation commissioning and maintenance will be in accordance with the British Standards from time to time accepted by the NSCIA and an NSCIA Certificate of Status and Competence will be issued upon the satisfactory completion of the installation of the Alarm System."

It should be supported by a specification which can be integral or attached to the Quotation but this specification should be clearly identifiable to prevent any confusion.

Terms and Conditions

These are important supporting clauses to any contract. The order and basic content of these are given but it is stressed that the precise wording and the need for any extra clauses is a matter which should be determined by the alarm company and its legal advisors. It is important however that the main headings given, and their numbers, are used so that the terms and conditions can be easily identified.

Part 1

Definitions
These should particularly indicate:
> The installation and alarm equipment in the specification which are the property of the customer
> Date on which the Servicing Agreement becomes effective
> Duration of the Service Contract and termination notice requirements

Part 2

Customer's obligations:
These should include:
> Consent to access for both installation and maintenance
> Provision of Post Office facilities and payment therefor
> Non interference with the system
> Report of defects
> Payment for call outs including emergency service

Part 3

Company's obligations
These should include:
> Routine maintenance in accordance with B.S. 4737
> Emergency service
> Terms of guarantee and the basis for any additional repair or other charges

Part 4

Extent of Company's liability
(a)   That which the Company accepts:
    (i)   death, personal injury arising from Company negligence
    (ii)   direct physical damage arising from Company negligence and any limit in this respect
    (iii)   breach in respect of Sale of Goods Act 1893 (as defined by the Unfair Contract Terms Act 1977)
(b)   That which the company will not accept.

Part 5

General
Examples
> Termination for breach of agreement
> Notification of increased maintenance and/or service charges
> Force majeure (engineers industrial action etc.)

Acceptance Document

The customer's acceptance should be on the same piece of paper as a copy of the Quotation to avoid any error or misunderstanding of what the customer believes he has agreed to buy. A copy of the specification and any other supporting contractual documents shall also be attached to the acceptance. It is further recommended that customers are asked to initial individual sheets to protect against possible substitution.

FIGURE 1–2   NSCIA guideline agreement; Annex B to NSCIA Code of Practice Section 3

The final design will produce detailed, integrated drawings and specifications. The design should not be compromised by architectural expedience or whim without the client being fully aware of the consequences.

At the conclusion of the final design, a decision on specification and tender or single contractor negotiation must be made.

Whichever decision is taken, if the client is to finally

obtain the system presented, an accurate specification must be written. Working in close association, the design team and the client will be very clear on what they want. The contractors to provide and install the system may not have been privy to all of the liaison meetings. These contractors are not mind readers and if they are to provide a system acceptable to the client, a clear description (and sometimes drawings) must be provided along with the contract documents.

Security specifications are still reasonably rare. Traditionally, clients would describe their approximate requirements to each bidder and rely on the good reputation of the company to ensure that they got what they wanted. This is still the case with companies that are interested in purchasing a small security system. The increase in intruder detection system costs and complexity have made this approach untenable on larger, commercial projects.

Like all other businesses, security installation companies exist to make a profit. To make a profit, they must first ensure that they win the contract. To win a contract most companies take the myopic view that their potential clients are also myopic and the lowest price will win the contract. All too often, this is true. Hence, to reduce costs, some companies are constantly searching to produce or buy cheaper products. This behavior can reduce standards, but what do they care—you probably are a one-time client, and you will not pay for a full service contract so they will not have to service the shoddy goods!

To prevent this downward spiral and to ensure quality and value for your money, a performance and quality specification should be written into the contract. The specification should refer to standards of materials, workmanship, etc., and give clear descriptions of where devices are required, how they should operate within the context of the building, etc.

The major weapon to ensure high quality is to demand the use of standards and insist on compliance with these standards. It is important to realize that many specifications are written and installations carried out or supervised by persons who have never read these standards. On larger projects it is essential that clients employ an enforcing agent or adviser to ensure compliance.

Selection and Use of Security Companies.

How can you tell that the company you are about to invite to advise you is reputable, knowledgeable, and provides high-quality security services? Before employing any professional or expert to advise you on security matters, check their references and verify their qualifications with appropriate sources.

Advisers.   By far the largest and most well-respected organization of experts in the field of security is the American Society for Industrial Security (ASIS). ASIS has more than 25,000 members, 13 standing committees

advising members on particular fields of security based on activity (e.g., banking), and 12 committees for general security topics (e.g., fraud, fire). There are various membership levels. The highest level of membership is demonstrated by *active members* and, more specifically, those who have successfully passed Parts I and II of the Certified Protection Professional (CPP) Program Examination. These examinations can also be taken by non-ASIS members.

The qualification *CPP* is not a license to practice, but it is a good indicator of the general level of competence and professionalism of the adviser. CPP also requires recertification by gathering credits, which prevents CPPs from resting on their laurels! Out of all of the ASIS members, only 12% are CPPs.

ASIS-stated aims are to "advance professionalism in the field of security." To its credit, the CPP qualification does this. It could, however, try harder! As a society, it is following the old role model of the European Engineering Societies and adopting a we-don't-want-to-get-our-hands-dirty approach. Even the venerable British Institute of Electrical Engineers (IEE) has now abandoned this rather insipid approach and is *directly* influencing and lobbying governments.

From my investigations, I conclude that ASIS does not

lobby for common, USA security licensing/standards
produce guidelines for employers, the general public, or clients of ASIS members regarding the use of security companies or professionals and typical conditions of contracts engagement and payment for the services offered by ASIS members
publish minimum design or installation standards, nor review quality of advice given, design solutions, or working practices

Like members of most learned societies and institutions, ASIS tend to forget that it is their duty to use their learning in a *direct and active* manner to protect their less-learned brethren, the public. Similar institutions in Europe, with far fewer members, have a direct effect on legislation and produce national minimum standards that can be enforced by law.

Installers.   The listing of burglar alarm installation and service contractors from the burglary department at Underwriters Laboratories (UL) is the most universally recognized standard of competence for installers. The company wishing to achieve this listing may apply for

- local alarm service, Grades A or B
- central station service, Grades C, B, A, AA, BB, or CC

Achieving any standard required to become initially UL-listed is easy for a company, as the incentives to be listed (and, hence, able to issue certificates on UL's behalf stating that installation meets standards) are very high.

Maintaining these standards against regular inspection

by UL inspectors is far more difficult. When selecting an installer, you should

ask UL if the company is listed and at what service level

ask UL how many installations with certificates the company has carried out to date

ask how many installations have been inspected for compliance with standards and how many have failed

ask whether the company has ever had its UL authorization withdrawn

ask the company how many installations and of which types it has completed. Compare this with the number of UL-certified installations. If the first number is high, it may be that work never passed inspection and no certification was given.

Remember to check references. While most companies with a UL listing issue certificates for 25% of their total business, this is not a guarantee that the company did the work in the stipulated manner, without undue disturbance, within budget, without delay, or without putting faults to right before completion of the job. A former customer of the company will tell you far more about the good and bad points of the company and individuals within it than will the company sales team.

UL systems installed by UL-listed companies using UL products currently account for only 15% of the USA's total intruder alarm and security business. This 15% applies to the high-value, traditional risk sector, such as chain stores, fashion shops, jewelry stores, etc. As explained earlier, consequential losses on less visible or concentrated valuable goods can be enormous and the use of full UL-approved product installation and maintenance packages should always be considered. At least 40% or more of the USA's systems could be more effective for a minimal extra cost.

A UL listing usually determines

installation competence (e.g., trained staff)

head office and central station organizational excellence in finance, work force management, maintenance service, record keeping, fault response, and repair

annual, random, unannounced system inspections and audits by inspectors

annual unannounced testing of central station response to an alarm

lower false alarm rates

insurer's approval

It is essential, when making inquiries or selecting contractors, that you show that you know UL certification procedures and that you require the submittal of these certificates. To ensure this, a simple certification specification should be included in the design documents and issued to the prospective installer. Written

confirmation of their acceptance of these procedures and conditions should be obtained before awarding the contract. The certification specification might include requirements to

issue signed copies of the company's UL-approved documents with contractor response

confirm, in writing, that the installation will be a UL system, using only UL-listed parts

require bidders to confirm they are part of UL Certificate Verification Services (ULCVS)

list all parts of the installation and appropriate levels of protection that will be provided (alongside of their definitions)

provide staged completion certificates (larger projects only)

issue lists of grades of service proposed for each part of the system and the whole system

list any deviations from the specification or design

list all system parts, including the part number, maker's name, and a net trade price schedule of rates (this schedule can be used for agreeing to fair prices for variations in the work or increases or decreases in quantities during the contract)

confirm that system completion and final payment will not occur until all signed UL certificates, maintenance manuals, installation drawings, and instruction manuals have been issued, checked, and accepted by the client

confirm that the contractor will issue annual inspection notices signed by UL inspectors, stating the system has been inspected, tested, and faults have been listed and corrected. If no certificate is issued, ensure they reimburse you for all maintenance and service charges.

make it the installer's legal duty to inform you of any *downgrading of the installer by UL*

The UL system of listing and categorizing is an attempt to document minimum levels of security provision. In doing this, UL is to be applauded. They must, however, be used as a *baseline* from which to consider and improve the security of an individual property.

UL publishes classification charts UL 601 and UL 1076 that enable individuals and insurance companies to consider the minimum appropriate security package. These charts are broken down into four grade categories:

Grade of central station guard response—the speed with which the guard arrives at your premises

A   =   15 mins.
AA =   5 mins. + line of communication security
C   =   30 mins.
CC =   30 mins. + line of communication security

Note that grades B and BB are now defunct, and grades C and CC are seldom used as response time is too long.

Extent of alarm system protection — alarm systems can be one of three installation types:

1 — highest — all sides (walls, floors, roof) protected
2 — enhanced — openings protected
3 — maximum — either all openings covered, all openings plus some movement detection, or door openings plus full-volume space detection or full sound detection to all areas

Type of systems — systems are classed as

local alarm
central station connected
direct connected
proprietary

Local alarms may be grade A (new work) or grade B (extension to older systems). Grade A is the higher grade as the alarm and control panel have most of the recommendations made in later chapters of this book.

Central station–connected systems have a local alarm and a digital communicator that allows transmission of an alarm condition message to the nearest alarm company's operating central station via telephone lines. This type of system forms the bulk of medium-priced cover/response provision for commercial retail and industrial premises.

Police station–connected systems have a local alarm that has a 5-minute delay device built in. During this delay period, the alarm message is transmitted *directly* to a constantly operated police desk. The delay period gives the police a chance at catching the intruder before he escapes as he will continue to enter the building assuming no alarm bell equates to no detection of his presence. This catch-the-culprit-risk-the-stock approach is different from the normal insurance company approach, which is to insist the alarm rings immediately when the intruder is sensed in order to frighten him away before any damage to or theft of stock or buildings is caused. While this method is understandable, it does nothing to improve the arrest rate of criminals. Police station–connected systems are becoming rare as police power is stretched with other commitments. In Europe, these systems are now found only in government or strategic buildings.

Proprietary systems are in-house systems with their own local control and immediate response force, and are mainly used on large sites or buildings where a permanent security staff is justified.

Using this system of classification, a building might have a Grade AA central station response, a Type 2 installation protection, and a UL-listed central station connection with a local bell.

Finally, do not forget to check for liability coverage. All experts, consultants, and installers should carry full liability insurance for providing advice, design, and/or installation.

Competitive Tenders and Contracts. If your scheme is to go to competitive bidding, ensure that all contractors know the exact type of systems, sensors, wiring, etc., they are to provide. Never issue a pure performance specification, as the interpretations that can be made are too wide and the result may be a cheap, flawed system.

Never accept bids until you are satisfied that they comply with the specification in every respect. Lists of sensors with purely descriptive names (or even catalog numbers) are not sufficient. A wall guard or motion detector might sound as though it meets the requirements, but it may be totally inadequate.

Unless you are absolutely sure that an alternative suggested by a bidder will perform the required function *better than the original,* never accept any deviation from a bidder. The deviation is usually a cost-saving device in order to offer a lower bid than a rival. If the deviation is acceptable in principle, ask other bidders for their prices and reactions to this alternative.

Never accept a tender or quotation simply because it is the lowest. Always negotiate; ask why one system costs more than its rival. Never let salemanship stifle reason.

Never, never let a main building contractor or electrical subcontractor employ a security company as a domestic subcontractor. Always appoint, under a totally separate contract, with guaranteed access, or as a "nominated" specialist subcontractor.

Remember in your budgeting that intruder alarm cables should be housed within conduits separate from the normal electrical service. If you do not include the installation of these cables in the security contract, you or your client could have a rather expensive extra or variation order to pay later to any electrical subcontractor. Conduit work is expensive and security companies often exclude it in their proposals so that overall system costs appear lower.

Do not split the responsibility for installation. Make sure your specification makes it clear that the security company contract is a turnkey contract and includes 110V supplies, etc.

INSTALLATION

Introduction

The installation of an alarm system within an existing building or as part of a new structure should be viewed in the same light as any other mechanical or electrical service. No architect would dream of allowing an electrical services installation to proceed to completion without specification of standards of workmanship and materials, design drawings, inspection, and site supervision. Architects have a duty to their clients to supervise and check installation standards. It has become common practice for some architects to look at security system wiring as something secret and incomprehensible, and to place it in the same category as telephone wiring (i.e., something done by others). Unlike telephone wiring, security wiring must be completely concealed, secure, safe, and free from defects (patent or latent).

Standards of Workmanship

While there are many excellent standards for the quality of equipment produced for security systems, there are far fewer standards for good quality of workmanship and installation practice. In general, UL and other American standards tend to be less particular about the standards of items, such as the protection and concealment of interconnection wiring, than European standards.

The following list gives sources of information that can be used to define "good" workmanship in either a specification or as an installation proceeds.

- UL 1641
- UL 1076
- UL 681
- National Electrical Code (NEC), Chapter 1
- SIA White Paper on Interpretation of NEC
- British Standard 4737
- United Kingdom National Supervisory Council for Intruder Alarms Code of Practice for Design Installation Operation Maintenance and Servicing of Intruder Alarms
- Alarm Industry Quality Control Manual: A Comprehensive False Alarm Reduction Program (NBFAA)

The standard of workmanship on the installation is critical to the future reliability of the system. Security systems are very intolerant of even minor faults. If architects or supervising officers see any work being carried out that they think is shoddy, not to specification, or dangerous, they must challenge the installer. Bad workmanship will put clients at risk, reduce the reputation of the installation company, and result in endless repair calls. At best, most security systems will only maintain adequate security for a period of 10 years. During this time, sensors, wiring, etc., will need replacement. If the system has been designed and installed properly, with enough flexibility, sensor replacement or even complete rewiring will be relatively easy. If sensors are difficult to access (e.g., wiring has been stapled), this inflexibility will cost the client a great deal of money.

Both the security industry and the electrical contracting industry have become more aware, during the last decade, that poor standards of design specification and installation mean lack of client confidence and, hence, lack of work.

Training Standards

Security systems will only work correctly if they are properly installed by trained personnel. These people are often not electricians. Electricians may train as security system installers, but do not expect an electrician to know as much or perform the task as well as the trained security installer.

Most reputable companies train their installers properly, so these personnel understand the importance of proper, secure, faultless wiring; properly mounted sensors; etc.

In the USA, reputable companies are normally members of national or state organizations, such as the National Central Station Operators, State Association, National Burglar and Fire Alarm Association (NBFAA), ASIS, etc.

In Britain, alarm installation companies are approved by the National Supervisory Council for Intruder Alarms (NSCIA) and subject to NSCIA Code of Practice and British Standards. They are inspected by the Security Systems Inspectorate (SSI) for compliance with these documents. The joint organization is known as the National Approved Council of Security Systems (NACOSS).

Quality of Workmanship

What should one look for to determine quality of workmanship? The first step is to *know the standards*. By reading the standards listed in the "Standards of Workmanship" section in this chapter, you will find that they are not very precise and leave a lot open to the interpretation of quality of the individual installer, thereby leaving you vulnerable and without possible redress. Be aware that the person installing your system may *never* have read the standards or may not have read them for years. They simply think they are doing things correctly as that's the way they have always done them!

These factors emphasize the importance of using a reputable company and, on larger jobs, of setting out your own specification with a section on quality of workmanship written by a security consultant who specializes in systems installation.

UL Standards

Wiring. Generally, UL requires that the interconnecting wiring between alarm panels, alarms, and sensing devices

follow contours
be located in a place where it is *likely to suffer the least chance of accidental damage*
provide mechanical protection where it *crosses* a joist

be protected by two layers of tape where it passes around sharp corners

be fixed at intervals of 2 ft (0.6 m), 4 ft (1.2 m), or 6 ft (1.8 m), depending on the material

have spliced then soldered joints, and washer or lug terminations

be segregated by at least 2 in (5.1 cm) from pipes, conduits, grounded objects, or signal wires, *or* be protected with two layers of tape

pass under, not over, steam or hot water pipes, etc.

be sleeved where it passes through wood floors and walls; taped is acceptable through concrete or masonry

be segregated by 2 in (5.1 cm) from power and lighting circuits or one of them be placed in electrical insulation or conduit

be fire-stopped where it passes through walls

These definitions are far too loose. European standards are more rigorous and their interpretation more uniformly followed on all but residential systems. For example, consider these standards:

Cables must follow contours, but must be at least 4 in (10.2 cm) clear of any likely fixing point (e.g., at the corner between the walls and the ceiling, where studs might be fixed or pictures hung) and not immediately beside door frame uprights. Cables must travel in straight lines horizontally or vertically (i.e., they must *not* travel diagonally across a flat wall).

"Likely to suffer damage" is a bit meaningless, particularly when wiring is permitted to be stapled directly to the side of wooden joists. These often have protruding nails, rough surfaces, etc. Wiring through roof spaces, above false ceilings *between* joists, etc., should be protected by a tough sheath or, as is common in Europe, encased in a high-impact polyvinyl chloride (PVC) or metal conduit throughout its length. This significantly reduces line faults, tampering potential, etc.

Jacketless (unshielded) cables should not be used; they are too prone to damage.

Any cable buried in plaster or passing through *any* wall or floor material should be sleeved in conduit or trunking and internally/externally fire-stopped to the same *fire resistance as the building material* (e.g., ½ hr, 1 hr, 2 hrs, etc.).

Cables should pass nowhere near steam or hot water pipes, if possible, and should always be sufficiently distant to prevent any rise in surface temperature above the general ambient temperature for which the cable jacket was designed.

Cable Types and Sizes. For cables, UL sizes are the *minimum*. However, the appropriate size might be greater. Always use cables that are jacketed and multi-

stranded, rather than solid drawn copper. This reduces likely bending fatigue (crack resistance increases, developing 6 months after cable installed), makes cables easier to pull into conduits, and reduces possible stress on the cable core. Also, use the largest cable core size permitted by the situation. Cable sizes are often used by habit. A larger core size costs very little extra (the installation labor dominates the price). The larger core reduces volt drops, increases future additional load capacity and sensor potential, and increases reliability due to strand redundancy and increased mechanical strength.

Terminations, Joints, and Splices. UL acceptance of soldered joints, and particularly washers, to clamp cables to termination studs is poor. Of all the sources of potential fault, poor terminations and joints are very high on the list. Good systems use insulated terminal blocks and connection boxes with sealed lids (not splices and solder), flat ring eyelets and crimped-on termination ends (not screws pinched directly onto the cores), or cores wrapped around the pillar screw and held down with a flat washer and a nut.

British Standards

In Britain, security industry practice requires that those involved ensure that there is no division of responsibility for system installation. In new construction, it is common for security conduits to be provided by the general electrical contractor as part of their normal subcontract. The security alarm installer gives the routes to the on-site electrician in the form of a drawing. This method is advantageous to the security company, as it is hiding the labor cost involved in providing conduits and thereby making installation appear relatively cheap. For the installation to proceed to a satisfactory conclusion, one of the following approaches should be adopted:

1. Make the security company provide a quote for the complete installation inclusive of conduit installation.
2. Allow the conduits to be installed by the electrical subcontractor, but ensure that the security company inspects the conduit work and confirms, in writing, that it is totally satisfactory before wiring is started. The security company is then responsible for the standard of workmanship and conduit provision given to them.

Ensure that

all wiring is totally segregated from other mechanical, electrical, plumbing, and heating services

wiring (unprotected or in conduits) is not exposed to hostile environmental conditions unless suitably rated or protected. Damage to or danger from the system could result from high or low temperatures (ambient or contact); moisture; water vapor; steam; corrosive liquids, vapors, or gases; and inflammable or explosive solids, liquids, or gases.

wiring is never run within lift shafts, dry risers, gas trenches, etc.

wiring is never run under carpets unless of a type specifically designed for such an application (flat armored or similar)

all parts of the system are properly grounded and bonded in accordance with NEC or IEE regulations

all power supply circuits are properly protected by fuses, miniature circuit breakers (MCBs), etc.

where cables are run underground, they are buried to a depth of at least 12 in (300 mm) (although this is commonly 3 ft or 900 mm), protected, and provided with a buried warning tape

no more than ten sensors of any type are installed on a single detection circuit. This reduces load and lack of cover under faulty conditions, and eases fault finding.

all conduit box covers are permanently fixed (e.g., secret key, allen key), access for rewiring is allowed, and draw wires are provided

circuits are provided with adequate numbers of test points in concealed but accessible positions

wiring is identifiable by means of a color-coding or cable-tagging system *known only to the installation company*. This will reduce tracing time should faults develop or the system require extension.

Remember that persons installing the wiring do not have to be trained electricians and often are not. They should, however, install systems at the same standard of workmanship as a fully trained, approved electrician.

BS 4737, Part 1, 1986 is applicable to all internal complete systems and states that the system shall comply with the British Standard in the environmental conditions to which it is likely to be subjected at the protected premises. These conditions include mechanical damage, weather, heat, dampness, corrosion, oil, electrical interference, or adverse industrial atmosphere. This standard is very stringent on the installer. If the company finds it is continuously cleaning sensor lenses or replacing sectons of corroded conduit, it cannot complain or charge the client unless the conditions have changed significantly since the installation date. It is, therefore, essential that room data sheets include environmental records.

INSPECTION, TESTING, COMMISSIONING, SETTING TO WORK, AND HAND OVER

The inspection, testing, commissioning, setting to work, and handing over of a system should be exhaustive. These elements are essential for reduced false alarms,

client satisfaction, and profit. The electrical, mechanical, and security services industries have long recognized the fact that these matters are poorly covered by standards or specifications or adhered to, even when they are issued. Systems should be subjected to the following tests:

- mechanical, visual, and audibility
- operational
- electrical
- communication

All systems should be inspected, tested, set to work, and commissioned in accordance with UL, ASIS, or British NSCIA Code of Practice guidelines. The work should be carried out in a continuous, logical progression by skilled personnel. The person performing this work *should not* be the person installing the system. The inspector should act as the devil's advocate and seek weak points in the system that could affect its operation or effectiveness. The inspector should be senior in rank to the installer and be able to demand clarification of any points the inspection raises. There is no reason why the appointed client representative should not accompany the person testing and commissioning the system in order to ensure that this individual is not simply going through the appropriate motions.

Poor contractors avoid proper testing; low bid prices on this element of the work often reflect this.

Testing, setting to work, and commissioning should include the following elements and the results should be listed.

Inspection (Visual)—Record Findings

Check wiring for compliance with UL standard or BS 4737, Section 3.30, 1986.

Check wiring for damage to cores, missing insulation, and stressing.

Check wiring for joints outside junction boxes and unapproved joint methods.

Check wiring for continuity of color-coding/segregation from other services, provision of screening, interference suppressions, etc.

Check wiring for encasement in conduits in which physical damage could result; remove from areas in which manipulation could take place.

Check wiring for suitable surrounding environment (e.g., no more than 86°F or 30°C).

Check wiring for bridge loops and correspondence with drawings, routes, etc.

Check conduits for grounding; continuity; glanding, bushing, and burr removal; periodic pull in boxes; and fixed/secure lids.

Check transformers for fusing, ease of access, and surrounding by noncombustible materials provided with permanent connection.

Check power supply unit (PSU) batteries for stated

capacity, battery type, recharge time, mechanical security, venting, ease of access and maintenance, and electrical protection and connection.

Check electrical mains supply for locking provision and fusing size.

For system sensors, refer to Chapters 2–4 of this book for issues dealing with applications, number of sensors per loop, security of fixing, and correctness of model type against specification.

For system alarms, compile a defects list and list of corrective actions taken; record all measures.

Electrical Testing

Disconnect all parts (particularly electronic parts) that could become damaged by testing and test

mains supply wiring and adherence to NEC regulations. Provide test certificate with circuit insulation resistance, ground loop impedance, continuity, etc., from the mains to the power supply unit.

grounding and isolation of transformers

voltage at primary side of transformer

continuity of interconnecting wiring with at least a 20,000 ohms per volt instrument

For the power supply unit,

measure the output voltage and current for both high- and low-charging voltage

measure individual battery outputs when practical; ensure they are fully charged

measure battery resistance with the system set and the mains supply disconnected

For the sensor connection point,

measure voltage and current supply under quiescent and alarm conditions, particularly alarms and movement detection sensors

Record all test results on system log book test sheets. Identify satisfactory readings, take corrective action, and retest the element. Carry out normal electrical testing of

- isolation
- polarity
- resistance of circuits
- continuity of conductors
- insulation resistance
- ground continuity and impedance (loop)

Record all of the results for every part in the system.

Measure the full load current demand on each circuit and the total current demand of the entire system. Check that when this value is divided into the rating of the standby battery (in ampere-hours), it complies with the required standby. Check that it also complies with the recharging time demanded by the standard. Check that the automatic switching of the system from the mains to the battery is working correctly.

Functional Testing

Testing the individual functions of system parts *is not* the same as commissioning. *Commissioning* is testing of the *entire system* in the modes in which you wish it to work to ensure it does so reliably. An analogy may be drawn using an air-conditioning system. Testing of a fan within the system will only tell you the fan works; it will not guarantee that you achieve the correct airflow rate, temperature gradient, humidity, etc., within a room or building.

Sensors. The overriding criteria for testing sensors should be that they

sense the presence of an intruder reliably repeatedly every time they are challenged

they are not affected or raise a false alarm when subjected to the actual working environment

To ensure this, sensors should be tested for correct operation against the *specific criteria* stipulated in the relevant section of the appropriate standard, under normal working conditions. Walk or field tests should be at the limits of range, speed of movement, etc. *Anticrawl zones* on movement detectors must be challenged and shown to work!

Sensors should be tested under *working conditions*. This could involve making telephones ring in the vicinity, slamming nearby doors, etc. There is little point in starting a *proving period test and record* if possible problems can be identified early.

When the preceding tests are completed, arm the movement detector in the system, but connect it to a fault analyzer for a 7-day period. If, after this period, no faults are found in the system's operation, the detectors may be connected to the control panel. If faults *are* found, corrective action should be initiated, followed by a further 7-day period of analysis as necessary.

Any device incorporating an antitamper alarm should be challenged to prove its operation.

Alarms. Use a sound level meter to check the audibility of internal and external alarms and buzzers, in all areas in which they are expected to be heard, under the worst ambient noise conditions that might reasonably prevail (road traffic noise, etc.). Sound levels should be checked against requirements of the appropriate standard.

Test the operation of self-powered bells to ensure all modes are operational. Bell delay, time-out, and rearm functions should be tested and recorded.

Communication Tests

Soak test—Before any security system, including vibration or volumetric sensors (e.g., beam, microwave)

can be directly connected to the communication medium, it should be operated under normal conditions on the premises for a period of 14 consecutive days without a single fault of any sort occurring. If any fault occurs, the test must stop and the fault must be traced, analyzed, recorded, and rectified. The 14-day test period must then start again.

Telephone line test—Check the number of lines, line quality, and message transfer integrity.

Ensure the tape dialer/digital communicator is within the secured area and is fixed and locked.

Check the wording of the message to be sent and the clarity of the message at the security company's central station or police station.

Ensure system functions are operating correctly.

Verify receipt of an alarm audibility test certificate, state licenses, telephone company certificates, etc.

Operation and Maintenance Instructions

The client should receive an operating and maintenance manual that includes

- a general description of the system
- information sheets on the types of detectors installed (the manual should not identify where these are installed)
- a clear, concise series of operating instructions for daily setting and closing procedures and for alarm call procedures
- the everyday and emergency maintenance service telephone numbers

This manual is often included in outline form as part of the security company's proposal. However, it often requires expansion and modification once the installation is complete.

UL 681, Appendix A provides an in-depth explanation of the certification processes for UL-installed systems. The UL also issues a book entitled *Procedures for Maintaining Burglar and Fire Alarm System Certificates*.

UL and other standards require installation and operating instructions to include record drawings. UL certifications should be included in manuals, as they give a synopsis of the system.

BS 4737, Parts 1 and 4.2, 1986 require that the installing company provide and securely retain records of the installation. The documentation referred to in Part 1, Section 3.2.4 of the British Standard is normally referred to as the *as-built drawings* in other electrical industries. These drawings are a record of the exact location, circuit routing, etc., as established at the end of commissioning and they show any changes made to the original specification or drawings. The drawings should provide the following information in order to comply with the British Standard:

- name and address of premises and subscriber
- type and location of detectors, sensors, processors, power supplies, warning devices, control equipment, signaling devices, etc.
- details of any client-isolated circuit facilities (e.g., sensor or warning device)
- details of any entry/exit routes

A copy of these drawings should be given to the subscriber and they should be secured. These drawings *should not* form part of the normal operation and maintenance manuals, as the details are in excess of those required for an engineer who might use the manual on a day-to-day basis to set and unset the system.

Hand Over

At hand over, the client or client's representative must accept partial responsibility for the alarm system. The client must also pledge to carry out certain maintenance duties not within the alarm company's remit, but absolutely essential to the efficiency of the system. Before hand over can occur, the client must be satisfied that

the system is as specified and the client understands its operation

the system has been installed based on a high standard of workmanship

the system has been fully tested and commissioned

the system is reliable, effective, and simple to operate

To this end, the client should insist on demonstrations, test certificates and drawings, and records.

Demonstrations.   The installer should demonstrate system operation and the effectiveness of all sensing devices by challenging the system. The client should be shown how various sensors can raise false alarms (e.g., shifting stock, installation of electrical heaters) and should be invited to try and "beat the system" within a given time frame in order to gain confidence and expand personal knowledge. If clients believe in their systems, they are more likely to take measures to ensure that their employees operate the system correctly.

Test Certificates and Drawings.   The client should examine signed copies of

- UL certificates (see Figure 1-3)
- certificates of status and competence
- NEC Regulation Test Certificates
- an alarm record book (to be given to the client)

**CENTRAL STATION
BURGLAR ALARM SYSTEM CERTIFICATE**

File No: BP9999 CCN: CPVX
Service Center No: 0
Expires: Mar 23, 1991
Issued: Mar 24, 1990 (Renewal)

UNDERWRITERS LABORATORIES INC.
NORTHBROOK, IL   SANTA CLARA, CA
MELVILLE, NY   RESEARCH TRIANGLE PARK, NC

an independent, not-for-profit organization testing for public safety

THIS CERTIFIES that the *Alarm Service Company* is included by Underwriters Laboratories Inc. (UL) in its *Directory* as qualified to use the UL Listing Mark in connection with the certificated *Alarm System*. This Certificate is the *Alarm Service Company's* representation that the *Alarm System* including all connecting wiring and equipment has been installed and will be maintained in compliance with requirements established by UL. This Certificate does not apply in any way to the installation of any additional signaling systems, such as: fire, smoke, waterflow, burglary, holdup, medical emergency, or otherwise, that may be connected to or installed along with the Certificated *Alarm System*.

LIMITATION OF LIABILITY: Underwriters Laboratories Inc. makes no representations or warranties, express or implied, that the *Alarm System* will prevent any loss by fire, smoke, water damage, burglary, hold-up or otherwise, or that the *Alarm System* will in all cases provide the protection for which it is installed or intended. UL may at times conduct inspections of the *Alarm Service Company* including inspections of representative installations made by it. UL does not assume or undertake to discharge any liability of the *Alarm Service Company* or any other party. UL is not an insurer and assumes no liability for any loss which may result from failure of the equipment, failure to conduct inspections, incorrect certification, nonconformity with requirements, failure to discover nonconformity with requirements, cancellation of the Certificate or withdrawal of the *Alarm Service Company* from inclusion in UL's *Directory* prior to the expiration date appearing on this Certificate.

ALARM SYSTEM DESCRIPTION:

System Grade: AA
System Type: Premises
Keys to Property: No   Line Security: Employed
Method of Alarm Transmission: Multiplex

Alarm Investigator Response Time: 15 minutes
Extent of Protection: 3
Alarm Sounding Device: None

Protected Property:

ELECTRONICS COMPANY INC
154 MAIN RD
NORTH, NJ 07647

Alarm Service Company:

AN ALARM SERVICE CO
333 PFINGSTEN RD
NORTH, NJ 07621

SN: **BC51145372**

Alarm Service Company's Representative

Date _____, 19____

© 1989 UL Form CS-C1

FIGURE 1–3 UL certificate

- police notification of registered key holder and system connection date
- as-installed drawings
- state licenses

Records. The installation company should keep records under the following headings:

System — specification, as-installed drawings, operation and maintenance instruction manual, etc.

Historical — date of every visit to the system by any company representative, time of visit, signature of responsible person on premises confirming visit, fault and cause, unit action taken, time of departure, etc.

Preventative maintenance — historical data along with details of tasks carried out and faults found (system or client operation)

Corrective maintenance — historical and preventative data along with a full description of the reason for the call out, fault found, and conclusive actions taken

Temporary disconnection — all preceding data, but with details of disconnections made, reasons, date to replace/reset, etc.

The normal standards indicate various lengths of time for this information to be retained, but, in general, no record should be destroyed before 2 years from the event to which it refers.

Maintenance

Intruder detection and alarm systems are, of necessity, sensitive responsive systems. If they were not, they would never detect an intruder. (Sensitivity should not, however, be equated with instability.) All sensitive systems require comprehensive regular maintenance. To a client, this maintenance can appear expensive and difficult to justify. It is also difficult to see positive results while it is happening.

Maintenance falls into two categories:

1. *Preventative* — routine servicing on a scheduled basis
2. *Corrective* — emergency responses to faults or false alarms

Unfortunately, in the past, many smaller companies regard the former maintenance method as simply a means of "checking the batteries," with no investigation or identification of potential faults. This has been encouraged by lower fees for preventative maintenance and higher fees for fast response, corrective maintenance. Poor attention to preventative maintenance results in large profits, but very dissatisfied clients.

A properly designed system will have been designed to a specific

- system life span
- mean time between failure of certain components
- level of noncritical redundancy/safety margin

These factors influence the overall system price and quality and, of course, consequential maintenance costs and the frequency with which they are charged.

Section 6 of the British NSCIA Code of Practice stipulates the minimum standards of maintenance that should prevail throughout the British industry. These standards must be complied with by the contractors listed on the NSCIA roll of approved contractors. Before signing any agreement, the client or client's representative should check that this standard prevails and that there are no hidden maintenance costs. Even with the best system in the world, regular maintenance and emergency maintenance will be required (as a result of vandalism, accidental damage, etc.). Labor costs are expensive, so check the works included in "normal service."

Most reputable alarm companies will offer a client a choice of maintenance agreements, depending on whether the client is buying, renting, or leasing the security equipment. If the cost of the installation is likely to be high, a client is, in most circumstances, best advised to rent or lease the installed equipment. If the equipment costs less; was bought or leased from a smaller, less well-known company; or is installed in a difficult situation, the maintenance agreement should offer *full* or *comprehensive service*. This equipment is more likely to give problems, and the service and maintenance agreement should include at least

- full maintenance every 6 months
- a limited number of *no charge,* normal working hour repair calls per year
- emergency repair service outside working hours and chargeable on an agreed-fee-and-expenses basis

If the system is leased, the cost for all materials included in any repair should not be charged to the client.

If the system costs more and was bought from a large, reputable security company, it usually will be less troublesome. In this instance, *limited cover maintenance* (at reduced cost) is usually sufficient. This type of coverage includes

full maintenance every 6 months or less, in accordance with NSCIA or other codes. Labor and materials required for repair of defective or worn equipment should be provided at no cost to the client.

all other repair calls at any other hours being charged to the client (i.e., labor and materials)

Remember that maintenance includes testing and inspection. In-house maintenance is secure and effective only if the employee is trustworthy and properly trained.

Grades of service exist for emergency call-out corrective maintenance programs. Many companies will agree to full comprehensive or partial security system service, but they are often very shy when committing themselves to a *grade of service* for emergency situations.

A grade of service brief should be established and turned into a statement of requirements that both parties can sign. The system owner should compile the document by taking *each system in turn* and deciding

the maximum time the system can be out of commission

the maximum time each part can be out of commission

the worst-case failure scenario

From this, the owner should determine

how long it would take to contact the repairers

how long it would take the repairer to reach the premises

how long it would take to fix each component

how long it would take to recommission, test, and complete each repair

what provisions should be made for on-site or repair company stock replacement

which components are quicker and more cost-effective to repair than replace

the costs for overtime and additional guarding that would result from failure to respond

A written grade of service requirement may then be compiled along the lines of, "System A—Attend site within 1 hour of receiving authorized instruction. Provide all spares by holding replacement stocks of items A, B, and C. Complete repairs under any circumstances within 2 hours."

Companies often ask, "What if traffic jams occur?" That is not your problem, it is theirs. For a fee they take a risk and compensation is payable if they fail!

Corrective maintenance should be initiated for all conditions, emergency responses to a client, or central station calls to correct faults or false alarms. The response time (the amount of elapsed time from call received to engineer on the doorstep) should not exceed 4 hours (unless the system is on an offshore island or is a local, audible alarm system only).

If the 4-hour period is impractical or excessively expensive for the client, a longer response period could be entered into a written agreement, but only with the approval of the insurers.

Always ensure that you see certificates of training and competence of the staff that the company sends to you. Make sure you have a photograph, personal details, description, etc., and keep these with your on-site security staff or key holder. Many crimes have been committed by persons arriving at the site, unrequested, to "fix a fault."

Preventative maintenance should be carried out within 12 months of commissioning on all local, audible-only, mains supply, and rechargeable battery standby power systems. Maintenance should also be carried out within 6 months of commissioning on all local audible or remote signaling systems using primary-only or rechargeable battery systems.

Preventative maintenance includes

a check of all positions, parts, routes, etc., against the records

a satisfactory operation check, including challenging tests of all sensors, not a simple panel function test

a check on flexible connection conditions

control panel function checks

audibility and functional audible checks and tests

These tests should be completed within 21 days.

It is essential that clients ensure that maintenance is properly carried out by a *trained* staff. It is very easy for both trained and untrained staff to give the appearance of carrying out routine preventative maintenance when they are, in fact, doing nothing, because they are bored or because they really do not know what they should be doing.

REVIEW OF SYSTEM

Do not rest on your laurels! You now have a system well made, correctly specified, and expertly installed. But remember, risks, technology, and people all change with time. Two years may have elapsed between specification and final system hand over. In that time, management plans, operational responsibilities of staff, and equipment and goods to be protected may have changed. Review the situation. How long will the system be adequate? Start planning for the future!

Standards. *Standards* protect the foolish, advance the knowledge of the wise, and give good companies baselines from which to improve. America has very few national *mandatory* laws, statutes, or standards relating to security in the commercial, industrial, and private sectors. Current minimum standards drives are being led by litigation cases based on "unfitness for use." This, however, means a very long and piecemeal progression toward what should be acceptable as a minimum.

Insurance companies set standards, but not "too high," in case individuals decide to go elsewhere for a cheaper quote. Trade and professional societies protect their own members. Some individual states have mandatory licensing and regulation, but minimum standards vary enormously. To be honest, the whole arrangement is a mess in which the system user has no real guarantees at all. "Buyer beware" is your best policy.

The organizations that stand out from the crowd in this sorry state of affairs are the American Society for Testing Materials (ASTM) and UL. The mandatory regulation of security systems and services by the government, on a nationwide basis using ASTM and UL standards as a starting point to write new codes of practice, is well overdue.

Appendix 2 lists most standards relevant to the security industry. Standards fall into the following categories:

manufacturing and testing specifications for component parts and individual elements of systems [UL, Loss Prevention Council (LPC), and British Standards]
specifications for sensors, alarm panels, etc.
Codes of Practice for selecting, planning, installing, and maintaining systems
guides for potential end users or novice specifiers in applying security equipment to particular building types
statutory regulations issued by the government, state and local authorities, and licensing authorities (radio regulation, noise abatement, etc.)
industry standards, such as self-regulating, self-policing standards issued by associations, trade policies, and company policies
insurance requirements and standards

Manufacturing Standards. One of the most important things to realize when looking at security products is that standards vary enormously. As in most other areas, you pay your money and make your choice. It is not self-evident that, although an item may claim to comply with "Manufactured Standard Specifications," this is primarily an únpoliced, claimed performance in use specification and is not a seal of approval. Many USA-manufactured products, such as CCTV systems, may, in fact, have a body from Ireland, camera and tube from Italy, lens from Japan, monitor from Britain, and motion detector from Australia. The *whole* assembly may never have actually been *tested* against the standard. *The claim is only a claim.*

Often, the component parts of a device might be well made, but the assembly may be poor. Fortunately, there are some key points for which to look among the detailed specification sheets issued by companies to assess quality. Some standards organizations within the United States follow.

UL Standards. UL carries out testing, inspection, and certification in many fields of engineering. The security industry, its companies, and components are among them. UL's main roles are

to certify, by UL listing, the quality, standard, or manufacture and performance of components (e.g., safes, motion detection sensors)
to certify, by UL listing, complete systems
to list companies that demonstrate competent levels of expertise and standards of service, repair response, etc.
to police the preceding actions by field inspections

The selection and use of UL-listed company products and systems will enhance the quality and reliability of that installation. However, remember that

it is up to you to check credentials. Claims of listing may be false.
UL-approved installers do not have to use UL-listed components or systems (i.e., they may offer cheaper products)
UL products may be good and safe, but may not be the best or state-of-the-art (many European standards are equal or better). Check the standard if you are not sure of performance.
insurance companies and UL do not warranty the installation. Insurance appraisal of the merits of a device is based on limited feedback. If your insurance company recommends a device and you are

robbed, the insurance company will pay (usually not in full) and will then demand larger premiums next year to recoup their loss, while insisting you pay for more security to cover failings in their specification!

the UL listing covers far more than manufacturing standards and later sections of this book show the value of the UL label or certificate

people selling devices often have no *real* knowledge of the product. They may have been selling cars the day before. Therefore, check that the product is stamped as complying with the latest revision of the appropriate standard and has the features you need and want. UL-listed performance criteria may not be the criteria you need in your device.

UL publishes lists of firms that have had products tested and lists the use of every product type tested each year.

ASTM. The ASTM F-12 Committee on Security Standards and Equipment is involved in the harmonization, extension, codification, and production of new standards for the protection of people and property. It is probably the only organization in the security field that is taking a *global* view of minimum manufacturing, installation, and maintenance standards for systems. It is also involved in producing guidelines for the methods by which protection can be provided.

National Institute of Technology and Standards. This organization (formerly called the *National Bureau of Standards*) publishes standards and technical reports, which are available to the public, that cover specific products, such as control units for intruder alarm systems.

Department of Energy—SANDIA, TAPIC, and so on. These laboratories test and establish standards for specific installation types or users, primarily military and federal agencies. Although they have spin-offs in generating discussions, extending the boundaries of the possible, etc., they publish few standards directly applicable to commercial security.

Security Industry Association (SIA). The SIA is a nonprofit trade association of more than 200 larger manufacturing, distribution, and sales companies operating in the security industry. SIA membership shows commitment to compliance with standards, understanding of the need for proper training, and a real understanding of products and their quality. The SIA issues its own white papers on various products and developments, and helps to develop national codes and standards.

International Electrotechnical Commission (IEC) Standards for Testing. Although there is little European harmonization on the tests that *must* be applied to particular security components, any literature that states that completed parts or systems have been tested to vari-

ous IEC environmental test procedures and standards will be worth reading (e.g., IEC 68 and IEC 839 for humidity and temperature, respectively).

British Standard (BS) Specifications. These are the most well-known European documents in security literature. These documents specify the *general requirements* of devices (e.g., an ultrasonic detector shall operate at a frequency of xHz with a coverage of ym^2). Remember, although a product may be marked "Complies with BS 4737 . . ." certain areas of compliance are very weak in most British Standards, so if you feel that you might have a problem, ask or specify the conditions. For example, an interference requirement might say, "shall be protected adequately against any interference that may normally be expected to occur in the working environment at the detector." What is adequately? How was your company supposed to know about your 100V/m radiation field generator next door!

British Standards may be amended—probably more often than manufacturers care to revise their labels!

BS 4737 is by far the most used document within the European industry. It was, however, published more than 16 years ago and, although amended, is at risk of being overtaken by new technological developments. False alarms in components and systems that comply only in spirit with BS 4737 are still far too common. The LPC's NSCIA (an installer's supervisory body) and insurance companies now operate a UL-style system. Insurers demand NSCIA-approved installers. NSCIA companies only use LPC-tested devices that meet BS 4737.

British Standard Codes of Practice. British Standard Codes of Practice supplement specifications by providing a shorthand description of many of the factors that should be considered with the application of British Standard specification devices or systems. 1987 saw the publication of *BS 4737, Part 4: British Standard Code of Practice for the Planning Installation of Intruder Alarm Systems in Buildings*—a useful document somewhat overshadowed by the publication of the first edition of this book in the same year!

British Standard Guides. The year 1986 saw the first guides for security against crime with the publication of BS 8220, Part 1, *Dwellings*. The year 1987 saw publication of BS 8220, Part 2, *Offices and Shops*. As their titles indicate, they are guides to the general security measures recommended for particular premises. They were published principally at the request of the British Home Office to provide free, basic security advice and, hence, increase the use of security products in the battle against crime.

Both guides are of excellent quality, although perhaps limited in scope and detailed advice. Many of the items of advice relating to physical security are expanded in *Security of Buildings* by Graham Underwood. The main bulk of the advice given in these guides relates to the prevention of intrusion by careful design, selection, and installation of fixings and locks to readily accessible doors and windows.

Government and Local Authority and Licensing Authority Regulations (Statutory). These regulations are generally restricted to areas of public concern, such as competence, gun carrying, noise nuisance from sounders, radio communication interference, etc. In some states, regulations and licensing procedures have been adopted that implement the findings and recommendations of the *Law Enforcement Assistance Administration (LEAA) Task Force Private Security Report 1976*. These guidelines outline the recommended public policy toward the advancement of joint, private, and public security industry initiatives to defeat crime. The report provides goals for the industry to achieve including

- qualifications for personnel (i.e., examinations, tests, experience, and screening)
- training of personnel at all levels and for various skills
- codes of ethics

The *Security Letter Source Book*, published by Butterworth–Heinemann, lists all state licensing authorities, giving their contact telephone numbers and a résumé of requirements. (See Appendix 3 for US and UK associations' short form lists.) The variations in the required standards and the amount of enforcement of the standards from state to state are enormous. Never take it for granted that a person, system, or installation company accepted in one state is accepted in the next state!

2
System Composition

SENSOR CLASSIFICATION

Initial Considerations

When considering the individual parts that make up the system, it is helpful to use an anatomical analogy in which the sensing devices are the eyes and ears of the system, the system wiring is the blood vessels or nervous system, the heart is the power supply, the control/processing panel is the brain of the system, and the signaling devices are the voice of the system.

If these parts are not properly assembled, installed, maintained, and, most important of all, operated correctly, a monster will be created that nobody will love and whose life will be very short!

All alarm systems are comprised of at least some of the following elements:

the protected area, volume, material, etc.
the sensing devices (i.e., *sensors* or *detectors*)
the signal cabling (field or loop)
the opening/closing (i.e., arming) device
the signal processor, analyzer, and control unit (i.e., alarm panel)
the "annunciator" (i.e., bell, alarm, or siren)
a remote signal-posting device (e.g., tape dialer, digital communicator)
the communication and transmission channel (e.g., telephone link, radio link)
the central facility or control room
the reaction force (police, private security organization)

It is essential that the roles of each element are fully understood. These elements are links in a chain of protection. If any link fails, the system fails. It is just as important for the annunciator to be correctly located as it is for the sensors. On a simple system using only an external bell, if the bell is removed before intrusion takes place, the intruder does not need to worry about detection—no alarm can be given to the reaction force. The system is not deaf and blind, but it is certainly dumb!

Surveying the Protected Area

The selection of a particular sensing device to cover an area or protect a particular surface requires

special knowledge of the latest devices available (e.g., their advantages, limitations)
practical installation experience
a very logical approach to the problem at hand to see whether the sensor is suitable, applicable to the risk, etc.

All of these services can *only* be provided by a trained specialist in the security field. However, it is a sad reflection on the security industry that no national licensing or governmental regulating body exists to police the industry with statutory instruments. Also, many of its members have little practical, hands-on experience of the installed systems. This is particularly true of many sales personnel. Unfortunately, the industry is fragmented. There are small companies that are very specialized in a particular field and some that have broad expertise and are large and reliable. Most, however, are small, financially vulnerable companies using untrained operatives, sales staff, etc. Many have little insurance coverage, most do not manufacture the equipment they install, and some even subcontract installation and maintenance.

With these facts in mind, buyer beware! When considering an installation or installation method, have the system, the sensors, the mode of operation, and the system's limitations explained to your satisfaction. Ask for alternatives and alternative price quotes, but never bully a company into using systems or sensors in situations that are totally inappropriate. Use the material in this book to help unravel the jargon, quiz the salesman, and gain a better understanding of the effect the system will have on the day-to-day operation of the building, the appearance of the building, and the protection the system will give.

Sensors

The sensors used in your system will depend, to a great extent, on the types of threat posed, the materials to which they are affixed, and the environment within which it is expected to work. It is important that the person designing the security system be given accurate information on all aspects of the building, the materials of construction, etc., and be notified of any changes during the detailed design phase. For instance, in dealing with an interior area, the designer must know, among other things,

- the likely threats
- the asset value
- the materials of construction
- the method of heating, cooling, and ventilating the room
- the purpose of the room
- room contents—furniture, machinery, equipment
- surfaces in the room—glass (e.g., laminated, annealed), walls, partitions
- room finishes—hard surfaces, carpets
- room size and shape
- the positions of the doors and windows
- the depth of false floors and false ceilings
- the periods of operation

The variety of available sensors, the number of different types within sensor groups, and the ways in which they interact, will at first seem totally bewildering to the layperson. It is this apparent complexity that has discouraged many clients from questioning the suitability of a sensor for a particular application, even when the client has information on the room or structure that may possibly make the sensor troublesome or useless. The attitude "leave it to the experts" is not a delegation of responsibility, but an abandonment of it!

In order to simplify the choice of a particular sensor, it is necessary to classify sensors in some way. Fortunately, this is easy and the range of *appropriate* sensors falls exponentially after a few simple questions are answered.

Sensors can be classed in a manner analogous to biological nomenclature (see Figure 2-1). Sensors fall into two *classes*—internal sensors and external sensors. Internal sensors are greater in number, variety, and purpose,

and are generally simpler, more reliable, and more efficient than external sensors. External sensors are more complex, more robust, more expensive, and require more maintenance. So, the first question to ask is, "Do I really need external sensors?" Would sturdier, higher fences, walls, etc., be simpler and cheaper in the long run? An external detection system can, like any other, be breached or bypassed given enough time. Does the time delay introduced by the external sensor significantly increase the probability of detection any more than that imposed by a fence or a wall? Would security lighting be better?

Sensor Nomenclature

Within each *class* of sensors there are three *orders* that generally describe the type of coverage given by each sensor:

- point
- linear (or area)
- volumetric

A magnetic contact door switch is an example of a point protection sensor. The signal is sent only if *that point* recognizes an attack. An attack 2 inches (5 cm) away may not be detected. A photoelectric or active infrared beam is an example of a linear detector. Breaking the line of the beam triggers the alarm, but ducking under the beam does not! Volumetric detectors fill the volume protected and recognize human characteristics of movement, noise production, body heat, etc.

The next division of sensor nomenclature is the *family* of sensors within an order. These are sensors that operate on the same principle, but are different in their physical appearance and functional capabilities. This difference is often reflected as the difference in the level or risk against which they will protect. Examples of families include

- magnetic protective switches (e.g., single, double, reed)
- passive infrared detectors (e.g., curtain, 360°, coverage)
- ultrasonic detectors (e.g., long range, broad coverage)

Let's take magnetic protective switches as an example of a family. The family can be further divided into actual sensor types:

- nonhazardous low risk—single pole magnetic switch
- hazardous medium risk—double pole magnetic switch
- extremely hazardous high risk—biased triple pole magnetic switch

The traditional method of selecting a sensor is to consider the item or area to be protected (e.g., windows, doors, safes) and match it to a normal solution.

This approach is rather simplistic, as it assumes there is only one solution. For example, it is common to find

Classes	1.	Internal	2.	External
Orders	1.1 1.2 1.3	Point Linear (area) Volumetric	2.1 2.2 2.3	Point Linear (area) Volumetric
Families	1.1.1 Magnetic switch 1.1.2 Pressure pads 1.1.3 Personal attack buttons 1.2.1 Geophones 1.2.2 Wiring, etc. 1.3.1 Ultrasonic 1.3.2 Microwave 1.3.3 Passive infrared, etc.		2.1.1 Inductive pad 2.1.2 Geophone (single), etc. 2.2.1 Microphone cable 2.2.2 Active infrared, etc. 2.3.1 Capacitive field effect 2.3.2 Microwave 2.3.3 Ultrasonic	
Sensors	1.1.1.1 Single contact (reed) 1.1.1.2 Double reed 1.1.1.3 Triple reed, etc.		2.3.1.1 Single line 2.3.1.2 Double line 2.3.1.3 Multiple line with dummy	

FIGURE 2-1 Sensor nomenclature

feasibility reports that refer to each *zone of protection* being provided with a device that is normally associated with the zone. For example, if you are interested in the *perimeter fence zone,* you may select from a fence vibration sensor, a capacitive field effect sensor, or active beam barriers. If you are interested in the *building grounds zone,* you may select from buried line sensors or CCTV motion detection. If you are interested in the *building fabric zone,* you may select from foils on glass, break glass sensors, taut wires, or vibration sensors. If you are interested in the *building interior zone,* you may select from volumetric sensors, PIR/microwave sensors, etc. If you are interested in immediate item protection, you may select from contact switches, antitamper switches, or small volumetric protection. Why is the zone approach too simple?

Let's consider the perimeter zone as your sole means of security protection. It is easy to see how ineffective this is, as it comprises your entire security program. Fences are rarely continuous (they have gates), they are often made of different materials in different areas, they are set up at various distances from building roads (which might upset some devices and not others), they could be on flat ground or very uneven ground and spaced 20 ft (6 m) apart. The best sensor is the one that is most effective for the *exact conditions* faced in a *particular area.*

Remember, it is the environment in which the sensor operates, the effects it must sense, and, most importantly, the level of skill that must be used to overcome the sensor that are the primary considerations.

Although the zone method of selecting sensors is beneficial from the security surveyor's point of view of focusing onto a product his company might produce or one he has used before, unless he has vast experience and a wide knowledge of the latest and best devices available across the market, mistakes can be made — alternatives can be missed, modifications to the structure to remove the need for a sensor can be disregarded — before real consideration. The best approach is to consider, in turn, each category of sensors and gradually focus on a family and type of detector suitable to meet your security requirements. This will remove any bias *and,* from the client's point of view, will highlight any unnecessary or unsuitable devices.

Figure 2–2 shows the common types of sensors avail-

FIGURE 2–2 Scale of security

Class: External   Order: Line   Perimeter Boundary Detection

Type/ area	To protect against or detect	Grades of security 1	2	3	4	T	Device or sensor	Star rating	Problems	Not appropriate or not recommended	Comments
1. Any perimeter with sturdy chain or weld mesh fence	1. Intentional trespass 2. Climbing 3. Cutting through 4. Lifting	L1 L3 L4	L1 L3 L4	L1 L3 L4	L1 L3 L4	L3	Fence mounted devices L1 = geophone, L3 = microphone 'cable', L4 = fibre optic cable	*** **** **	1. In any urban area all must be protected by an outer un-alarmed fence to prevent nuisance/ vandal alarms	1. Any fence which might be banged, climbed by children, etc 2. Not for protection against tunnelling under fence	1. Geophone most widely used fence mounted device 2. Microphonic cable is excellent sensor but 'Hi-tech' and expensive 3. All systems are effectively ground hugging
2. Any open field site or site with 3–4m free area around perimeter	All the above	L2	L2	L2	L2	L2	L2 = capacitive field effect	****·?	1. Difficult to commission. Takes far longer than any of the other fence mounted devices 2. Requires cordon sanitaire 3. Blown debris and snow	1. Small sites under 200m boundary length 2. Low or medium risk sites	1. Capacitive field effect fences have 'depth to the zone of detection unlike L1, L3, L4. this makes them very difficult to bridge or climb in any way and hence they are really volumetric in nature

Class: External   Order: Line   Perimeter Boundary Detection

Type/ area	To protect against or detect	Grades of security 1	2	3	4	Device or sensor	Star rating	Problems	Not appropriate or not recommended	Comments
1. Any perimeter where space is limited i.e. town centre site say under 3m clear between building and boundary	Trespass on to site intentional or accidental	N1 N2 N3 N4	N1 N2 N3 N4	— — — N4	— — — —	Active infrared sensor N1 = single beam separate transmitter receivers, N2 = single beam transceivers, N3 = multiple beam stack parallel beam, N4 = multiple beam stack variable angle beam	** ** *** ***	1. Attenuation by fog 2. Flat ground applications only 3. Stack-type large and open to vandalism 4. Requires cordon sanitaire	1. In areas where vandalism a problem unless behind good perimeter fence 2. On undulating sites 3. For direct-to-police communication systems	1. More stable but less sensitive than ultrasonic or 'microwave/fences' 2. No 'depth' to detection zone unlike ultrasonic and microwave 3. Single beams prone to false alarms caused by birds, etc 4. Single beams usually shorter range than microwave 5. 'Modulated' beam essential
2. Any perimeter with more space than (1) and a perimeter fence	Intentional trespass	01 02 03 04	01 02 03 04	03 04	04	External microwave sensor 01 – short range beam, 02 – long range	*** *** *** ****	1. Must be kept away from public highways and glazed areas of	1. In city centre or dense urban areas 2. On building plots which do	1. Generally higher level of security than active infrared sensor 2. More prone to false alarm than active

FIGURE 2–2 (continued)

		beam, 03 – strip-line (vertical fence), 04 – slot-line (horizontal fence)	secure building as field could 'spill' into these areas 2. Requires cordon sanitaire	not have a good perimeter fence 3. Not for site subject to annual drifting snow	infrared 3. Less effected by environmental conditions than active infrared 4. Waves penetrate most materials so tunnelling under 'fence' less likely than infrared sensor

Class: External   Order: Line   Grounds and Approaches

Type/ area	To protect against or detect	Grades of security 1	2	3	4	T	Device or sensor	Star rating	Problems	Not appropriate or not recommended	Comments
Any open area or ground free of plants, vegetation etc, and not close to any road (public or private), railway line, etc, which would carry traffic when system armed	Intruder ap- proaching a building facade or theft within space, i.e. goods yard	M1 M2 M3 M4 M5	M1 M2 M3 M4 M5	M1 M2 M3 M4 M5	M2 M3	M2 M3	Buried in ground sensors M1 = fluid pressure sensor, M2 = geophone, M3 = PLE20 electronic cable, M4 = 'Audio' geophone, M5 = strain magnetic	** **** *** *** **	M1 prone to puncture/ leak M2 requires gravel bed and path M3 must be proofed against moisture M4 Expensive M5 Must no run near buried electrical cables	1. For ground with high watertable which might freeze 2. Ground near intermittantly used roads 3. Near tree roots, telegraph poles, pylon foundations, sub-stations, generators, compressors (unless differentiating types)	1.Geophone and fluid pressure sensor are oldest and most widely used devices 2. M1 not as sensitive as M2, M3 or M4 3. M1 acts on local pressure only 4. M2 reacts to vibration over a longer distance than M3 which is essentially a 'near field' sensor 5. M4 will give better indication of direction intruder is taking than any other sensor

Class:Internal   Orders: Point & Line   Doors and Doorways 1

Type/ area	To protect against or detect	Grades of security 1	2	3	4	Device or sensor	Star rating	Problems	Not appropriate or not recommended	Comments
1. Ex- terior or interior Solid hardwood, or	Breaking or cutting through	A1* A2 A3 A4 A5	A1* A2 A3 A4 A5	— — A3 A4 A5	— — — — A5	A1 Taut wire A2 Wire in tube A3 Open-span wiring A4 Filigree screen A5 Purpose wired board	* * ** ** ****	A1 Stretch A2 Inward opening window A3 Damp surface A4 Accidental damage A5 Delivery time	A1 Other than domestic A2 Windows A3 Expensive finishes A4 Heavy traffic doors A5 Where punc- turing could occur	A1 Low security A2 Looping easy A3 Damage before alarm A4 Very ac- ceptable A5 Mesh form best
2. Glazed portions sheet glass	Cutting or breaking panel	B1 B2 B5	B2 B5	B5	B5	B1 Foil on glass B2 Film in glass B5 Breakglass monitor	* ** **	B1 Damage B2 Con- nections B5 Visible	B1 Displays B2 Any grade below 4 B5 Not on laminated	B1 Poor in- stallation common B2 Ex- pensive B5 Reliable
3. Plate glass	Cutting or breaking panel	B3 B4 B5	B4 B5	B5	B5	B3 Acoustic breakglass B4 Vibration strip B5 Breakglass monitor	* **	B3 'En- vironmental noise B4 Vibration B5 Visible	B3 Heavily cur- tained areas B4 Domestic B5 Not on laminated	B3 Free win- dow space B4 Low security B5 Reliable
4. Win- dows or doors	Removal of whole win- dow or door pane	C1 C2 C3 C4 C5	— C2 C3 C4 C5	— — C3 — C5	— — — — C5	C1 Vibration sensor C2 Internal filigree screen C3 External meshes C4 Knock-out bars C5 Internal fixed bars	* ** *** ** ****	C1 Sensitivity C2 Damage C3 Aesthetics C4 Ageing C5 Ventila- tion only	C1 Thin surfaces C2 Heavily used doors C3 Daytime at- tacks (?) C4 Physical security C5 Where limited fire escapes	C1 Careful selection re- quired C2 Covert device C3 Deterrents C4 Easy to defeat C5 Best barrier
5. Attack on entire door	Jemmying of lock prep Frame spreading Lock removal (drill around area) Ramming Pulling Lock open- ing or picking	N/A				Steel panel long staple/ bolt dog bolts Dog bolts Rebate frame Steel inner Espagnolete Lock Rack bolts Espagnolete Lock Reinforced staple No trough door fixed accessories Locks: see section 6	Proper consideration of resistance to violent attack/cutting, etc, requires co-ordinated consideration of all aspects, wall/frame/fix- ing/ironmongery/lock/material. For advice see reference (1)			Reference (1). The Security of Buildings, Grahame Underwood (Architec- tural Press)
6. Whole door opening. (Also ap- plies to windows)	Entry into area follow- ing suc- cessful operation of/picking of lock	E1 E2 E3 E4 E5 E6 E7 E8	— — — E4 E5 E6 E7 E8	— — — — — E6 E7 E8	— — — — — — — —	E1 Mechanical micro- switch E2 Single-reed surface E3 Single-reed flush E4 Balanced/ double- reed flush E5 Coded double reed E6 E5 with security ring E7 Triple reed E8 E7 with security ring	* * * * ** *** *** ****	E1 Obsolete E2 Domestic only E3 Minimum standard E4 Magnetic bridging easy E5 Normal commercial standard E6 High security, E4 E7 Alarms before opening E8 Best available device		

FIGURE 2–2 (continued)

Class: Internal   Orders: Point & Line   Window and Window Openings

Type/ area	To protect against or detect	Grades of security 1	2	3	4	Device or sensor	Star rating	Problems	Not appropriate or not recommended	Comments
Items 2, 3, 4, 6 of 'Doors' also apply	Window opening, vandalism, smash/ grab, whole pane removal	N/A				For further advice on protection of windows (particularly non-alarmed) see Reference (1)				Reference (1) *The Security of Buildings*, Grahame Underwood (Architectural Press)
Window pane	Break-glass detection	F1 F2 F3 F4	– F2 F3 F4	– – F3 F4	– – F3 F4	Break-glass Sensor F1 vibro switch F2 foil-on-glass F3 break-glass F4 acoustic break-glass	* ** *** ***	F1 Unreliable F2 obtrusive F3 not near phones F4 not on laminated glass	F4 not near fans	F1 virtually obsolete F2 easy to bridge F3 very small, possible and therefore a deterrent F4 more than one pane cover

Class: Internal   Order: Volumetric 'Movement' Detection

Type/ area	To protect against or detect	Grades of security 1	2	3	4	Device or sensor	Star rating	Problems	Not appropriate or not recommended	Comments
Any enclosed area with relatively stable environment and no infrared heat generating equipment Office, store, warehouse, shop	Intruder walking through door or window	H1 H6	H1 H6	H1		Passive infrared sensors H1 – H5 – H1 curtain type ceiling mounted H2 narrow long-range multizone H3 narrow, long-range twin zone H4 broad-spread wide angle plus lookdown H5 ceiling mounted 360° coverage 2 zone H6 active internal, infrared beam	*** **		1. Areas above 27°C ambient temp 2. High security areas 3. Areas where access to zone immediately under wall-mounted devices possible	1. All volumetric detector should be considered in light of an environmental check prior to final selection 2. Might not be acceptable to insurers 3. Very reliable and trouble free 4. All tenants must leave at same time, i.e. before alarm in corridor set. No good in multi-tenancy offices 5. Number and position/distribution of zones of detection critical 6. Less troublesome but less sensitive than ultrasonic and microwave devices 7. H5 cheap, non-intrusive method of 'out of hours' protection of high value goods in cases on racks, etc
	Intruder entering through curtain wall or number of windows	H1 H2 H6	H1 H2 H6				*** * **		Easy to mask	
	Intruder walks down long corridor	H3	H3				****			
	Intruder walking across long corridor	H2 H6	H2 H6				** ***			
	Intruder moving in open plan area of office	H4	H4	H4			***	Sensitivity must be adjusted for space		
	Space with many partitions, desks chairs, etc, 'Bureuland-shaft'	H5	H5	H5			****	Can see heat rising from radiators and create false alarm		
	Spot protection of cases, cupboards, etc	H5	H5				***			

Class: Internal   Order: Volumetric 'Movement' Detection 2

Type/ area	To protect against or detect	Grades of security 1	2	3	4	Device or sensor	Star rating	Problems	Not appropriate or not recommended	Comments
Any enclosed area which is environmentally stable and there are no draughts, fans, fast-response warm air heating systems	Intruder walking through door or window line	I1 I2 I5	I1 I2 I5	– I2 I5	I1	Ultrasonic sensor I1 wall mounted tube transceiver I2 corner mounted disc transceiver I3 ceiling mounted ring transceiver I4 master–slave system I5 any of I1 I2 I3 multihead system	*** ** ***	1. Blinding by solid objects 2. Multiple interference if multiple units used and not 'crystal controlled' 3. Penetration of badly sealed or open doors to occupied areas, causing false alarm	Where 1. Fan coil or unit/fan convector heater used 2. Where large temp/R.H. changes occur (i.e. above 15°C/30°% RH) in short periods of time 3. Audio systems in operation 4. Extensive areas of soft furnishings, bedding, etc (attenuation of signal)	1. Not suitable for air conditioned areas 2. I1–I3 small areas say 8 × 4m I4 for medium sized areas (13 × 7m) requiring possible future extension I5 for large areas warehouses bonded stores etc coverage up 180–200m2 pair coverage with systems economic from 20 to 100 pairs 3. More sensitive than PIR usually less troublesome than microwave 4. Not recommended for areas which are 'noisy' during 'alarm hours', i.e. machines/telephone rings/computers in action, etc
	Intruder walking down long corridor	I2 I5	I2 I5	I2 I5			*** ***			
	Intruder moving in open-plan area or large space	I1 I2 I3 I4 I5	I1 I2 I3 I4 I5	I1 I2 I3 I4 I5	I3 I4 I5		*** ** **** **** ****			
	Intruder in space with many shelves/cupboards, etc	I3	I3	I3	I3		***			

Class: Internal   Order: Volumetric 'Movement' Detection 3

Type/ area	To protect against or detect	Grades of security 1	2	3	4	Device or sensor	Star rating	Problems	Not appropriate or not recommended	Comments
Any enclosed area with solid walls	Intruder breaking through	J3	J3	J3	J3	Microwave sensor J1 general purpose coverage wall	***	1. Some devices sensitive to	1. Coverage perimeter must not be within	1. For large areas more economic than PIR or ultrasonic

FIGURE 2–2 (continued)

and block/brick partitions or inner walls with little or no metal racking surfaces	windows (detection prior to attack)					mounted J2 ellipse coverage J3 long-range narrow coverage (overlapped)	****	fluorescent lamp interference 2. Can penetrate lightweight partitions, giving rise to false alarms	3m of any large area of metal sheet mirrors or tinted windows, etc	2. More sensitive and hence less stable than PIR or ultrasonic 3. Select units with false alarm, precheck sensitivity adjustment and coverage overlap capabilities.
	General movement in long corridor	J3	J3	J3	J3		****			
	General movement in large area	J2 J1	J2 J1	J2 J1	J2 J1		**** ***	Do not use in too confined a space. Check coverage not too great		
	General movement small area shop/cellular offices, etc	J1	J1	J1	J1		****			

Class: Internal   Order: Line (or area)   Walls and Floors and Ceiling

Type/ area	To protect against or detect	Grades of security 1	2	3	4	Device or sensor	Star rating	Problems	Not appropriate or not recommended	Comments
Heavy-weight homogeneous walls of concrete, brick or block over 114mm thick	Cutting through before penetration	G5 G6	G5 G6	G5 G6	G5 G6	Structural penetration sensors G1 = inertia, G2 = acoustic, G3 = strain, guages, G4 = continuous integral wiring G5 = geophone, G6 = piezoelectric	*** ****	1. All devices should have 'counting', 'summing' or 'sampling' circuits to remove false alarms caused by single vibration 2. G1 does not detect high-speed drilling as well as G5 or G6 3. G3 requires careful installation and commissioning	1. Walls under stated thickness 2. Wall subject to natural vibration caused by trains, underground heavy traffic, etc 'environmental check' required 3. G3 not for any building subject to 'settling in', particularly new buildings	1. Piezoelectrics have better sensitivity adjustment than geophones. 2. Geophones primarily designed for vertical movement (fence lifting etc) not horizontal 3. Little sensitivity adjustment possible. 4. G3 must be permanently fixed to the structure
As above, but under 114mm thick	As above	G1	G1	G1	G1		***			
As above	As above	G3	G3	G3	G3		***			
Any partition, ceiling or floor	Cutting through only. Detection after penetration	G4	G4	G4			**			5. G4 expensive to install and not particularly difficult to detect. Not suitable on any surface which might become damp or suffer from surface or interstitial condensation
Any room insulated from effects of ambient noise, i.e. saferoom, strongroom basement store, etc		G2	G2	G2	G2		***	Extraneous noise	Other than for high security heavyweight structure rooms	

Class: Internal   Order: Point   Protection of cases, cupboards, etc.

Type/ area	To protect against or detect	Grades of security 1	2	3	4	Device or sensor	Star rating	Problems	Not appropriate or not recommended	Comments
Any cabinet in any carpeted area	Direct approach to case/cupboard etc	P1	P1			Pressure pad or pressure mat. P1	**	Case position must allow approach only via the mat	Any situation as sole method of protection	Very reliable but simple to defeat if found by intruder. Not for public traffic areas
Glass cases or cabinets	Smashing or cutting of glass	P2 P3 P4 P5 P6	P2 P3 P4 P5 P6	P2 P4 P5 P6	P4 P6	P2 foil-on-glass P3 laminated glass P4 vacuum glass P5 continuous wiring P6 break-glass monitor	** **** *** ** ****	P2 intrusive cleaning P3 Cost P4 Cost P5 Intrusive P6 Visible	P2 display cabinets P3 weight P4 installation P5 display cabinets P6 laminated glass	Foil-on-glass to detect breaking only Wiring must be secure
Any	Opening of (in addition to the above)					P7 protect switches or contacts See sheet 'Whole door opening'				In addition to detection measures all cases/cabinets, etc, should be 1. Bolted down to prevent removal. 2. Fitted with a good lock of key design (Assa, Bramah, etc) 3. In the case of filing cabinets plan chest, etc, if high value secret information stored should have combination lock.
Any	Any approach to cabinet, cupboard, etc	P8	P8	P8		P8 ceiling mounted 360° passive infrared sensor	****	Other IR sources		

*Not recommended **Fair ***Good ****Excellent

able. For the purposes of initial suitability and comparison between devices, Figure 2–2 uses a five-point scale of security:

Grade 1—deterrence of vandals and opportunist criminals, prevention of mild confidentiality, unskilled criminal

Grade 2—protection of low-value goods against deliberate preplanned criminal act, skilled or knowledgeable criminal

Grade 3—protection of moderately valuable goods against deliberate professional criminal, highly skilled individual

Grade 4—protection of high-value goods against organized criminal gang with equipment, highly skilled with technical backup

Grade T—defense against terrorism

After considering the tables, refer to Chapters 3 and 4 for further guidance. You may then be in a position to act as the informed *agent provocateur* in any discussion on proposals put forward by a security adviser, surveyor, or salesperson.

Sensing devices have not been listed or classified under the type of common applications for which they are best suited. They have been classified by whether they are for internal or external use and by their principle of operation.

Only by appraising all viable options can you select the most suitable sensor. By listing sensors by applications, as many books do (e.g., fence-mounted sensors), tunnel vision can develop early in your selection process and alternative methods or sensors fail to receive proper consideration. Security system selection requires logical reasoning and lateral thinking!

Although it provides a comprehensive explanation of most of the commonly used elements of security systems, this book is by no means exhaustive. To cover all devices, specialized tools, and company services, would require four volumes, not one. However, as most of these omitted subjects are *specialized,* general advice could be misleading. Expert guidance will always be required in these areas of security. Advice will often be provided from within the client's organization or by a government agency. Security company services (such as watchmen's logging or warden systems) have not been covered in this book, as these are primarily physical systems and are simple to understand.

Remember you are not alone when you select a suitable security system. The police, the reputable security companies, and the insurance industry all have a vested interest in ensuring that the system you finally select will be effective and reliable. Their services and guidance are often free. Use them, but use them wisely. The number and complexity of security systems installed in all types of buildings are increasing dramatically year by year. Fortunately, the security industry is beginning to learn the lessons of past mistakes and is now concerned with giving truly professional service. More and more regulation of the practices of the industry's companies is being introduced. This can only benefit the industry and its customers. Even small companies will now have no excuse not to be approved by a responsible body.

SENSOR CABLING AND SIGNAL TRANSMISSION METHODS

Standards of Workmanship

The sensor of the intruder detection system must pass its response to a central control processing unit. In order to do this, a communication network is required. We will see, in future sections, that most sensors provide good protection and are often fitted with antitamper circuitry that makes it difficult to bridge them or the materials to which they are affixed. Because of this, there has been an ever-increasing trend toward attacks on the part of the system that has traditionally received the least attention during design and installation—the *sensor wiring* (or *sensor loop*).

Many manufacturers produce similar sensors. Competition is fierce and prices are very, very competitive. Hence, in this market, there is little scope to undercut a rival company on the capital cost of a series of particular sensors. Cabling, however, is very labor-intensive and, therefore, costly. All too often, the lowest bid is a reflection of the lowest amount set aside for wiring labor. Once a job gets to a site, the architect should not be too surprised if the wiring is stapled to beams or finishes, badly protected, and unconcealed!

Very often, the first real involvement of the consultant or architect occurs when the system does not work and he is summoned by the client. In many, many cases the cause of the problem is poor cabling. Even worse than this is the case where the system works on commissioning, but is either breached or fails (due to tampering, damage, heat affecting insulation, etc.) before the expiration of defects liability, and the client suffers a loss. It is not sufficient to specify that the electrical subcontractor will provide an occasional conduit and a "provisional sum" of money to cover all security installation responsibilities. If, as recommended, you have selected a UL-approved security installation company, then at least you can expect the installation to comply with NEC and other regulations. If you have not selected this type of company, then standards are far more likely to be poor. Even approved companies sometimes use the minimum standards stipulated by UL, as though they were a high standard of materials and workmanship specification. They are not. The following section summarizes the recommendations of NEC/UL standards. An inspection of these soon shows that wide interpretations are possible, resulting in many standards of cabling and means of protection that may be totally unsuitable for the level of risk faced. When scrutinizing the security proposals put forward by any company, it is wise to

know the type of system used and its security

establish the exact means, materials, and methods of mechanical protection to be used during wiring installation

UL Standards

UL standards (previously discussed in Chapter 1) and, in particular, UL 681, *Mercantile and Bank Burglar Alarm Systems,* require connection wiring to

be generally no less than no. 22 "Authorized," or American Wire Gauge (AWG) (0.005 in^2 or 0.32 mm^2) copper and no. 16 AWG (0.02 in^2 or 1.3 mm^2) between battery or power supply and any sounding device

follow contours

be located where least subject to damage

be protected by conduit or other metal protectors where it crosses joists

be tape-protected from abrasion on sharp corners

be fixed using insulated fixings if it is jacketless (unsheathed)

have fixing instances specified

be spliced or joined with a device acceptable for the purpose, or soldered

be spaced at least 0.5 in (1.3 cm) from all other services or be tape-protected

be run *over* piped services

be spaced at least 2 in (5 cm) from main cables (110V or above) or protected by conduit or insulator

be protected by tape where it passes *through* wood floors or walls

In general, UL requirements are the *minimum* needs of a system. For improved security and reliability, wiring should be conduit-protected through its entire length and not joined or spliced between terminal points.

The cabling of the sensors should be more secure than the sensors themselves. If a single sensor fails, detection is still possible if other devices are present. If, however, a complete loop of sensor wiring is bridged, no sensors will operate.

Wiring should

use the shortest, most economical, and best-concealed route available

not be placed near an electromagnetic interference (EMI) field

not have identifiable markings on the sheath and not be labeled (color-coding can be used, but only the installer should know the code)

be secure throughout the entire installation

Cable Specification: UL

UL specification of wiring simply refers to a gauge size and "applicable requirements for burglar alarm wiring." However, *applicable* is not defined, nor is *burglar alarm wiring.*

The specifications for low-voltage wiring, AWG installation, etc., are contained within the NEC. Consideration of British Standard 4737 gives a few more clues to good cable specification.

Joints and Terminations

UL and the NEC both cover methods of joint wiring to ensure electrically and mechanically sound joints are made. These include

- wrapped terminal joints and splices
- crimped joints
- soldered joints
- clamped joints

All joints should be covered with an insulating material or contained within a junction box. However, sound though this advice may be, it fails to make the point that *wiring joints should be avoided at all costs* for the following reasons:

1. Joints increase system resistance and, hence, voltage drops and reliability is reduced.
2. If used, joints provide ready-made points at which the system may be bridged, tampered with, etc.

In general, wiring should be one, unbroken length between devices, wherever possible. If joints are made, they should always be made with a steel joint box and have a permanently affixed lid. Standards below these reduce security.

Cabling Methods: British Standard 4737

The previous European edition of this book attacked British Standard 4737 (and UL standards) and their lack of specification on standards of workmanship or interconnection, cabling types, and installation methods. Fortunately, since the last edition of this book, BS 4737, Part 1, 1986 and Part 4.1, 1986 have been published and improve matters considerably. Installation methods are, however, still ill-defined.

BS 4737, Part 1, Section 3.2.3 requires that the entire system be protected from all likely damage, including mechanical, electrical, environmental, etc.

BS 4737, Part 1, Section 3.3.3 requires that all interconnecting cables be adequately supported and their installation conform to good working practice (*whose* practice, it does not say). It further states that interconnecting wiring shall not be run in the same conduit or trunking as mains cables "unless they are physically separated." Physical separation could be read as an air gap and not a steel or PVC segregation plate. As IEE and many international regulations do not allow the mixing of extra, low-voltage cables with mains cables, unless the former has the insulation resistance of the latter or a segregation plate is provided, this requirement is too imprecise.

BS 4737, Part 4.1, Section 4.5.2 requires wiring to be screened and protected from electrical and radio interference. As most people would agree that electrical services within buildings are now complex and dominated by the requirements of the mains voltage wiring regulations, then, in all nondomestic buildings, the requirements for mains voltage wiring should prevail. Adoption of this policy would mean that interconnection wiring would always run in separate conduits or trunking compartments segregated from Category 1 (mains voltage) or Category 3 (fire alarm) circuits. This would ensure the British Standard was met, the cables are properly protected from all extraneous environmental effects (or deliberate tampering), and the cables are properly supported.

Cable or Wiring Specification: British Standard 4737

BS 4737, Part 1, Section 3.30, 1986 is the *Specification for PVC-Insulated Cables Used for Interconnection Wiring in Intruder Alarm Systems.* Its publication was a significant step in the right direction as the standard can now be used to established reasonably easily whether or not a contractor is complying with normal standards and practices. Unfortunately, the advent of *multiplex bus wiring* and *addressable alarm systems* means that this section is already in need of amendments to allow for high-grade, coaxial or shielded transmission cables.

Despite this, the most common (non-UL) alarm systems should have all parts connected in accordance with BS 4737, Section 3.30 wiring, which is

only for use in internal locations and not for use in
    locations open to the weather, unless suitably
    protected
only to be used when suitably segregated and rated
protected or shielded from interference in accordance
    with Part 4.5.1, Section 4.1, of the British Standard

Conductors should be comprised of solid, annealed copper (manufactured to BS 4109 Condition 0) conductors of strands each plain or tinned surface coated. Solid conductors should have a minimum cross-sectional area (CSA) of 0.002 in^2 (0.2 mm^2) (24 AWG) and to Table 1 of BS 6350, 1988. If under 0.008 in^2 (0.5 mm^2) CSA max-

imum (20 AWG), resistance at 68°F (20°C) shall be 95Ω/0.6 mi for 0.003 in^2 (95Ω/km for 0.2 mm^2) and pro rata to 0.008 in^2 (0.5 mm^2). Stranded conductors should have a minimum CSA of 0.003 in^2 (0.22 mm^2) (23 AWG) to BS 6360, Table 3, 1981, Class 5.

For *cores comprised of conductors,* solid or stranded conductors should be a minimum of 0.3 × 0.008 in (7 × 0.2 mm) or 23 AWG in diameter. Each core should be covered with Type 2 or TI1 PVC insulation manufactured to a BS 6746 minimum thickness of 0.006 in (0.15 mm). The insulation should not adhere and should allow for easy stripping without damage to the conductors. Cores should be able to withstand a 1500V RMS spark test.

A *cable* is comprised of a number of cores or wires. Cables should be sheathed with Type 6 or TI1 PVC manufactured to a BS 6746 minimum thickness of 0.015 in (0.4 mm). Cables should include a rip cord to allow for easy removal of the sheath without the need for excessive cutting, which could damage conductors or cores. A cable should be able to withstand an insulation resistance test of 500V DC for 1 minute between *each conductor* and for all remaining conductors in the cable with an insulation resistance of no less than 50 megohms for 100 minutes at 68°F ±5°F (20°C ±5°C).

In practice, each conductor actually means *each core,* as with stranded conductors this would be impracticable.

Compliant cables should be used on drums and spools and should indicate the manufacturer's name, length of cable, number of cores, minimum CSA of the conductors in in^2 (mm^2), number and diameter of strands in a stranded conductor (e.g., seven at 0.008 in or 0.2 mm), British Standard number, and date.

Wiring Methods

Wired pairs and twisted pairs are the traditional names given to small-diameter, single-sheath electrical conductors. Although these two wire conductors, which resemble *bell wire,* are still found on domestic scale systems, they are not suitable for commercial or industrial systems, which should always use multistranded, four-core insulated and sheathed cables as a minimum standard (BS 4737, Part 3.30). Systems wired using these cables are often referred to as *hard-wired DC systems.* All circuits wired in this manner are usually 12V or 24V DC. There are four basic wiring configurations that use this type of cable.

A *normally open, open loop,* or *unsupervised system* is a passive system, in that the sensor works as a switch that only allows current to flow through cables when an alarm condition arises and the sensor closes. It is, inherently, the least secure, as the sensor monitors the area or material, but nothing monitors the cables for faults or tampering (see Figure 2–3). The system is seldom, if ever, used today.

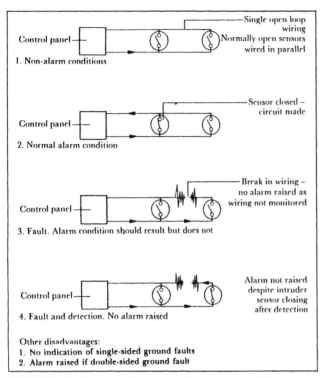

FIGURE 2–3 Unsupervised, open loop sensor wiring

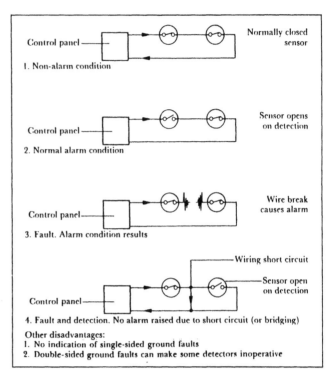

FIGURE 2–4 Supervised, closed loop sensor wiring

In a *normally closed, closed loop,* or *supervised circuit,* the sensors act as a series of switches that open in the event of an attack and cut off the current supply that normally flows through the circuit. The circuit is said to be *supervised,* because any attack on the cable that causes the current to be stopped or interrupted (e.g., cutting or primitive bridging) will raise an alarm, even if the sensor does not (see Figure 2–4).

BS 4737 recommends that minimum cable security standards include

continuous monitoring of an open circuit or short circuit (24-hour)
raising a local, audible tamper alarm within 20 seconds
a latching indicator on every processor or sensor, including electronic processing, to indicate that part of the sensor or interconnection wiring that is in an alarm condition

The UL recommends similar measures and military standards specify the current variation percentages allowed before alarm.

In a *double pole* or *closed loop* detection circuit, a positive pole core and negative (or neutral) pole core are taken into and out of every sensor. If the adjacent conductors or conductors of opposite poles are connected, or if a short circuit occurs, an alarm will be raised. This method makes bridging more difficult. Only double pole–supervised or double pole–monitored circuits offer

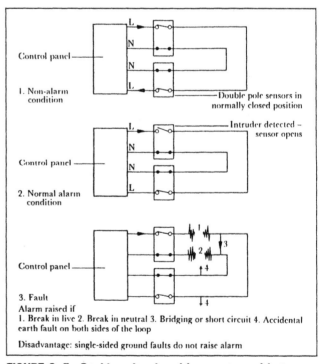

FIGURE 2–5 Double pole, closed loop sensor wiring

acceptable levels of security in most situations (see Figure 2–5).

Monitored double pole or *closed loop circuits* with an *end-of-line impedance* are circuits of the normally closed variety in which the current flowing in the circuit is continuously measured and analyzed by a device within the control panel. With most previous methods, ground faults (either

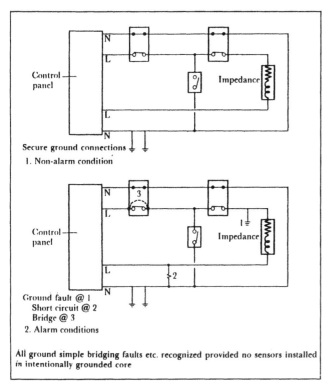

FIGURE 2-6 Monitored, double pole or closed loop with end-of-line impedance

intentional or otherwise) were not recognized and could compromise the system.

To overcome this, the negative (or neutral) pole core is deliberately grounded and the current in the live (or hot) wire and to ground via neutral are monitored.

Any change in the current magnitude, direction, etc., causes an alarm to be raised, even if the cable is not cut. This is the best system for most low- to medium-level security risks (see Figure 2-6).

The advantages of the traditional methods of hard wiring listed above are

• cables are small, flexible, and easily concealed
• installation is fast and easily carried out
• low voltages and current result in inherent safety in use and a low-battery backup capacity requirement

The disadvantages are

• fixed installations (wiring and sensors)
• criminals are familiar with these systems and, hence, methods of defeating them

Line Carrier or Mains Signaling

Until 5 years ago, mains signaling (see Figure 2-7) was primarily restricted to large electrical or telecommunication networks. Nowadays, it is found in many buildings, controlling everything from fans and lights to refrigerators and cooling appliances. Mains signaling consists of sending a signal through the normal wiring of a build-

ing (i.e., the wiring that supplies sockets, light fittings, etc.). This is done by superimposing other frequencies on the normal 60Hz frequency, which will activate switches coupled to receivers that identify the frequency used. The mains signaling intruder alarm system consists of a master transmitter, located at any convenient socket outlet or other electrical outlet, and similarly placed receivers. The transmitter is usually hard wired to a sensor (e.g., break glass, vibration) and will send a pulsed signal to any receiver to which it is matched. The receiver will then act in a predetermined manner. It may ring an alarm, operate a digital dialer, or lock or unlock a door. The master unit can be used to reschedule frequencies, pulse codes, etc.; maintain security; or change the pairing of transmitters and receivers.

It is important to remember that with this system, any outlet can "talk" to any other (i.e., lighting outlets can communicate with socket outlets). This communication occurs because most systems operate on a signal transmitted through the neutral and ground connections to the outlets. As these are all connected *en masse* at the mains intake position terminal blocks, there is always a signal route through the wiring. Usually, the only block to the signal will occur if small transformers exist on any circuit. The block exists because transformers are physical breaks in the circuit across which the signals cannot readily translate (see Figure 2-7).

The advantages are

very cheap installation for little extra capital cost expenditure on devices

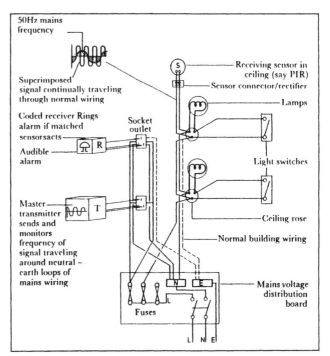

FIGURE 2-7 Mains signaling

adaptability—this can be very important in large buildings with tenants on every floor

comprehensive retrofitting of existing premises is possible

systems are easily extended

The disadvantages are

to prevent malicious or accidental removal of transmitters or receivers from the mains outlets, they should be hard wired into proper electrical junction boxes with allen key fixings to lids

cables used in the mains wiring should preferably be screened from electromagnetic interference (EMI) to prevent loss or distortion of signal. This usually means circuits should be mineral-insulated, copper-covered (MICC) cables or PVC-insulated, single-core cables in steel conduits.

the security of the line of communication is dependent on the continuity of supply. Hence, if a neutral or ground link is disconnected at any time, the signal will either be lost or undetected and a false alarm will be raised.

security can be compromised by sharing mains lines with signals used for other switching purposes (e.g., energy conservation) and access to the system being permitted to false "electricians"

lower overall security than hard-wired systems

not many systems are in use yet, so industry experience and expertise are low

Radio Transmission (Wire-Free)

With low-powered, internal radio transmission systems (see Figure 2–8), the normal, hard-wired connection between sensors and control panels, and control panels and setting devices is removed. In its place, each traditional sensor has, coupled to it, a radio transmitter and battery pack. Radio signals (or occasionally ultrasound) are used to transmit information to a common, remote receiver at the control panel. The most obvious advantage of this form of communication is the removal of

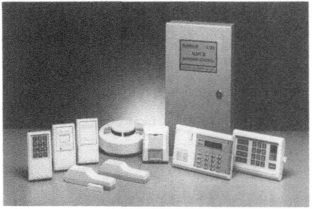

FIGURE 2–8   Wireless supervised control panel

very high labor and materials costs associated with hard-wired-in-conduit systems. The price advantage has resulted in the rapid expansion of the radio, wire-free system during the past 2 years, primarily in low or moderate risk areas.

While there are general rules in various states governing the use of *internal* radio transmission systems, there are no universally recognized USA standards. UL does not have specific requirements for wireless installation; they simply ensure that the hardware used meets normal panel-testing and proving procedures.

In Britain the picture is clearer. Responding to the expansion of sales and the need to classify and specify a wide variety of devices on the market, British Standards introduced BS 6799, 1986, *Code of Practice for Wire Free Intruder Detection Systems.* Because these systems are relatively new to the normal commercial security market, there is not as much case history or established knowledge of the systems within the industry. Buyers should, therefore, deal only with known experts in the field and should exercise great care in selection. Many exaggerated claims are made and BS 6799 emphasizes that any trade description that claims their component of the system conforms to BS 6799, 1986 is, in fact, false! This is because BS 6799, 1986 is a code covering the completed installation within a protected premise or location to a high quality of workmanship. Only the completed installation can claim compliance, not components.

BS 6799 classifies systems in ascending order of integrity (or security) of the transmission method and the level of sophistication of the system's self-monitoring capability.

A *Class I* system sensor transmits on alarm or deliberate activation. It also gives local or remote warning of low voltage or high impedance in its battery-charging circuits. The level at which it does this must correspond to voltage/ampere-hours remaining that are sufficient to support a further 7-day quiescent condition followed by the alarm signal duration (this does not apply to deliberately operated panic button–type devices).

A *Class II* system is the same as Class I, but, in addition, each receiver or control panel indicates the sensor that is transmitting.

A *Class III* system is the same as Class II, but, in addition, the system identifies any blocking (deliberate or accidental) of the radio transmission with a duration greater than 30 seconds. Identification will generate a fault indicator and an agreed action procedure will ensue.

A *Class IV* system is the same as Class III (with the exception of the panic alarm devices being temporarily removed from the premises) plus

a return-to-normal signal within 30 seconds of alarm or fault, *or* interrogation of the device state from the control panel receiver automatic reporting of the state of each transmitter (e.g., battery voltage), by each transmitter, at intervals not exceeding 8.4 hours

a separate low-battery voltage warning *before* the 7-day, capacity-remaining time limit

a fault indication on the panel for each and every transmitter that fails to send (or receive) any automatic report within the 8.4-hour period and the alarm condition initiated if more than 2.5 hours elapsed since the last automatic report was successfully received from any sensor

a fault indication of any low-battery report from a transmitter that lasts longer than 8.4 hours and displays a continuous fault indication until the battery is replaced, recharged, or the system is manually reset.

A *Class V* system is the same as Class IV, but automatic report periods are reduced to 1.2 hours (maximum) with alarm, if they have failed to report within a 3.6-hour period.

Class IV and Class V systems are referred to as *supervised* or *monitored systems.*

In Britain, equipment manufactured for the purpose of wireless intruder detection systems is regulated by the Department of Trade and Industry (DTI) Radio Regulatory Division. Equipment is tested to several specifications, depending on the equipment's *type of use.* All parts of approved systems display a type-approved label. This will indicate the specification against which it was tested. In the USA, approval may be needed if transmitters emit more than 100 milliwatts (mW) of power.

British radio transmission frequencies reserved for these wireless devices are 173.225MHz (narrow band) and 418MHz UHF (wide band). Radio-paging devices and social call (e.g., senior citizen distress) systems also operate under these frequencies, but no part of these may be incorporated or substituted into a BS 6799 system, as DTI specifications differ. The normal type specifications in the UK (and hence on UK-exported products) are MPT: 1309, MPT: 1340, and MPT: 1344 for land-based fire and security systems; and MPT: 1265 for offshore or portable marine systems.

In the USA, always check for compliance with FCC regulations and ensure systems are FCC-approved.

The general rules that apply to the application of sensors and their advantages and disadvantages, apply to wire-free systems. They do, however, have some particular features. Advantages include

panic alarm devices can be portable and hands free

no difficulty or expense in relocating devices

devices may be added or removed from the system at will

vastly reduced installation time, and wiring and labor costs

reduced impact on building structure (very useful in refurbishment of sensitive buildings, such as art galleries)

individually addressed and annunciated sensors at low cost

unlimited numbers of sensors (50 or more common) provided the panel can identify each sensor separately

Disadvantages include

system is dependent on both primary and secondary (transmitter) battery condition

transmitter battery capacity limits range and period of renewal (i.e., the longer the distance, the more energy used and the shorter the battery life). In practice, range is usually up to 180 ft (55 m) for a battery life of 6 months.

sensors must be accessible and there are high labor costs on large schemes with high-sensor positions for changing batteries on each transmitter every 6 months

no condition reporting for Class I, II, and III systems; do not use

noncontinuous monitoring on any class of system; the best is Class V and is monitored once every 1.2 hours

frequencies used are not exclusively reserved for wire-free systems and blocking (deliberate or accidental) can occur. The most common blocking is accidental and is caused by frequency drift to other radio signals in the vicinity.

cannot be used in the area of other high-activity radio transmissions (e.g., police transmitter) or similar systems of same type

range is limited by massive concrete and mesh-structured walls or by presence of large amounts of grounded metal

lower overall security level than hard-wired system

can suffer from building radio transmission "dead spots" caused by multiple signal transmission reflection

Better systems

are frequency modulated (FM), not amplitude modulated (AM)

are Class IV or Class V supervised; only these apply to normal commercial security levels

have subcarriers or digital coding of the signal to provide interference protection; simple amplitude or frequency modulation is not sufficient (This also prevents mutual interference.)

are wide band receiver–based

use complex, 16-level signal encoding to address sensors individually

can simultaneously decode more than one transmitter message arriving at the receiver at the same instant by "wait and store" or "duplicate reception" techniques

are supported by full information. Many manufacturers are overfond of quoting "free space range" distances for radio transmission; however, as the devices are only made for installation within real buildings, which contain many articles that dramatically reduce the ideal free space range, the claim is meaningless.

Applications. A full radio transmission path and interference survey must be undertaken by the sales company before any claims for its use can be made. This survey should tell you *all* of the building's null points and will often show that, in certain areas, the system cannot be used. Seek out the views of the insurer. Approval is normally given only for lower risk, small, commercial or domestic premises. Remember that bells, digital communicators, etc., must all still be hard wired.

All quality devices should have lithium or manganese-dioxide batteries if quality transmission is to be maintained. Lithium batteries are primary cells and cannot be recharged. They are also relatively expensive, so maintenance agreements should include free replacement. Alkaline batteries are inferior. Lithium batteries have a very long life, high ampere-per-hour output, and a high voltage per cell. This leads to overuse and shelf life abuse in many instances. Batteries should always be supplied with a manufacturer's date stamp. Otherwise, how do you know if the battery has been sitting on the wholesaler's shelf for the last year?

All metal or metal-backed objects can cause problems with signal alteration. It is fortunate that devices are somewhat portable, as these objects include foils, metal blinds, sliding screens, metal partitions, shelving foil insulation boards, vapor-proof lining papers, chicken mesh reinforcement, etc., and all of them could be introduced at any time. Do not position devices near pipes, mains wiring, etc.

Locate receivers at the geographical transmission center and never in a basement (due to ground attenuation of the signal). Receivers should be placed close to the control panels to minimize the hard wiring between them. Many systems have receivers built into the panel.

Transmitters on *all* devices require thorough and exhaustive functional testing in their final, *fixed* position. Commissioning engineers are fond of obliging the installing electrician by confirming transmission strength and recording it before it is fixed to save him from having to fix it twice. By merely *handling* the device, the commissioning engineer is altering its transmission characteristics.

Once all the devices are fixed, it is not good enough to simply test each transmitter and receiver. What you really need to know is

If more than one transmitter is signaling, can a signal clash occur, thereby canceling or corrupting the signal?

Are there any null points on the installed systems?

Does the supervision system work?

Have any radio frequencies been introduced during the purchase and installation periods that might upset the system?

Transmitters should never be used in a cold environment. Transmitters have a high *current draw* on the battery (hence the need for lithium batteries). As temperature of operation falls, lithium battery output capacity dips significantly and causes frequency drift on the signal modulation. This dip, however, can and should be used to an advantage—as a test that all supervised transmitters are sending a low-battery signal, which raises a fault condition at the control panel. If good, reliable power is needed, use *gelcell* lead-acid batteries; they surpass lithium in these applications, if they are used with trickle chargers.

Most transmitting devices have very, very small internal aerials. However, in problem transmission areas, short external aerials are sometimes needed.

Because receivers are small, architects are very tempted to completely conceal them (even multichannel receivers are only coffee-table-book size). This should be avoided. You must be able to receive the strongest available signal and see the panel status constantly.

It is better to have as many unique channels as possible on the receiver, with each transmitter having its own channel and unique address. This greatly helps response forces identify the actual point and path of intrusion, and facilitates false alarm identification.

When costs allow, use systems with high-gain antennae to allow for a substantial increase in the attenuation of the radio signal by moving the object in the building before full relocation or reevaluation of the system is needed.

Aerials on or in transmitters should be at least 6 ft 6 in (2 m) above ground level or higher. Do not place them near sensitive electronic equipment (e.g., computers, fire alarm panels).

CONTROL PANEL AND CENTRAL EQUIPMENT

General

The control panel is the heart of the intruder detection and alarm system. A poor-quality control panel will mean a poor-quality alarm system. A typical traditional control panel may contain the following:

- mains electrical supply connections
- batteries (either primary, nonrecharging, or recharging)

- transformer/rectifier
- battery trickle charger (if rechargeable batteries are used)
- zone control, isolation, maintenance, opening and closing, arming, and testing switches
- master-routing, bell time-out, and day/night switches
- fault, alarm, low voltage, zone indicators (LEDs) and audio alarm generator (buzzer on panel)
- automatic changeover switch or floating battery circuitry
- mimic diagram and legend

It is important that the control panel is not treated as a "black box" that performs complex rituals and only occasionally bursts into life. The control panel must not be selected based on appearance, but on its fundamental quality and useful functions!

The following sections give sufficient information for those of you selecting a system to be able to distinguish between high- and low-quality apparatus.

UL 1076, National Institute of Justice (NIJ) Standard 0321.00, and Corps of Engineers Guide Specifications (IDS) are the most commonly referred to quality standards. They are all very comprehensive and explain the actual, specific requirements of the individual parts of the control panel and the testing of the whole panel.

Of the three standards, the NIJ standard is the most easily understood and is simple for field engineers to test to if problems or disputes occur. UL 1076 (and other similar UL codes) have greater detailed explanations.

Power Supply

The system must obtain its electrical power from somewhere and the type of supply, the way in which it is provided, etc., will give the first clues to the quality of the system and control panel. Most systems are powered by one of the following means:

- transformed/rectified mains supply voltage plus rechargeable (secondary) cells, power supply unit, or uninterruptible power supply (UPS)
- transformed/rectified mains supply voltage plus nonrechargeable (primary) cells
- primary cells alone

For the majority of the time, the first two methods rely on the electric company's supply, until mains failure. The quality of this supply, means of provision, and continuity should be carefully considered, as this will partially determine the level of required standby battery capacity. The mains supply should be

provided by a source that will not be disconnected by

the electric company for more than 8 hours in any 36 hours, or more than twice per year

provided from a supply that is not isolated at night, on weekends, etc.

taken from a switch fuse with a metal case with lockable lid and lockable means of isolation (in the "on" position), and fed directly from and wired (without interruption) to a bus bar or distribution panel with lockable lid

free from voltage spikes or current surges (arresters and zener diode protection circuits should be incorporated within the panel)

within the tolerance ranges of 115V ±10% or from 100V to 130V with a transition time of less than 10 msec, and frequency variations no greater than ±4% of 60Hz. (Although these are within the normal supply limits, if these conditions cannot be ensured by the electric company, the control panel might comply with any standards, but will still fail to operate correctly.)

supplied directly to the control panel (or other system supply point) and not via switches, plugs and sockets, external fused spurs, etc.—all of which could fail or be switched off either intentionally or accidentally

Primary Supply

It is wise to remember that the primary power supply for the alarm system is the electricity supply to the building. Criminals rarely cut the mains supply, as this attracts too much attention. For most of the time, your normal building power supply will support the system. This normal supply is 110V, 60Hz AC. Most alarm systems work on 12V or 24V DC, so the primary power supply must be transformed and rectified. This may be carried out in a power supply unit PSU or UPS unit (which includes other functions), or in a simple transformer/rectifier unit.

Whichever the case, remember that all security system devices are change-of-state devices. *Any* change of state might cause an alarm to be raised. The sensors and system sensitivity is voltage- and current input–dependent. Hence, transformer quality is important.

Transformers should comply with UL 310, UL 1310, UL 1585, or UL 506, as appropriate. Transformers should be placed where they derive the most stable input mains voltage that is unaffected by other building loads. They should be secure from interference. The supply voltage and current at the proposed position should be trace tape–monitored for a few days, prior to installation, to prove stability. Transformers should be enclosed, ventilated, and safe. They should not be placed on any flammable surfaces and they should not have exposed terminals.

Transformers should include a secondary voltage regulation that will produce the secondary 12V DC needed, irrespective of normal primary input voltage. Also, the transformer should, whenever possible, be placed at the geographic center of the sensor loop wiring to shorten 12V cable routes and hence reduce cable voltage drop.

Transformers should be UL standard–approved. Their voltamp and amp output must be 200% greater than the total system load. Primary and secondary sides of the transformer should be properly labeled with the polarity clearly marked *live* (hot) or *positive, negative* or *neutral*, or *ground* or *earth*.

The AC-to-DC *rectifier* associated with the transformer should include an overcurrent draw, thermal cutout and a current trip with a manual reset. The rectifier should be rated between 150% and 200% of the total system DC load to allow for future expansion and reduce false alarm potential.

Secondary Supply

As the name implies, the secondary supply on most systems only takes over when the primary supply fails—whatever the reason for the primary supply failure. Secondary supplies are provided by battery charger/changeover battery bank units, with secondary cells or nonrechargeable primary cells with changeover units. Whatever the battery system used, certain aspects of selection are fundamental and common to both types of secondary supplies:

> the number of amps the secondary supply can produce throughout its life as required by the system in *alarm* conditions (e.g., UL 1076 requires a maximum intended load support for 24 hours, as the minimum)
>
> the number of hours the secondary supply can maintain the alarm condition and quiescent condition output (i.e., amp hours)
>
> the rate of reduction of amp output toward the end of battery charge life
>
> the minimum voltage level for the required output amp

Remember that battery output is temperature-dependent. At 70°F (21°C), battery output is normally 100%; at 20°F (−6°C) battery output is 80%. Batteries should never operate below 32°F (0°C) or above 130°F (49°C). Also, remember "Murphy's Law"—"If it can happen, it will happen." Just because your building's heating system normally keeps the building temperature at 70°F (21°C), does not mean the heating system will operate during the next blizzard with temperatures below 32°F (0°C). If your battery's temperature falls, your system could fail. Remember to use battery packs that are complete with links and connections, not loose with cells.

Battery Types

General.   There are five main battery types found in security systems:

1. Lead-acid (L-A) recombination cells—by far the most common rechargeable cell (i.e., gelcells)
2. Nickel-cadmium-sealed (NiCad) cells—once the favored rechargeable cell for larger or more critical systems, but now seldom used
3. Leclanche cells—carbon-zinc type; nonrechargeable
4. Manganese (i.e., Mallary or Duracell) nonrechargeable cells
5. Lithium, manganese-dioxide nonrechargeable cells—seldom used

In general terms, primary (nonrechargeable) cell systems (either primary only or primary plus normal mains) are looked upon as less inherently secure than secondary (rechargeable) systems, because of the constant worry and monitoring for low or no power output.

NiCad Rechargeable Cells.   The advantages to NiCad cells are they

> are economical over long life of system (3–5 years)
> are robust
> are nongassing/dry
> have a wide temperature tolerance
> have long shelf lives
> are smaller than most other types
> are rapidly rechargeable
> are voltage regulated (i.e., they dip far less than lead-acid batteries under a heavy continuous load; they are more stable on large systems)

The disadvantages to NiCad cells are they

> are expensive and are now seldom used on security systems for this reason
> have a lower voltage per cell than lead-acid batteries (1.25V versus 2V)
> have a higher current discharge potential (lower internal resistance) and, therefore, are more energetic than lead-acid batteries. Their low current draw makes them an overspecification for all but large systems.
> are easily overcharged and require a charge-limiting device in the battery charger to prevent damage
> must be *warranted new* when installed, as their shelf life is long
> should be completely discharged three to four times per year or their capacity will diminish

Lead-Acid Rechargeable Secondary Cells (Gelcells). The advantages to gelcells are they have a

> lower cost than NiCad cells
> higher volt output than NiCad cells and, hence, fewer cells are needed

The disadvantages to gelcells are they

are larger than NiCad cells and less robust

have a shorter life than NiCad cells (3–4 years versus 3–5 years)

have a steady deterioration in rated capacity after rated life

have a larger voltage dip at end of their rated output time

Note: Open- or *wet-type* lead-acid batteries should not be used on security systems. Only sealed or gelcell nongasing types are suitable.

UL 1076 requires secondary battery sources to recharge in less than 48 hours (if fully discharged) with trickle charging and the battery hour capacity cannot be less than 2 times the recharging time and never less than 24 hours.

BS 4737 is slightly more explicit as to the load the battery should be able to support and under what conditions. BS 4737 requires that the battery be able to supply all systems parts with sufficient electrical energy to maintain their continuous operation for a minimum of 8 hours if the mains supply fails and requires that the battery trickle charger/rectifier transformer unit be able to supply the normal system load while *also* recharging the battery to a full charge from a zero charge in a period not exceeding 24 hours. BS 4737 also requires that the battery be automatically and immediately introduced upon mains failure or when the mains voltage falls to a level below which the output voltage of the panel would be insufficient to run the system (in practice, this means when the mains voltage falls below 216V AC) on 240V AC systems.

Where systems rely on a *rechargeable battery only* supply, BS 4737 requires a sufficient capacity to provide normal load, including 4 hours in an alarm condition, for the maximum specified period before recharge.

For mains voltage systems that rely on *nonrechargeable* (primary cells) systems, BS 4737 requires a 36-hour normal load capacity (quiescent state), including 2 hours of operating under a full load alarm state. In other words, the battery must be capable of supplying 2 hours of alarm capacity after 34 hours of mains failure, with no alarm.

BS 4737 requires that the battery be automatically and immediately introduced upon mains failure or if the mains voltage falls to a level below which the output voltage of the panel would be insufficient to run the system. Also, BS 4737 requires that the primary cells be replaced at intervals not exceeding 75% of their expected shelf life. In practice, a good-quality system maintenance agreement will quote the replacement interval and it should not exceed 50% of the actual shelf life. Installers should mark the installation date indelibly on every battery installed.

Primary cells used in security systems are usually either Leclanche cells (of the carbon-zinc type) or manganese cells. Of the two, the manganese cells are indicative of a better-quality system.

For systems that rely on *nonrechargeable (primary cells) only*, BS 4737 requires that the battery supply comply with the requirements of BS 397 and BS 1335, and that the batteries have sufficient capacity to support the quiescent state system load for the total period they are installed, plus 4 hours in an alarm condition. BS 4737 also requires that the batteries be replaced at intervals not exceeding 75% of their expected shelf life and that the date of installation of every cell be indelibly marked on the battery.

Most alarm systems operate on 12V DC. Hence, a number of cells (eight or six) is required to provide the supply power source and standby power source.

It is extremely important that the method of cell interconnection be well designed and use custom-made battery packs. It is also important that battery types and manufacturers are not mixed, and that the alarm system manufacturer quotes the manufacturer of the batteries used. Many poor-quality batteries are beginning to appear on the market. Only the better manufacturers such as Duracell, Saft, and Eveready should be used.

UL 1076 requires all systems to be installed to NEC, ANSI, or NFPA 70-1987 specifications. UL 1076 refers to primary cells as "storage batteries" and is far less specific regarding system needs, simply referring to a 24-hour, maximum-intended load period the battery must support.

Battery Chargers and UPS Units

In quality intruder alarm systems, it is now very unusual to have separate transformer, rectifier, and charger units. Systems are normally supplied with either a battery charger unit (BCU) or a UPS. The principal difference between the two devices is that the UPS has a far higher ability to remove interference, surges, etc., on the mains supply, which cause false alarms. UPSs are commonly used as backup power supplies for computers.

Whichever unit you select, it is important that the intruder alarm company provide and install it to prevent divided responsibility for the system. It is not uncommon for service engineers to falsely blame battery chargers or UPSs for false alarms when they are called out to a site. If the UPS does not belong to the installation company, the repair charge will become the responsibility of the client. It is very difficult to disprove the claim.

Battery Chargers. Battery chargers should comply with UL 1236 or LESP Report 0202.00 and at least

be able to recharge all batteries to a full charge within a specified period (normally less than 24 hours while maintaining the system alarm load) be hard and permanently wired into the normal electrical service (i.e., not plug and socket)

be *internally fused* on both primary and secondary sides (external fusing could lead to tampering)

be free-floating across the main and not switch over on failure

include audible and visible alarms for mains failure, secondary side failure, and current and voltage regulator failure and shutdown

have a DC or AC voltage trigger to activate digital communication and pass on failure messages to a central or local response force

have an antitamper device within its cover protection

have a grounded negative on the secondary DC mode for safety and reduced false alarm rate

include voltage and current meters to allow visible inspection of battery condition and to facilitate the maintenance of battery records

have short-circuit protection to prevent permanent damage

Good, UL-approved, battery-charging units will have lifetime replacement warranties.

UPS Units.   UPS units generally offer more protection from interference and more monitoring and recording capabilities.

The most commonly quoted standards for the UPS and control panels are

- UL 603 power supplies for use with burglar alarm systems
- UL 634 connectors and switches for use with burglar alarm systems
- SIA 1987 NEC interpretation (white paper)
- BS 4137, Part 1

Most good-quality security panels derive their power supply from one or more UPSs instead of primary cells. A good-quality UPS will have a

low-heat-output, fully rectified 110V AC to 24V DC transformer

solid-state voltage regulator

solid-state linear current regulator (some manufacturers offer foldback regulators, but these are inferior)

high-temperature (resettable) cutout device

continuously monitoring, low-voltage alarm circuit

solid-state, main input suppression filter for removal of transient high-voltage spikes

trickle charger for the standby battery

If the system uses a UPS with NiCad batteries, it is usually of better quality than a system that relies on lead-acid batteries, *provided* that all the preceding capabilities are included.

BS 4737 Part 4.6, Section 4.1, 1986 requires that the UPS

be of sufficient capacity and recharging rate to cope with any prolonged isolation of the main supply for fire safety, normal isolation, and isolation to work on part of the electrical services

be located to promote easy maintenance

be provided with sufficient ventilation to prevent gas buildup (on the vented battery), which could cause damage or injury

be marked with date of installation

not be exposed to corrosive conditions

have fully restrained cells to prevent accidental fall, spill, etc.

BS 4737, Part 1, Section 7.4 requires any UPS that uses the main supply to

have a safety isolating transformer built to BS 3535 specifications

provide the required output plus recharge requirement under any construction of +/− rated supply voltage and +/− rated supply frequency at temperatures between 14°F and 104°F (−10°C and 40°C)

UL 603 requires similar standards but is far more comprehensive than British Standards; UL 603 standards should be used.

Power Line Conditioners or Surge Arresters.   Power line conditioners or surge arresters prevent distortions of voltage, current, and frequency caused by lightning, load switching, etc. They are often needed on CCTV access control and alarm control systems, and should offer at least

- ferroresonant design
- no moving parts
- isolation primary to secondary sides
- 125% of full load capacity
- ±1% output volts at 85% load, with −25% to 15% of nominal input volts
- voltage change on load change ±3% for three cycles and 95% correction for two cycles
- high-speed, leading edge suppression
- high-energy resistance, metal oxide resistor

The Institute of Electrical and Electronics Engineers (USA) (IEEE) 587 (previously ANSI/IEEE C62.41, 1980) classifies the normal transient waveforms for protection and standards. A transient/surge protection device wired in parallel with the building main supply, with a much smaller local security system supply surge protector, will normally remove all problems.

Interference from the Environment.   All modern intruder alarm control panels (and most sensors) contain sensitive electronic components. These components are generally very reliable, but will only remain so if properly protected from a variety of environmental electromagnetic influences. Failure to provide adequate protection will result in increased false alarms, intermittent faults, blown fuses, destroyed microchips, etc. This can be costly in both maintenance and replacement, but even more costly in terms of user confidence in the system. The major influences are as follows.

Lightning Storm Damage. Lightning is probably the largest single destroyer of system components or instigator of false alarms caused by other than human incompetence. The levels of risk faced due to lightning obviously varies from area to area (Florida is high risk). It is possible to carry out a simple calculation to determine statistically the chance of lightning affecting your system. Why bother! Unlike lightning conductor systems, protection of alarm systems to reduce the risk of damage is inexpensive and should be included on all systems of quality. Essentially, any system design should include

surge arresters on the secondary side of transformer wiring and sensor wiring to prevent excessive currents or voltages

grounding of the system's main transformer to a separate consumer grounding roof or copper earth tape and connection by means of heavy gauge multistranded wire (insulated)

bonding (i.e., physical connection of all EMI and heavy, walled conduits to the system grounding point)

avoidance of exterior wiring

NIJ Standard 0321.00 includes a simple lightning surge test that can be used to test systems. Also, the Corps of Engineers Guide Specification 1672S (IDS) gives very clear guidance on surge protection.

Radio Frequency Interference (RFI) and EMI. There are two major types of interference of intruder alarm systems—RFI and EMI. RFI is caused by radio sources (e.g., transmitters, CB radios, police radio transmitters, discharge lighting). EMI is caused by either a radiated source of electromagnetic energy (similar to RFI) or a conducted energy source on the mains or sensor wiring. EMI sources include welding tools, motor starting and stopping, lightning, etc.

Both types of interference distort the normal current flow (usually DC) within the system or its components.

The main methods of EMI and RFI protection are

- filtration (primary)
- screening and shielding using grounded conduits or coaxial-type cables
- filtration (secondary)
- segregation
- earthing and grounding

Filters are essentially electronic sieves that only permit certain types of currents or signals to be passed. The filters should be included on the primary (control panel)

and secondary (lines or sensor wiring) to remove conducted interference.

Screening and shielding involve encasing electronic components in metal boxes or conduits, which are then grounded. Radiated electromagnetic or radio energy hits the shield first and is conducted harmlessly to the ground before it can affect the system. Screening normally refers to cable screening (i.e., the outer woven screen of fine wires in a coaxial cable) and to metallic enclosures of a more solid type. *Segregation* is simply the physical avoidance of a potential problem by rerouting cables or repositioning controls. One of the most common problems found in segregation is that of common cable ducts, which carry data, communication, or alarm cables in parallel within normal power voltage cables. Long runs in parallel can result in induced harmonic currents from the radiated energy of the power cable flowing in the alarm cable. The most successful way to overcome these harmonics is to reroute cables so they occasionally cross at right angles, but do not travel in parallel. Where this is not possible, shielding by installation of a grounded metal plate, conduit, etc., between the cables will improve matters (0.8 mm or 1.6 mm aluminum).

Buildup of Static Electricity. Static electricity can build up in materials, the atmosphere surrounding control panels, or the people working on control panels. Static electricity, discharged well below the level at which the average human feels discomfort, can severely damage electronic components, particularly electronic microchips. This damage may be permanent, temporary, or intermittent. The latter is particularly worrisome as it causes intermittent, repeated, random faults and false alarms.

In areas where control panels are located, do not allow static electricity. Change carpets or wall finishes or move the control panel. Static should not damage components, provided the cases containing them (and any persons working on them) are properly grounded.

To summarize, with the increased use of electronic component circuitry within all alarm panels and the increase in number of different voltages and frequencies used in buildings, panel circuit protection has become very important. It is essential that the control panel and wiring be

screened from the effects of radiated RFI and EMI

screened to prevent them from generating RFI and EMI, and hence damaging other equipment

filtered to prevent them from generating mains interference

correctly fused and grounded

protected from damage due to lightning strikes on any part of the system earthing, supply lines, or equipotential bonding

Many good manufacturers hold test certificates from independent testing stations to indicate that their panels are resistant to all known interference tests currently applied. Good control panels are interference- and sensor loop resistance–tolerant (i.e., the electronic and electromagnetic circuits within them are not oversensitive to system or supply variations).

Control Panel Location and Housing

Location. The location of the control panel within a building should be carefully considered. The control panel should be

- in a position where it will remain secure under all conditions, even when the area is unmanned. Often, panels are placed near reception desks—in open areas where they are totally unsecured. Whenever possible, the control panel should be located in a room with a lockable door that is wired to the alarm system, where it is not visible from the outside, and where no building occupant or visitor can tamper with it.
- positioned where it will be possible for a guard to monitor it, while also watching the main entrance to the building and vehicular approaches
- positioned at the economic center of the wiring. Although most security devices do not suffer greatly from voltage drop, excessive cable distances increase the cabling costs. Excessively long runs or runs that cannot be grouped for at least part of the route will have a high installation cost.

Guard/receptionist room areas in centrally located, ground floor foyers are good locations for a control panel, as they can be secured and they provide sufficient privacy for the system controller. If a room of this type cannot be provided, the control functions should be separated from the indication functions and the latter placed in the entrance/reception area. This can be provided by an exit route repeater circuit and indicator panel. This allows the control unit to be activated, deactivated, or monitored by means of a remote indicator board with secure key switch. This is very useful where exit/entrance areas are not secure enough to locate the actual panel.

The Control Panel Housing. BS 4737, Part 1, 1986 calls for

- one or more containers to be securely fixed and fitted with antitamper protection
- containers to be constructed of either 0.05 in-thick (1.2 mm) mild steel, 0.04 in-thick (1 mm) stainless steel, or 0.012 in-thick (3 mm) polycarbonate at the minimum. Other materials are permitted only if

they provide equivalent or better durability and security of fixing and resistance to attack.
- have a 200 differ key reset antitamper circuit on the housing lid to prevent unauthorized entry. When the system is unset, an immediate, audible alarm should be given if tampering is detected; the alarm should only be silenced by using a key. When audible devices are disconnected at the panel, the panel shall give warning via a panel indicator and not allow setting (arming) of the panel. When the system is set, the panel should give a normal audible alarm, just like any other sensor or detector.

In addition, good control panel housings should

have lockable front panels
be metallic and/or offer EMI screening
be recessed flush to structure for improved security (with ventilation allowed)
have a plain-faced, locking cover plate
have NEMA 12 metallic/NEMA 250 internal housing
have NEMA 4 metallic/NEMA 250 exterior housing
have NEMA 4× /NEMA 250 enclosures where corrosive atmospheres are present

UL 1076, *Proprietary Burglar Alarm Units and Systems,* is far more exhaustive and comprehensive than most European standards for both the control panel housing and the contents of the housing. For example, UL 1076 lists at least 15 different conditions that a housing must meet to be approved and also lists 26 specific tests that the housing and its contents must survive. There is little doubt that while UL documents on this subject make for very tedious reading, UL-approved control housing and contents are a good indication of high quality.

Control Panel Facilities

If the control panel is to operate and be operated correctly, it must be

- simple to use (fool- or user-proof)
- flexible
- reliable
- secure

In addition to these fundamentals, the control panel must offer facilities (see Figure 2–9) that are commensurate with the risk, but do not overburden the system or overcomplicate its operation. On small, domestic and commercial systems, the facilities offered by complying with standard codes normally suffice.

Setting Devices. Every time the alarm system is required for operation it must be set up. This involves three distinct phases:

FIGURE 2–9   Controller and LED keypad

- energization
- fault testing
- system alarm loop arming (or setting)

No matter how sophisticated the setting device of the system, it must carry out these functions. Energization of the system is usually carried out by means of a key-operated switch located on the control panel. Turning the switch to the first position provides power to all normally de-energized circuits. If the power supply is sound, an indicator will usually confirm this fact. Testing of the circuits to ensure that all devices are in their proper state and that there are no wiring faults (accidental or deliberate) is accomplished by turning the switch to the next position. If a system fault is detected, the control panel should never permit the operator to progress to the arming phase, regardless of which type of setting device is used. This makes the system foolproof and is extremely important as many false alarms are user induced by arming when faults exist.

The methods of arming a system vary. In a system that has a key-on-panel-only arming device, a single, three-position, key-operated switch on the panel provides energization, testing, and arming capabilities. This system is common on panels for residential premises and on small, low-risk, commercial panels. A disadvantage of this system is that a person who gains entry to a structure other than through the legitimate final exit door, can turn off the system without having the key to that door.

Many systems on the market still offer what is commonly referred to as a *shunt* or *setting switch,* which is a microswitch fitted within the door or lock of the final exit. The key switch on the panel is used to energize and test the system. If testing is successful, the local alarm bell will ring or, if it is a silent alarm system, a signal will be sent to the central station. The user must now go to the final exit door, leave, and lock the door. Only when the key turns the final exit door locks to the closed position will the alarm system become armed. If, at this point, the alarm stops, all is well.

Although these switches are reliable and offer easily understood closing and opening regimes, the security they provide is limited by

- the mechanical protection afforded to the microswitch
- the type and arrangement of the microswitch
- the security offered by the lock

If the lock is picked or left unlocked, a copied key used, or the microswitch cut from the door, the system is compromised. Better systems reduce these dangers by

enclosing the switch as an integral part of the staple of a very long-throw, dead-locking mortise lever or mortise cylinder lock

controlling only the volumetric detector in the final exit corridor from the switch (all other circuits having been previously activated by the control panel enabling switch)

putting the lock switch into the door with a covering plate that locks into place over the keyhole thus doubling the keys required to defeat the system

ensuring that the lock and key used are of the high-security type with a special key blank and are registered or have a similarly restricted key issuing policy

putting a timer on the switch to set and disarm the panel within a given time, thereby reducing the time the intruder has to locate and disarm the panel. This is the simplest form of "setting timer."

Externally mounted switches located within secure containers, rather than locks, should never be used. In general, the simplest shunt switch device does not offer the security or flexibility required of most medium-to-high security systems. The ubiquitous microchip has made much more sophisticated setting regimes possible. These more modern devices put the final, exit-arming switch on the panel itself (within a secure room, within a secure boundary), thereby removing all the dangers facing the exposed shunt switch. These devices are normally referred to as *setting timers* or *digital event timers.*

Setting Timers.   Setting timers are antitamper, programming switches that allow the following features to be programmed:

opening time—only permits the panel to be switched off during a predetermined time slot. Hence, if the key is lost or stolen, the system is less likely to be successfully deactivated by an unauthorized key holder.

normal closing—raises an alarm if system is not set by a predetermined time. This promotes efficient closing procedures and reduces the risk of personnel not setting the alarm.

FIGURE 2-10 LED keypad

bypass — allows deactivation of opening device circuit in noncritical, lower security risk areas for established periods before and after closing and opening procedures in order for cleaners, etc., to work

overtime working — sets a time limit override of the normal closing time to allow for occasional overtime working by staff

a *delayed action or silent alarm timer* for the ringing of an alarm bell on use of a personal attack device — immediate alarm energization would put the user under increased risk

sounder duration timer (bell time-out) — can be set to allow bell to ring for a set duration only, thus reducing nuisance caused by false alarms

entry timer — increases security by requiring a person to complete the opening procedure within a preset time and, hence, increasing the risk of detection for those unfamiliar with the opening procedure (see Figure 2-10)

In general, opening time and normal closing time capabilities are only applicable to medium-to-high security systems that are linked to alarm company central stations. With these systems, the alarm is raised *only* at the control station, not locally, allowing for the possibility of the criminal being apprehended or reducing nuisance if the user sleeps in! These systems are often referred to as *timed* or *timetabled silent alarm systems.*

Bell time-out circuits are now required by most city, state, and local authority bylaws. When selecting a bell time-out circuit, ensure that

the time-out period is the minimum possible to comply with bylaws

the bell rearms itself if a loop or sensor is reset. For example, an intruder might gain entry, let the bell

time-out, and then work in silence. While other sensors register his presence, the bell is still disarmed. during the time-out period, an externally mounted warning light over the final exit shows that the system has not been reset and that an intruder might, therefore, be within the building. This will warn the response force.

Control panels with isolating switches, which offer the ability to de-energize individual alarm circuits for maintenance and fault isolation, should never be chosen lightly, since they might invalidate the client's insurance coverage.

On larger systems, a *verify* switch is useful, as it prompts the operator to check that all alarms and timers have been properly initialized before activation. To comply with UL standards and BS 4737, the panel must *not* allow arming if a circuit or sensor is faulty.

The choice of a particular setting device and facilities on the panel depends almost entirely on the client, who determines the level of risk faced and the value to be protected. Setting timers are of more interest on buildings that have regular opening and closing hours (e.g., banks, shops). In general, the order of setting devices, from least secure to most secure, is

• key-on-panel-only arming
• shunt switch (bell set)
• shunt switch, but with timed access route alarm loop
• local audible alarm only and setting timer with preset opening and closing times
• silent alarm to central station and setting timer with preagreed opening and closing times

Alarm Circuits.

Night Alarm. Night alarm sensors are only required to detect at night and are connected to night circuits. During the day, a day/night switch is used to select *day circuits,* thereby deactivating all loops connected to night-only sensors. These would normally be volumetric detectors, break glass, vibration, etc., covering areas heavily used during the day (see Figure 2-11).

Day Alarm. During the day, the user selects "day" on the day/night switch. This keeps all circuits that are linked to that position energized during the hours of opening and any sensor on that circuit armed. If any attempt is made to bridge wiring during the day, or if circuits are damaged or persons intrude into areas covered by the day sensor, an alarm will be raised. This is an extremely useful function for premises such as banks, drugstores, etc., which might want store doors, rear exits, rear windows, etc., alarmed 24 hours a day to prevent intrusion, pilfering, etc. Day alarm circuits are usually only connected to the alarm panel buzzer to notify the user that an action or repair is required. This prevents nuisance calls and false alarms to be sent to the response forces.

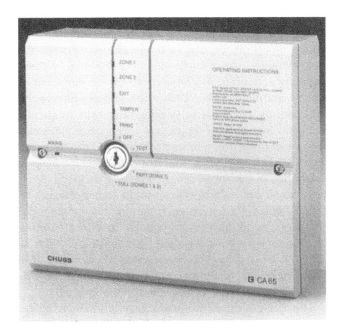

FIGURE 2–11 Two-zone domestic control panel

Personal Attack Alarm. This type of alarm has a silent or delayed action bell.

Antitamper Alarm. This type of alarm is used on all circuits/housings.

Access Route Alarm. The access route circuit is controlled by a setting timer and allows persons to enter the building by a predetermined route. If the panel is accessed within a given time, no alarm will be raised. Some panels allow for an audible buzzer on the access route to be raised as soon as the unsetting switch on the entry door is opened. The buzzer continues as a reminder that the system must be unset at the control panel. If the panel is not unset within a specific period, a full local alarm is given. After a second predetermined period, if the system is still unset, a signal will be sent to the remote central station.

Open Circuit. The open circuit alarm circuit is used to connect sensors (e.g., pressure mats) that normally open the circuit only, thereby closing an alarm condition. Better systems have a completely separate loop for the devices, without any terminal common to open and closed circuit sensor loops.

Passage Alarm. A passage alarm is used in remote areas of the building that are normally monitored by the system, but are occasionally passed through. This system allows a local buzzer and panel buzzer to sound as the person enters, but inhibits full alarm signaling. If the person has not vacated the area within 5 minutes (maximum), a full alarm is signaled within 5 seconds.

Indicators. Indicators should be comprised of bright LED lamps and should have an audible signaling device. Most good panels include

- condition and status of sensor indicators
- system set indicator
- self-testing, healthy/faulty indicator (electronic circuitry)
- low voltage or current input, low output, and disconnected indicators
- logos on individual zones showing the type of device connected (e.g., personal attack) so that the reaction force knows the type of intrusion
- a full mimic diagram (showing building layout with indicators for zones, rooms, or individual sensors) to assist the reaction force
- alarm indication and reset required for each sensor, processor, etc.
- alarm indication for subsequent alarmed sensors and processors

Indicators should only be visible and audible during alarm setting and unsetting or test procedures, as appropriate.

Printers. Many modern panels include a memory and printer output port that allows a small printer to be attached to the system. The printer provides a hard copy printout of the information contained in the memory. This usually includes data on

- hours run
- setting and unsetting times
- alarm times
- mains voltage failure time

This feature is extremely useful in pinpointing the time the system was attacked or went into a fault condition and also provides maintenance system reliability records.

Signaling and Communication

Introduction. There would be little point in installing even the simplest intruder detection system if it did not raise an alarm that could be heard immediately, and recognized and responded to by a reaction force.

Before considering the technical side of signaling, consider these questions:

Could police officers in a passing patrol car clearly read the name and street number of your building (day or night) from the car?

Is the name and address on your message tape the same as that on the building?

Is it clear to the police or response force where the main and subsidiary entrances to the building are (day and night)?

If you answer "no" to any one of these questions, do not be surprised if the police or the security company does not catch an intruder on your premises.

In the design of the overall system, the signaling and communication method is the *Cinderella element*—scant attention is paid to it. However, this method should be the first element considered and established with the client. If the communication and signaling equipment is not suitable, the client may

suffer losses due to its failure

be forced to disconnect it due to nuisance and/or false alarms

pay unnecessarily high operational costs

In order to reduce the risk of any of these occurring, it is prudent to ensure that

the client knows to whom the signal is given, by what means, and how quickly they are likely to respond

the method of communication is clearly understood by the client. Clients are often amazed that neighbors, instead of calling the police, telephone to say that an alarm bell has been ringing for 30 minutes and is causing a disturbance.

police, local authorities, and neighbors be notified of the proposal to install a system. The former two organizations should always be consulted for advice on local conditions, bylaws, response, etc. The police offer good, free advice.

the control panel includes diagnostic and checking facilities that check circuitry and the validity of a signal before raising an alarm

the system is installed correctly and maintained properly by trained personnel on a routine basis

parts are renewed regularly, particularly if subject to heavy use (e.g., door contacts)

the system operation is checked thoroughly, and commissioned and tested for a period of at least 14 days before it is handed over

it is made extremely difficult for communication channels to be blocked, cut, bridged, or otherwise interfered with. They should never be accessible to unauthorized persons, including caretakers, electricians, and plant engineers.

A fundamental paradox that exists within the philosophies of different parts of the security industry is the type of alarm. Insurers would like to suffer zero losses and, as they have leverage with the client, their views can predominate. In order to minimize loss, many insurers wish to prevent entry or scare off intruders by using a very loud, local alarm. This was the traditional means of alarm and, theoretically, the intruder was supposed to run straight into the arms of a policeman on the beat. As the police moved off the beat into cars, another means of catching thieves was required. This involved a silent alarm being sent to the police station and a delayed local alarm to which patrols could respond, thereby catching the criminals on the site. In the transition between the two methods, the security industry boomed without applying enough of its resources to the development of *reliable* equipment. The result was a massive increase in false alarms. Local alarm systems rang bells frequently when no intrusion or attack had taken place. This created a bad noise problem, and the public began to equate all alarms raised as "false" and stopped reporting them to the police as they occurred. Arrest rates dropped. Similarly, direct silent alarm systems to the police called them out on many wasted visits, stretching the police's patience. Local authorities responded by either banning local alarms or reducing their maximum audibility or duration of ringing. This made them less effective. The police response to direct-connected systems (from private companies) was to have them disconnected.

The result of this action was a huge increase in the use of *middleman systems*—central alarm response stations (manned 24 hours a day) provided by security companies. These companies offer the ability to either check the system's alarm before relaying the alarm to the police or provide direct response to the alarm followed by police notification. These systems have been, and continue to be, a very lucrative field for the industry. They have not, however, cured the problems that brought about their introduction and in which the client is more interested—namely, false alarm generation, increased response times by the police, and security of communications. Although many of the communication capabilities offered by private security companies are excellent, the technology involved often obscures the overriding concern of the client—how quickly will your system get a response force to my buildings? If police respond directly, the question is, perhaps, unfair, as the security company will have no control over police priorities. If, however, the alarm company provides the response force, the question is fully justified and clients should know where they are on the security company's list of priorities.

Bearing this in mind, the common signaling and communication systems can now be described. The security industry normally divides systems of signaling and communication into two fundamental types:

local, audible alarms, which require no interface or interaction with out-of-building signal transmission systems (The alarm signals and communicates!)

remote signaling and communication systems, which require a device for communicating and a signal transmission system

Local, audible alarms are well known and used in many small, low-risk, low-value premises. The second category of systems is divided into subtypes, differentiated only by the type of communication signal sent and the medium used. Any system in the remote signaling category that requires sending a message to a remote response force, must have either a

- tape dial unit
- disk dial unit
- digital dialer (communicator)

The most sophisticated, secure, digitally encoded signal transmission system in the world will be as useless as a local, audible alarm without a bell gong, if the tape dialer or other device fails to send the "help" message. The "Remote Signaling" section in this chapter considers these devices in depth.

Local, Audible Alarm.

Principle. The sensors are linked via the control panel directly to the system alarm, which is usually mounted on the facade of the building. No direct link with any response force exists.

Types. System alarms are divided between bells and sirens (see Figures 2–12 and 2–13). Bells are gradually going out of style within the industry because they are easily confused with a fire alarm, shift change, or other audible indicators or alarms.

Uses. Local, audible alarms are only used alone in low-security risk situations, such as small retail outlets and domestic premises. They rely on the cooperation of the local populace and the frequency of passing police patrols for their effectiveness. For this reason, they are not recommended for multitenancy offices; inner cities; areas of small warehouses, factories, etc.; or large, isolated industrial estates where they are unlikely to be acted on, even if heard by the public. They are still used extensively due to low cost and absence of operating costs, telephone line rental, etc.

Selection. BS 4737 lists the following minimum standards for alarms:

    must be two-note type, continuous, or one-second-on one-second-off

    the alarm must emit at least two fundamental frequencies in the band 300Hz–3kHz

    must *not* be of police, ambulance, or other emergency service type

    must be totally enclosed and weatherproof, with a case offering at least the protection that would be afforded by 0.047 in (1.2 mm) of steel

    must produce at least 70dB(A) mean sound level *and* 65dB(A) in any one direction at 10 ft (3 m), with the security case or cover in place

    must not sound during normal opening and closing procedures

FIGURE 2–12 High-security alarm and beacon

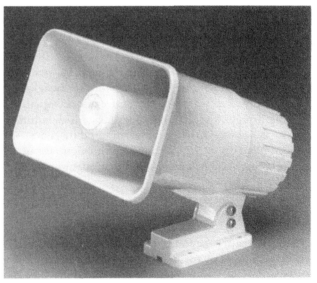

FIGURE 2–13 High-intensity siren (before boxing in)

    must be self-powered by a battery source that offers either a 30-minute (secondary cell) or 2-hour (primary cell) alarm state duration with maximum recharge time on the secondary type of 24 hours a short circuit of a secondary battery charger should not result in the secondary cell discharging

local control-of-noise orders give guidance on methods of preventing noise disturbances in sensitive areas. A 20-minute automatic cutout of the alarm and automatic rearm is usually required under many local bylaws.

must include an antitamper alarm circuit

In addition to these minimum standards, good, local, audible alarms incorporate

adjustable bell-ringing times from 0.5 minutes to 30 minutes in duration, with automatic resetting (i.e., bell time-out rearming circuit)

noncorroding cases weatherproofed to NEMA standards or IP55 (polycarbonate polypropylene or stainless steel)

vandal-resistant case and workings (bells can be damaged by impact on outer case). High-security alarms should be placed in a solid, stainless steel case. Bells with double-skin cases and drill detection offer good protection against vandalism and "professional" attack.

low-current draw, electronic, motorized bells of a minimum diameter of 6 in (150 mm)

good temperature resistance. BS 4737 calls for bells to be able to operate in the range −18°F to 122°F (−10°C to 50°C) 10–95% relative humidity. It is not unknown for temperatures to drop below −18°F (−10°C) and for bell mechanisms to freeze solid. The better the weatherproofing, the lower the likelihood of this resulting from weather (or intentionally created low temperature!).

adequate sound power level in watts (L_w). Decibels are a relative level measurement. A 65dB(A) bell in a 65dB(A) ambient environment will not cause a disturbance as it will be barely noticeable. In situations of normally high sound level (e.g., factories bounded by motorways, flyovers, or where buildings are surrounded by others), a loud siren might be much more noticeable than a bell of equivalent power and decibel rating.

low-input voltage monitor and automatic alarm sounding device

Many cheaper sounder housings include grills or louvers within them to allow sound to escape more easily. If alarms of this type are used, it is essential that a freeze or setting foam detector is included within the housing to prevent alarm disablement. The detectors are normally two-beam, double-knock infrared beams. The foam agents used in attacks cause both beams to break and the alarm to be raised immediately.

Sounders are subject to being pulled from walls, hammer blows, foam attack, fire, tampering, short circuiting, drilling, and muffling. UL 681 specifies conditions to avoid or reduce these risks by stating that alarm housings should be visible from the public street and placed as high as possible on a building but be still maintainable—not normally above the fourth floor unless an auxiliary alarm is provided.

Advantages.

- simple, cheap, and reliable
- no third party or communication network for which to pay or maintain
- easily maintained
- always available, even during telephone strikes, power cuts, etc.

Disadvantages.

- no direct link with response force; chance of apprehension of intruder is low unless on-site guards are present
- not suitable for medium-to-high security risks
- false alarms can cause direct disturbance of neighbors

Installation.

- The alarm unit must be as inaccessible as possible (preferably at least two stories from the ground)
- The alarm must rest on a solid, monolithic structural base of concrete, brick, or hardwood if some of its power is not to be lost in vibrating the building structure
- The area in direct line of sight to the bell should, if possible, be free of high, dense hedges; rows of trees; etc., that might attenuate the sound. (Sound reduces by 6dB for every doubling of initial measurement distance on flat, open sites and by up to 10–12dB in built-up or tree-planted areas.)
- Bell units should always be rear entry with no exposed conduits or wiring of any type.
- With a mean background noise level of 25–30dB(A), a siren will be audible at 0.5–3 mi (0.8–4.8 km) depending on terrain, building, etc. A bell would be audible 328–1640 ft (100–500 m). In general, sirens are used more commonly on very remote, isolated buildings or very high security installations, and are often referred to as *howlers*. When used in cities, they must have a short-duration, bell time-out circuit to ensure no nuisance is created. An audibility test should always be carried out (under normal conditions) after installation. Pick the worst direction in terms of surrounding attenuation, the nearest busy road or path, and listen to check that the alarm is still clearly audible.

Remote Signaling. Whatever form of remote signaling is considered, the system always includes two basic elements:

- a means of communicating with the response force
- a system through which a message can be transmitted—the signal transmission network

The first of these elements can be carried out by

- a tape or disk dialer (automatic telephone-dialing unit)
- a digital dialing unit

Tape and Disk Dialers. With tape and disk dialers, a microchip memorizes the emergency number as a series of dialing tone electronic pulses. When the alarm condition is raised, the system automatically rings this number in the same way a conventional push-button telephone would. Within 5 seconds of the last digit being dialed, the system begins to transmit a verbal message that has been prerecorded on a disk or endless tape loop. The system will transmit this message irrespective of whether the

- alarm has stopped
- telephone line has been successfully connected
- line has been cut

The tape will repeatedly transmit the message for between 1 and 5 minutes total duration. After this time has elapsed, the system will assume that it has been

- connected
- received
- understood
- successful in initiating a response

Most devices have no way of confirming any of these assumptions. Once the tape has started transmitting, it continues to the end of the 5-minute period and then ends the connection. The tape will then lie dormant until a fresh alarm signal is received.

Types. The *disk* was the gramophone disk and player, and it has gradually been superseded by the tape device. The *tape* is a continuous loop tape, sometimes multitrack, allowing different messages to be sent at different times. Most tape units have a delayed bell/sounder ringing device that inhibits local, audible alarms for up to 15 minutes to allow the police to arrive (if the message is transmitted successfully). This will increase the chances of the intruder being apprehended.

Uses. Disk or tape dialers are only used on low-level risks. They are less reliable than digital dialers (sometimes called *digital communicators*), due to the fact that voice transmission can be garbled, the line may be bad, and there is no way to interrogate the system remotely to ascertain that it is actually ready and available to work. (It could have been disconnected to mute the system.)

Selection. Mechanical automatic dialing units that use a mechanical device to hook up the telephone lines are almost obsolete and require excessive maintenance. Electrical automatic dialing units, which have the telephone number (or numbers) encoded on the same tape as the verbal message, are more reliable, as the mechanical hookup line device is removed and the number is called electronically.

Good tape or disk dialers should have the following characteristics:

The equipment must be acceptable to the telephone company and installed, powered, and protected in accordance with British Telecom Technical Guides Nos. 26 and 30 (see also NIJ 0322.00).

The equipment must be within a secured area, not visible from outside the building.

The dialer housing must have the same resistance to attack as a control panel and must comply with the same standards; the housing must be securely fixed.

The housing must include antitamper devices that prevent it from being removed from its fixing without alarms sounding and prevent its lid being opened without alarms sounding.

It must include a rechargeable (secondary) battery of capacity sufficient to transmit five full alarm call durations, and to recharge to this capacity within 24 hours.

The telephone line to which it is connected should be an underground, not overhead, supply.

It should operate within 5 seconds of alarm; continue to transmit, even if the alarm is reset; and not transmit alarm conditions that fail to last longer than 200 msec.

A line monitor device shall also be installed, if a delay alarm device is incorporated, to look for telephone line fault or disconnection at the exit end; if found, it should sound the alarm within 15 seconds (overriding the previously set time delay).

In addition to these requirements, good tape or disk dialers should offer

a built-in programming unit that allows the person responsible for the system to record the message and change the message on-site

programmable emergency call numbers of at least ten digits

telephone line fault and line cut alarm

adjustable message length

repeat-attempt dialing (e.g., three attempts, each of 2–3 minutes duration) to increase the chances of the message being received

lightning damage protection (surge arresters)

Advantages.

- easy to install
- low maintenance
- cheap
- reliable in operation (though sometimes not in results)
- uses normal telephone system and lines
- no cost of line rental if the system does not have a dedicated line connection

Disadvantages.

- essentially low security risk is its only application
- no way of determining whether the call was received or

understood; no record of message at premises or receiver's premises

- slower than digital dialer
- not suitable for trunk calls to remote response forces
- can be more easily compromised than a digital dialer
- should only be used on outgoing-only lines
- line reversal can be problematic
- busy line seizure is not possible

Installation.   Before installation, the telephone company should approve connection. The unit must be direct wired (not via a jack plug and socket) and secure, and the telephone line located away from the unit, preferably in a locked services cupboard. The unit should be connected to, in increasing order of security,

- one of several normally used lines (not recommended)
- an outgoing-calls-only line
- an ex-directory line

A common method used to try and foil simple systems is to *phone block* the outgoing line by calling on all lines while an entry is taking place. A new phone company service, *caller identification,* records the origin of all incoming calls and the time at which they are made. This assists the police on all types of transmission systems and helps with bomb scare and nuisance calls.

Digital Dialers and Communicators.   The digital dialing unit, as pictured in Figure 2–14, connects the telephone line in a similar manner to the tape dialer (i.e., a series of pulses imitate the phone dialing action and a prerecorded number is rung when the intruder detection system goes into an alarm state. Unlike the tape dialer, the digital device can sense whether the call has been successfully received. This makes it more likely that the response force will arrive at the scene. Hence, the device is now commonly used on the majority of low- and

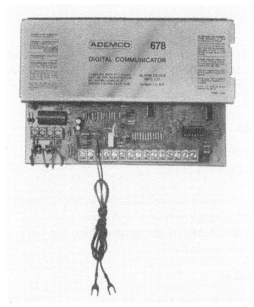

FIGURE 2–14   Eight-channel digital communicator

medium-risk intruder detection systems that use the public telephone transmission system.

The second advantage the digital dialer has over the tape unit is that no voice communication or transmission is needed. The message is sent as a series of encoded pulses that form a type of binary code. This action is carried out by a transmitting device that is linked to the dialer, rather like the nonerasable memory in a calculator. The pulses are received by a similar device at the police or security station, decoded, and transformed into an alphanumeric printout or video display unit (VDU) display. Because no voice communication is involved, bad lines or garbled messages that can occur on the tape system are virtually eliminated. By digitally transmitting the message, there is no speed limit on the message; unlike a voice message, which must be heard and understood. The digital binary code is converted to a tone code (by a device called a *modem*) before it is transmitted and reconverted at the central station by a similar modem.

Devices differ in the capabilities they offer, but most operate in the same basic sequence:

1. An event, such as closing and setting the alarm system or sensing an intruder, signals the intruder alarm panel circuits to initiate a trigger signal.
2. The control panel passes the trigger signal to the digital dialer transmitter control unit on the premises.
3. The control unit within the transmitter either connects the digital communicator to a reserved, outgoing-calls-only line or seizes a line away from its normal connection to a telephone handset. Having obtained a line on which to attempt communication, the control unit allows a dialing pulse generator to send a dial pulse to a preset number via the telephone line.
4. At the alarm company central station, the dial pulse sensor awaits such signals.
5. On receipt, the sensor prompts the transmitter to acknowledge with a handshake signal. This transmits an acknowledgment that the communication link is made to the dial tone recognition unit.
6. The dial tone recognition unit awaits the handshake. If it is received, the control unit encodes the information to be sent. If it is not received, the unit waits for a preset period and carries out a sequence of repeat attempts, which may include repeat calls to be made on the same line to the same number, alternative lines to the receiving station, alternative numbers, or alternative signals. Usually, ten attempts are made before a hang-up and failure report is generated.
7. Assuming the correct handshake is received, the encoded message is sent via the message transmitter.
8. The message is received and acknowledged and a signal to this effect is sent back to the transmitter. This generates an off signal that stops the message

repeat signal. In some units, the message is decoded and checked before the off signal.

If no acknowledgment is received, the first ten messages will be exhausted, the transmitter will stop, and the unit will repeat the calls to

- the same number on the same line
- other emergency numbers
- the same number using a different outgoing line to the telephone exchange

9. As the message is received, an audio/visual alarm alerts the central station that a message is being transmitted.
10. The message is decoded.
11. The message is checked.
12. The message is displayed and printed.

Types. The types of digital dialers are defined by the basic functions they provide:

- single number, single line, repeat call
- different number, single line, repeat call
- single number, multiple lines, repeat call
- multiple number, multiple lines, repeat call

Obviously, the greater the permutations, the greater the chance of successful connection. Hence, multiline, multiple number units give the highest security.

Uses. Digital dialers should be used on all intruder systems using the public-switched telephone network. They offer higher security at little, if any, extra cost.

Selection. Primary selection should be based on advice given by the police or security company, which will match the security risk to the types of previously listed units. Materially, the unit should have the same features as those listed under selection of tape and disk dialers. Better devices include

encoded signal transmission—*encryption*—(time division or frequency division multiplexer) to at least Data Encryption Standard (DES) Publication 46

silicon chip, memory message board (not wired circuit cards)

programs that allow the users to periodically change the signal encoding or message content by prior arrangement with the response force

listen-in, line seizure, and release capability

priority interruption, which allows messages to be changed (e.g., from "window broken" to "vault open") in response to individual sensor circuits

line fault, line cut buzzer, and instant alarm, respectively

manual and automatic transmission reset capability

individual premises message-acknowledged code (e.g., Browns Limited transmits "1010"; on receipt of message, "a message sent from 1010" is recorded automatically in the receiving station log. The acknowl-

edged signal sent to the premises is similarly encoded.

prearranged message content recognition. Only a restricted format message will be sent (e.g., "help required address") before the message-acknowledged signal is sent back to the premises. The receiving station checks to see if all of the message is present by comparing it with the system's memory. If the message is not complete, it asks for a repeat and will not acknowledge or let the transmitter hang up prematurely.

UL 1635, *Digital Burglar Alarm Communication Systems Units,* is the most common selection standard referred to in the security industry. NIJ 0323.00 is the alternative.

Problems. There are several potential problems with digital communicators that should be considered before purchasing one—public telephone network reliability [using the public service telephone network (PSTN)] and special information tone clash. As we will see later, there are many ways to transmit an alarm signal to a central station. By far the most common is the everyday public telephone company voice grade telephone line. One major problem with these lines is that they are not monitored for disconnection and many users have found themselves with no path to a central station when the telephone company decides to repair their system. Parallel alternative lines should be considered if this is a problem and the risk warrants an alternative.

One way this problem can be eased is the linked or looped digital dialer. With these devices, there are two outputs—one links the direct, normal line to the central station, the second links a neighboring digital dialer, which has a separate similar line, to the same station. Should the normal line fail, the digital dialer will pass calls through all the dialers to which it is connected until it finds a good line over which to transmit the message. This is not as quick or secure as a multiline, multinumber dialer located on the same premises, but is far cheaper than a direct, dedicated, monitored, private telephone line.

Special information tone clash can occur when a digital dialer makes an attempt to ring a number on the PSTN. If it finds that the line is busy, the dialer will stop and try to ring again. Unfortunately, in an effort to promote a good public image, telephone companies have been introducing special information tones (SITs). These tones precede a recorded message that details the congestion problem. The difficulty is that these tones can be similar to those used by a remote station to indicate that your message was successfully received, *when, in fact, it never got through!* What are known as fast and super-fast dual frequency dialers rarely have this problem, but slow-format dialers do. Therefore, be aware of the type of dialer you have or will have.

BS 4737, Part 1 requires the following of dialer units:

compliance with BS 6320 modem connection

battery capacity for five separate alarm call sequences (or 20, if supply to the recharger cannot be guaranteed)

a battery recharge within 24 hours

a 1-second initiation of call sequences from alarm recognition

the communication channel open signal to be received within 1 minute. If this does not happen, the device should hang up and restart the sequence (e.g., different line, different number)

once the communication line is established, up to ten messages may be sent in an attempt to get an acknowledged signal. If, after ten attempts, acknowledgment is not received, the line shall be released and the whole sequence repeated for a maximum of three lines. If, after the third attempt, the message-acknowledged signal is still not received (and the system has no deliberately operated devices), a local alarm should ring within 10 seconds

if a fault, off-hook, phone-in, or other block in the monitored line is detected (and the system has no deliberately operated devices), then the system will sound a local alarm within 30 seconds.

Advantages.

- message-received verification and validation of correctness
- number and/or line change
- no voice transmission, no garbled messages
- line cut and line fault sensing and alarm
- will transmit over very long trunk lines with several exchanges
- higher security offered than tape dialer
- two-way message and command possible

Disadvantages.

- slightly more expensive than a tape dialer

Signal Transmission

Introduction. The final leg in the long journey of the intruder alarm signal is from the building to the response force. The ways in which transmission can take place are numerous, but fortunately, due to various factors, the range of options can be narrowed quickly. The process of narrowing the options usually only requires answers to a couple of fundamental inquiries:

What system of transmission are the building insurers prepared to accept?

What system of transmission and connection do the local police favor for the type of building and risk involved?

If neither the police nor the insurers are particularly rigid in their recommendations, an analysis of the economics of the various systems proposed by security companies will narrow the field.

Although the *ways* of transmission are varied, the *means* of transmission usually involve at some point the use of telephone services external to the building. When considering any of the alternatives, it is important to remember that

signals that are encoded are more secure

the fewer the interruptions to the signal on its journey, the more secure it will be—direct lines to the police, using encoding, offer the highest security

systems using nondedicated public telephone lines are unsuitable for all but the lowest risks

the further the signal travels, the greater the cost and the slower the response time

any system connected to a normal, nondedicated section of a telephone exchange can be interrupted, disconnected, etc., at any time, without notice being given. This frequently occurs when the system exchanges are in danger of collapse due to overload caused by many simultaneous calls. This can occur during local disasters, bad weather, Christmas morning, etc.

any comparison of "line hire" cost should be priced over the entire system's life. Telephone line rental charges rise annually and what may at first seem an expensive multiplex system, can soon look very favorable when compared with a private, dedicated line (e.g., using discounted cash flow methods over 10 years).

with any system using telephone lines, the building user has no control over the length of time it takes to repair faults and cannot make any claim for loss of service

Clients are often surprised at the operating cost of line rental. Invariably, the negotiation for the line will take place between the security adviser, the intruder alarm company, and the telephone company. It is essential that clients be aware of the quality and reliability of the line they are hiring and that they feel it is commensurate with the risks they face. If the line quality is too high, they will pay dearly for the next 20 years! If it is too low, they could find themselves disconnected at the very time an alarm signal should be sent. Ask the security company adviser to provide details of the comparative quality and reliability, and costs of line hire for the various systems available. In general, these lines are

- unspecified, ungraded, unknown loss lines—no guarantee of service
- point-to-point on various quality and tariffs
- tandem on various quality and tariffs
- omnibus on various quality and tariffs
- multipoint on various quality and tariffs

The common transmission systems available are explained in the following sections. The trend in the industry at present is away from direct lines to police stations and toward private security company central stations (either manned or unmanned). The exponential rise in criminal theft and limited police resources are forcing a change in police and government attitudes toward private security forces. It is very likely that within the next few years, we shall see a great increase in local area networks for security, with manned stations provided by security organizations. These stations will provide a signal relay point and an immediate response force. The police would then respond to any confirmed incident as and when time and manpower permitted. Figures 2–15 through 2–18 show systems without private security organization involvement.

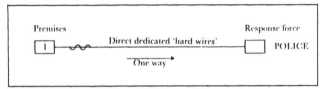

FIGURE 2–15 Individual direct line

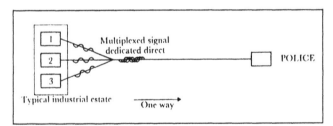

FIGURE 2–16 Shared direct line

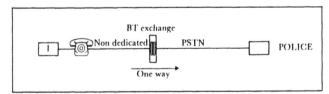

FIGURE 2–17 Indirect, nondedicated exchange/police

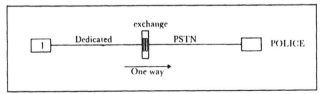

FIGURE 2–18 Dedicated exchange/police

Remember that signal transmission systems may be tailored to the local and wide areas needed. For example, a large university campus might run all its alarm signals from various buildings using the existing data network of its computers or the existing telephone line. Off-campus buildings on the other side of a multilane freeway might send messages directly across the freeway using a microwave or infrared link to prevent long-distance telephone or direct line routing under the freeway, both of which may be expensive.

Signals must be transmitted under the worst likely scenarios, including attack on the system, bad weather conditions, etc. The signal must be received, without corruption, during these scenarios.

Remember that transmission systems encounter common problems, such as attenuation, effects of RFI and EMI, and on-line or conducted interference. Also remember that the telephone line termination block, the digital communicator, etc., must all be housed in strong, environment-proof, tamper-proof secure enclosures or rooms that are wired to a security system.

Direct Line Systems: System Types

Individual, Direct Line. A single, dedicated line (see Figure 2–15) is taken directly from each property and without interrupting the response force premises. The line does not pass through any exchange. This system usually operates in conjunction with a communicator, which transmits a continuous encoded signal. Failure or change of signal raises the alarm.

This type of system is usually only considered for the highest security risks—banks, military premises, jewelers, government establishments, etc. If a system of this type is being considered, a specialist's advice must be sought. Good systems incorporate varied code or random code transmitters and matched receivers. These ensure that even if one code is broken, the intruder cannot be sure which code is used on the premises or, in the case of a random code unit, which code is being used at a particular moment.

Advantages.

- direct action
- no intermediaries involved
- no switching or manual actions required
- immediate warning of alarm at police station
- fastest response possible

Disadvantages.

- line is in continuous use and is used solely for the alarm system
- both installation and maintenance costs are high
- many police forces will not accept direct line connection due to limits of space for termination equipment and waste of manpower due to false alarms

- lines of more than 8 miles are usually prohibitively expensive
- only normal, alarm, and address messages can be reported
- a reassurance capability at the protected premises is sometimes available to monitor the line condition and receiver, but this is more expensive

Indirect Systems Using Telecommunication Company Lines

Indirect systems are identified by the fact that signals are either routed via a telephone exchange or a private security organization's central station. The main types of indirect system are divided between those using and those not using private security company central facilities or stations.

Systems Not Using Private Security Company Central Stations.

Nondedicated Line via Telecommunication Exchange to Police. A nondedicated line via telecommunication exchange to police (see Figure 2–17) is probably the most common system for low security risk installations. A tape dialer or digital communicator is connected to either

- a normal telephone line on the premises
- an ex-directory line
- an ex-directory call-out-only line

The first two methods are too easily breached and only the last is recommended (even for low security risks). Tape dialers are used to send messages to the emergency operator at the exchanges and the operator then takes the appropriate action.

Advantages.

- cheap line rental and user costs
- normal calls can be made on-line during deactivated intruder system time

Disadvantages.

- no check against false alarms
- no resistance to cutting, bridging, etc.
- all the disadvantages of tape dialers (see "Signaling and Communication")
- should only be used on underground transmission lines

Dedicated Line via Telecommunication Exchange to Police. A dedicated line via telecommunication exchange to police (see Figure 2–18) is used when security of the line to the exchange is required, but overall distance to the police station is high, thereby preventing the use of a direct line to the police due to economic reasons.

This system is seldom used because the line between the exchange and the premises returns very little revenue to the telecommunication company, unless the line rental charged to the property owner is very high.

In general, all indirect systems that do not use private security organization central stations are more costly to run and more prone to false alarms.

Systems Using Security Organization Central Stations

Systems that send an alarm signal to private security organization central stations (see Figure 2–19) are now the norm within the industry for most types of risk. Many security organizations make extravagant claims for their central stations in how they will improve security. If the purpose of security is to reduce crime, the criminal must be caught at the site of the crime to *show* that crime doesn't pay. Catching the criminal requires a fast response to an alarm signal—any delay gives an exponential increase in the chances of escape. Indirect transmission via central stations will always be slower than direct lines to the police. This is, however, their only real disadvantage and the growth of such central facilities is a testimonial to the one big advantage indirect transmissions *do* have—optimization of police resources. Central station personnel can make value judgments, interrogate systems, distinguish line faults, etc.—none of which can be done on normal direct-line systems. The result is a reduction in the false alarms transmitted to the police and, hence, a more effective use of manpower.

Shared Direct Line to Central Station. A shared direct line (see Figure 2–16) is basically the same format as an individual direct line, but the difference is that a single *cable* carries signals directly from up to, say, ten different premises, all in the same locality. The signals are multiplexed onto the line so that the line can carry ten different signals simultaneously.

Advantages.

- lower line rental per individual property

Disadvantages.

- not as secure as an individual direct line

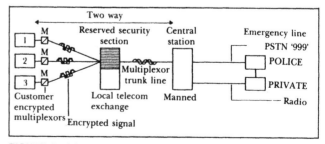

FIGURE 2–19 Unmanned satellites

- limit to the amount of encoding, varied code, or random coding that can be carried out
- can only operate on-site when several high-risk properties are in close proximity

Shared Line to Central Station.

Using Satellites. *Satellite* is the term applied to an unmanned room within a building or enclosure located near the geographic center of a number of premises. The alarm system on each of the premises is connected to the satellite via a single, dedicated, private telephone line. The satellite has within it a multiplexing unit that transfers all the incoming signals onto a single, outgoing (or trunk) line that is leased from the telephone company by the alarm company (see Figure 2–20). The trunk line connects the satellite to the central station, which might be 50 miles away (although usually within the same police force area).

Advantages.

- reduced line rental cost due to line sharing for up to 95% of the journey to the central facility (most satellites are within 10 miles for properties located in cities)
- the line from the premises to the satellite is monitored
- most links between satellites and central stations are multiline, hence reserves exist if a trunk line becomes damaged

Disadvantages.

- only available for premises in larger towns and cities
- satellite failure can cause a prolonged loss of security facilities. Temporary guards might be needed for the period of failure. Some security organizations offer these guards as part of their system guarantee and some companies house equipment for satellites in unmanned offices, which, in the event of automaticreceiving equipment failure, become manned "manual" satellites. This reduces their fixed guard manpower commitment to clients.

Remember that you pay a considerable amount of money to rent a telephone line connected to a central station and to hire their services. The quality of this service, just like any other, should be questioned before signing an agreement. You should determine line capacity and grade of service.

Central Station Line Capacity. A perfect central station would have facilities to accept every incoming call and act on them all simultaneously. Most stations do not attempt this due to the high capital investment in equipment and technology it would require. Instead, they use fairly questionable statistical analyses to show the likely peak traffic flow into the station in units of *line hours of occupation* (or *"erlangs"*). These analyses seldom incorporate civil unrest, riots, major fires, and other catastrophic factors. So, inquire:

How many lines are connected to the central station?
How many connected users are there?
How many seconds does it take to acknowledge your call under various scenarios?
Can you visit the station to which you will be connected and see it in action?
Can you challenge the system randomly and monitor the response times?

Central Station Grade of Service. *Grade of service* is a term often used to tell you the odds that your call will *not* be accepted and acknowledged. It is often expressed as a simple percentage (e.g., 10%). This percentage is meaningless. Ten percent of what? For how long? You need to know,

What is the assumed peak call input? Ten thousand calls?
What is the *first call* grade of service? If it is 10%, then 1000 calls may not be answered and yours could be one of them.
What are the various repeat call grades at the peak time? For example, the second call grade might be 8% (not too bad). If the grade of service is still 10%, things are not looking very good and yours could easily be one of the unlucky calls.

Using Telephone Company Exchanges. In cities and major towns, the choice between satellites and telecommunication local exchanges must be made. In rural areas, the local telecommunication exchange is the only facility available. Systems signaling via local telecommunication exchanges take lines on each property from an area of the exchange that is reserved for security companies' outgoing lines. Instead of using a multiplexing unit within a satellite, each site transmits a unique, multiplexed signal. At the exchange, the various signals are fed onto a single, shared line that is connected to the central station. Each signal from each property is integrated in the overall signal transmission pattern for that area.

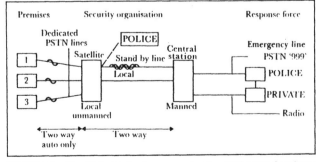

FIGURE 2–20 Using telephone company exchange in place of local satellite

Advantages.

- telephone exchanges are nearly always closer than satellites and, hence, over the same total distance to the central facility the exchange system is usually cheaper to run
- no possibility of satellite failure
- no penalty incurred by the client for being the only user in the area (satellite cost can be high if only a small portion of the capacity is used)
- extremely cost-effective for shopping malls, factories, estates, etc.

Disadvantages.

- greater dependence on telecommunication facilities
- usually no standby line to central facility from local exchange

Telecommunication Systems

In the early days of message transmission, most premises' alarm systems were linked by direct line to the local police. Because all alarm systems suffer periodic faults, false alarms became a nuisance to the police. A way of combating the faults caused by malfunctioning automatic dialing units or preventing line faults from occurring was needed. British Telecommunications' solution to this problem was to develop two security signal transmission systems known as *Alarm by Carrier (ABC)* and *Communicating Alarm Response Equipment (CARE)*. Many American telephone companies offer very similar systems under different names.

Alarm by Carrier.   ABC systems (see Figure 2–21) use a normal telephone line to the local exchange, which is continually monitored for failures, faults, or alarm signals. All of these signals are "carried" on the top of local, audio, telephone line signals that simultaneously use frequency division multiplexing (see "Multiplexing"). At the local exchange, faults are reported to a central reporting station run by the telephone company. Alarm signals are gathered onto a dedicated part of the exchange, known as the *local processor*. Any valid alarm signals are transmitted to the area exchange processor, which routes calls to the appropriate police station after rechecking the line, message content, etc. Once the signal is sent, the police force sends an acknowledgment back through the system. Many clients make use of this capability as it offers the benefits of monitoring (fault and failure) and semidedicated lines, without additional line rental charges or installation costs.

Communicating Alarm Response Equipment.   CARE systems (see Figure 2–22) are extensively used by letter

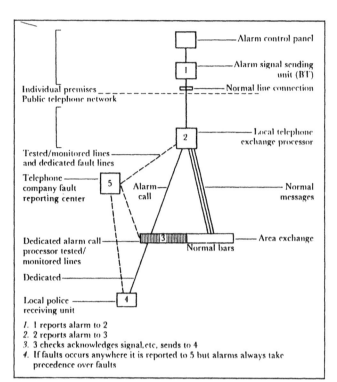

1. 1 reports alarm to 2
2. 2 reports alarm to 3
3. 3 checks acknowledges signal, etc, sends to 4
4. If faults occurs anywhere it is reported to 5 but alarms always take precedence over faults

FIGURE 2–21   ABC system

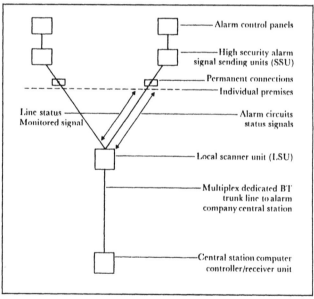

FIGURE 2–22   CARE system

security companies that operate central station facilities. A CARE system is basically a sophisticated, high-security ABC system that reports directly to the central station and not to the police. It offers

- line monitoring—fault, cut, and other failure reports to a local scanner unit (LSU)

status reporting of alarm circuits whenever the telephone is not being used for normal voice communication

very highly reliable services (very little chance of lines being disconnected at any point in the system)

nondedicated, but continuously available, telephone lines

full encryption of a signal from the premises to the security company computer

dual redundancy alternative lines, in case of line failure

remote resetting of alarm system compability

automatic polling or searching for line faults

separate identification (and, hence, speedier correction) of customer and customer network faults

All of these capabilities make systems like CARE an exceptional value. This is particularly true when connection by annual subscription is available (irrespective of line distance) to remote users (rather than annual line rental).

Multiplexing

Multiplexing means the *simultaneous transmission* of multiple signals over a single transmission system, with each signal being uniquely identified by the receiver. Multiplexed systems exist in two basic forms—time division multiplexing (TDM) and frequency division multiplexing (FDM). One or the other is used on virtually all hard-wired, remote signal transmission systems.

Why use multiplexing?

Telephone line rental costs are very high. Hence, dedicated telephone lines that are directly connected to central stations (or other transmission methods) are very costly. Your shared line and path of transmission with others is cheaper.

Very few premises need the physical security of their own dedicated lines with the available signaling and encryption methods. Alternative path switching and fail-safe alarms make shared line use very secure.

Sharing lines with others allows you to use higher data transmission rate lines (e.g., optic fiber lines) than you could economically purchase on your own.

Frequency Division Multiplexing.

FDM (see Figure 2-23) is true, *simultaneous* transmission. Instead of each building receiving and transmitting a response in turn, all buildings simultaneously transmit over the line. The transmissions are individually identified by the exact frequency and wave shape of their transmission signal (e.g., 9.100Hz, square wave form).

Telephone lines, microwave transmission, etc., are only permitted and possible within strictly specified frequency limits. These limits are termed the system *band-*

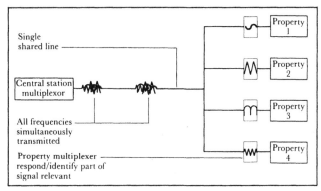

FIGURE 2-23 Frequency division multiplexing

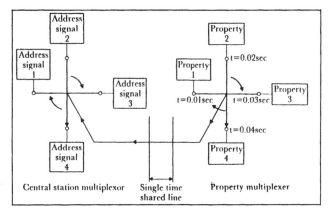

FIGURE 2-24 Time division multiplexing

width. As the bandwidth tends to be narrow, very few discreet frequencies can be achieved on the same line. Hence, while having your own, special transmission path that is in *constant* contact with the central station sounds better, it is almost always more expensive than TDM, where frequencies or tones may be similar, but separated in time.

Time Division Multiplexing. TDM (see Figure 2-24) is *not strictly simultaneous* transmission. The central station transmitter sends a series of coded signals over the line to the various premises in cyclic pattern, which is repeated time and time again. Each property must answer the signal with its own encoded response signal in a similar response cycle (i.e., within a *time envelope*). The entire cycle of transmission and reception of signals for ten buildings, for example, might only take 10^{-3} seconds before it is repeated. If a response is not received from a particular building, the alarm is initiated at the central facility. If the signal response is still not present after another 10^{-3} seconds, for example, the alarm will be confirmed and lights will start flashing at the alarm station. The number of failures to respond and the cycle time depend entirely on the line length and the equipment. Because the system is essentially two-way interactive and is continually monitored, it offers considerable advantages over normal, unidirectional alarm signal transmission.

Coded Time Division Multiplexing. Coded TDM is the version of basic TDM that is most commonly used for communicating between alarm control panels and alarm company central stations. Security is improved by ensuring that the signals sent to or received from individual premises are coded by means of a pulsing Pulse Code Modulator (PCM) or Stream Burst Modulation (SBM) device. These devices send signals either as a binary code (1 = particular frequency, 0 = another frequency) or as a pattern of particular frequencies. With coded TDM, each building has a unique address code, remote management and maintenance function codes, etc. For example

- property address code = 7018
- arm system code = 17018
- disarm system code = 27018
- test system code = 30718

Frequency Shift Keying (FSK). Most TDM (PCM or SBM) systems operate by transmitting information to the telephone line as a digital binary code comprised of 1 and 0, which is represented electronically as 1 and 0 volts. In order to be transmitted over the public telephone system, the simple 1V or 0V must be converted to AC tones. The conversion is carried out on an FSK modulator/demodulator (modem). In order for the information on the line to remain secure and recognizable, the 1 and 0 levels are also represented by different frequencies. Hence, the signal information will shift frequencies depending on the message content. The message is transmitted with amplitude (1 and 0) discrimination, frequency discrimination, and real-time discrimination, making it very secure.

TDM is an *analog-generated* signal. That is, information is represented as *either* of two states or levels. This could be current-on and current-off or 1A and 0A. The 1V might represent the binary number 1 and the 0A, the binary number 0. There is no intermediate state. It is clear that a simple switch will generate these two states easily and that a very fast switch will allow for a string of 1V and 0V states to form a code to be transmitted very quickly. A group of eight of these states (or bits) is referred to as a *binary word*. Note that binary words are not *alphabetic* words—no letters are used and they do not necessarily represent any particular letter.

So how is a message sent? Well, let's say the alarm system is looking for answers before it lets you into a building.

Are you male or female? 1 = male, 0 = female
Are you married or single? 1 = married, 0 = single
Are you 21 years old? 1 = 21 or older, 0 = under 21
Is your hair curly? 1 = yes, 0 = no
Can you swim? 1 = yes, 0 = no
How is your father? 1 = well, 0 = not well

By only accepting the binary six-digit word 100101, the code only lets unmarried males, under 21 years of age, with curly hair, who cannot swim, and whose fathers are healthy into the building! As you can see, even part of a binary word can transmit a lot of information. Most TDM systems operate with *at least* 300 binary words (bits) per second!

Simple 1V/0V DC signals are not very useful. If they were used, a hard-wired pair of wires would be required and the distance of transmission would be very limited due to voltage drop on the lines. In practice, this is overcome by turning each of the 1V/0V states into an audible frequency (or tone) or into two distinct AC signals. This format allows a whole range of transmission paths to be used, including data grade telephone lines, voice grade telephone lines, infrared or microwave (frequency) links, etc.

The device that converts the DC voltage signals (1V and 0V) to these tones or frequencies and vice versa is called a *modem*. Modem stands for *modulator/demodulator*—indicating it can perform two-way conversions (i.e., from AC signals back to DC voltage signals in real digital form).

There are many types of modems. All have different facilities and abilities, different rates of data transmission, etc. Modems are usually used in conjunction with multiplexers. The multiplexer allows many modems to hook up to a single four- or two-line telephone line and for all operators to use this line simultaneously.

TDM Interrogator/Responder Units. If each group of buildings is given its own address or each group of buildings within a given tone transmits a different binary word message as a pulse or stream burst, it is easy to see that real-time clashes could occur if every building tried to transmit a message at the same time or in different directions over the same line. Some frequencies might add to, subtract from, or simply corrupt the individual signal. To prevent this from happening, some means must be used to synchronize and schedule the time at which signals are sent. In very, very early devices, this device was, literally, an electromechanical clock that opened relays on the lines.

A better and more secure means of preventing clashes uses a system known as the interrogator/responder (IR) unit. This device is used on over-and-out (half duplex) or simultaneous talk-and-respond (full duplex) telephone lines. Each building's TDM multiplexer is able to transmit and receive binary tone codes. The central station sends a signal to a building multiplexer. This signal has special characteristics particular to the building and is called the *interrogator signal*. On receipt of the special tone with a unique code of binary digits ("address"), the modem receiver checks the validity of the code. If the signal is valid, a handshake is made and the receiver then transmits a "hello" status signal, over the line, to the central station interrogator. If no handshake is made after three interrogation signals have been sent (within a short time), the alarm is raised. If the handshake is made, the "hello" signal gives a full report to the central station, as the alarm system digital communicator is programmed to do.

Encryption. Multiplexers used on security systems must prevent false information or substitute commands from being injected into the system, as they may allow an all-normal state to be displayed when, in fact, an intrusion is taking place.

True information is turned into a code to prevent this from happening. This process is known as *encryption.* Only persons or systems with the code key will be able to read or decipher this code. The only reasonable standard of encryption that is universally adopted is the National Bureau of Standards (NBS) Data Encryption Standard (DES). This standard has within it several, actual, encryption methods. The minimum acceptable encryption method for security systems should be DES pseudorandom encryption for both interrogation and responders using automatic encryption key shifting—preferably on a full duplex four-line private dedicated network (or two-line trellis, full duplex) system.

Selection. In pure security terms, there is currently little to choose from between TDM (PCM) and FDM. On the whole, FDM hardware is simpler, smaller, cheaper, and proven over a number of years. TDM is slightly more costly (first cost), but for larger buildings it is a very flexible system, as it allows one pair of dedicated lines to the local exchange to carry multiple bits of information (e.g., security, fire, direct debit banking). TDM permits the division of signals into high-priority and low-priority signals. These may range from "general perimeter intrusion detected" to "personal attack in bullion area." TDM's only real disadvantage over FDM is the length of the time period during which it cannot transmit because other buildings are using the line. In general, the shorter this "dead time" and the longer the building's transmission time slot, the better the system. Overall, TDM systems are probably more suited to the needs of security and offer greater potential.

The type of multiplexing and modem you select depends on

the level of security required

the amount of system flexibility required (e.g., levels of access, information transmitted, time taken to transmit)

an analysis of the installation and operating costs for each system

Whichever system is selected, it is essential that

it is acceptable to the line provider

the equipment used is approved (e.g., correct modems)

the system is fully compatible with the intruder alarm system. The multiplexing systems should *always* be provided and installed by the contractor who provides the intruder alarm, *under the same contract.*

Of the various TDM systems available, Simplex's digital clock synchronization is the cheapest, slowest, and least secure. The half duplex IR is best for medium-risk security systems transmission. Full duplex, two-way, simultaneous transmission with an IR is the most commonly used system.

Multiplexer Information. To properly select a multiplexer for a security (or data) application, you need to know whether it has X25 packet-switching, statistical-multiplexing, or TDM transmission techniques. Then, you need to know

- the maximum number of channels
- whether it can accommodate both synchronous and asynchronous data
- the maximum aggregate input speed (kilobits/second)
- the channel speed for asynchronous and synchronous channels (kilobits/second)
- the number of voice channels and the voice technique used
- the maximum composite link and its speed
- whether it supports point-to-point linking; onward linking; and drop, insert, and multimodal facilities

From the client's point of view, the type of multiplexer used will be irrelevant, provided it

- is reliable
- is cost-effective
- has sufficient capacity and security

Transmission System Selection. What should you look for in the transmission system?
Telephone Line/System.

- shortest geographic route (actual distance)
- least number of telephone exchanges
- best line quality possible
- least amount of bridging or switching in the telephone exchanges
- electronic exchanges (e.g., system)
- alternative switching paths during bad weather
- data compaction rates on the line of no greater than 4:1

Central Station Routes.

- direct to a local, manned central station (not an unmanned local or remote satellite station)
- central station with a fast, real-time overall response (determine the maximum guaranteed response time)

- local compaction of multiusers onto same line to central station

Multiplexer/Modem.

- full duplex
- FSK/TDM
- IR
- 16-port data entry
- automatic line switching on line failure
- NBS data encryption (pseudorandom) on two-way IR with automatic key shifting

Modems. Modems should be selected based on three fundamental questions:

1. How fast do you want the modem to transmit your information? (Line time = money, the longer the distance, but the shorter the transmission time, the lower the cost.)
2. Should the modem support synchronous or asynchronous transmission?
3. Should the modem be half or full duplex?

The difference in capital cost between modems is relatively small. Select the fastest modem you can afford that is compatible with the size of your building, complexity of your system, and amount of information you have to transmit. Common speeds are 1200, 2400, 9600, 14,400, 24,000 (1 kilostream), and 2,048,000 (1 megastream) bits (or binary words) per second. To give an example of the speed of transfer, a 9600 bits/second modem will transfer data from a word processor faster than the fastest typist can type the data. The data on an A4-sized sheet of text or a fax, transmitted over a 9600 modem, will take 14 seconds to reach its destination. An automatic drafting system for transferring full architectural plans would need at least 64,000 bits/second and *live* television pictures would require at least a 10 megastreams transmission rate.

Be aware that synchronous transmissions tend to be used in security work and a full duplex transmission is better, by far, for little extra cost.

Modems should also have the following features:

- automatic dialing
- automatic answering
- autoseek for fast response to longer data
- an adoptive equalizer to recognize telephone line quality and improve reception quality
- automatic disconnect and follow call protection
- multipoint input and number store
- automatic retaining
- part contention
- diagnostics

Statistical Multiplexer (Statmux). Statistical multiplexers work on the TDM principle. The difference between statmuxes and normal TDM is efficiency of data transfer and, hence, line rental cost. A statmux allocates a time slot to a building's alarm system *only* if it has data to transmit. The statmux design is based on the idea that not all buildings on a system need to report in at the same time, with similar data, or with similar time slot lengths. High-security buildings might need to report a lot of information very frequently. Low-security buildings might only report opening, arming, closing, and alarm states. The former should not dictate the average line rental cost for the area. A statmux allows each building or group of buildings to pay the correct cost contribution. Most good computer data transmission systems use statmuxes.

X25 Multiplexing. *X25 multiplexing* refers to an International Standard Dialing Network (ISDN) international switching standard that allows for the very fast information transfer of security data over the public data network (PDN) system. The X25 multiplex allows packet compression of data (i.e., *packet switching* or *PSS*) over PDN lines. About 50% of all data transmitted in the world uses PSS over the X25-linked PDN system. The ISDN standard is common to the United States, Europe, and Asia.

X25 multiplexing is really cost-effective when members of a *closed user group* want to relay data among themselves at very fast transmission rates. Remote security central stations can use this system to provide a very good service to their customers. A particularly useful development in this area is the use of the extremely fast – 10 megabit (20,480,000 bits/second!) – real-time transmission of television pictures from on-site CCTV cameras to remote, manned central stations.

Signal Transmission Medium

There are essentially three types of signaling media:

- electrical energy – McColloh grounded system, direct current, alternating current, and multiplexed digital signal
- light energy – fiber optic transmission
- free space transmission, using microwave, radio, infrared, or other electromagnetic spectrum wavelengths

Electrical and audio (voice grade) signals are still the most common systems. These systems use simple telephone line link cables (with stranded copper wire) provided by the local telephone company for line rental. These signal systems are commonly referred to as *hardwired* systems (although optic fiber systems could also be referred to as hard-wired systems).

Regardless of the medium used, it is essential that the signal get to its destination quickly

be clear when it arrives

be secure from false injection

convey the information needed (e.g., normal, open, closed, bandit attack, intruder alarm, line fault, line cut)

be obtained at the optimum price commensurate with the level of security of the intruder alarm system and the protected property

be reliable

be "clash"-proof

McColloh Loop Systems. The McColloh Loop System was the first TDM system invented. A single transmission line of cable links various buildings within a neighborhood under different or the same ownership. Each building has its own transmitter. The transmitter sends a pre-coded signal to the single loop of cable at a predetermined time. The signal is a code comprised of a sequence of direct current "switch offs." Pulses are generated either by a mechanical or transistorized switch (mechanical switches are almost obsolete). A limited amount of line monitoring and signal encoding is possible. Systems *must* include pseudorandom test signaling to the central station and simultaneous, signal clash prevention techniques.

Advantages.

- cheap to use and install
- very useful for on-site systems with a secure, perimeter fence or individually protected transmitters

Disadvantages.

- lower signal security than DC transmission (or any other common form)
- low-grade security applications, as methods of defeating are well understood and relatively easy for skilled intruders
- must be very well grounded to work properly
- ground must be free of ground loop currents within it
- restrictions on the length of cable to be looped for any given area (commonly 25 miles or 40 km)
- restrictions on the number of users connected (normally a maximum of 50) and hence the shared cost within an area is higher

Direct Current (DC) Transmission Signals

Direct current transmission systems will only operate on a hard-wired, dedicated, two-core cable between the control panel on the premises and the alarm company central station (or proprietary station). This makes the use of DC transmission expensive for its relatively poor signal security. DC signaling is now quite difficult to obtain over the public telephone network, as it ties up too many

lines for a single purpose, which could be let for line rental to many users using multiplexers.

The only time a DC transmission signal system should be considered is when using a continuous DC voltage and current that is balanced bridge-monitored and tone burst principle. The *tone burst* is, in fact, two or more superimposed AC signals. The DC signal is the normal signal and any predefined variation in its level activates an alarm. *Bridge balancing* reduces the ease with which a false current may be injected to fool the central station. The AC signals enhance this level of line supervision. A person wishing to compromise the system would need to monitor, record, and reinject all signal currents and voltages.

Advantages.

- more secure than the McColloh Loop System
- direct, dedicated lines make it fast and reliable

Disadvantages.

- very expensive on nonproprietary, off-site central stations
- not as secure as multiplexed, encrypted signals
- can be seriously affected by line inductance pickup, ground leakage currents

Alternating Current (AC) Transmission Signals

Alternating current transmissions are produced as a series of tones of frequency in the audible range for humans (i.e., between 25Hz and 16kHz) (mains AC is 60Hz). As they are *audible* tones, they may be transmitted over voice grade telephone cable systems (private or public), through exchanges, etc. These use a bandwidth of 30–3400Hz. The AC signal tones can be monitored by a variety of techniques for amplitude, frequency, and period between tones to reduce signal injection risk, send information, and monitor the line. The telephone lines on which the signals travel can be monitored using a reactance-balanced bridge or carrier frequency technique.

Advantages.

- uses PSTN and private telephone systems, where available, giving great flexibility
- allows for voice and data transmission over the same lines
- more secure than DC or McColloh loop signals
- cheap system for a central station connection on lower security risk properties that are relatively close to the central station
- a commonly used, well-understood replacement for DC systems

Disadvantages.

- not as secure as multiplexed signals with encryption
- not as fast as a DC transmission, as transmission is not direct but occurs through exchanges
- line maintenance by telephone company can (and quite often does) upset the signal transmission, causing a fault to appear on the control panel

Remember, regarding two-line versus four-line telephone cables, four lines are twice as expensive to rent for an equivalent distance, but are more reliable and give a clearer signal (signal-to-noise ratio twice that of a two-line cable).

Also, the bandwidth of the tones the signals send should be at least 2% of the total available, to allow for variations in temperature, supply voltage, and current without damaging the signal's integrity. At the end of the tone sequence a silence/reset period of 4-10 seconds should always be allowed to reset system and optimize false alarm versus security considerations.

Fiber Optic Transmission

Principles of Operation. A fiber optic is comprised of cores or filaments of extruded glass in continuous lengths. The glass is extremely pure and has a very high transmissivity to light along its length. This property of linear light transmission is used to transmit data in *optical* form. This means that energy in the visible part of the electromagnetic spectrum is used to transmit data, instead of energy with frequencies, like microwaves or infrared. Light travels at 984 ft/sec (300 m/sec), is very easily controlled and generated, and, hence, is an ideal transmission medium in many ways.

A fiber optic transmission system (see Figures 2-25 and 2-26) is comprised of a converter/transmitter, fibers optic, photo receiver/converter, and, often, a reception amplifier. The transmitter takes a normal input voltage signal, usually from the information transmitter, and transforms the coded signal into a modulated beam of light that varies in waveform and/or frequency, depending on the code. The light is then focused into the individual optic fiber cores and transmitted at the speed of light to the photoreceiver.

The photoreceiver is comprised of a very light-sensitive chemical receptor/converter that converts the coded light energy into an output signal via an amplifier.

The frequency of transmission and bandwidth depend on the application, but are usually in the range of 10Hz to 10MHz for CCTV transmission and 10Hz to 20kHz for audio transmission. Data transmission is usually quoted in kilobytes per second and nanometer wavelengths.

Uses and Application. Fiber optic links will transmit the following *simultaneously* over the same cable in either uni- or bidirectional modes:

- audio signals
- video signals
- alarm signals (contact closure data)
- pure data
- telephone voice grade or any combination of these, depending on the type of cable and transmitter/receiver used

In the security industry, fiber optic transmission systems are extremely useful due to their multifunctionality. A single cable around a building will transmit guard intercom signals, CCTV signals, audio signals, logging data from patrol points, control and switch alarm system signals, lighting system relays, and access control system data.

The predominant use in the security field is for either direct, point-to-point CCTV transmission (particularly outdoor in harsh environments) and direct, buried, multiplexed transmission cables connected to central stations.

Fiber optic systems have a very high immunity to practically all environmental conditions and are inherently safe to use under virtually all conditions.

In CCTV links, fiber optic cable is far superior to coaxial cable, the former having a fixed, uniform, transmission loss across all frequencies and no common-phase restriction.

As fiber optic cable is relatively expensive compared to other types of cable, careful thought must be given to the actual need for it versus other cables. Fiber optic cables often win over other cables due to their interference resistance, safety, and transmission speed and quality.

Losses. The primary calculation carried out by cable system designers is that of assessing the total transmission loss (TTL). TTL is the sum of the linear attenuation of the cable; the attenuation of joints, splices, and connectors; the coupling loss at the transmitter and receiver; and an aging loss allowance. The transmission losses in fiber optic cables are measured in decibels (dB).

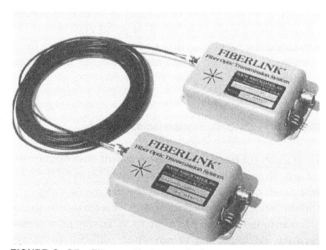

FIGURE 2-25   Fiber optic transmission system

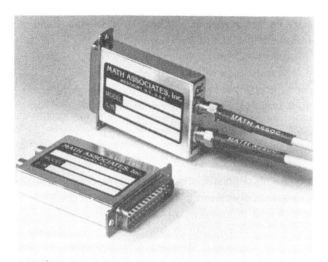

FIGURE 2–26 Fiber optic transmission unit

As light is a very concentrated form of energy and photosensitive detectors are now extremely sensitive, an optic fiber system will continue to function with 90% *loss* of transmitted light energy.

In any system, TTL calculation is essential to prove that the calculated dB loss is less than the *maximum allowable loss* (MAL) for the system. Unlike many normal cable systems, this calculation is critical because joint and splice losses are huge in comparison to cable losses. If too many joints are present in the system or too many joints are subsequently introduced to the cables, the system may not work.

A simple equation will demonstrate this. Assume the following:

maximum allowable system loss = 10dB
cable used = direct, buried, unjointed armored, 50 micron, four core
operation wavelength = 850 nm
cable attenuation = 5.0dB/km
cable length = 1 km
coupling losses, transmitter = 0.2dB
coupling losses, receiver = 0.2dB

Total transmission loss = (5.0dB/km × 1 km) + 0.2dB

+ 0.2dB = 5.4dB

If the MAL is 10dB, this system is acceptable. However, let's say the length of the cable was measured incorrectly *and* a mechanical splice is introduced. If the

new cable length = 1.5 km
mechanical splice dB loss = 4dB

Total transmission loss = (5.0dB/km × 1.5 km) + 4dB

+0.2dB + 0.2dB = 11.9dB

This system is unacceptable. However, if a better fusion splice is used, the splice loss is reduced to 0.1dB and now

TTL = (5.0dB/km × 1.5 km) + 0.2dB + 0.2dB + 0.1dB

= 8.0dB

This system is now acceptable. So always use elastometric splices or fusion welds (0.1–0.2dB loss) in preference to epoxy glue or mechanical connectors (2–4dB loss).

Equipment Types and Selection. The fundamental features to consider when selecting an fiber optic transmission system include

- distance point-to-point
- whether cable is buried, overhead, or on the surface
- whether your application is indoor or outdoor
- the system transmitter wavelength
- the system bandwidth
- the environmental conditions it must withstand
- fiber grade
- fiber size
- fiber quality
- cable makeup and type
- cable size

Grades of Fiber. Cables have various grades of fiber. The grade of fiber determines the bandwidth of data transmission frequencies the cable will carry and the attenuation loss limit within the cable (i.e., the limiting length of cable without amplification stations). The grades commonly used are

- multimode step index fiber (SIF)—narrow bandwidth of 20MHz for 3281 ft (1000 m) lengths
- multimode graded index fiber (GIF)—wide bandwidth of 20MHz for 12–18 mi (20–30 km) lengths
- monomode fiber core (MFC)—very wide bandwidth of up to 2000MHz per 0.6 mi (1 km)

SIF cables are seldom used, except in pure low-capacity signal or data transmission. GIF cables are the industry standard for fiber optic links for television or CCTV transmission systems with full control, telemetry, and voice, plus picture (color or black-and-white) and, hence, security work. MFC cables are used primarily for high-speed data transmission.

Fiber Sizes. Cables can be purchased with many fiber diameters. The most common are 50 micron, 62.5 micron, and 100 micron (10 micron and 8 micron are for MFC cables). Different diameters of fiber have

- different total maximum optical attenuation (dB)
- different maximum input power limits (microwatts)

For example, the industry standard for CCTV links tends to be 50 micron, 10dB optical attenuation, 10 microwatt power, 3dB maximum attenuation per 3281 ft (1000 m), and a 58–65dB signal-to-noise ratio.

Fiber Quality. The "quality" of the fiber is defined by its *numerical aperture number (NA)*. The NA is a measure of the fiber's ability to absorb light for transmission purposes. A high NA indicates more light absorption and lower losses at coupling splices.

Cable Makeup and Type.   Cables are made to suit the environment in which the optic fiber core will operate. The materials used to encase the fiber are primarily used to prevent mechanical damage to the fiber. Cable types include

- general indoor
- heavy-duty indoor and outdoor, unarmored
- UL approved for plenum, low-smoke emission, indoor
- aerial, catenary, duct-strained
- direct, buried, armored

Cables are normally comprised of indoor cables: fiber optic core, loose buffer tube or sleeve, braided Kevlar strengthening fibers, PVC outer sheath) and outdoor cables: steel central member, interstitial filler, buffer tubes, fiber cores, inner Kevlar fibers, outer PVC sheath armor, outer PVC sheath, respectively, radiating outward from the core.

Cable Sizes.   Indoor cables are normally 0.15–0.5 in (0.38–1.27 cm) in diameter, with a minimum bending radius of 1–10 in (2.54–25.4 cm). Outdoor cables are normally 0.3–0.6 in (0.76–1.52 cm) in diameter, with a minimum bending radius of 9–11 in (23–28 cm).

Limitations.

- cost
- bending radius
- connection and splicing limits

Advantages.

- wide bandwidth—makes any signal transmission possible
- little attenuation on cable (fixed attenuation across frequency range)
- up to 3 miles of cable length without amplification
- no interference problems from lightning, RFI, or EMI
- no radiated energy (RFI or EMI)
- no explosion risk, fire risk, or spark risk
- difficult to access
- does not corrode
- cables made to high standards to withstand harsh environments and installation practice (including automatic burial, self-supporting catenary, etc.)
- huge data and simultaneous transmission capability
- small diameter for capacity
- strong tension
- very useful in EMI areas (cheaper than heavy metal conduit plus shielded cable)
- very useful in potentially explosive environments

Disadvantages.

- cost

- limited bending radius
- can be *crushed* and damaged (250–400 lb/in or 113.4–181.4 kg/cm normal limits)
- limited temperature for operating−14°F (−10°C) indoor, −40°F (−40°C) outdoor cable types
- care in terminating (expensive to do properly)
- does not work in nuclear radiation environments

Microwave Transmission

Principle of Operation.   The principle of microwave transmission operation is identical to that of the internal and external microwave intruder detection sensor (see Chapter 3, "External Microwave Sensor," and Chapter 4, "Microwave Sensor"). Microwave transmission is comprised of a transmitter and a receiver. The transmitter emits microwave energy (i.e., energy between the wavelengths of light, heat, and subsonic sound) at frequencies in the 10–25GHz range and 4–10-watt power output. This energy is passed by line of sight, directly to a receiver, which takes the energy pattern transmitted and converts it back to a voltage signal, thereby recreating the signal sent. The beam shape is usually far broader than the normal external intruder sensor microwave.

Uses.   Microwave transmission links are commonly used to transmit large quantities of digital information over distances 2–50 mi (3–80 km) with 10 mi (16 km) being the normal, optimum jump. As infrared, free space line-of-sight transmission has a range up to 3 mi (5 km), microwave links take transmissions to the next stage—large sites or long distances across large, open line-of-site spaces. Microwave links are widely used by telephone companies to digitally transfer information point-to-point within cities or on major trunk routes where optic fiber cable or other large data capacity systems would be difficult to install. Security uses normally include long-range multiplexing of big sites to central stations and simultaneous television signal (color or black-and-white), speech communication, and telemetry command and control for television surveillance systems.

Current expansion of microwave transmission technology focuses on the area of portable, cableless, television surveillance equipment. Typically, this includes

mobile transmitters connected to mobile receivers (e.g., helicopter or fixed-wing plane to vehicle on ground)

mobile transmitter to fixed receiver (e.g., soldiers or patrolling guard with portable, low-light video camera/power pack)

fixed transmitter to a fixed receiver point with switching capability (e.g., pan, tilt, zoom, cameras at remote fixed location to fixed receivers; multiple overhead cameras on football stadia roofs, prison roofs, etc.)

Application.

Never buy a microwave transmitter system without a proving test and full field survey report of the likely line-of-sight transmission routes.

Always have an alternative microwave transmission path available.

Always include full commissioning *by the manufacturer* of the equipment, when possible.

Always check claims for compliance with standards. Ask the regulatory or licensing organizations if they have tested and inspected the manufactured item. Do not accept claims by salespeople that the equipment complies. Compliance may be to the *letter* of the rule and not to the *spirit*. The difference is quality.

Equipment Types and Selection. Microwave transmission equipment is specialized and there are few manufacturers of quality equipment with experience in this fast-expanding, new transmission medium. There are, however, many retailers and installers who see the advantages of microwave systems over cable systems and are prepared to sell substandard microwave transmission equipment and install it based on their limited knowledge.

Systems should

- be FCC-approved
- have a bandwidth compatible with the video equipment used (normally 5.6MHz or less)
- be able to operate in temperatures that range from −4°F to 113°F (−20°C to 45°C)
- have at least four or five frequency range products to which to tailor your needs and keep costs down
- have at least two selectable frequencies
- have at least two subcarrier frequencies (50Hz to 4kHz or 50Hz to 20kHz) that can be selected on-site

Transmitters should have

- a maximum of 8MHz deviation
- better than 50 parts per million frequency stability
- audio and video inputs that match the equipment used
- weatherproof, rugged enclosures to prevent damage
- a suppression capability to prevent local interference (better than −56dB/W)

Receivers should have

- capabilities similar to the transmitters and have a sensitivity that is better than −82dB for a 30dB signal-to-noise ratio

Advantages.

- high integrity and high picture quality
- real-time, very fast, data transmission
- simultaneous video telemetry and voice transmission
- very secure signal (difficult to inject false picture)
- high degree of multiplexing possible (i.e., 32 channels per receiver)
- signal weather condition immunity very high

Disadvantages.

- line-of-sight only
- expensive

Free Space Infrared Transmission

Principles of Operation. Free space infrared transmission systems are comprised of a transmitter and a receiver. The transmitter is placed beside, and is hard wired to, the devices that transfer data to and from the receiver. The data transmitted may include

- video television
- telemetry (instructions)
- raw data
- sound

The transmitter is comprised of a number of large, optical-focusing lenses in a case. Behind each lens, an infrared LED generator or a *gas laser* generates infrared energy in the 350 terahertz (350×10^{12} Hz) frequency. This frequency is approximately midway between microwave energy and the visible light spectrum and, therefore, has some of the penetration power of microwave energy and is invisible to the naked eye.

The lenses focus the infrared energy into a beam (or beams) of approximately 1° width and very accurately direct the beam (or beams) toward the receiver. At the receiver, a lens and photodetector similar to those used inside a low-light CCTV camera produce a video output analogue of the received infrared energy frequency, amplitude, and modulation.

The system is a *free space* transmission. There is no wiring or guiding between the transmitter and the receiver, and it is a line-of-sight system. The system range is limited by the power of the LED infrared generator. These systems are normally only operated up to 5000 ft (1524 m), with most operated at 2000−3000 ft (610−914 m). It is, therefore, a medium-range transmission system that fills the gap between cable and more expensive, long-range, microwave or radio free space systems.

Equipment Types and Selection. Free space infrared transmission systems are normally defined by

a range of link and frequency (i.e., 350THz over 1.86 mi, or 3 km)

the information carried (e.g., video television signal and telemetry, opposite directions only; television signal plus two-way telemetry; television plus telemetry plus sound)

beam spread and number of individual paths. Short range (i.e, 109–347 yd, or 100–400 m) devices normally have wider single beam paths.

Selection. Units should include

- modular electronics
- IP65- or IP66-rated enclosures (or NEMA equivalent)
- sun shields/demisters (heater to remove mist or condensation)
- range-finding and alignment devices
- reception signal drift or signal attenuation monitors and alarm devices at the transmitter and receiver
- multiple-beam paths each 1° wide. Sixteen beams are fairly common, with a 2:1 compaction beam ratio at the receiving station. This arrangement reduces interference and noise, and improves picture or data quality and signal integrity.
- integral pan/tilt heads and mounting platforms
- 20mA telemetry standard (where required)

Applications. Infrared transmission is commonly used to replace 22GHz microwave or optic fiber cable links in transmission of CCTV signals. Like optic fiber cables, infrared transmission is capable of high-quality real-time picture transmission. Infrared transmission is commonly used in instances when hard wiring would be difficult or dangerous (e.g., in built-up areas; links across rivers, railways, runways, highways, and cliffs; portable situations; military exercise grounds). Picture quality is dependent on the camera used, the frequency of the link, and the range required.

Systems can be used to transmit CCTV pictures, data only, or both. Color CCTV transmission is standard.

Multiplexing of many camera signals onto the single transmission path is possible, but slightly problematic. When trying to achieve multiplexing, it is essential that cameras, multiplexers, and the infrared link are all purchased from and installed by *the same company*. Definition of the extent of responsibility for compatibility is extremely difficult to define properly if a bad picture or data results when separate suppliers are involved. Similarly, this procedure must be adopted when data telemetry is required, to complement the video television signal.

The problem of structures being erected that obscure the line of sight is a real one and contingency planning is needed to identify alternative paths that may be used, such as using relay transmitters on other premises. Mounting of these units must be rigid and vibration-free.

Advantages.

- no cable
- flexible
- can be redirected easily
- can cross any terrain and area, where cables cannot

- requires no license
- produces no radio interference pattern or readily detectable signal
- invisible, with good resistance to signal jamming or false signal insertion
- low cost compared with cabling; often lower cost than microwaves up to 1.8 mi (3 km) transmission distance
- speed of installation
- no other contractors involved (e.g., electrical, civil)
- gives real-time picture where only slow scan previously could have been used due to poor telephone line transmission speed
- available for telemetry-plus-picture transmissions
- color picture transmission inherent in system
- penetration properties of wavelength are good

Disadvantages.

- must have very stable mount as devices are line of sight and vibration affects quality
- buildings or structures may be erected in the transmission paths, thereby cutting the link without prior warning (regulations do not prohibit this)
- range, performance, and quality are affected by weather conditions and the nominal specified range of distance required to receive good transmissions during bad weather must be derated to 50–60%

Long-range, External Radio Transmission

Principles of Operation. Long-range, external radio transmissions (see Figure 2–27) of information between protected premises and a central station is controlled by the FCC. The system is comprised of

- an alarm reporting center
- a receiver, encoder, and transmitter
- a coaxial cable connected to the transmitter
- a transmitting antenna

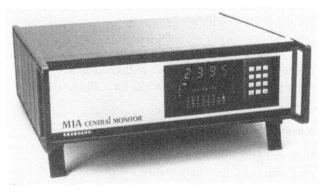

FIGURE 2–27   Long-range, radio telemetry receiver/decoder

- a radio path through the atmosphere
- a receiving antenna
- a coaxial cable connected to the receiver, decoder, and monitor
- a printer and annunciator
- an uninterruptible power supply at both the transmitting and receiving ends

The transmitting range varies between 20 and 50 mi (32 and 80 km); 30 mi (48 km) is the optimum distance for the power levels used.

These systems use ultra high frequency (UHF) or very high frequency (VHF) radio frequencies of either 150–170MHz or 450–470MHz (the most common range is 460.9–466.0MHz).

Long-range, external radio transmission systems use either amplitude modulation (AM) or frequency modulation (FM). FM is far superior to AM as it is *almost* unaffected by atmospheric conditions, bad weather, etc., and does not drift. AM should not be used.

Antennae for receivers and transmitters should be within the line of site and are commonly placed on top of large buildings. The antennae are usually 30–50 ft (9–15 m) high.

Full reception and transmission tests are essential prior to making any decisions on purchasing this type of system. It is important to remember that radio transmission systems are passive, not active, and normally only transmit during fixed, automatic reporting time intervals or alarm conditions.

Advantages.

- no cables to lay or bury
- cables cannot be cut outside protected areas, unlike telephone lines
- flexible and portable
- many transmitters for one receiver
- speed of message transfer
- not as prone to suffer storm damage as cables
- no standing charges or rental charges for telephone lines
- very useful in remote locations
- speed of installation
- simple range extension with additional devices
- little maintenance required
- very wide range of applications
- can be a very cost-effective solution
- very good for moving to temporary areas for protection
- each transmitter is uniquely addressed; many addresses and zones available at a relatively low cost

Disadvantages.

- basically a line-of-sight system unless "jumper" stations are used

- range can be dependent on atmospheric conditions, topography, antenna height, etc.
- antennae are very visible and vulnerable to attack
- clear signal paths can become blocked by future building
- signal jamming is possible, although difficult
- requires operational license from FCC
- multipath reception caused by reflection of the signal from nearby objects. The multiple paths can cancel one another causing a dead point on the transmission. Reception tests prior to ordering are essential.
- *synchronicity* — caused by a clash between two nearby transmitters with a single receiver, which corrupts the transmission. Systems should always include timers to prevent this event.
- not continuously monitored

System Selection. Good systems use

supervised signaling and automatic reporting at half-hour intervals

FM radio signal not AM, as the AM signal is very prone to RFI and EMI

reception improvement threshold amplification to boost signal integrity over greater distance

dual frequency transmission (i.e., parallel paths) with an alternating message transmission path

repeated message sending

high surge protection on antennae and coaxial connection

package or turnkey installation to prevent split responsibility (e.g., power supplies, masts, antennae, coaxial units, transmitters, receivers, printers)

UPS units (24-hour minimum reserve)

automatic alarms for signal jamming, signal interference, low voltage, and power loss

automatic frequency seizure to transmit alarm situations

narrow band filters and crystal locking

UL-tested standards/FCC-type acceptance

good signal-to-noise ratio (12dB or better)

two-part signal transmission — *address* followed by *message*

high-frequency stability (not greater than 0.0005%)

VHF on open field transmission sites; UHF in town or built-up areas

high-gain antennae should be used

FSK signaling with digital transmission, rather than analog

6000 bits per 0.5 second transmission burst rates for status reporting capabilities

Alarm Transmission Using Computer Data Networks

Based on reading the preceding sections, it should be clear that digital dialers, multiplexers, modems, etc., are all also commonly found in computer systems. Similarly, computer systems transmit data over short or long distances using the same media (i.e., PSTN telephone lines, infrared and microwave links). So if microprocessor-controlled alarm systems and computer-controlled alarm systems use common parts, why not combine them and share links and devices by using private (or closed) user group computer data cables?

Advantages.

- shared data lines reduce the capital and line rental costs associated with high security systems' separate, dedicated, direct lines over long distances; risks are not significantly increased if the lines are private (between companies in different locations), such as supermarket outlets to the headquarters building
- data lines generally have a very large spare capacity for additional data and the additional security data uses very little of this
- conversion or adoption of systems to carry security data is relatively inexpensive
- multiplexers, modems, etc., will already be in place and will normally have the speed to support the data transmission line; speeds are far better than those normally justified for alarm-only systems
- most data transmitted within companies (e.g., wired cash transfers, marketing data) are sufficiently sensitive to warrant some form of high-grade encryption to prevent industrial espionage. Hence, security signal transfer will often be sufficiently encrypted if it uses a similar format.
- data lines are often more important to companies than alarm systems, as their loss for a single hour will cause companywide shutdown. Their reliability, maintenance, capital expenditure, budget upgrading, etc., get a far higher profile than the lowly alarm system (where costs are difficult to value until a major crime occurs).
- data systems commonly have spare routes, channels, backup equipment, and generator-driven and UPS power systems that are difficult to justify for alarm-only systems

Types of Systems. There are, potentially, as many system types as there are computer data network types. However, the most common systems are

- derived or parallel transmission
- bus
- "token ring"
- tree and branch
- X25 packet switching (PSS)

Derived or Parallel Transmission. Derived or parallel transmission systems simply separately inject alarm and data signals, using separate (or derived) channels or signals to identify each signal, into a common modem on the premises.

These two signals (derived and parallel) share a common telephone line to a local telephone exchange, where they are *both* entered onto the normal, direct, data line to the headquarters building. Here, the reception modems divide the signals back into separate alarm and data signals. If the response force is within the headquarters building, the alarm signal goes directly to the response force. If the response force is not in the building, the alarm signal is remultiplexed and sent to the alarm company central station over another data or PSTN line.

These data systems are commonly referred to as *star systems*, as the data lines all collect at a central node before being retransmitted. The most common system name is *Arcnet*. These systems have the disadvantage of slow transmission rates and usually very little backup (resilience) on the main transmission link between nodes. So for high security links or links needing to transmit CCTV pictures, diagrams of site, etc., they are not recommended. (Typically, the transmission speed is 15 times *slower* than good-quality, direct PSTN lines or private lines with alarm data only.)

Bus Systems. Bus systems are connected using the straight line radial principle—a single, high-speed data line reaches each building, which plugs into this line with a direct, radial connector. Again, the channels are separated, multiplexed, etc. The common system names are *Ethernet* and *DECNET*. These systems use what is known as the CSMA/CD protocol to listen and send data to prevent collision on the line.

Ethernet systems tend to be the de facto industry data transmission method for all non-IBM computer networks. They are low cost and allow many types of machines to be integrated with the system (including IBM), unlike the IBM "token ring" system, which actually tries to make all devices IBM-compatible devices. This costs a lot of money, particularly when dealing with alarm system panels.

Bus systems are faster and preferred to the star systems.

Token Ring. Token ring is the IBM trade name for its own transmission system. Blank tokens, which are constantly moving around a ring, which has spurs into each building or area. When data are transmitted, the system hijacks a passing token and rides it to its addressed delivery point.

There is no possibility of data clash or collision, as the data transmission speed is always fixed by the token flow rate, not by the amount of data traffic on the line. This architecture makes the system secure, reliable, and fast, but costly. Off-site links are set up as star connections to the exchange. The IBM system uses a unique Synchronous Network Architecture (SNA) protocol to allow a separate alarm system token ring to be established within a building in parallel to the data ring. Both data and alarms report to the same building controller and modem. All lines from local modems are radially wired to a concentrator that then prioritizes the information into alarm and data signals and sends the signals down separate lines to their appropriate destination. The bottom line with the token ring system is that it virtually locks in the customer to using one company's equipment.

Tree and Branch Systems.   Sometimes called *hierarchical multiplexing* systems, tree and branch systems are fast, but of lower intrinsic security than the other systems discussed due to the number of stages, points of multiplexing, or onward transmissions that often occur before the high-speed, direct link is reached.

X25 Packet Switching (PSS).   X25 packet switching is becoming the worldwide telecommunication standard transmission format. Essentially, PSS allows any system of any type (e.g., data, security, fire, IBM, EPC, Ethernet, Honeywell) to connect via a device called a packet assembler/disassembler (PAD) to the international, national, or company wide area network (WAN) or local area network (LAN) computer data transmission system. The PSS system offers exceptional benefits to users, as it will link and run anything, reducing line and data costs considerably by increased sharing of resources and using very fast transfer rates. The system works by creating a *packet* of data on the premises and addressing it to a destination. The packet may take many routes to its destination, all of which are fast and secure. The basic advantages of the system are

- low cost
- wide compatibility with hardware
- high security
- many routes to destination, hence nondelivery very, very unlikely
- many different transmission rates can be selected and alternative routes used for economy/security trade-offs
- local central stations (manned or unmanned satellites) may route (or post) packets for companies that have no local presence. Two or more companies in a state may both use onward posting of each other's packets, giving very high reliability to catastrophic failure of one central station.

Ethernet local systems connected to X25 packet switching systems over private WANs or PSS telephone company systems are probably the best systems available.

COMPUTER-CONTROLLED SECURITY SYSTEMS AND INTELLIGENT SENSORS

Introduction

The application of the ubiquitous microchip is bringing about a gradual revolution in the security industry, as in many other industries. This revolution is occurring on two fronts—central data gathering, control, analysis, and action; and addressable, intelligent, analogous multistate sensors.

The first of these developments is already well advanced. Many security systems, particularly for large buildings, hotels, and industrial complexes, have the normal control panel (with its zone indicators) replaced by modern computer control panels that often incorporate a status printout capability or video display unit (VDU). Because the capital cost of the computer hardware is still relatively expensive, most small- to medium-sized companies needing to purchase the systems outright will continue with the traditional control panel for a few years. The sharp end of the development of cheap computer control is in the on-line access control systems market and "smart cards."

However, many companies, foreseeing the advent of addressable, intelligent sensors, have introduced intermediate, computer-based control panels as direct replacements for all traditional, small- to medium-sized panels. These panels are modular, to easily enhance system capabilities, and include a powerful microprocessor that has the capability of running a fully addressable, intelligent sensor system. In their basic form, these panels usually include the following capabilities, all of which are controlled by a keypad:

- improved detection of faults in the system that are potential sources of false alarms. This is achieved by monitoring circuit resistance characteristics.
- multizone capacity of 50–250 individually programmable zones
- 30-character liquid crystal display annunciator that gives the panel fully interactive programming capabilities on a do-you-want-to/yes-no format
- a continuous, clock *event log timer*

An *event log timer* is basically a memory that stores real-time information about the state of the system. This feature is extremely useful for police, insurers, and maintenance crews, as it allows instant recall from its memory to display

- times and sources of system faults
- times and sources of system alarms
- time each zone was last programmed (to allow for checks to see who programmed zones)
- permanent record of all tests carried out, by whom, and when
- a paper printout of events, by means of plug-in devices
- multilevel, multiperson, secure system access for code settings, programming maintenance, etc.
- interactive two-way communication with alarm company central station

Although computer control developments help the client by offering reduced maintenance and fault-finding times, development of intelligent multistate sensors will have the biggest impact on the industry over the coming years. At present, all detection system sensors tend to be either totally passive or active, two-state ("alarm on" or "alarm off") devices. These two states will only operate effectively if their sensitivity is relatively high. Unfortunately, this leads to instability and false alarms. False alarms are the *bête noire* of the security industry and have led to its poor public standing. Intelligent sensors offer the promise of overcoming many of the false alarm problems.

The *intelligent sensor* consists of an *analogue* sensing element to which a uniquely addressable microprocessor is affixed. The combined sensor and microprocessors are wired in loop fashion and the whole loop terminates at the central processor. The central processor sends a signal down the loop to each sensor every few milliseconds. The signal is unique to the microprocessor on a particular sensor, and addresses and interrogates the sensor, asking it to report its status at that particular instant. The central processor then looks at the state of the sensor on a sensitivity scale of $1-100$, for example. If the sensor sensitivity is between 1 and 20, the sensor can be said to be in good order and not approaching alarm. If, at the next interrogation (100 msec later) the sensor reads 50, the central processor will recognize that a prealarm condition has been reached. If, however, the sensor was to display 50 for several interrogations, the prealarm will be changed to give a printout stating a maintenance requirement due to an aging sensor, warped door frame, general abuse during nonalarm hours, etc. If, 100 msec later, the sensor displays 75, the processor will recognize an alarm condition and raise the alarm.

The overall effect on the system that uses intelligent, multistate, interactive sensors is the improvement of system security and reliability. Initial development of this type of system has already occurred in the fire alarm industry, where it has passed rigorous examinations and has already had dramatic results in the reduction of false alarms. Analogue, addressable fire alarm sensors now dominate the European market for commercial fire detection systems. A big advantage of the uniquely addressable sensor is that there is now no need to zone the system or to provide local indicators, as every detector becomes its own zone and the place, pattern, time, etc., of alarm can be recorded and stored.

Central Computer Control

Large buildings or industrial sites pose severe security problems. These stem from the clash between right of entry and intruder detection. In some establishments, notably military bases, security is maintained primarily by massive use of manpower—stationary guards, patrolling guards, pass-issuing procedures, vehicle restrictions, etc.

Within commercial complexes, these security methods are not usually practiced or accepted. Computer-controlled security is used to supplement manpower. This method of security restricts entry, challenges a person's right to enter an area, monitors security points, and carries out preprogrammed responses to unsecured situations. *Computer speech* has inevitably found its way into computerized security systems, often confusing specifiers or alienating potential clients. The use of the language is unavoidable. The following explanation should clarify the subject.

All computer-based security systems are comprised of

- the central computer-controller device (hardware and peripherals) (see Figure 2–25)
- the communications wiring to the sensors, card readers, etc. (multiplex wiring)
- the program of actions the computer carries out under any set of conditions or instructions (i.e., the software program)

Central Computer.   The central computer should

be dedicated to security system control, only. Systems that incorporate building management, energy monitoring, and fire and alarm detection tend to be overcomplex and confusing to the user when confronted with a "panic" situation. Remember that the person sitting through the night in front of your sophisticated computer may not be the most intelligent person in the building. This individual must be presented with clear, concise instructions, which appear on the screen.

be of reputable manufacture. Ask who actually makes the computer and its microprocessor. Less reputable companies will try and associate their name with "badge name" equipment. The computer should be a standard model that has been modified and made into an expert system under the supervision of the security company.

have at least 256Kb of main memory capacity and dual disk drives with at least double this storage capacity for each drive. One drive should be used to store the software programs controlling the system and the other to store the information gathered and generated during system operation. Some systems house the former disk within the body of the computer.

have dual control. Many systems offer dual control computing and processing capabilities as standard. The second control unit is an automatic backup device if the first control fails. Failure of the primary system most often results from deliberate attack on the system or main electrical supply distortion or interruption. The latter can be avoided if the system incorporates a UPS and a mains filtration/suppression device. With dual systems, it is essential that all sensor circuits be permanently hard wired to both computers, thus eliminating the possibility of the failure of any automatic switching device.

only become active when an alarm condition or access control verification is needed — known as *stand off*. Systems of this type employ *distributed intelligence* that allows detectors and control units to have some autonomy of action under normal conditions, thereby reducing the data-processing load on the central computer.

Multiplexing. Multiplexing is the method by which signals are simultaneously sent down a single line of communication. Each signal differs from its neighbor and is coded to a particular circuit or device that it either addresses or interrogates when the signal is required to report its status.

Multiplexing communication systems will be most useful to the client if they are modular in concept. Good systems allow for easy expansion as the site or complex, and thus the system's requirements, grows. Systems of this type usually include devices known as loop or line controllers and sensors or point multiplexers. The loop controllers send the multiplexed signal around a loop and look for status reports. The sensors are connected to the loop or alarm sensor circuit centers (e.g., individual area, building) and only monitor the sensing devices spurred off the loop at that point.

Multiplexing is usually either TDM or FDM (see page 61).

Software Program. The usefulness of the computer control system will be determined by the computer software. The software program supplied with the computer should be

- easy to use
- flexible
- applicable
- adaptable

If the system is difficult to use or understand, it will be a burden, not a help. If the system allows for the addition of 100 sensors at some later date, but the software cannot handle the additional load, the system is inflexible and unadaptable. If the type of sensor, its position in the building, and its status are not easily identified, the software is not appropriate.

The computer system exists to provide useful information to help identify intruders. This information must be provided in either complete written or graphic format on a VDU. Codes or shortened descriptions (such as "IR level 4.A") are useless, if the guard familiar with the codes is absent from work on the day the alarm message is transmitted!

A picture is worth a thousand words and, hence, systems incorporating VDUs that can display actual floor plans, sections of a building, etc., are far better tools than simple printers.

A good system software package will include the following:

A comprehensive database that allows the user to draw and manipulate plans of the buildings, the system, etc., and designate high-risk areas on the VDU and store them for display. These systems are usually referred to as computer-aided design *(CAD)* systems. The best systems allow drawings of the building to be placed on the screen by simply running a "mouse" over the plan and keying in the required scale.

The incorporation of good maps or on-screen drawing capabilities is of enormous benefit. Visual and graphic displays are far more efficient at quickly presenting large amounts of data in a more readily understandable manner than pages of text. Map types include

local area maps — that show the local road layout, approaches to the building, traffic congestion spots, and barriers. These are useful in directing response forces to routes of escape and to your building when they are not familiar with the area.

perimeter of grounds — that show blind spots, positions of cameras, positions of detectors, guard patrol progress around the grounds, positions and status of gates, etc.

floor plans — that can zoom to any area at any level of detail.

sections of plans — that show entry and exit routes, service ducts, lift positions, and status

Point programming to allow the user to program a series of actions that occur when any particular sensor is activated. The system should be able to support a response for every sensor if necessary. Point programming can be extremely useful when, for instance, perimeter fence detection is subject to vandalism. The system could be programmed to warn of vandalism or intrusion but not raise a general alarm. Should the beam detector that backs up the fence protection register an alarm condition just after the fence sensor raises a warning in that area, an intrusion is confirmed and the response is to raise the general alarm.

Simple alarm processing. The computer keyboard should have a number of keys, dedicated to alarm processing, located to one side of the normal alphanumeric keys. The alarm keys should be arranged vertically or horizontally, in logical order of use, and should indicate their functionality. A typical set of alarm processing keys might be

Alarm Display — causes the alarm to be listed on the printer and shown in position on a VDU floor plan by a flashing symbol. The VDU should also give a message, such as "Alarm #1, infrared detector, ground floor entrance."

Alarm Acknowledged — automatically stores a complete log of information regarding the incident, including the sensor number, time activated, time silenced, etc.

Next Alarm Display — shows the next sensor that went into alarm (if several were activated within a short period of time)

Command — allows the user to tell the system what to do next, such as typing in "timed manual override on card access door 2; stay closed." Commands may also be preprogrammed and only require that the command button be pressed.

Status — initiates a status report on the whole or part of the system. This can be used to check that things are back to normal or to check the quiescent status.

The software program should allow the user to specify the following for each point or sensor in the alarm or reporting system:

priority level — gives priority of display and processing to alarms generated in sensitive areas over those in less sensitive areas, should several alarms be activated simultaneously

message — allows the user to define the message to be printed or displayed when that point of the system is activated

response action — is treated as a message

description — describes the point on the system and the loop to which it is connected. The area protected can be entered by the user

audit trail — allows the point on the system to be included or excluded from the audit trail when listing system activity over a specified number of hours

password protection — in addition to the actual key/lock on the system, passwords or codes should be used to grant access or prohibit personnel from accessing the information contained within the computer at various security levels

shunting — allows points on the system to be deactivated. Good systems allow for manual shunting at the keyboard, single time period shunting (e.g., 1–30 minutes), or timetable shunting (e.g., off 0900–1830 hours Monday through Saturday). On large sites with multiple entries and exits, deliveries, shift-working, or cleaning shifts, flexibility of shunting can be the difference between a useful system and a useless system

access recording — gives a record of *all* persons who have accessed the system and states the person's name, clearance, etc.

There are many other functions offered by the security companies selling computer-controlled security systems. When selecting or ordering any system, you should bear in mind that computer-based security systems use many watts of power. They are also very prone to corruption from accidental or deliberate EMI, RFI, and main carrier interference. They do not work when the main power is cut off either deliberately or by a storm! It is, therefore, essential that high-quality UPS and conditioning filtering units be purchased to support the whole system and its peripheral devices in the event of mains failure. The normal mains wiring should be taken via this device, so it floats across the mains supply and is always instantly available and always filtering the normal supply. A normal battery backup unit for a traditional intruder alarm system is *not* suitable for computer-controlled systems.

The Corps of Engineers Guide Specification (CEGS) 16725, *Intruder Detection Systems,* gives excellent advice on the overall needs of the client. It advises on the product, the technical data package, testing, training of staff, etc., and should be carefully considered before making any purchases.

Central Station Facilities (Also see page 58)

Security company central stations are comprised of three key elements:

- people
- computers, printers, telephones, or other hardware (as seen in Figure 2–28)
- software systems

People and Buildings.   People come first! A central station manned by a staff with a combined IQ of 10 will defeat the most user friendly computer! Staff must be well trained in *all* aspects of security work and not simply be telephonists. Buildings must be fire and intruder resistant. If they are not, then major loss to you and many others could result from a relatively minor accident or unskilled attack on the building.

FIGURE 2–28   Computer-controlled security system

Hardware. Computers and other hardware should be of sufficient capacity, modular, and developed by a recognized manufacturer (e.g., IBM). Most good central stations use IBM or IBM-compatible machines, each with at least 40 or 65Mb capacity, that are linked by an internal network with spare terminals and central processors. Subscribers to central stations should get written confirmation of the level of backup a station provides during a catastrophic event. Complete backup records in hard copy, tape, and floppy disk form should be kept of all events. Full graphic displays based on CAD systems should be used. Color screens on VDUs are essential for efficient operations.

Standards. Probably the most comprehensive document on the requirements for central stations is the recently revised British Standard 5979 (1991). This contains a wealth of points to be checked before choosing a facility in America. Central stations should *always* comply with and exceed the requirements of UL 611 and UL 1610, and have software programs capable of handling data and recording events in a manner that complies with these standards and grades of service standards.

Software. In a central station, the software program really runs the show. It is, in effect, an expert system for handling data, determining the responses required, and issuing instructions to human beings to carry out the instructions! The software must be fast, simple to operate, and flexible. It should always incorporate the following features, at the minimum:

automatic and verbal call pass code entry and verification by subscriber

record of false or failed attempts to enter personal identity numbers (PINs), codes, or passwords

access denial after two unsuccessful attempts

reminder to operator to check status of premises

alarm handling—separate VDU and signal channeling for emergency events and routine events (such as logging off/on)

automatic priority ordering of alarm type by personal attack/bandit alarm, fire, UL grade of service response time, property intrusion, perimeter intrusion, etc. Check your priority order.

response data, such as event handled by and action taken by responding agencies' verification codes, comments on event and action types generated by pressing a single button or by operator direct input

subscriber details that automatically appear on-screen for alarm-essential data only (e.g., address, telephone number, first and second key holder, alarm company responsible, nearest police station and details, nearest subscriber, security response force details)

backup subscriber details, such as people to call for

particular problems or buildings, telephone numbers, addresses, emergency authorized building repair or salvage companies and their telephone numbers and addresses

automatic dialing capabilities

automatic dialing in the event that the first key holder or response force is not responding

automatic dialing of next telephone number if the first person is not contacted

automatic logging of calls made, calls connected, calls not connected, and responses made

instructions and temporary instructions (e.g., "Call Mrs. Smith if fire or alarm systems are triggered; telephone number is" or "Do not log off until incident verified by the following persons")

holding for response confirmation. When an alarm is raised it may be a false alarm. To verify a true alarm, the station will ask for a response to a signal. This response confirmation may be put on hold if further alarms are received. An automatic system must always require operators to check the response signal after a set time so it does not remain permanently on hold.

shift change logging and information exchange. The staff at the central station should personally log on and off the system. It should not be possible for a staff member to cover for another. Failure to log on/off should result in nonpayment of wages. Staff should not be permitted to log off while there are calls still awaiting action. Handing over in-process calls can cause confusion, loss of confidence in the central station, and often result in delays or no action at all!

lists and reports, including properties listed by type, alarm installation company, single aimer, company, key holder, subscriber, and address; alarm company details, such as contacts and numbers; and authorized persons details, such as age, height, weight, sex, etc. Lists of these types allow for very detailed statistical analyses of problem areas, alarm systems, user misuse, etc.

event timetables for opening, shutting, and arming schedules; unscheduled UL time window opening and shutting; next expected event versus next actual event; duress codes for opening and shutting codes; and long-term, short-term, daily, and hourly time tables.

cross-linking and referencing capabilities. For example, if you change one telephone number in one register, the system automatically changes the same number stored elsewhere in the system.

search capability that searches data by company name, street name, person's initials, codes, etc.

maintenance and guarding services—Will the central station facility construct a maintenance testing inspection program for your alarm system? Will it construct a guard resources, guard patrol, and logging schedule?

Other very important factors to consider before subscribing to a central station are

number of operatives on duty
extent of geographical area covered
familiarity with your property type and operation

local knowledge of public order problems, traffic congestion, etc., held by operators

real-time telephone and television surveillance or time-lapse video recording

security of the station from sabotage or fire—Is there a backup station in case of these events? Would their system tell you if they lost control or would they let you find out 3 days later?

3
Sensors for External Use

INTRODUCTION

Before considering any form of exterior sensing device, clients and their advisers should stop and ask, "Do I really need this? Would the money be better used elsewhere in our overall security strategy?"

Some food for thought:

Exterior sensor installation costs for even modest-sized sites can be very expensive when the full cost of trenching or ancillary works are included.

The cost of installation often restricts external sensor use to military, strategic, research, explosive, or very high-value stock areas. These sensors are sometimes used to protect large stock compounds or marshaling yards, car lots, stud farms, racing stables, etc.

Very few systems will defeat a professional criminal who puts in the research, practice, and effort needed to attack or circumvent the system.

None of the systems could really be recommended for use on unmanned or nonpatrolled sites. These systems may only really be considered for use as a manpower reduction/increased productivity tool.

Patrolling guards (particularly with dogs) are far more feared and a greater deterent than most external systems.

External sensors are potentially more troublesome to use, and more expensive and more difficult to select, than interior sensors. *For this reason,* far more effort should go into ensuring that external sensors are carefully selected, specified, and installed. The following chapters will highlight the specific problems that can arise in the application of individual sensors. The following general topics should be considered *before* focusing on a particular sensor. If they are not, the result will be false alarm malfunctions and subsequent reduction in user confidence.

While care in selection is needed, exterior sensors offer some extremely attractive functions. They

provide *time* to react to intrusion (before, during, or after entry). This greatly increases the likelihood of intruder detention compared with internal sensor only systems.

minimize the risks to guards or building occupants

reduce the likelihood of criminal damage, vandalism, etc.

provide an enormous deterrent value when used overtly

Environmental Considerations

Vegetation.

- grass more than 4 in (100 mm) high in fenced areas (electronic or physical)
- bushes, shrubs, etc., blowing into path of devices
- autumnal leaf accumulation
- movement of tree branches
- movement of ground due to tree roots or tree sway

Weather.

- snow accumulation
- wind
- lightning strike or interference
- fog or mist
- very hot or very cold (86°F to 5°F or 30°C to −15°C) air temperatures
- frozen surface water
- dust
- shifting and settling of sandy subsoils
- drying of clay soils

Human.

- children playing near devices
- vandalism to devices
- intrusion (microwaves) or annoyance to others (views spoiled by fences, etc.)

Ambient Conditions.

- electromagnetic interference from overhead lines, buried services, radio transmissions, welding equipment, etc.
- vibration caused by traffic, generating equipment, transformers, etc.

Security Considerations

Position of Equipment.

- Does device need outer fence protection?
- positions of processing and signaling equipment
- positions of sensors, blind spots, etc.
- security of buried cables
- security of fence-mounted cables

Security of Power Supply.

- type of standby supply—batteries, generator, both?
- hours of standby capacity—1 week, 2 weeks?
- means of system isolation—position and security of isolators
- types of conduits

Operational Considerations.

- Who will operate the system? How competent will they be?
- What type of service agreement is likely? Cost?

Economic Considerations.

- lengths of zones of detection per sensor
- management and guarding of zones; guards per area
- cable routing lengths
- voltage used, cable sizes, transformer positions

FLUID PRESSURE SENSOR

Introduction

Table 3–1 summarizes the types, uses, typical dimensions, and advantages/disadvantages of the fluid pressure sensor.

Principles of Operation

Pressure is force per unit area (lb/in^2 or N/m^2). The fluid pressure sensor works on Pascal's principle of pressure equality—if a given mass of x lbs (kg) is resting on a piston of, for example, an area of 1 yd^2 (1 m^2), which is linked to a piston of an area of 2 yd^2 (2 m^2), via an incompressible, leak-free, fluid-filled tube, the piston will exactly balance a mass of $2 \times x$ lbs (kg). This is the princi-

Table 3–1 Fluid Pressure Sensor

Types	Single line.
	Single loop.
	Double loop.
Typical range area enclosed by loop	Single line 328 × 5 ft (100 × 1.5 m).
	Single loop zone of detection 957 yd^2 (800 m^2 or 200 × 4 m).
	Double loop zone of detection 656 × 5 ft (200 × 1.5 m).
Activated by	Pressure differential across loop zone of detection
Uses	Open field site/perimeter protection "in depth."
	Tank "farms," military installations, etc.
Not to be used	Under hard (road) surfaces or in areas of standing water.
Typical dimensions	Pipe diameter 0.6–1 in (15–25 mm) outside.
	Pipe loops 328 × 5 ft (100 × 1.5 m) typical.
	Analyzers/processors 20 × 12 × 12 in (500 × 300 × 300 mm).
Associated work	Trenching.
Budget cost	Labor cost predominates.
Advantages	Totally invisible, difficult to detect.
Disadvantage	Wide perimeter zone free of traffic required.

ple on which the hydraulic ram operates. The fluid pressure sensor is a form of hydraulic ram.

In the fluid pressure sensor, a small-diameter tube, hose, or pipe that is sealed at one end and that has *flexible walls*, acts as the first piston of the ram. If a load is applied to the walls, they compress, causing pressure to be applied to the liquid. The force is transmitted through the liquid to a pressure-sensing bellows and transducer at the opposite end from the seal. The pressure change is sensed by the bellows transducer and transformed into an electrical signal that is sent, via a cable, to an analyzing unit that will raise an alarm if the pressure change is significant.

Most commercial sensors consist of two loops in which each loop has its ends joined to a sensing bellows or transducer to form a figure-eight shape. Each tube will sense the pressure change wherever it occurs and will relay it to the sensor. By having two loops, the sensor can compare and reject pressure changes that are uniform across a large area and common to both loops (such as the pressure caused by snowfall) with those caused by a point load on one loop or one side of a loop. The device is sensing the *pressure differential* across the loop's zone of detection.

An additional refinement is achieved by placing a self-compensating bleed valve at the midpoint of each loop.

If a large load is imposed for a long period of time, such as a truck parked in the area, the pressure the truck exerts on one side of the loop is equalized by bleeding fluid into the other side of the loop. This equalization maintains the sensor's status quo and sensitivity. If, in this example, the truck driver now moves the truck, the alarm will still sound.

A pressure sensor transmits *everything* it senses. It is the function of the analyzer or trained operative to raise the alarm after determining what has been sensed.

The zone of protection for a single system's loop is approximately 219 × 4 yd (200 × 4 m).

Uses

Fluid pressure sensors are most commonly used as an alternative to the geophone or buried electromagnetic cable to detect persons crossing open ground. They are seldom used for commercial developments, but are often used to protect tank farms, military bases, etc., where normal pedestrian traffic is low and space is not restricted. These sensors are useful on very large *greenfield* developments (where there have been no previous developments).

Application

Although burial requirements vary slightly between manufacturers, the following measurements are common good practice for estimation of cost purposes. A trench 1 ft (0.3 m) wide × 1 ft (0.3 m) deep is required in which to bury the pipes. As the pipes or hoses are usually spaced 3–4 ft (0.9–1.2 m) apart, it is usually more economical for a common, 4 ft (1.2 m) wide excavation trench to be dug in one operation. The bottom of the trench must be clear of large or sharp stones. A 2 in (5 cm) layer of fine sand should be placed in the trench and firmly compacted.

The pipe should be laid on a bed of fine sand and surrounded with the same before backfilling the trench with soil. Great care should be taken with the preparation and location of the trench. A sample pit should be dug to test the draining properties of the soil. The trench can easily become a sump or water conduit and may even have voids created by the sand washing away *within* the trench. This is particularly true of waterlogged sites or clay subsoils. The problem with these sites is that a water-sodden trench can freeze to a solid mass of ice, preventing any sensing of pressure changes. Areas of very high seasonal rainfall or a highly variable water table can also cause problems.

After filling the trench, some companies prefer to see the trench materials artificially compacted, so immediate field trials and commissioning can commence and their

fee can then be placed in the bank! Avoid artificial compaction! If the trench is properly prepared, it will take 1–2 months for the materials to settle. During this time, normal traffic over the area should be encouraged. After this period, depressions should be filled and tests commenced. Tests should be active and a variety of types (e.g., people running and jumping, vehicles parking and moving) should be undertaken over a number of days, during different weather conditions. If the tests are satisfactory, you can then start sowing grass seed!

Installations under nonelastic surfaces (e.g., concrete) are slightly less sensitive and need more care and planning than installations under gravel, sand, asphalt, or grass.

A good precaution is to bury a pipe warning tape 8 in (20 cm) below the ground's surface to indicate the pipe's presence. The system can suffer false alarms due to natural ground movement resulting from the roots of trees, foundations of pylons, etc., that move during high winds. (It is often recommended that no system pipe be closer than half the vertical height of the tree, pylon, etc.)

Equipment Types and Selection

The type of sensor depends, to a great extent, on the application. Most buried-in-ground sensors are liquid filled (often with antifreeze or hydrocarbons). Some are single-line sensors, but most are double-loop (see Figure 3–1) or four-pipe, double-loop sensors.

The reliability of the system depends on two factors:

- the robustness and durability of the buried sensor and pipe
- the faith one has in the ability of the analyzer to raise only "real" alarms and not to "cry wolf" at every falling leaf

Before final selection, consider the following.

Pipes.

From what are they constructed? What is their life span? All pipes that bend, suffer from fatigue cracking. This type of failure can be advanced by corrosion or by erosion contact with soils. Try to select either sheathed or double-walled pipes. Avoid materials that corrode due to saltwater or that suffer electrolytic corrosion. Neoprene or thin-walled polyethylene are probably the best types of materials, although some plastics are extremely good.

Most systems operate with the liquid within the pipe or hose pressurized to 2.5–3.0 atmospheres. At this pressure, even slow leakage over a long period of time can appear as background noise. Systems should include fluid pressure drop transducers to monitor for leaks during quiescent conditions.

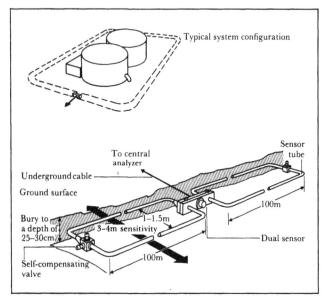

FIGURE 3–1 Typical fluid pressure sensor configuration

Fan-pipe (double-loop, same side) systems increase the width of detection and reduce the risk of ground heave false alarms.

Sensitivity is a function of *impact* force on the ground above the pipe and the degree to which the ground transmits this force to the pipe wall. Well-compacted loam soils are the best transmitters; loose, sandy, waterlogged soils are the worst.

Sensor-analyzer link cable lengths are normally limited to 0.75 mi (1.2 km).

A distance of 25 ft (7.6 m) between roads and pipes and 60–120 ft (18–36 m) from railway lines or aircraft runways is advisable.

Surface fences may be required in some locations to prohibit large animals (more than 50 lbs, or 23 kg) from walking into detection zones. Similarly, buried fences might be needed to protect pipes from rabbit or mole damage.

The pipe must register the pressure change and relay it accurately. Hence, the walls must be as flexible as possible to improve sensitivity (at whatever pressure) and all joints to valves or sensors must be extremely rigid and leakproof. The manufacturer should be able to test the joints to show their reliability.

All joints and sensors will be buried, hence they should be of bronze (preferably phosphoric bronze) or high-quality stainless steel.

Single-pipe systems are poor in operation compared to double-pipe systems.

Cable. The cable taking the signal from the buried sensor to the analyzer should be encased, screened, coaxial cable. For some installations, the cable will need to be monitored to ensure it is not cut, bridged, etc., and false signals are not being transmitted to the analyzer.

Sensor. Mechanical bellows, electrical and mechanical contacts, etc., fail more easily due to moving parts than sensor elements comprised of piezoelectric material.

Analyzer. The analyzer should incorporate

- a direct-reading pressure meter with maximum and minimum stops to show normally expected environmental pressure loads
- a sensitivity-setting switch to enable alarms to be generated at preselected levels of pressure
- a memory capability so that it can store the characteristic pressures of objects that are normally present and filter these out during preset times (e.g., vehicular traffic moving during the night for collection purposes, parked vehicles). This is only feasible on very large installations.
- power failure, fluid loss, and pipe-cut alarms
- standby power capability (at least 8 hours)
- amplitude and frequency analysis capabilities with preset characteristic comparisons

Limitations, Advantages, and Disadvantages

Limitations.

- cost
- ability to install with ease

Advantages.

- simple
- invisible
- can be used on most terrains and gradients
- can be zoned to track intruder paths or approaches to building
- self-compensating
- *depth* of detection better than all fence-mounted or barrier, electronic fence devices
- low false alarm rate and high probability of detection. The false alarm rate is normally one in 24 hours. The detection probability is normally quoted as 85% to 100% for people and 95% to 100% for vehicle movement.
- very useful in environmentally sensitive areas where security fences would be unacceptable. Similarly, the system can be considered for potentially dangerous or explosive environments.

Disadvantages.

- can be bridged by ladders, jumped over, etc., if single-line type
- should be protected by fence to prevent persons accidentally straying into area
- cannot be used under concrete or paving slabs, sets, etc., that do not flex

- good, clear route critical to success
- can be punctured and dug up, expensive to repair damage
- requires space in which to operate
- needs good maintenance—annual at the minimum

ELECTROMAGNETIC CABLE SENSOR (MICROPHONIC OR ELECTRET)

Introduction

Table 3-2 summarizes the types, uses, typical dimensions, and advantages/disadvantages of the electromagnetic cable.

Principles of Operation

Any conducting material will accept an electrical charge. The greater the charge it can store, the higher the potential to do work. The capacity of the conductor (commonly called its *capacitance*) is the ratio of its stored charge to its potential charge. If two conducting materials are separated by a material that stores a large charge before passing it (a *dielectric*), a capacitor is formed, and its capacitance is the ratio of the charge on one conductor to the potential difference between the conductors. Some dielectric materials have the ability to store a

charge for an extremely long period of time, while maintaining a very high potential difference (voltage) between opposing surfaces. The electromagnetic cable uses materials that have these properties to create a continuous, capacitive microphonic cable (see Figure 3-2).

The cable consists of an outer insulating sheath, a layer of braided conductor, a precharged dielectric filler layer, and an inner conductor. If a current is applied to the outer and inner conductor cores, an electromagnetic, capacitive field is created that is interrupted by the dielectric layer. If a dielectric material moves within the field, it will cause a change in the current flowing in the outer conductors, which can be sensed. In practice, the movement takes place in the outer conductor as pressure is applied *to it,* but the result is the same. The current change will directly reflect the pressure/vibration on the outer conductor in both frequency amplitude and waveform, and hence is called a *microphone*—a device for faithfully transforming mechanical energy in the form of pressure into electrical energy for transmission and reproduction purposes.

Provided the cable is not of excessive length, it will not lose any appreciable energy. No matter where the disturbance is caused along its length, the cable will produce the same strength signal. If any attempt is made to cut or tamper with the cable, an instantaneous signal is transmitted due to the physical disturbance to the outer conductor or to a change in its electrical resistance.

With electromagnetic cables, it is important to remember that alterations to sensitivity cannot be made at the source (i.e., on the cable) other than by altering the material to which it is fixed, so that it does not vibrate so readily or transmit vibration. The cable will always reproduce every vibration or deflection sensed by its outer core. In order to distinguish between vibrations caused by wind, birds landing on a fence to which the cable is attached, or

Table 3-2 Electromagnetic Cable

Types	Electret (registered trade name) (sometimes called "Microphonic").
Range	Up to 328 yd (300 m) of fence per processor unit/cable length.
Activated by	Vibrations characteristic of a person cutting, lifting, or climbing a fence.
Standards	None (covered by USA patent).
Uses	Mainly fence-mounted perimeter detection where space is limited.
Not to be used	On any fence easy to bridge or to tunnel under.
Typical dimensions	Cable: 0.126 in (3.2 mm) diameter. Processor (fence mounting): 8 × 10 × 5 in (200 × 250 × 120 mm). Audio display unit: 1 × 1 × 1 ft (300 × 300 × 300 mm).
Associated work	Protection of a supply cable to the fence-mounted processor unit and trenching of cable.

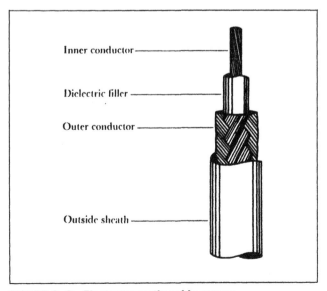

FIGURE 3-2 Electromagnetic cable sensor

cutting or lifting of the fence, a signal processor is required, which analyzes the frequency, amplitude, waveform, etc., of the signal and makes a discriminating decision.

Uses

- for perimeter protection when mounted on *inner* chain-link or mesh fences. This is a device *par excellence*.
- buried in-ground as a ground cover vibration device. In this application, it is normal to bury the sensor in a *cordon sanitaire* consisting of a layer of pea gravel that looks like a path, but, in fact, acts to amplify any surface footfalls on the path to the cable.

Application

Cable. Fence-mounted cables are primarily used and the cable is affixed to the inside of the fence. The fixings are usually comprised of a noncorroding, tough, nonremovable tie and the cable is supported at intervals of approximately 20 in (0.5 m). On some fences of height in excess of 10 ft (3 m), additional stiffening using horizontal stay wires might be required. Most cables are installed in discrete 328–984 ft (100–300 m) sections, each with its own secure, signal-processing unit or relaying unit, which is fixed to the fence at the same height as the cable.

The cable must touch the fence in order to work, so the more frequent the fixings, the better.

For additional protection, cables are occasionally run within non-ultraviolet-degrading, PVC conduits that are affixed to the fence (with the cable loose within it). This practice is of questionable benefit.

Processor. Processors require a 12V DC power supply and battery backup packs are usually included inside the processors. The power supply to the processor should be protected within a high-impact PVC or alkethene (or polyethylene) conduit throughout its length.

Audio display units usually require a 240V AC power supply with an integral battery backup and must be located in a secure room, guard post, etc.

Buried Cable. With buried cables, the manufacturer's advice should always be sought regarding the trenching detail required.

Equipment Types and Selection

The electromagnetic cable is a patented device. Only a few cable types exist, all of which are extremely similar in performance, although many security companies offer the cable as part of a system under their overall trade name for the systems (see Figure 3–3). The registered trade name for this type of cable is *Electret*. Electret is one of the newer sensing devices and its successful application depends, to a great extent, on the abilities of the processing/analyzing unit. For this reason, it is prudent only to invest in equipment manufactured and installed by one company and not to mix cables and processors. Most security companies that offer these systems do not manufacture the equipment; they only install it. Always ask the name of the original manufacturer and determine the soundness of that company. Otherwise, future parts, servicing, etc., could be a problem. After a few inquiries of this type, you should notice that the same names keep cropping up. Any oddities should be eyed warily, particularly if there is any appreciable difference in system costs. A good system will incorporate

cable sensors of the coaxial type that are constructed to international standards and include lightning protection, surge arresters, high operating-temperature ranges (212°F, or 100°C +), high-stability outer cable sheaths, and custom-made, hermetically sealed connectors

permanent stainless steel or nylon attachment kits

processing units that include normally open or closed contacts with a minimum rating of 0.5A, variable alarm pulse width capabilities (0.5–3 sec at a minimum), printed circuit, modular boards for easy repair, and a memory capability to memorize deliberately introduced signals to be ignored in future (e.g., wind). This "learning" capability should be programmed during an extensive field test, with some preset levels entered at the manufacturer's plant to reduce commissioning time. Most good systems already include processors that have selected band pass filters. These filters each recognize the signature of common, nonintruder vibrations and erase them from the signal transmitted. This action increases the signal-to-noise ratio, making true detection more likely and reducing false alarm rates. Processing units should also include antitamper, cut, and fault alarms (instantaneous); full EMI suppres-

FIGURE 3–3 Microphonic fence external cable

sion of the power supply to the processor/analyzer; custom-made cable connection and shroud units; and site-adjustable, preset band pass filters to reduce commissioning time, increase reliability, and reduce the likelihood of environmentally induced false alarms.

an end-of-line-balancing, antitamper resistor to allow for continuous monitoring

cables with very tough, robust, outer sheaths. Cable operation and life expectancy will suffer if these are damaged accidentally or deliberately.

Most systems offer the processor as a direct, alarm-generating device and, as an option, an audio output/display unit. The display unit is only of use when a trained operator is in residence who can make value judgments on the nature of the signals received by the unit. When this option is offered, the manufacturer should be able to assure the customer that memory capabilities within the processor have not been sacrificed based on the assumption that the operator will act as a filter! Operators change and what might seem like a minor wind buffet to one operator, might be read as a major intrusion by another.

Limitations, Advantages, and Disadvantages

Limitations.

- seldom used on unmanned sites
- should never be used on publicly accessible fences, which are subject to malicious damage, nuisance alarms by pranksters, etc. Wherever possible, an outer fence should be used on perimeter zones facing public pavement, roads, etc.

Advantages.

- no moving parts or mechanical elements to service
- easy to install, repair, and maintain
- very effective and reliable
- low capital cost for the cable—ideal for large installations such as power stations, petrochemical plants, prisons, airports, and high-tech industrial sites
- tamperproof passive materials and low current consumption
- no loss of sensitivity due to length of cable
- not sensitivity dampened at source—will see all frequencies, unlike inertia, which gives preference to high frequencies, or geophones, which favor low frequencies

Disadvantages.

- complex signal-processing electronics
- field test, and lengthy and exhaustive commissioning recommended for future reliability

- outer fence required to protect from mischievous individuals
- persons must climb, lift, cut, or otherwise come into contact with the fence to trigger the alarm. If they crawl under the fence or jump over it, the cable will not detect them.
- does not normally identify *where* the intruder has entered; could be anywhere along a 328 yd (300 m) length of fence corresponding with a zone

FIBER OPTIC CABLE SENSOR

Introduction

Table 3-3 summarizes the types, uses, typical dimensions, and advantages/disadvantages of the fiber optic cable sensor.

Principles of Operation

A beam of pulsed light is transmitted through the cable core. A receiver at the other end of the cable receives the transmitted light and registers its magnitude, wavelength, and any other preprogrammed, coded information that the transmitter encodes. If the cable is cut or interfered with, the receiver recognizes the change and initiates an alarm.

Uses and Application

Fiber optic cable sensors are principally used for perimeter fence protection in hostile or hazardous environments where conventional, detector cables and sensors or those carrying or generating current and potential differences would be unsuitable. Typical applications include protection of gas stores, munition stores, sites in very exposed areas, or sites subject to high EMI levels. In most commonly encountered situations, however, fiber optic cable sensors are usually bettered by Electret or geophone sensors.

Some high-security fence manufacturers include a fiber optic cable core within the hollow strand of their normal fence material. Hence, what appears to be a barbed wire fence is, in fact, also wired to an alarm system.

Equipment Types and Selection

All cables are of the same type—stranded or single core with an outer envelope and protective sheath. Receivers and transmitters differ in the complexity of the information encoded on the cable. Most use the light/no light

Table 3–3   Fiber Optic Cable Sensor

Types	Various specified by quality of transmission.
Range	Virtually unlimited.
Activated by	Physical break or tapping.
Standards	None in security application.
Uses	Covert within fences surrounding hazardous areas or in strong electromagnetic field strengths.
Not to be used	Where less expensive alternatives viable.
Typical dimensions	Under 0.16 in (4 mm) diameter.
Associated work	Special terminations.
Budget cost	Individual applications determine.
Advantages	Very reliable, very secure.
Disadvantage	Cost.

Table 3–4   Capacitive Field Effect Sensor (CFES)

Fence systems Types	1. Single field wire, grounded post: cheapest, least reliable. 2. Double wire (field and ground wire): better. 3. Triple wire: high and low-level field wire, central grounded sensing wire, "dummy" conductors; best system.
Range	Usually either contact only (simple system) or depth of field detection 20 – 39 in (0.5 – 1.0 m). Most fence systems operate with a single processing unit controlling up to 547 yd (500 m) in length.
Activated by	Disturbance to electrostatic capacitive field caused by person in field of detection.
Uses	Perimeter fence, wall top, or roof level; relatively high security system.
Not to be used	Where grounding is poor or vandalism/nuisance alarm possible, and difficult to prevent.
Typical dimension	As standard of security fence adopted (usually up to 16 ft, or 5 m).
Associated work	Cable trenching, fence erection and anchoring, auxiliary grounding independent of electricity supply.
Budget cost	Part of overall fence cost.
Advantages	Takes up very little space, more difficult to overcome than "beam fences." Comparable in security performance, etc., to electromagnetic cable fence protection.
Disadvantages	Expensive; "false alarm" more frequent than beam fence protection.

change, while others actually transmit a secret signal of which anyone trying to bridge the cable would be unaware and unable to match, even if they succeeded in forming bridging connections.

The cables are extremely small in diameter and can be contained within the hollow core of a section of barbed tape, interlaced within or strapped to fence wiring. Always select a company with an extensive past record of successful installation. Fences with "collapse and cut" outriggers offer additional protection.

Limitation, Advantages, and Disadvantages

Limitation.

- normally a covert device, hence little, obvious deterrence

Advantages.

- very safe to use
- unaffected by almost all environmental conditions
- almost impossible to tap into or tamper with
- very unobtrusive; can be completely hidden

Disadvantages.

- expensive to install and repair; repairs are difficult and can indicate fence type
- can be avoided if position is detected; fence can be lifted or climbed without causing alarm (unlike Electret and geophone)
- a direct contact device
- can only replace existing fence

CAPACITIVE SENSORS

Introduction

Table 3–4 summarizes the types, uses, typical dimensions, and advantages/disadvantages of the capacitive field effect sensor (CFES).

Principles of Operation

Capacitive field effect sensors operate on the principle of electrical resonance. If a capacitor (C) and an inductor (L) are wired in parallel, and a voltage (V) is applied across their ends, a current (I) will flow in the circuit. This current will be the sum of that flowing through the capacitor (I_C) and that flowing through the inductor (I_L).

If the *frequency* of the alternating current in the circuit is gradually raised, the current flowing through the inductor will decrease and the current flowing through the capacitor will increase. At a certain frequency, the currents will be exactly equal. This frequency is called the *resonant frequency*. When resonant frequency is reached, current surges backward and forward between capacitor and inductor, which alternately store and discharge energy. While this is happening, very little current is drawn from the supply.

The oscillating current set up in this way may be very large, many times that normally drawn from the mains supply, and, hence, any change in the inductance or capacitance of the circuit will cause loss of resonance and a large, sudden draw of current from the mains supply to one device or the other will occur for an instant. Capacitive field effect sensors sense this change in current caused by a change in the capacitance or inductance of a *tuned circuit,* as a resonant circuit is frequently called. The change may be caused by a person walking in a capacitance field or touching the capacitive part of the tuned circuit.

In practical capacitive sensors, the capacitor is formed by either the material of the object to be protected or by a layer of air between two electrodes. The inductor is part of the *analyzer circuit* that is wired to the object to be protected or forms part of the signal processor. There are three main categories of capacitive sensors:

- capacitive relays
- capacitive effect cables (CECs)
- capacitive field effect sensors (CFESs)

Capacitive relays or *capacitive proximity detectors* (internal use only) rely on the ability of the protected object or material to conduct electricity and, hence, to form a physically recognizable part of the electrical circuit. The most common use for these devices was the protection of metal safes, cabinets, barred enclosures, etc. These devices, although extremely sensitive, suffered greatly due to stray electromagnetic fields, static electricity, and variable resistance of the protected object, and, hence, were extremely difficult to stabilize without great loss of efficiency. This led to their rapid demise and they are now seldom used.

Capacitive effect cables are simple, capacitive, static electricity–generating cables arranged in triple configurations on fence-top outriggers. The wires are provided with a low-voltage AC signal with a frequency of approximately 3kHz. This signal is insufficient to create an air field of capacitors, but is sufficient for the air to have a measurable capacitance between the wires and the grounded outriggers to which they are attached. The total capacitance of the circuit is comprised of two capacitive cables of equal length and the air capacitance. The cable capacitance dominates. An intruder touching,

cutting, or otherwise disturbing the cable affects the directly measured capacitance and causes an alarm to be raised.

These systems are basically low-price, inferior versions of CFESs—true field systems—and should not be used if CFESs can be afforded. If they cannot, then very careful comparison of Electret or other fence-mounted cables with CECs should be made; they will not normally win!

Capacitive field effect sensors (often referred to as *electric field* or by trade names such as *E-field*) make use of the ability of air to store a charge and, in effect, become a capacitor. The "field" in the name of the sensor describes the pattern of static electricity distribution created when a capacitor is formed using air as the dielectric material between two electrodes. In order to form the capacitor, a normal fence is provided with horizontal wires on outriggers. These wires form continuous electrodes when supplied with a very small current. In simple systems, a single wire is used to form the positive electrode, while the fence and posts, which are metal, are grounded to give the necessary potential difference (voltage) across the air gap between them. In better-designed systems, three wires are used—one as the positive wire, one as the ground wire, and one as the sensing wire. Four-wire or multiwire systems offer further protection.

The *positive* or *field wire* is supplied with an alternating current of perhaps 10kHz in frequency, by an oscillator. This current creates the capacitive electrostatic field around the field wire. The field created is "cut" by the sensor wire mounted near the ground and a current is induced in it. A current now flows in a closed loop through the field wire, the air capacitor, and the sensing wire with its attached inductor. When this circuit is tuned, very little current is drawn from the mains. If the air capacitor field is interrupted by an intruder, the tuned circuit will no longer resonate and a comparatively large current will be drawn.

The system processor unit determines whether the disturbance to the field was caused by a human or an animal by the magnitude of the current drawn, the speed of change from the norm, the duration of the change, and whether the change was continuous in one direction or another. This ability to site-adjust sensitivity for the mass rate of movement and the duration of the disturbance gives CFES systems a very high detection probability (98–100%).

Uses

CFESs are used primarily on *secondary* perimeter boundary fence protection (behind and attached to a deterrent fence) for establishments requiring fairly high levels of security. They are extremely useful on very large sites and are rapidly gaining ground on the more conventional methods, such as geophones, etc. CFESs offer some excellent advantages to the user and are a strong challenger of the electromagnetic cable as the "best" form of sensor available for most high-security fence applications.

Simple systems can be very effective if installed in roof eaves or on flat roof parapets to detect anybody trying to climb onto a roof to force an entry. This method of detection is extremely attractive to larger, urban factories, warehouses, etc., that have very little boundary space.

Simple fence, wall, or parapet systems offer a secondary advantage of increasing the effective height of any fence to which they are added. Many of the world's most sensitive installations are protected by CFES systems.

Application

For successful operation with minimal false alarms, it is essential that the CFES system be installed correctly. Wherever possible, the system installer should be asked to install the complete perimeter fence and sensor system to remove any friction between fence installers and their standard of workmanship, and security system installers and their fence requirements. Where a fence already exists or it is not economically possible to get the system installer to install the entire fence and system, you must ensure that

- the fence is high enough and of a type that will provide reasonable protection, even without the detection system
- it is not easy to bridge or crawl under the fence; anchor or bury the bottom of the fence in structural concrete
- the fence posts are substantial and rigid, and the fence is taut and will not deflect appreciably in the wind (additional tension wires may be required)
- a *cordon sanitaire* exists on each side of the fence. Sensor wires are placed at 12–16 in (300–400 mm) above ground level and no plants, shrubs, etc., should be allowed to grow under them or for a distance on either side that is equivalent to the depth of the sensing field.
- tree branches, signs, cars, or other objects that might move occasionally are not present within the sensor field

Most systems are installed in discrete lengths of 984–1640 ft (300–500 m), each with its own processing unit. This gives the benefit of *sensitivity zoning,* which allows lower sensitivity settings to be used in unstable or less than perfect environmental zones of the perimeter without greatly affecting the overall detection potential.

The one major drawback of this system is that it can be severely affected by objects or materials touching the sensor wire and by snow and ice. Weeds, blown paper or debris, drifting snow, or ice collecting or forming on the sensor wires will all affect the circuit resistance and give rise to false alarms. These false alarms can be reduced by using

- multisensing, dummy wire systems instead of two-wire systems
- catch trenches on each side of the fence to stop drifting snow, water buildup, etc.
- simple processors with good-quality, accurate band pass filters that only forward certain frequencies characteristic of human disturbances to the control unit
- central controllers, which compare processor information from different and adjacent areas of the fence for signal amplitude, duration rise and decay, etc.
- signal processors that self-compensate for disturbance effects as intruders approach or withdraw from the area, keeping signal strength steady
- central controllers that have several different current generation circuits to allow several zones to be used or operating frequencies to be changed on a single zone, if problems are encountered

Systems should include at least three band pass filters. The lower frequency filter rejects vibrations caused by low-mass, slow-moving objects such as windblown debris. The higher frequency filter rejects vibrations caused by wind vibration of the field wires. The third filter is normally a lightning protection electromagnetic surge arrester.

Most systems operate on the *coincident variance principle*—if a frequency that passes the band pass filters exhibits

- an amplitude that reflects the mass and likely proximity to the field that a human would create, *and*
- a frequency change caused by net forward or backward physical movement of a human walking, shuffling, crawling, etc., *and*
- a change of signal present for a time period that would be needed to cut a wire, climb or cut a fence, etc.,

then, and only then, will the system sound the alarm.

Another environmental hazard that can cause a false alarm is the accumulation of salt. This makes the system difficult to use in seaside applications where salt-laden air can cause corrosion, electrical tracking of insulators, etc., giving rise to false alarms, and cleaning and maintenance problems. Selection: Consider the following and ask salespersons

- What happens when a flock of sea gulls decides to roost on the field wires?
- What effect would the system have on the radio reception or the patrolling guard's receiver?
- Would electronic data disks or tapes, carried by people as they pass through or near the system, be corrupted?

Sensible answers will tell you a lot about the competence of the salesperson selling the system!

The most critical area governing the successful application of CFESs is the availability of a good, electrical ground (i.e., less than 1 ohm). In the past, many system failures have, in fact, been attributable to grounding failure or lack of low ground resistivity. Before any installation on made-up soil, shillet, shale, clay, etc., is considered, it is best to have an electrical contractor, and/or better still, a lightning protection specialist, measure the electrical system ground and the soil's ground resistivity. This costs very little compared to the cost involved in obtaining even a relatively poor ground as a remedial measure.

Equipment Types and Selection

CFES systems are relatively specialized and, hence, not many manufacturers exist. This has the disadvantage of restricting comparisons on reliability, cost, etc., but is outweighed by the advantage of dealing with companies that only manufacture one product and, therefore, know its capabilities and drawbacks very well and are unlikely to saddle you with a troublesome system (as their reputation will suffer). This can be a strong bargaining counter and the client should ensure that the company selected will provide guarantees above those expected for less expensive systems (e.g., unlimited warranty, repair response within 8 hours, complete set of spares for the system, liability for system failure). Leasing arrangements for systems can often prove to be an attractive proposition.

Equipment types can be divided into security categories and installation categories. In security terms, the poorest (and hence cheapest) systems are *single-wire, grounded post systems* for fence protection. These consist of a single wire fixed to an outrigger on the top of a normal chain-link or welded mesh perimeter fence that has metal posts. These systems are often unreliable and troublesome. A better system is the *double-wire system,* for which one wire is the field wire and the other wire is the grounded-sensing wire. This is probably the most common system in use, as it provides adequate sensitivity and reliability at the lowest price for most commercial risks.

The next level of security is provided by the *triple-wire system,* which has two field wires, one on each side of a central-sensing wire. This system provides a truly "shaped" field of detection and variable sensitivity. The most secure system is the *multiwire system* (see Figure 3–4), which uses multiple pairs of field and sensor wires with intermediate dummy wires to shape the field produced by the system. The multiwire system, by its configuration, creates a true fence of detection, unlike some of the wall or fence top-mounted, two-wire systems that only provide anticlimbing protection.

During installation, the systems are usually divided into

- freestanding fences (see Figure 3–5)
- fence or wall top-mounted
- fence-mounted, secondary multiwire fences
- internal roof space protection systems to protect against roof penetration in very large warehouses, hangars, etc.

When selecting the system, look for

multiwire systems that have field-shaping capabilities that limit the charged atmosphere range

systems that are offered as part of an overall fence construction package—are matched to the fence and not bolted on

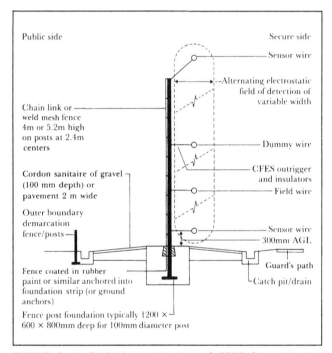

FIGURE 3–4   Typical arrangement of CFES for perimeter fence

FIGURE 3–5   CFES E-field system

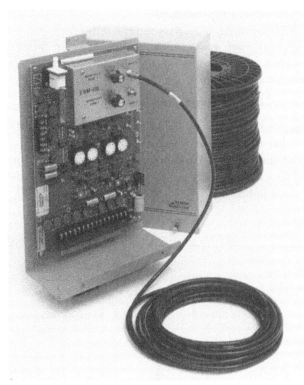

FIGURE 3–6   CFES E-field system analyzer

FIGURE 3–7   CFES E-field system central processor

systems that give good depth of detection without excessive spillage (typically, 31.5 in, or 0.8 m, deep) and have fields that hug the ground contours, preventing intruders from crawling underneath

systems that offer a wide range of tuned circuit frequencies that can be used to counter changes in atmospheric conditions, circuit aging, or resistance change—all of which affect sensitivity

systems that have frequency-locking capabilities to prevent drifts in frequency caused by power supply conditions, circuit anomalies, standing wave phenomena, etc., allowing full sensitivity to be maintained

a system with a central processor controller analyzing unit, as seen in Figures 3–6 and 3–7, that offers as many capabilities as possible. This unit will actually activate alarms based on what it sees. It must be able to monitor the circuit conditions, change the sensitivity, and foresee potential maintenance problems.

modular components to allow for easy replacement. It should be possible to pinpoint faults on the field lines (by means of a Wheatstone bridge/mimic diagram arrangement or similar) without extensive searching.

systems using high-tensile, low-resistance stranded wiring, which is lightly insulated to prevent corrosion or hardening. The wires should be kept taut by spring-tensioning devices located at regular intervals.

systems that have electric company–standard insulators to prevent tracking or shortening of the field wire to the ground. These insulators should be strong and tough, preferably porcelain, glass, etc.

systems that incorporate some form of antitamper circuit on both the processing and fence wire circuitry

systems that will allow individual, mutually exclusive frequencies to be set for each tuned zone or timed circuit when more than one zone of protection is used. This will reduce mutual interference that might regularly occur if frequency locking is not incorporated.

systems with warranties on parts and labor linked to the mean time between failures (MTBF) quoted by the manufacturer. If, for example, 25,000 hours is quoted and the system is in operation 5000 hours per year, the warranty should be at least 5 years.

systems that permit field and sensor cables to be run in conduits across or under gates in fences to allow for a single, continuous zone and reduce costs or the need for separate zones on each side of the opening

systems that are not affected by RFI or EMI

Limitation, Advantages, and Disadvantages

Limitation.

must be second inner fence or out of public areas to avoid vandalism

Advantages.

low space requirement, typically a *cordon sanitaire* of approximately 3.3–4.9 ft (1–1.5 m) in width (total)

follows terrain (ground hugging)

not greatly affected by wind, animals, etc.

extremely difficult to bridge electrically

very sensitive, but has good reliability and fairly low false alarm rate if correctly installed

can provide depth of field; harder to tunnel under or physically bridge than fence-mounted geophone or Electret cables

can be mounted at almost any angle without loss of performance

wires are physically difficult to avoid and have deterrent effects to the uninitiated, as they look like high-voltage, electric fences

long coverage per zone, typically at least 820–984 ft (250–300 m)

can be used in combination with other systems, e.g., simple, two-wire CFES plus fence vibration sensor

sample two- or three-wire CFESs and ground-buried systems will provide preattack warning and increase reliability and security

Disadvantages.

effective grounding; this can be a problem in many types of soil and, when this is the case, a ground wire should always be used and a clean ground provided to that wire

requires a fair amount of routine maintenance and inspection to look for mechanical damage to materials and insulators, corrosion, etc.

sensitivity can be affected during thunderstorms and in areas of high EMI (particularly long-wave radio signals)

in order to reduce false alarms caused by birds sitting on the wires, ice forming on the wires, and debris distorting their shape, the wires must be kept taut. This can induce fatigue if good materials (e.g., multistranded or stainless steel conductors with outer sheath, tensioning secondary catenary wires) are not used.

some fence-mounted systems (particularly single-wire or fence top-mounted) will not detect rapid, fence fabric lifting

GEOPHONE SENSORS

Introduction

Table 3–5 summarizes the types, uses, typical dimensions, and advantages/disadvantages of the geophone sensor.

Principles of Operation

Geophone sensors operate on the principle of electromagnetic induction. If a magnet is passed through spring coils made up of fine wires that conduct electricity, it will cause an electromagnetic force (emf) to flow through the wire. This, in turn, will induce current flow in the wire. The magnitude of the current produced is dependent on and proportional to the strength of the magnet's magnetic field, the number of coils the magnet passes in its travel through the spring, and the speed at which it travels.

To produce a current in the coil, only relative motion is required; the spring may move over the magnet, the magnet through the spring, or both. If the coils are fixed in position and the magnet is suspended above them on a spring plate, anybody standing on the plate or anything causing it to vibrate up and down will generate a current in the coil. This current can be used to signal an alarm condition. A device that acts on the principle of ground-, wall-, or fence-born vibration to induce a current is known as a *geophone*.

Piezoelectric seismic sensors operate on a similar principle, but in a slightly different manner, relying on pressure or stress to produce the current. The names of these devices are often interchanged or confused by specifiers and manufacturers, and it is essential that both parties understand what is actually required and provided.

Uses

mounted on inner perimeter fences (protected by an outer fence) to detect climbing, lifting, cutting, etc.

budied in ground to detect movement across open space

fence-mounted type mainly restricted to large-scale perimeter protection sites (e.g., prisons, industrial processing plants, generating stations)

wall- or floor-mounted types will protect against most forms of structural attack (including internal tampering); often used as an alternative to inertia sensor

fence-mounted devices should only be considered on green-field sites and should never be the only means of detection. These devices are the first line of defense.

wall-mounted devices should be carefully considered before being used on party walls in offices, shops, etc., or on external walls in built-up areas. False alarms caused by neighbors fixing pipes, shelves, etc., are a frequent event; avoid if possible.

fence-mounted types should not be connected directly to alarm stations or police stations. They should always include a listen-in facility for a local guard to make a judgment as to whether assistance is required.

Table 3–5 Geophone Sensor

Types	Buried, fence-, wall-, or window-mounted models (other geophones are in fact devices operating on a different principle).
Range	Buried: 20 ft (6 m) diameter circle usual per sensor.
Activated by	A ground, structural, or fence-born vibration characteristic of a person walking, running, or crawling across the ground; in the case of fence-mounted, a person climbing, cutting, sawing, or lifting the fence; or structural attack in all forms.
Standards	BS 4737, Part 3, Section 3.10.
Uses	Fence type: primarily rural or green-field sites or on second fences behind a boundary fence. Most useful against unskilled, untrained intruders only, a device easily circumvented and obvious. Buried sensor excellent deterrent against even the most skilled, but must not be in area of easy access or where nuisance could be caused by persons inside the site. Wall types guard against removal of masonry.
Not to be used	Fence-mounted: near gates, on loose or badly maintained fences, near trees, etc., that might vibrate the fence-mounted sensor. Buried: underneath or near large trees, near vibrating equipment, heavily used roads, etc. Wall- or fence-mounted: on surface that could vibrate or rattle in normal conditions.
Typical dimensions	Geophone, including mounting plate/junction box: 4 × 4 × 1.6 in (100 × 100 × 40 mm). Control, rack mounted inside secure premises: 20 × 12 × 20 in (500 × 300 × 500 mm) (for up to 10 "chains" or radial circuits of detectors). Window attack types: 2 × 1.5 × 1.2 in (50 × 40 × 30 mm). Wall types: 4 × 4 × 2.4 in (100 × 100 × 60 mm).
Associated work	Fence-mounted: all carried out by installer. Buried: trenching and special preparation/backfilling with sand layers.
Advantages	Prepenetration alarm. Follows ground contours. Not badly affected by weather.
Disadvantages	Fence-mounted: can be relatively easy to circumvent. Buried: *Cordon sanitaire* free of equipment, trees, pylons, etc., required. Window type/structural type: not as good as break glass and inertia, respectively.

Application

Fence-Mounted. See Figure 3–8.

Sensors should be mounted at the manufacturer's recommended spacings and heights (usually 16–33 ft, or 5–10 m, spacing and 5 ft, or 1.5 m, height). Most geophone systems are sold in chains of up to 20 to 50 sensors per cable.

Always choose fence-mounted geophones that are mounted with the vibration axis as the vertical axis and are desensitized to horizontal movements caused by winds.

Always install the connecting signaling cable (with its antitamper circuit) inside a continuous conduit permanently affixed to the inside of the fence (stainless steel cleats). Ensure that the junctions and terminations are completely watertight or hermetically sealed if possible.

Ensure that the method by which the geophone is affixed to the fence is rigid and will resist attack and vandalism. No relative movement should exist between the fence and the geophone.

Make sure that the fence is strained correctly, fence posts do not move at all, and tree branches, bushes, gates, and barriers do not cause false alarms.

Anchor the fence base securely into the ground or bury it at a depth of 20–39 in (0.5–1 m) to make tunneling more difficult. Although costly, a continuous ribbon foundation holding both the posts and the fence base is best.

Make sure the fence is high enough not to be easily bridged (minimum 10 ft, or 3 m).

Ensure that the geophone will trigger an alarm if a ladder is placed against a post and climbed.

Always ensure that any signal amplifiers (required on long perimeters) are secure.

Buried. See Figure 3–9.

Always bury the sensor strictly in accordance with the manufacturer's instructions. Never cut corners. Installation usually involves digging a trench along the line to be protected. The trench is usually 12 in (300 mm) wide by 6–12 in (150–300 mm) deep. If a flat-based geophone is

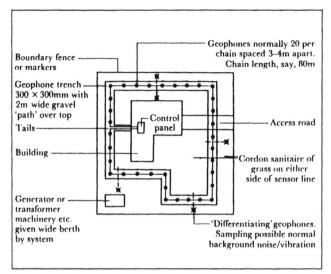

FIGURE 3–8 Fence-mounted geophone

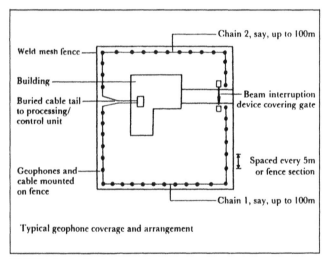

Typical geophone coverage and arrangement

FIGURE 3–9 Buried geophone

used, a 12 in (300 mm) deep trench will be backfilled with 4 in (100 mm) of sand or pea gravel. The geophone and cable are then laid down and another 4 in (100 mm) layer of sand installed; 4 in (100 mm) of sifted earth is then tamped to just below the original ground level. The surface is either planted to match the surroundings or leveled with gravel to a width of 65 in (2 m) to produce a *sound box*. If spiked-based geophones are used, these are pushed into the base of a trench that is 6 in (150 mm) deep until the spike is completely embedded. Layers of sand, earth, then gravel are placed onto the detector. Trenching and backfilling are often carried out by a building contractor, but in order not to split responsibility, it is often better to allow the building contractor to excavate only, and to contract the security company to carry out the backfilling, as this is often considered part of the detection system. Even if the security company subcontracts the work back to the building contractor,

responsibility rests with the security company and no three-cornered arguments can develop!

When sensors are to be buried in asphalt, the manufacturer's instructions must be carried out *to the letter* or damage and false alarms will be a continuous problem.

Never be tempted to extend a line of sensors or to install too many per line.

Never bury a line of sensors near trees, lampposts, or gateposts, or close to heavily used roads, transformers, generators, etc., that could cause the ground to heave or vibrate.

Wall- or Floor-Mounted. See Figures 3–10 and 3–11.

Never affix to unsound, lightweight, or vibrating structures.

Always affix to any surfaces likely to be a target.

Never select for use without having person selling equipment carry out an environmental check.

Always specify the area to be covered and the materials of construction, not just the number of sensors you think are required. Never change the structure without notifying your security adviser.

Equipment Types and Selection

All true geophones are of the same basic type — an encapsulated, electromagnetic current-generating switch of approximately 4 in (100 mm) in diameter. Different manufacturers produce devices suited for particular applications, making the device particularly sensitive or insensitive to lateral, backward and forward, or up and down motion. True geophones are manufactured for three particular applications:

- buried
- fence-mounted
- wall or floor window guards

Buried Geophones. See Figure 3–9.

A buried geophone system is comprised of a number of sealed sensors evenly spaced along a signal cable in radial fashion. The whole line is buried underground in the area to be protected. Buried geophones and their associated signal-processing and control panels are designed to detect low-frequency, ground-transmitted vibrations that occur over a relatively long time (e.g., 2 seconds). These geophones are used primarily on sites that have a perimeter zone of grass, gravel, or asphalt (e.g., 10 ft, or 3 m) and are run parallel to the fence, often with two zones of protection.

Fence-Mounted Geophones. See Figures 3–8 and 3–12.

Fence-mounted geophones are similar to the buried

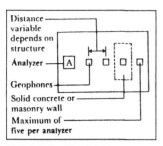

FIGURE 3–10 Wall- or floor-mounted geophone

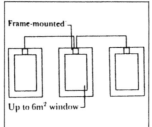

FIGURE 3–11 Window-frame-mounted geophone

FIGURE 3–12 Fence-mounted geophone

type, but as they are often subject to much greater horizontal disturbance than buried sensors (caused by wind, animals, etc.), they are made less sensitive in this direction. This is done by either manufacturing the sensor slightly differently or by using more sophisticated and flexible signal-processing units.

Wall-Mounted and Floor-Mounted Geophones. See Figures 3–10 and 3–11.

Structural attacks on walls or floors produce many more disturbances and characteristic vibration patterns than the movements seen by fence or ground-buried sensors. Wall-mounted and floor-mounted sensors also face far fewer problems from environmental false alarms (e.g., wind). This means that, in general, these geophones can be more sensitive and reliable than their counterparts.

It is important to remember that a floor-mounted geophone is not necessarily suitable for wall mounting. When selecting a geophone, consider the following:

Always choose devices that are connected to a site-adjustable analyzer unit, because normal, structural vibrations cannot be anticipated. A factory or road could be built next door!

Always ensure that the geophone has alarms against internal interference (circuit tampering) and attempts to pry devices off walls or push pads between wall and sensor surfaces.

Never install without first having a suitability test carried out.

Always use devices that will detect drilling, scraping, hammering, etc. Some geophones will only sound an alarm as a result of percussion drills or hammer and chisel attacks.

Ensure the device is a closed-circuit wiring configuration to give added security.

The choice of a particular sensor in any application is extremely limited, as there are very few manufacturers (although many retailers). The quality of the system is more easily established by investigating whether the processing of the incoming signals is tailored to a particular application or whether the same processor would be provided irrespective of whether a fence-mounted or buried system was adopted.

Signal Processing. In fence-mounted sensors, most devices limit the horizontal disturbance caused by wind, etc., within the actual device, leaving the processor to identify signals of a frequency range, duration, and amplitude characteristic of a person climbing, lifting, or cutting the fence. This detection is usually achieved by a band pass filter, which filters out certain high and low frequencies. Those signals that are admitted are checked against the background, electronic noise in the system. If the signal-to-noise ratio is above a preset limit and occurs for a predetermined duration, an alarm is generated. This is a sensitivity adjustment, rather than true processing, as detailed later.

The buried sensor responds to low-frequency sounds transmitted by the ground, which vibrates with the intruder's footfall. A band pass filter and signal-to-noise-ratio system is still used to look for the characteristic signature of a human moving over the ground. However, because ground vibration can be caused by traffic, machinery, etc., the better devices also incorporate a dual backup or comparison signal in order to verify that the vibration is caused by a person. This is usually provided by either *differentiating sensors* or *split or dual channel signals.*

Differentiating sensors consist of a number of sensors placed in areas of the site that are prime suspects for the generation of false alarms (e.g., near loading bays, doors, generator enclosures, transformers). Signals from these sensors are sent directly to the processor where they are compared with and subtracted from the signals received by the true geophone protection line. The only disadvantage of this system is that, with some processors, the presence of a large differentiating sensor signal can reduce the sensitivity of the actual geophone line.

Split or dual channel signals are the result of the signal from each of the geophone lines being divided and passing through a separate band pass filter. The first signal goes through a filter that looks for specific frequencies characteristic of a person in very close proximity to the geophone line. The second filter looks for signals of lower amplitude or frequencies that are characteristic of vibrations being transmitted from a distance well outside the protected area (e.g., a passing truck). If the first sensor signal is subtracted from the second, or vice versa, and the signal-to-noise ratio is great enough, an alarm will be generated. Some systems use a second line of sensors inside a *cordon sanitaire* to detect trucks approaching a fence, for example, which have no right to be there and could be used to bridge the fence.

Some systems supplement automatic alarm-sensing processing and signaling with an audio signal output that can be used by a guard to verify the signal in safety before initiating a response.

Any system that offers either of these two primary methods will be of better quality than the one that simply offers band pass filtration. In operation, there is little to choose from between the abilities of either method and selection will be made on cost, space, and a particular problem.

Before selecting these devices, it is always a good idea to consider the alternatives. When correctly installed, maintained, and monitored, geophone sensors can provide a high degree of security, but in practice they are often troublesome. This happens because of untrained operatives, incomplete field trials, excessive sensitivity, or inadequate processing. In general, electromagnetic cables are better suited to give high-security fence and open ground protection. If devices are specified, they should always

be part of a complete system designed, manufactured, installed, tested, and commissioned by one company. Subcontractors should never be used and responsibility never split.

include controls that allow preamperage gain adjustment, feedback adjustment, and sensitivity adjustment, and contain a low-voltage warning indicator, an antitamper or bridging detector, and an impedance monitor

be fail-safe and normally closed, if possible

Limitations, Advantages, and Disadvantages

Limitations.

Cannot be used with ease in urban situations where traffic vibration, underground transport, road digging, malicious attacks, etc., might cause continual interruption or entail reducing sensitivity to an unacceptably low level.

Many fence-mounted devices are linked to a central control and analysis unit that requires a trained operative to be in attendance.

If standard cable is used, actual fence cable length is limited by the overall circuit resistance. In practice, preamplifiers are often needed for longer fence systems.

Advantages.

buried type not visible and therefore less prone to attack than beam interruption devices

not affected by mist, fog, etc., which affect some beam interruption devices

buried type can be arranged to provide positive path tracking of the intruder across open ground zones, reducing the level of CCTV that might otherwise be required

fence types are quick and easy to install (far more so than CFESs)

detects before entry is gained

fence types follow ground depressions, unlike beam systems

take up very little surface area

correctly applied, they are quite reliable

Disadvantages.

other devices often more suitable due to environmental disturbances

best application is for ground protection, but, in this role, snow cover can greatly affect sensitivity

on fences, can be bridged, tunneled under, or magnetically attacked until faith in system is destroyed

to prevent false alarms, fence-mounted devices have to-and-fro movement sensitivity deliberately suppressed and only up-and-down movement is given full rein. This means that, in practice, devices are less sensitive to cutting with bolt cutters than they might otherwise be.

Glossary of Terms

Geophone—a device operating on the principle of electromagnetic induction (magnet involved). Can be either fence mounted or buried in ground. Sometimes referred to as a *seismic sensor* to distinguish the vibrating medium as the ground, as opposed to the air. Also used on heavyweight walls to detect breaking through.

Piezoelectric sensor—a device that operates on the principle of stress or pressure producing an electrical current. These sensors can be window, buried, structure, or fence mounted and are often misleadingly called *geophones* or *seismic sensors*.

Buried line sensor—a name given to either of the preceding terms when these devices are cabled together and buried. Often confused with buried electromagnetic or Electret cables, which have no individual sensors, but are continuous, linear, sensing elements.

Inertia sensors—devices that use a sensing medium coupled to an enclosure and attempt to remove oversensitivity to normal motion caused by environmental effects. Most environmental movement conditions are fairly intermittent and variable in frequency and amplitude, but have very little acceleration or deceleration content. Many intrusion attempts, such as fence lifting, using ladders, hitting a fence, scaling a fence, etc., do produce rapid acceleration of force inertia, which inertia sensors detect. Inertia sensors are often simply desensitized versions of geophonic or piezoelectric sensors and should be considered with reservation. Inertia sensors will reduce false alarms and high noise levels, but can compromise real detection ability.

ACTIVE INFRARED SENSOR (AIRS)

Introduction

Table 3-6 summarizes the types, uses, typical dimensions, and advantages/disadvantages of the active infrared sensor (AIRS).

Principles of Operation

The AIRS is basically a fail-safe, beam interruption device that consists of three major components—the source, the transmitter, and the receiver. The source is designed to emit infrared light of an exact wavelength, when supplied with electrical energy. The source is contained within the transmitter, which takes the light energy created, shapes it, and transmits it through open space as a beam of energy to a distant receiver. The receiver recognizes and captures all ambient infrared energy of a particular wavelength and code falling on its collecting surface. The receiver then transforms (transduces) the incident infrared light energy back into an electrical current using a photoelectric cell, which is used to keep an alarm device in a quiescent state. If, for any reason, this supply of infrared energy to the receiver is stopped while the alarm is set, the alarm will be activated.

In order for the device to operate, one must ensure that the transmitted infrared light impinges on the intruder. Being opaque to infrared light, the intruder will absorb the energy of the beam, thereby stopping the energy flow to the receiver and causing the alarm device to sound. It does not rely on the infrared energy generated by the human body and, hence, is totally different in operating principle from the *passive* infrared detector described later.

The light source is usually a light emitting diode (LED)

Table 3-6   Active Infrared Sensor (AIRS)

Types	1. Internal single beam: range 49–984 ft (15–300 m). 2. External single beam: range 328–2625 ft (100–800 m). 3. External fence stacks: range 33–328 ft (10–100 m).
Activated by	Loss of 50% of beam energy for period greater than 40 ms. Should *not* operate for 100% loss of beam energy for a period of up to 20 ms duration (minimum requirements).
Standards	BS 4737, Part 3, Section 3.12, UL etc.
Uses	Across gates, door openings, window openings, or as flat terrain perimeter protection. Principally as low risk/value first line of defense.
Not to be used	In internal areas where accidental movement of curtains, stock, etc., might obscure beam or external areas where vegetation is above 4 in (100 mm) high.
Typical dimensions	1. Internal single beam: 3.5 × 3.1 × 1.2 in (90 × 80 × 30 mm). 2. External single beam: 8 × 4 × 6 in (200 × 100 × 150 mm). 3. External fence stacks: 14 × 14 × 39–197 in (350 × 350 × 1000–5000 mm).
Associated work	1. Fixings for mounting brackets. 2. Pole and foundation: or wall bracket holes. 3. 24 × 24 × 24 in (600 × 600 × 600 mm) mass concrete base with 4 × M10 rawlbolts, and cable sleeve, set into ground. 1. and 2. can be recessed by builder.
Advantage	Reliable.
Disadvantage	Flat terrain/line-of-sight use only.

completely encased within a protective enclosure. The source emits light having a wavelength within the range of 0.9–1 micrometers (9–10 × 10³ angstroms) at a carrier frequency of 500Hz and a frequency modulation of 20kHz. Individual transmitter and receiver current consumptions range from 20mA to 70mA. The beam of light is usually very concentrated, with a diameter of 1.2–2.8 in (30–70 mm) and 2.1° of divergence.

AIRSs can be divided into several different types of products:

- internal-use, single beam
- external, short-range, single beam
- external, short-range stacks
- external, long-range, single beam

Internal-Use, Single Beam.   Internal-use, single beam products usually have ranges on the order of 16–33 ft (5–10 m). They can be of the separate transmitter/receiver type (see Figure 3–13) or can use transceivers that use a single housing to transmit and receive the beam — the beam is reflected optically from a point opposite the transceiver. These devices are invariably mounted on a surface wall and have dimensions on the order of 3 × 4 × 5 in (75 × 100 × 130 mm). The transceiver type offers the advantage of single-point wiring to a single unit.

External, Short-Range, Single Beam.   An external, short-range, single beam AIRS commonly ranges from 98 to 492 ft (30 to 150 m). These devices incorporate various measures to ensure environmental protection and reduction of false alarms caused by birds, etc. They are usually pole mounted at 4–6 ft (1.2–1.8 m). This sensor is the cheapest external device and can obviously only offer very limited detection and protection capabilities (see Figure 3–14).

External, Short-Range Stacks.   External, short-range stacks are the most commonly used devices for *invisible fence* perimeter protection. Basically, they consist of several transmitters/receivers arranged in a vertical array within a weatherproof column. The result is a series of beams across an open space that form an invisible fence. The beams can either be parallel (for use where wildlife may be present) or diverging (where animals may not be a problem and economy is required). The "fence" provided can be up to 16 ft (5 m) in height (see Figure 3–15).

External, Long-Range, Single Beam.   External, long-range, single beam AIRSs are essentially high-power versions of the short-range, single beam AIRS and are used on industrial perimeters, airfields, etc. Beam lengths vary from 328 to 2625 ft (100 to 800 m). All have separate transmitters/receivers (see Figure 3–16).

FIGURE 3–13   Internal-use, short-range, single beam active infrared sensors

FIGURE 3–15   External, short-range stacks

FIGURE 3–14   External, short-range, single beam active infrared sensors

FIGURE 3–16   External, long-range, single or stackable beam active infrared sensors

Uses

Internal-Use, Single Beam.

window- or door-opening protection
particularly useful to guard against intrusion through skylights or across glass shop fronts
protection in storage racks/areas
very unobtrusive

External, Short-Range, Single Beam.

very cheap perimeter protection for use behind chain-link boundary fences to supplement boundary patrols or static guards
very useful at lifting or slewing gate entry points for out-of-hours protection

External, Short-Range Stacks.

for use behind boundary fence (*not* a wall) or where higher level of risk is present
very useful on medium to large industrial sites with static guards in guard houses
should be protected behind fence to prevent vandalism and nuisance alarms caused by children
can trigger CCTV to give very good perimeter protection

External, Long-Range, Single Beam.

perimeter protection of very large industrial sites or complexes where stationary and patrolling guards cannot hope to give adequate cover

All devices can be used to trigger other equipment (e.g., cameras for recording vehicles leaving or entering the site after hours, automatic gate-locking mechanism on leaving premises).

AIRSs are generally less secure and, hence, less suitable for high-risk and high-value applications, than microwave or ultrasonic sensors. They are, however, less prone to false alarms and, hence, more stable. They are cheaper than many alternatives.

Application

Figure 3–17 shows the typical AIRS beam patterns. When considering the adoption of a *stack* system AIRS, the following points should be noted:

The most common methods of defeating the system are climbing over the beam fence using ladders and streetlight repair trucks, tunneling or crawling under the beam using dips in the land, masking or attacking the device prior to intrusion and making it appear like an act of petty vandalism, and using the actual beam stack enclosure as a means to climb over the fence.

In order to reduce the risk of system penetration, certain features should be designed into the scheme:

- a fence to protect the device from vandalism and sufficient depth between the fence and the device to prevent attack from spray paints, stones, etc. Usually a small outer fence and inner fence with a 3 ft (0.9 m) gap are best.
- wall bracket–mounted devices where no fence is possible and perimeter security lighting to deter attack on the system components
- a low-level ground beam, for example 12 in (300 mm) above ground level
- an antitamper detector alarm on the stack
- an anticlimb device (e.g., spring-loaded switch) on the stack
- the outer casing of the stack should be independent of the beam sources, receivers, or their mounts and the trigger for the anticlimb device should only respond to vertical movements (up/down). If not, wind vibration can cause loss of beam aiming and will result in nuisance alarms.
- an antimask device

Design positions of stacks to give overlapping fields of cover so that one stack is in the protected zone of the next stack.

The range of the device should not be overstretched. The distances between the transmitter and the receiver, quoted by most manufacturers, are the maximum distances for good conditions. Ask whether this range is an underrun or true maximum range. If it is the former, you need not be conservative in your spacing of equipment. Infrared radiation is attenuated by thick fog, smoke, etc., so when deciding on the equipment to be used, be conservative as to its range capabilities.

To reduce false alarms on external systems always use equipment that can differentiate between moving objects by

- recognition of actual size (percentage loss of signal)
- speed of travel through beams
- dual sensor verification (or quadruple sensor)

External stack systems with parallel beams or programmable interruption times give reduced likelihood of repeated false alarms from wildlife, pets, livestock, etc. Systems that sound an alarm only when individual beam clockage is 75–90% for 230–295 ft (70–90 m) are more stable than those using the minimum 50% per 131 ft (40 m) BS 4737 loss values.

BS 4737, Section 4.1 recommends that multiple beams are spaced no wider than 20 in (500 mm) apart and that coincident treating of two beams may be required to prevent false alarms.

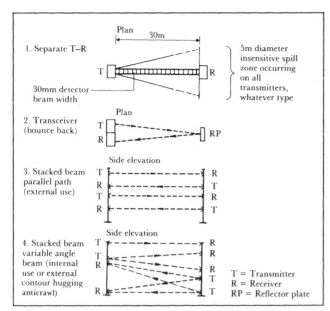

FIGURE 3–17 Typical AIRS beam patterns

Always use devices that include screening of the receiver to resist false alarms due to natural light and prevent system penetration by "blinding" from a remote distance with another infrared source shone from an angle.

Infrared beams are concentrated (usually 2 in, or 5.1 cm, in diameter) and they do not fill the space. Hence, great care should be taken to ensure they cannot be circumvented.

With stacked systems, the highest and lowest beams should always raise an alarm if they are broken. Only intermediate beams should operate on the multiple or successive break principle. This is a compromise between stability and the need to prevent intruders from crawling or climbing over the beam fence.

Being a line-of-sight device, it esssential that the beam from the transmitter always be in direct alignment with the receiver. Any drift of the beam will result in the receiver interpreting this as loss of signal caused by the presence of an intruder. For this reason, ensure that the beam aiming/alignment procedure carried out during commissioning is accurate and thorough, and that external beam sensors or stacks are mounted on substantial foundations that will not shift due to wind, ground action, etc. This is particularly important for clay soils (and even more so if there are also trees in the vicinity). Foundations are usually 2 × 2 × 2 ft (0.6 × 0.6 × 0.6 m), with long bars into the soil for clay or sandy soils.

Never allow branches of trees near beams; roots of trees near bases; vegetation, grass, flowers, or shrubs to grow under the lowest beam to a height greater than 4 in (100 mm); vehicles to be parked across the beam line; vehicles to be parked near the stacks, as

weight could cause beam drift or collision damage could result.

Chance of detection is greatly increased if stacks or single units can be recessed, concealed, or camouflaged into the natural surroundings.

Equipment Types and Selection

When an AIRS is used for *interior* detection, the capabilities offered by the device should include

ease of adjustment (bracket) and receiver test switch
robust casing; steel/polycarbonate and lexan lens cover
sensitivity adjustment; beam locking
antitamper/antimask devices; pulsed signal

For internal use, recessed, flush devices are far more secure than bracket-mounted sensors.

External AIRSs require far more extensive capabilities, such as

environmentally protected enclosure [built to ingress protection (IP) 55]
antivandal materials used in construction (polycarbonates)
corrosion resistant materials (polycarbonates, aluminum)
self-adjustment or alarm device to indicate transmitter or receiver covered with snow, ice, etc.
attenuation alarm indicator to indicate presence of fog. Thick fog can make the beam visible and can also eventually prevent the system from working.
integral transformer and battery backup system
antitamper, anticlimb, and antimask alarms
defogger, defroster, and heating elements
low beam dispersal over distance
ability to discriminate between falling leaves, wildlife, trees, etc., and intruders
infrared emitter (IRE) source modulation. These sources send a beam of IRE light in discrete pulse patterns that are difficult to reproduce with a blinding source and, hence, the device is more secure. The higher the frequency of modulation, the better (10–100kHz). In addition, modulation reduces the likelihood of unintentional interference from other infrared sources. (Also, look for *matched synchronization*.)
mirroring or angle adjustments to allow crisscrossing and beam direction changes to prevent potential intruders from becoming familiar with the beam pattern of any particular manufacturer

Wherever possible, avoid the use of transceiver devices that have exposed mirrors to reflect energy. Mirrors can go out of alignment, crack, become dirty, etc., all of which will reduce the overall sensitivity or increase the risk of false alarms. Stacked barriers, including enclosed mirrors, are not so troublesome, but the best devices do not resort to the use of mirrors.

Model Information.   Clients should ensure that they know the quality of the product they are considering by obtaining the following information. This is particularly important for external devices.

Enclosure.

- physical dimensions
- affixing details and whether the device is root or base mounted (requiring concrete base)
- materials of construction
- weight
- assembly details
- IP rating
- indoor or outdoor use

Application.

- number of beams
- beam width, range, spread, attenuation in fog
- typical applications
- typical items it would and would not recognize; at what speeds, directions, etc.
- Will AIRS go into an alarm condition if only one beam was broken? More than one? Successive beams?

Electrical.

- light source (LED, gallium arsenide, or gallium phosphide)
- response time in milliseconds
- alarm signal and time of reset
- contact rating volts AC, volts DC current
- supply voltages AC/DC
- current consumption — receiver/transmitter
- voltage regulation (±V, etc.)
- integral transformer–type, heat output
- integral battery type (e.g., nickel cadmium)
- backup rating (ampere-hours and actual hours of supply interruption)
- heater element rating in watts
- level of environmental protection for electrical equipment (IP rating)
- method of sensitivity checking and adjusting
- temperature operating range and humidity operating range
- light source wavelength and carrier frequency of transmission modulation frequency
- size and type of electrical termination

Limitation, Advantages, and Disadvantages of Stacking Systems
Limitation.

- low and medium security risks only

Advantages.

- takes up very little space
- relatively easy to set up and commission
- relatively trouble-free
- can incorporate CCTV and low-light CCTV (stack types)
- sensitivity adjustable
- low power consumption for long ranges
- invisible beam
- fail-safe
- safe to use
- very useful for giving protection to highly glazed curtain walls

Disadvantages.

- does not hug ground contours and, hence, either lots of posts or flat ground required (unlike some microwave systems or electromagnetic cables)
- requires direct, line-of-sight applications
- requires conventional fence protection
- can be affected by thick fog, although with careful equipment selection, this problem can be reduced, but not removed
- difficult to disguise the function of or to hide the device
- can be masked by an intruder
- can be tunneled under or climbed over unless you use stacks of considerable size
- can be blinded by other infrared sources (unless shielded and using a pulsed beam)
- stack type physically larger than microwave, fence protection devices
- requires 14-day probation period before police will accept calls

EXTERNAL MICROWAVE SENSOR (EMS)

Introduction

Table 3–7 summarizes the types, uses, typical dimensions, and advantages/disadvantages of the external microwave sensor (EMS).

Principles of Operation

The EMS is identical in basic microwave generation to the internal version, but various modifications in design make it more suitable for its external role. The major differences are

- beam interruption (line of site) or phase shift alternatives
- physical size

Table 3–7 External Microwave Sensor

Types	Long-range narrow beam or "fence" with either greater depth than height or vice versa.
Activated by	See Chapter 4, "Microwave Sensor."
Standards	BS 4737, Part 3, Sections 3.4 and 3.12.
Uses	Perimeter detection by creation of "beam fence."
Not to be used	Areas where moving objects such as swaying trees, cars, etc., could cause a continuous sensitivity/false alarm problem.
Typical dimensions	Beam type: 21.7 in (550 mm) high × 12 in (300 mm) wide × 10 in (250 mm) deep on pole 4–10 ft (1.2–3 m) high. Horizontal type: 20 in (500 mm) wide × 10 in (250 mm) high × 16 in (400 mm) deep on mounting 4–10 ft (1.2–3 m) high. Vertical type: 4 ft (1.2 m) high × 12 in (300 mm) wide × 12 in (300 mm) deep.
Associated work	Creation of *cordon sanitaire* 6.5–10 ft (2–3 m) high × 10 ft (3 m) wide, free from all moving objects.
Budget costs	Dependent on application.
Advantage	Not as affected by environmental conditions as all other external beam or fence-mounted devices.
Disadvantage	Very sensitive: requires careful commissioning to reduce likelihood of false alarms.

- beam shaping
- enclosure

The sensors differ in the principles on which they detect intruders. Internal microwave sensors invariably sense intruders by characteristic frequency or Doppler shift. A transmitter transmits a very high frequency (10.7GHz) electromagnetic wave toward the boundaries of the room. The energy is reflected from the surfaces and returns to a receiver at the same frequency. If an intruder enters, he is opaque to the energy and reflects that energy, causing a change to the normal pattern. If the intruder moves even slightly, he will compress the waves of energy traveling toward him and will change their frequency, which in turn changes the net frequency at the receiver. This frequency change is known as a *Doppler shift* and relies upon reflection from an opaque surface.

External sensors will also operate on this basis, but because there is generally very little effective "boundary" that would, itself, not move (trees, shrubs, etc.) and cause false alarms, the Doppler shift principle is seldom used. Signal processing would need to be extremely complex in order to filter out noise. In addition, in order to recognize a person, the sensor would need to create a field high and wide enough to ensure that the intruder was in it long enough to give a reliable characteristic signal. As most external sensors operate in situations where space is at a premium, this is often not possible. External microwave sensors that operate based on true Doppler shift and have a wide coverage are, thus, very expensive.

Most external sensors operate on a slightly less effective principle — *beam interruption*. The transmitter radiates one or more beams of microwave energy to a transmitter in a direct line of sight. Each beam is quite narrow and particles of energy within it travel at the speed of light. The receiver compares the level of energy and the waveform of the transmitted energy to that energy leaving the transmitter. If a significant difference in waveform, magnitude, or amplitude of energy is present for a specified duration and happens more than a set number of times within a given period, an alarm will be triggered. Hence, in principle, the external microwave sensor is very like an ultrafast, ultrasensitive, external, active infrared beam device. Some devices on the market operate on contained Doppler shift/beam interruption principles in order to provide a useful compromise.

Uses

As EMSs are commonly used as perimeter detection devices, the beam is shaped to present a fence of microwaves through which an intruder cannot pass without detection. In this role, EMSs compete directly with AIRSs. The principal difference between AIRSs and EMSs is the depth of the fence provided. AIRSs produce beams of approximately 2–4 in (50–100 mm) in diameter over normally used ranges and a fence is erected by standing one beam above another or by crisscrossing them. With EMSs the shape of the fence will be directly related to the type of the transmitter used, but the fence is often in excess of 6.5 ft (2 m) deep.

Application

The following guidelines apply to external microwave sensors:

EMSs are generally more sensitive than AIRSs, and require much more careful planning, positioning, and commissioning if false alarms are not to be frequent (i.e., two or more per year).

Microwaves are prone to spread and drift. Hence, in situations where the devices are very close to public paths, roads, etc., false alarms could result due to the device seeing cars and pedestrians as intruders, unless the correct type is selected.

Devices with "range cutoff" help prevent false alarms. Multipath, phase lock looping also helps by distinguishing between full or partial blockage and blinding attempts.

As with AIRSs, EMSs should always be protected by a substantial fence and be out of range of vandal attack.

Devices in which the transmitter and receiver are in totally separate enclosures (either remote or separated by 6.5 to 9.8 ft, or 2 to 3 m, within the same case) give much more reliable performance than small, unitary transceivers.

Devices that transmit in the X band with wavelengths of 12 in (300 mm) and frequencies of 10.5GHz are prone to mains-induced ripple interference and to false alarms caused by rainfall. Some manufacturers offer devices that operate in the L band, with wavelengths of 4.7 in (120 mm) and these are inherently more reliable than the X band types. Devices using "Q" Bands with wavelengths of 0.3 in (8 mm) are now becoming available. These will be useful for sites with limited space on the perimeter.

Installation Economics and Power Supply. Microwave detector systems operate at various voltages, from 10–60V DC. Although they consume relatively little power (0.2–0.5A), the distances of cabling required can result in large cables being required if excessive voltage drop is to be avoided. It, therefore, usually makes better economic sense to specify devices that have integral transformer/rectifier sets and to support the units at 240V or 110V AC. Being supplied from the building underground, all devices can be vulnerable to power failure caused by electrical fault or severing of mains cables (intentional or otherwise). For this reason, all devices should incorporate a mains failure battery and trickle charge unit of at least 8 hours standby capacity. For very large, remote installations, some manufacturers offer solar-powered and charged systems, with microwave signal transmissions.

Sensitivity. It is important to remember that external microwave detectors *can* be affected by snow, falling leaves, birds, dogs, etc., and these are all potential sources of false alarms (although these are far less of a problem than with AIRSs).

Any device that will allow easy recommissioning and resetting of sensitivity after an initial trial period will be better than a system that cannot. A competent security company engineer should be able to carry out this adjustment without reverting to replacement or invalidation of warranties. The device should be accompanied by charts that illustrate graphically and/or pictorially, the physical result of dropping or raising the sensitivity, so

that security managers or those responsible for the premises can readily understand the consequences of these actions.

Devices operating on Doppler shift techniques should always be site tested to determine the minimum single movement speed required to activate the system. Local ground conditions will change this from site to site.

Most vertical pattern detection devices incorporate only low- or high-sensitivity switching. Some better devices include up to five levels. The greater the number of settings provided, the finer the tuning available. Most horizontal pattern devices offer a maximum of ten sensitivity levels.

Equipment Types and Selection

There are three basic types of equipment — the beam type (cylindrical), the vertical type, and the horizontal type.

Beam Type. Beam types produce a field of energy that is cigar-shaped, the cigar being quite fat in the middle and pointed at each end. The exact shape is dependent on whether the transmitter was transmitting into and reflecting from a parabolic reflector (see Figure 3–18) or whether it was transmitting through a wave guide or horn fitted to the front of the transmitter.

The former device produces a much sharper cigar shape than the latter, but the width of the field at the central part, between transmitter and receiver, will depend on the range of the device. Both of these devices are distinguished from those that follow by the fact that they are physically more compact devices and their pattern of energy is symmetrical about a center line from the transmitter to the receiver in both the horizontal and vertical planes.

The basic beam subtypes are

• short range
• long range

Short Range. Short-range beams are commonly used to protect gates in fences and entrances to buildings, and have ranges from 66.6 to 164 ft (20 to 50 m). Most consist of a small transmitter dome that is mounted on a pole 3–6 ft (1–2 m) high and a similar receiving dome placed in direct line of sight with the transmitter. The pattern of

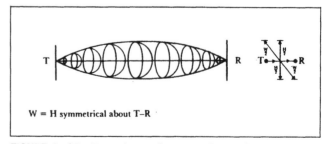

W = H symmetrical about T–R

FIGURE 3–18 Beam-type microwave transmitter

coverage is usually cylindrical and the unprotected zone beneath the transmitter is on the order of 20–40 in (0.5–1 m). The beam diameter is usually under 3 ft (1 m). Several cheaper alternatives, specifically for gate coverage, have the transmitter and receiver in the same housing (transceiver), but most are limited to ranges of under 65.6 ft (20 m).

Long Range. Long-range beam devices use separate transmitter/receiver stations and ranges vary from 197 to 984 ft (60 to 300 m). The increasing range is usually at the expense of beam width or diameter and this can reduce the level of security provided. With areas requiring long ranges — airfields, armed forces bases, warehouse complexes, etc. — the temptation is to use a long-range device for economy. This is dangerous. Often, the terrain is not flat, the narrow beam will not hug the ground, and, as remote areas might be infrequently patrolled, there is greater chance of intruder evasion. It would be far better to use several mid-range, 492–656 ft (150–200 m), devices.

Long-range devices really come into their own where boundaries are extensive and perimeter space between the fence and the building is restricted. In this application long-range devices can provide simultaneous coverage of fence and building openings, doors, windows, etc., without spilling onto nearby pavements, roads, etc., which might give rise to false alarms (see Figure 3–19).

Vertical (or Stripline) Type. The vertical type uses a very long microwave transducer that is mounted vertically and directs its energy into a parabolic trough. The result of this configuration is to produce a field of energy that is much broader than it is high (see Figure 3–20).

As the device transmits close to the ground, it provides almost total crawl protection. These are by far the most common devices marketed. They usually consist of a

FIGURE 3–19 External, long-range beam microwave transceiver

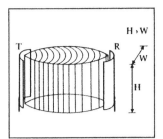

FIGURE 3–20 Vertical (or stripline) microwave transmitter

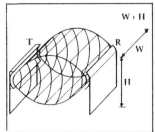

FIGURE 3–21 Horizontal (or slotline) microwave transmitter

triangular-sided box with dimensions of 12.5 × 8.6 × 8.6 in (320 × 220 × 220 mm), with a height of 21.6 in (550 mm), mounted on a pole. All have separate transmitter/receiver stations and variable range settings of 3–656 ft (10–200 m) maximum. All have greater sensitivity to movement in the horizontal beam plane than the vertical, and, hence, will be more sensitive to objects passing through the beam (e.g., persons, animals) than to objects falling through the beam (e.g., leaves, fruit, branches). Most of these vertical beam devices claim to filter out unwanted signals from animals, but this is usually at the expense of sensitivity. Most claim to distinguish a person from a bird, bat, or rabbit, but dogs and cats are not usually mentioned in the literature. A man on all fours can look very like an Alsatian dog to a microwave detector! This problem is usually overcome by mass, speed, and percent break of beam processing.

Horizontal (or Slotline) Type. The horizontal beam device is basically the vertical type mounted on a pole and laid on its side to give a field that is much higher than it is broad (see Figure 3–21).

In addition to the transmitter types, devices are divided into those with a single-head transmitter or receiver only, or those with a transceiver (i.e., containing a transmitter and receiver in the same assembly).

From the preceding transmitter illustrations, it can be seen that the potential Achilles' heel of the external microwave detector is the zone immediately beneath the transmitter. Here there is very little, if any, transmitted energy and a person crawling under the assembly might not be detected. To improve the likelihood of detection, it is common practice to

- overlap the fields of single-head devices by approximately 26 ft (8 m) (see Figure 3–22)
- use transmitters/receivers to cover the dead zone, giving an effect rather like a stacked, active infrared beam (see Figure 3–23)
- use transceivers and reflectors as an alternative to overlapping the fields of single-head devices (see Figures 3–24 and 3–25)

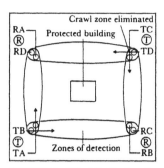

FIGURE 3–22 Single-head transmitter/receiver zone layout with overlap to eliminate crawling

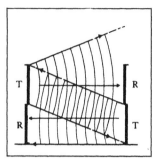

FIGURE 3–23 Phase and frequency separate transmitter/receiver stacks. No lateral separation of heads.

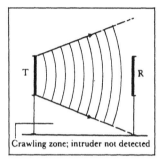

FIGURE 3–24 Normal, single-head device. No overlapping of field.

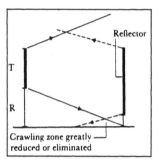

FIGURE 3–25 Transceiver/reflector

Beam-type transmitters are usually used when either the minimum field scatter or maximum range is required. Although it does not have a great linear range, the vertical transmitter offers the following advantages over the beam types:

> superior ground cover near transmitter with far less likelihood of crawling intrusion in any configuration
>
> a fence of much greater height, making scaling more difficult
>
> restricted field width allows use in schemes when perimeter boundary space between building and boundary, boundary and road, etc., is restricted
>
> reduced risk of false alarms caused by trees, wildlife, etc., near transmitter. The narrower field width near the beam-type transmitter means that even quite small birds can effectively block a considerable amount of the transmitted energy and, hence, appear equal in value to a human intruder.

The *horizontal* transmitter offers the following advantages over the others:

> greater field width gives greater depth of cover. As many devices detect based on a combination of blocked and Doppler shift energy, the greater the depth of cover, the greater the detection reliability

and discrimination of human from animal intruder. The device has time to check size and speed, net forward or net backward movement, etc.

> less affected by vegetation movement than either beam- or vertical-type devices
>
> superior beam-shaping capability allows usage in sensitive and confined perimeter zones where overspill could give serious false alarm problems. A 6 ft (2 m) clearance from an exterior fence will normally suffice to prevent passing traffic false alarms.
>
> sharp, face-down angles mean that the field of generated energy is slightly asymmetrical in the horizontal plane, giving a very small blind spot where crawling under the beam is a risk
>
> tend to incorporate a close field–standing wave trigger within the processing unit. This *standing wave* is a waveform that is constantly reflected from the ground near the unit. Any movement within this area (whether or not it breaks the wave) causes the standing wave to collapse and an alarm to be raised.

Specialized horizontal devices exist that allow minor undulation in the ground and still provide cover, usually up to 5 ft (1.5 m) in depth if gradually sloping.

Its obvious drawback is the free space width required and its overall size. In many urban applications, the space is not available and its size and radar-like appearance make it a target for vandals.

Where space is not a problem, these horizontal devices offer the best level of security of the three types. They are usually large—typically 18 × 16 × 12 in (450 × 400 × 300 mm) on a 3–6.5 ft (1–2 m) pole. The coverage patterns can vary, but most give a very high, wide fence—typically 13 ft (4 m) high × 10 ft (3 m) wide, with a range of 197–492 ft (60–150 m). This depth of fence allows devices to evaluate speed, size, and net directional travel, thereby reducing alarms caused by gale-blown trees, wild animals, and even dogs and cats. This all, of course, increases the cost of the unit, compared to beam or vertical detectors.

Another form of horizontal detector that can prove very useful on rough, undulating ground, is a chain-link microwave fence. This consists of a single transmitter at one end of the chain and a number of intermediate antennae (links), linked together with a cable, which continues around the boundary to the receiver. The transmitter transmits the signal across the final exit gate in the fence to the receiver, which relays the received signal around the antennae. If any link is broken, the alarm is activated. The antennae are affixed to the boundary fence and are separated by distances of 6.5–16.4 ft (2–5 m). Each antenna provides an ovoid volumetric detection pattern that follows the contours of the ground over its 16.4 ft (5 m) range. As the antennae are cheaper than transmitter/receiver stations, the whole installation can prove much more economical than conventional systems for high security on rough ground (particularly on double-fence applications when the device is mounted on the inner fence).

No matter which of the preceding sensor types is chosen, any device that incorporates at least some Doppler shift recognition will be more secure and free from false alarms than a device that is purely line-of-sight/loss-of-beam activated.

Microwave versus Infrared Perimeter Protection

On many installations, the actual choice of external perimeter sensor will be made between microwave and infrared sensors. This choice depends on many factors, not the least of which are the intended area of use, the competence of the user, and the level of skilled on-site personnel to carry out work associated with the grounds around the device. The final decision should always be left to the expert. However, consideration of past applications shows that

micetrowaves are not suited to dense urban site use; infrared beams are

the microwave sensor is not as greatly affected by the environment (e.g., snow, rain, fog) as the infrared device—it is an all-weather device

the infrared stack is usually better at detecting crawling intruders

the range of the infrared device is less than that of the microwave device

within the security industry, microwave fences are usually regarded as medium- to high-security devices, whereas AIRSs are seen as low- to medium-grade security devices

Limitations, Advantages, and Disadvantages

Limitations.

expense
care needed to prevent excessive beam spread

Advantages.

long range possible for low energy use and capital cost
good anticrawl properties with some manufacturers' equipment
depth and height of provided cover can prevent climbing or crawling, provide advance warning of persons *about to climb* a perimeter fence, and provide detection around and within temporary buildings due to the penetrating property of microwaves (see Chapter 4, "Microwave Sensor")
not affected by air movement, temperature changes, fog, or rain

devices can stand alone (see Chapter 4, "Microwave Sensor")
difficult to evade, mask, blind, or confuse
higher level of sensitivity than possible with AIRSs or fence-mounted vibration detectors
longer wavelength and greater penetration property than AIRSs

Disadvantages.

less stable than AIRSs
susceptible to false alarms caused by passing traffic, etc., and, hence, requires careful planning and aiming
less robust and more intrusive than AIRSs
still affected by drifting snow, moving trees, long grass, etc., just like AIRSs
requires 14-day probation period before police will accept calls
line-of-sight device; undulating ground cover difficult to protect
straight line fences or perimeters required without too many changes in direction for economy (to avoid multiple transmitters)
affected by reflections from any nearby metallic structures or water, particularly if these move periodically and intermittently
an outer, protective, 10 ft (3 m) fence is normally essential, with an inner fence of at least 3 ft (0.9 m) to demarcate the microwave field width and prevent accidental blocking

Remember, unless you have very high guarding levels and fast response times, any intruder could run through a microwave, ultrasonic, or AIRS system unhindered, achieve her aims, and depart! The physical fence *and* the beam detection give *you* the element of time, not the intruder.

TAUT WIRE SWITCH

Introduction

Table 3–8 summarizes the types, uses, typical dimensions, and advantages/disadvantages of the taut wire switch.

Principles of Operation

A cylindrical pendulum switch is fixed to every horizontal strand of wire on a perimeter fence. The switch consists of a central pendulum contact and an outer cylinder contact. Normally, the switch is in the open position and no current flows. The switch will close if the pendulum is moved sufficiently in one direction due to the move-

Table 3–8   Taut Wire Switch

Types	Pendulum (associated types: magnetic reed, weak link).
Typical range	Only limited in zone length by natural expansion/contraction.
Activated by	Displacement of suspended wire held in tension by more than a set distance (usually 2 in, or 50 mm, or less).
Uses	Low- to medium-risk perimeter fence protection in urban areas where space limited.
Not to be used	For high-risk areas.
Typical dimensions	Wire strand and terminators, various.
Associated work	Best used on purpose-matched fence (not retrofit) or wall brackets.
Budget costs	Labor costs predominate.
Advantages	Simple, stable.
Disadvantage	Easy to circumvent.

ment of the strand of wire to which it is affixed. To prevent an intruder from gently prying the strands of wire apart, the wires are held taut by end post terminators and carried on roller supports on other intermediate posts. Any natural creep or temperature-induced expansion or contraction is taken up in the terminators.

There are several common variations of the pendulum switch principle used on taut wire perimeter fences. Two of the most common are the floating magnetic reed switch and the weak link switch. The former is a type of magnetic reed switch (see Chapter 4, "Magnetic Reed Switch"). The switch is comprised of a magnet, which is physically attached to the fence strand under tension, and two reed contact switches of the normally open variety. If the fence is disturbed, the magnet moves from a neutral position toward a set of reeds. They adopt different polarities, are attracted to one another, close, and raise the alarm.

The weak link switch has an enclosed *weak link* in each fence strand. The strands are run on insulating bushes on outriggers and carry a very low, monitored current. The weak link consists of two extremely thin foil contacts with a central pin- or solder-type connection. In normal conditions, the weak link is just strong enough to cope with the tension imposed on the wires (e.g., wind). If the tension is increased or decreased, the pin or solder mechanism holding the foil contacts together shears or the contact itself shears. The monitored current then ceases to flow and an alarm is raised.

Although the floating magnetic reed and weak link switches are useful alternatives to the pendulum switch, they are not quite as stable or reliable—the floating magnetic reed switch having the edge over the weak link device.

Uses and Application

Pendulum principle taut wire switches are used primarily on fences surrounding low- to medium-security-risk buildings. They are commonly applied in areas where an intruder might want to remove large amounts of sizable objects from a site and to do this would have to cut or climb a fence. Hence, taut wire switches are suitable for storage depots, parking lots, marshaling yards, etc. In this application, their lack of sensitivity over more advanced sensors, such as fence-mounted geophones, is no great disadvantage. It *can be an advantage,* as these types of sites are often in inner city areas where vandalism and petty theft are the main risks and potential intruders would not possess the expertise to defeat the more expensive systems.

When installing these devices, it is essential that every area of the fence be protected. Only direct contact with a particular area will raise an alarm; an area of 16 × 16 in (400 × 400 mm) is sufficient for a person to gain entry undetected. For this reason, it is recommended that

- up to 6 ft (1.8 m) from the ground, horizontal wires and switches are spaced at a maximum of 6 in (150 mm), with the first strand no higher than 4 in (100 mm) above the ground
- above 6 ft (1.8 m), the fence should either be crank-topped and each strand on the crank arm protected or, if the fence continues higher than 6 ft (1.8 m), the maximum spacing should be increased to a maximum of 9.8 in (250 mm)
- the strand length between terminators be kept as short as economically sensible
- the fence strand deflection required to raise an alarm be repeatedly tested on every strand during commissioning and checked against the manufacturer's published, recommended deflections

Equipment Types and Selection

Taut wire devices have been used for many years in the industry and still have a very useful role. Although there are minor differences between manufacturers' equipment, true taut wire devices all act on the same principle. Good systems include

- totally encapsulated, hermetically sealed switches
- switches that are affixed to the fence using stabilizing or balancing tags (i.e., short lengths of plastic material affixed to each side of the switch) to allow them to equalize the position of the central pendulum contact, as the fence tension changes very gradually due to the temperature expansion and contraction not seen by the end terminators

noncorroding materials, rollers, terminators, etc.
sensitive and fully secure, adjustable, fence-tensioning devices

Limitation, Advantages, and Disadvantages

Limitation.

does not prevent tunneling or bridging of fence

Advantages.

only raises an alarm if the wire is significantly displaced, cut, pulled, etc.
reliable
stable; although less sensitive than many other fence-mounted devices. Most other fence sensors will give false alarms in high winds, when objects hit them, when snow accumulates near them, etc.
fewer are required to cover a specific area of fence

Disadvantages.

cannot be easily refitted
not very sensitive; hence, used for low- to medium-security applications only
very susceptible to damage by vandals and expensive to repair

Separate Taut Fence Systems

Taut fence systems use separate straining posts that are set back behind the main fence by a distance of 8–12 in (20–30 cm). Each zone of fence is approximately 164 ft (50 m) long by 12 ft (3.6 m) high, with 24 strands that are 6 in (15 cm) apart, with four vertical zones. Each 16 ft (50 m) zone of complete fence reports signal variations, on strand disturbance, into a *concentrator,* which has a unique multiplex address on a loop of wired concentrators. Signals are sent down the loop to an analyzing unit that actively interrogates each concentrator, in sequence, on the loop. Full interrogation of all concentrators takes less than 1 second. These systems are superior to fence-mounted systems. Costs should always include posts and trenching costs.

LEAKY COAXIAL CABLES

Principles of Operation

Radio waves have different transmission and penetration properties through different materials, depending on their frequency or wavelength. Figure 4–27 in Chapter 4

shows that VHF radio signals have longer wavelengths than microwave energy and less penetration power. This means that the quality of transmission is excellent, but the range for a given power is quite short, as attenuation through different media is quite high. Leaky coaxial cable systems use VHF radio signals and these attenuation losses to a great beneficial effect.

Two cables are buried underground and separated by a short distance—one is a transmitter, the other is a receiver. Radio waves pass through both the ground under and the air above the cables. The reception strength and characteristic signature is monitored. A person moving across the path of either the transmitter cable or receiver cable interrupts (or attenuates) the flow of radio energy to the receiver, causing interruption, attenuation, or phase shift.

The success of the system is based on the fact that humans are primarily composed of salt water. The amount of salt water in a person varies very little and is only slightly proportional to the physical size of that person (usually 0.3 cords, or 1.2 m^3). Salt water is also a *constant, dielectric* material—that is, its attenuation and reflective property to radio or other electromagnetic energy does not vary with signal strength, time, age, sex, etc. Because of these two combined constants, the system can readily detect humans and discriminate people from all but the largest similar mammals (bears might be a problem?!). In effect, the saltwater content of a human gives a unique radio signature.

System Components

Radiating Cables. These cables are usually equal-length pairs (transmitter and receiver). They are leaky coaxial cables made up of a solid, copper core; insulating dielectric materials surround the core, and an outer, braided, copper coaxial core (see Figure 3–26). The coax-

FIGURE 3–26 Ground-buried, leaky coaxial radiating cable

ial braid has regular, triangular holes formed within it (with centers of graduated size or cross-sectional area) along the length of the cable. The holes in the braid allow the radio signal of the cable core to escape the cable in a predefined field pattern both below and above the ground. The graduation of the hole size ensures the field width is constant, despite increasing the length of the cable from the transmitter power source. Simpler systems do not have this graduation.

Transceiver Modules.   Each transceiver module is connected to two pairs of cables that radiate out from the module to form two zones. The transceiver is the zone sensor; it provides the power to the transmitting cables and receives the signals from their matched receiver cables. The transceiver constantly switches or polls each pair of cables, searching for variations above a normal quiescent condition plus acceptable variance. Transceivers are normal, linked together via their sensor cables in a ring fashion with spurs for complete ground cover.

Control Module.   The control module provides power to the receivers, collects the data and signals from the transceiver, acts as an annunciation link, and differentiates signals from different zones and allows these to be displayed graphically or detailed by a printer.

Equipment Types

There are two dominant manufacturers of these cable systems—one American and one Canadian. These cables are marketed throughout the world under a variety of brand labels.

Systems differ in their processing (i.e., pulsed wave digital or carrier wave analogue), levels of coincident activity before an alarm is raised, and whether they are package or extendable systems.

Pulsed wave systems transmit VHF as a burst of encrypted pulses of energy toward the receiver cable. The attenuation caused by the intruder and the timing of the duration between sending and receiving a pulse, allow the high-speed digital processor to determine the intruder's position along the zone length of cable within a few yards (meters) in a zone of up to 492 ft (150 m), which is a very useful feature at night. This ability is commonly called *range gating*. Pulsed wave systems effectively use distributed intelligence or signal processing at the transceivers. These systems are extendable, but are far more costly than carrier wave systems due to the processing complexity, graduated cable holes, etc. Unfortunately, the wide bandwidth used by these systems can cause mutual interference with other radio devices and other zones unless precautions are taken.

Carrier wave systems require the distribution of a signal and data cable around the perimeter to each transceiver, which passes data to an analogue signal processor. Some smaller, packaged systems have combined central transceivers and a signal processor, and two radiating zones only.

Carrier wave systems use ungraduated-holed coaxial cable, simple analogue processing, and a narrower frequency band than pulse wave carrier systems. This means these systems are significantly cheaper, are less secure, do not pinpoint the location of an intrusion, and can be used more easily when other radio sources are present.

Both systems use VHF frequencies of 20–120MHz. The choice of frequency used on any particular cable type is usually set by the manufacturers, but is sometimes site adjustable. The ability to adjust the frequency is important, because radio waves at these frequencies are susceptible to variations in ground conditions (e.g., water content), ground absorption, ground-air interface attenuations, etc. If the frequency is too high and wavelengths too short, poor intruder detection fields will be provided. False alarm rates are low, but discrimination of humans from animals, birds, etc., is poor. As the frequency decreases, discrimination improves and attenuation and field width decrease. Most systems operate with frequencies of either 40MHz, 60MHz, or 66MHz. In general, for most applications, the 40MHz units are better, but more expensive.

Selection

With only two major production companies and two different types of cables from which to choose, selection is not a problem. See sales representatives and view demonstrations of each type of cable. Systems should include

- encrypted radio pulsing
- automatic gain control
- self-adjustment capability
- self-testing capability
- compound signature detection (CSD)

CSD makes the sensor even more reliable and stable as it means that the signal processor and transceiver will only raise an alarm condition when a certain level of coincidence occurs between

amplitude change—related to the intruder's mass and proximity to the radiated field

phase change—a blockage, in real-time, of the signal, causing a shift in the pulse receipt rate. It only identifies pulse rates typical of intruders crawling, running, jumping, etc., and no larger animals crossing the detection area.

time of sample—the preceding conditions must be coincident for a certain period of time

Some systems use bidirectional power and data transmission between linked transceivers, and radio frequency decouplers. This configuration is very useful, as if the cables are cut at any point, the system still functions on two halves. It also means that faults do not disable the system.

Uses

These systems are used for any situation when high-value goods or protection life outweigh cost. They are frequently used around military bases and versions that comply with military standards and field trials exist today. They are excellent systems, particularly on sites with a fence, where they are used to trigger other systems (e.g., security lighting and CCTV camera installations). They are particularly useful with covert-low and low-starlight CCTVs, when a *zone* quite often matches a field of view well and an intruder can be detected, observed, and detained often before recognizing any threat.

Similar systems described in "Guided Radar Systems" later in this chapter can be used for aboveground, temporary protection or for situations when burial is not possible.

The choice of these cables is often the result of a comparison with buried fluid pressure sensors, and electret sensors discussed in this chapter, buried geophones or piezoelectric cables.

Application

Cables are normally buried 4–24 in (16–61 cm) below the ground and 6–7 ft (1.8–2 m) apart. The field of sensitivity may be increased by bringing cables closer together. The width of the zone of detection is normally 3–4 ft (0.9–1.2 m) on either side of each cable, giving a total width of up to 15 ft (4.6 m). The height of the detection field is dependent on burial depth, soil condition, and transmitting strength, but normally is about 3–4 ft (1.8–2 m).

The zone of detection should be a *cordon sanitaire*, if possible, but grassed over to prevent any detection of buried cable.

Most literature avoids mentioning the need to provide a uniform trench material and, if possible, material across the whole zone of detection. While the system will work satisfactorily without this, its performance is greatly enhanced by having a graded uniform soil and sand backfill. It is also beneficial if large rocks, etc., are removed.

The cables are not armored and any damage to the outer sheath leads to real problems with water penetration, damage to the coaxial parts, etc. Where possible, cables should be run in and protected by telephone-grade, PVC underground conduits or ducts laid in the trench, chemically fused at joints, and air pressure–tested to prove watertightness.

When conduits are not affordable, a pea gravel (not sharp gravel) base to a trench laid to falls with intermediate sumps or soakaways, will prevent too much cable damage from sharp stones and help to take running and standing water away from the cable.

Cables should not be used in areas of high water table, salty subsoils, etc.

Cables should be kept a good distance (normally 3 ft, or 0.9 m) (seek the manufacturer's advice) away from cables buried in the ground, buried in metal objects, etc. Underground objects can cut the radiated underground (and aboveground) radio field, causing local distortions that vary with time. Chain-link fences with direct, rooted, metal fence posts and fence material sunk into the ground are commonly forgotten and must be at least 10 ft (3 m) away. Separation by a depth of 3 ft (0.9 m) can cause real problems, as many cables and most pipes are buried 2–3 ft (0.6–0.9 m) below the surface.

The most secure system is normally one that has the system in front of or behind a 9.8 ft (3 m) security fence. The former provides prepenetration detection as intruders attempt to defeat the fence, but the system can be defeated if ladders are used. The latter protects the cable and usually prevents bridging or jumping of the zone of detection.

All coaxial cables suffer from bend loss attenuation. Cables must not be bent any tighter than stipulated in the manufacturer's guidelines.

Lightning surge protection and ground current filtration are both required on these systems.

Limitations, Advantages, and Disadvantages

Limitations.

cost
needs manned site
should not be selected for areas that have a high water table, heavy snow melt, frozen ground, etc.

Advantages.

unaffected by most environmental conditions and weather
not visible
low maintenance
very reliable
very sensitive
low false alarm rate
can be used under concrete, pavement, etc. (with reduced sensitivity)
variable zone length
follows terrain

will detect most small-scale tunneling, jumping, crawling, and bridging attempts

will cover perimeter up to 2 miles quite easily

can pinpoint intruder entry

not affected by seismic activity, ground disturbance,
or vibration of passing traffic

Disadvantages.

must be clear which type of system used (e.g., analogue
or digital)

carrier wave H-field or similar system, while cheaper,
must still be used with fences and not as good at
recognizing and discriminating between humans
and animals

systems using carrier wave, nondistributed intelligence
tranceivers need separate trenches for signal, power,
and sensor cables

Guided Radar Systems

Guided radar systems are an aboveground (surface) version of the leaky coaxial cable system. Essentially, these
systems use radio and radar frequency bandwidths
similar to the coaxial cable, but produce a cylinder of
protection of 6–8 ft (1.8–2.4 m) in diameter. The system
is specifically designed for rapid deployment around
sensitive equipment, military camps, outdoor exhibition
areas, aircraft, parking lots, etc.

The system's cables are usually supplied in prepacked
dispenser reels of 300 ft (91 m). Each cable is affixed to
support stakes driven into the ground around the area to
be protected, approximately 40–50 in (102–127 cm)
between cables. An end-of-line analysis and alarm-
sounding unit in a weatherproof, tamper-resistant case
with primary or rechargeable batteries completes the
self-powered, self-contained package. Each analyzer unit
can normally support two 300 ft (91 m) zone lengths.

Guided radar systems are relatively new, but the
technology is well understood. They are reliable, but only
of real use on sites where rapid guard response is always
available. These systems are expensive but very, very flexible; adaptable; and can be used and transported very
easily.

INDUSTRIAL LASER BEAMS

Principles of Operation

Industrial laser beams operate on the same principle as
AIRSs—a number of line-of-sight beams of light are
transmitted to a receiver. An interruption of the beam
triggers an alarm.

Equipment Types and Selection

These systems are often offered as an alternative to AIRS
systems or other beam detectors for outdoor use. They
are, however, in a stage of development that makes them
more suited to interior use, as environmental effects,
such as mist and fog, can seriously attenuate and disperse
the beam if it is not of a high intensity. Hence, this system
is very expensive. Great care should be taken to determine actual performance as many other system alternatives are currently available.

EXTENDED PIEZOELECTRIC CABLE SENSORS

Principles of Operation

Chapter 4's section entitled "Piezoelectric Sensors" explains that piezoelectric materials exhibit the property of
producing a voltage that is directly and linearly proportional to the amount of mechanical stress, strain, or compression to which they are subjected. With static
piezoelectric sensors, a rigidly encapsulated piezo
material is required. This limits its use to "point" detection as metal containers are needed to surround the
material. Chemists have taken the basic piezoelectric
compound and created a tough polymer (normally a
variant of barium titanate or polyvinylidene fluoride)
that allows the material to be extruded and molded into
a continuous cable core.

Each type of piezoelectric material has a particular
ability to produce a given voltage for a given, subjected
force. This is called the *piezoelectric coefficient.* For a given
cable type or manufacturer, this coefficient may vary
considerably. Buried cables are typically 10pC/N (pico-
Columb per Newton); wall- and fence-mounted cables
are 0.25pC/N at the same temperature. Fortunately,
temperature in the range $-4°F$ to $140°F$ ($-20°C$ to
$60°C$) only varies the coefficient by $\pm10\%$ around the
mean. Therefore, the coefficient may be considered a
mark of sensitivity to intrusion—the higher the figure,
the better the sensitivity.

The linear gradient of the voltage output of the sensor
will be different for strain in the longitudinal direction,
strain in the cross-sectional direction, and direct compensation (or hydrostatic pressure). All of these gradients are normally reflected in a particular intrusion attempt in lesser or greater proportions. The system's
analyzer unit will be designed to recognize these strains,
put limits on the normal production due to hailstones,
minor ground vibration, etc., and trigger an alarm under
certain major variations of one element or under smaller
coincident variations in all three types of pressure.

There are three basic versions of the 0.1 in (3 mm)
diameter cable—a sensitive buried cable, a medium sen-

sitivity wall-mounted cable, and a lower sensitivity fence-mounted cable.

A typical system is comprised of a control unit for the whole system with four pairs of cables linked to each zone's analyzer unit (one per cable). The analyzer is normally mounted at the perimeter, as close to the sensor cable connection as possible. When direct connection is not possible, a coaxial cable and splice is used to link the sensor to the analyzer unit.

Equipment Types and Selection

Buried Cable. Buried cable is normally comprised of a cross-linked, polyethylene, outer sheath; a tinned copper, braided or silvered-tin (tin with small silver content), solid sheath electrical interference screen; a graphite-serving, piezoelectric polymer under that; and an inner core of a special, low-melting-point alloy.

The braided screen cables are more flexible and lighter than the others, but slightly less resistant to electrical interference.

The buried cable will recognize and usually discriminate between ground vibration caused by tree roots, vehicles, tunneling, movements of intruders, etc.

Fence- or Wall-Mounted Cable. With fence- or wall-mounted cable, the robustness of the cable is increased by removing the graphite layer within the cable and replacing the rather weak, low-melting-point alloy core with an electrically conductive, stranded, tinned-copper conductor. The increase in robustness is at the expense of sensitivity.

With either type of cable it is important to remember that the piezoelectric coefficient is not uniform for every cable batch produced. While there are limits to production quality, a good rule is to ensure that all cables come from the same production batch and are of the same manufacturer. While minor differences can be taken up by the analyzer units, not ensuring that cables are bought from the same batch simply creates problems from the start.

Selection. Selection is primarily based on the abilities of the analyzer associated with the cable, as the cables are produced by only one or two companies. The analyzer must be digital with a powerful microprocessor that allows

- positive discrimination of false alarms from real alarms
- self-testing
- pinpointing intruder crossover points by balance bridge techniques
- self-adaptation for variations in ground hardness, compressibility, etc., under various ground conditions (e.g., moisture content, soil variation)

Uses

- All forms are of perimeter-buried protection and detection.
- More expensive than Electret and leaky coaxial cables, and less well developed and tested.

Application

Buried Cable.

- Zone lengths should be restricted to 984 ft (300 m) due to increases in signal noise.
- A full site survey *by the cable manufacturer* is essential. While the cable is simple to lay down, cable positioning, trenching, etc., are critical to the system's success. This is not a job for the installer's surveyor.
- A trench 6 in (15 cm) wide is required in which to bury the cable. The trench is normally 6 in (15 cm) deep. The bottom of the trench is tamped down, stones are removed, and a layer of soft builder's sand 2 in (5 cm) deep is placed in the trench and packed down. The cable is laid in the trench and pegged in place with plastic pegs every 6 ft (1.8 m). An additional 2 in (5 cm) of sand is placed on top of the cable and compressed. Soil is then placed on top of the sand and turf replaced to hide the cable route *after* the trench has settled for a few days.
- Burial depth must be constant or the width and sensitivity of detection will vary considerably along the cable lengths. Cables should never be laid less than 2 in (5 cm) below the surface; use 4 in (10 cm) for a safety margin.
- The cost of trenching and careful installation to prevent damage is high. The potential for damage or incorrect trench preparation is enormous. Trenching preparation, pegs, soil removal, two-stage sod replacement, and paving should *always* be carried out by the cable installer; never subcontract these tasks.
- Cables buried under paths, roads, and parking areas must be installed as separate zones, due to the slightly different compression characteristics from soil and potential damage. When placed under a path, the cable is normally zigzagged for 5 parallel crosses, each spaced 10 in (25 cm) apart.
- While the cable is very flexible, tight bends and variations in burial depth should be kept to a minimum.
- The cable, if properly installed, can be used under water, which makes it almost unique in this respect and very useful where rivers form a natural boundary.

Wall-Mounted Cable. For wall-mounted cable configurations, the piezoelectric sensor is extremely useful. The applications include buried within the top layer of brick or block mortar on perimeter boundary walls to detect persons climbing them

buried into brick mortar joints, plaster, or concrete coating; directly buried into concrete walls; or built into studs or timber partitions as constructed. All of these provide detection of climbing, attempts to break through, etc.

attached to roofing materials

As the cable is passive, it is totally safe to bury it. As it is screened, it is unaffected by electrical cables that run near it.

With both ground- and wall-buried cables, it is extremely important to keep accurate records of burial locations to complete all initial testing for damage *before* the final surface is reinstated!

Fence-Mounted Cable.

Installations on fences should only be attempted if the fence is in first-class order, buried in the ground at the bottom, and very taut. The system, like most other fence-mounted types, is less prone to spurious alarms, when installed on weld mesh or palisade fences. The types of fencing in any zone must not be mixed.

Cables should be permanently tied to the fence using nylon (non-ultraviolet-deteriorating) ties, which are at least 0.4 in (10 mm) wide, every 8 in (20 cm). At every post, slack of approximately 1 in (2.5 cm) should be available to allow for fence post movement and to prevent chaffing of the cable against the post. Every 33 ft (10 m), a service drop of 6 in (15 cm) should be tied to the fence as slack for future repairs.

When fences are more than 13 ft (4 m) high, two parallel cables may be needed.

Lightning protection and a local grounding rod are always required.

Limitations, Advantages, and Disadvantages

Limitations.

expensive

new

in later stages of signal analysis development

Advantages.

buried type very, very difficult to detect if installed correctly

passive

safe in most environments

potentially very reliable and discriminating

flexible

virtually any zone length tailored to meet your needs

not sensitive to most radiated energy

not affected by most environments (except freezing of all surrounding ground)

can be used under water

wall-buried devices give pre-penetration alarm

Disadvantages.

careful consideration of real performance through use and testing of a prior installation is recommended

covert, so no deterrence value

PASSIVE MAGNETIC EFFECT CABLE SENSORS

Principles of Operation

Passive magnetic effect cable sensors are buried. The cables are either pure magnetic effect or combined magnetic, strain, and seismic effect cables. Both types include a pure, magnetic disturbance element in their detection principle.

In the *pure magnetic* cable, a ferromagnetic alloy is used in the core of the cable. This material has what is known as a *high residual magnetism* or ability to hold on to a magnetic field into which it has been introduced for an almost indefinite period. The cable, when buried in the ground, creates a unique magnetic field *coupling* or pattern (both below and above the ground) with the earth's normal magnetic field. The lines of invisible magnetic field are known as the *magnetic flux paths* and, depending on the particular environmental conditions and the earth's magnetism at the installation, these paths may have varying densities. The cable is passive—it requires no energy input to radiate a field; it merely reacts with the earth's normal background magnetism.

An intruder carrying any ferrous metal object or having any of these on his clothing (or in his body!) inadvertently cuts the magnetic flux paths between the cable core and earth's magnetic field that induces a voltage difference onto a coil of wire within the cable, which is wrapped around the core. The voltage difference causes a current to flow in the normally passive cable. This current is detected and, along with the voltage, is greatly amplified, filtered, and analyzed.

The level of disturbance caused to the magnetic field and the strength of the signal produced depends on the

amount of ferrous material

speed with which the disturbance crosses the cable (the faster, the better)

number of passes over the cable (right angles are best; parallel is normally worst)

The combined magnetic effect/strain effect/seismic effect cable is identical, except that an added level of detection is provided by introducing a second coil of wire surrounding the ferric alloy core. When an intruder walks on the cable, she creates a local pressure on the cable, thereby disturbing it. If she attempts to jump over or tunnel under the cable, she causes a seismic or ground-transmitted vibration pressure on the cable. These pressure variations cause the ferric alloy core to strain or compress very slightly, but enough to disturb the normally constant, very short, magnetic flux coupling pattern between the core and the surrounding coil. This relative movement of the core induces a voltage onto the coil and a current flow. This current is separately analyzed. Coincident currents in both coils or an exceptionally large disturbance in one coil causes an alarm to be generated.

Equipment Types and Selection

The primary differences between the systems are

buried or surface-mounted cable specification. Buried cables should not be used for surface-mounted applications.

purely magnetic or combined seismic magnetic. For the most part, the combined cable is better and the extra expense is justified in terms of increased security and reduced false alarm rate.

most cables have flexible, stranded, ferric alloy cores, a layer of insulation, the sensing coil, outer armor, and cable protective sheath. Cables using a stainless steel tape-armoring and grounding connection are preferable to plain, steel wire, armored cables due to increased EMI resistance, lower corrosion problems due to damage to sheath, and superior protection to the sensitive, continuous-sensing coil conductors.

combined magnetic and seismic sensors. These sensors should always have separate, initial signal amplifier channels, their own individual frequency prefilters, and amplifiers. The amplified signals should be separately analyzed, each by its own signal processor, before combination, and introduction to the processor, which determines whether the coincident criteria or individual element of the signal requires the alarm to be raised.

systems that have hermetically sealed, ground-buried, signal amplifiers that are permanently connected at the factory to a set length of cable. These systems are more robust and more reliable than site-assembled systems.

Uses

This system is a very reliable, relatively cheap, medium-security measure for perimeters in stable, uniform soil conditions around sites 100–2000 ft (30–610 m) long.

These devices have been developing in reliability for more than 25 years and are well known in military circles.

Application

Cables are normally purchased in sets. These are comprised of a fixed length of sensing cable; a permanently attached signal processor, filter, analyzer, and amplifier unit; and a weathertight signal cable connection box.

Cables are normally 100 ft, 300 ft, or 500 ft (30 m, 91 m, or 152 m) long. They are buried in the ground where the magnetic permeability has been already assessed to determine correct burial depth.

A trench of the recommended depth (normally 2–14 in, or 5–35.5 cm, deep) is dug. The trench is usually 6 in (15 cm) or 12 in (30.5 cm) wide. All sharp stones and the soil are removed from the trench. The cable is laid in a surround of 2 in (5 cm) of soft builder's sand (not sharp sand, which will wash away!), and the whole area is then gently vibrated to assist settling. The topsoil is replaced after commissioning trials prove successful over a given cable length. The topsoil is gently compacted, allowed to settle for 2–4 weeks, and then reseeded with grass or covered with the original surface material.

Signal cables are taken away from the amplifiers and sensing cables, at right angles to the latter, to prevent mutually induced interference. Signal and sensing cables should *never* be run parallel to one another, or in common or close trenches.

Commissioning tests should include tests such as minimum ferric material carried, effects of standing water over trench, intrusion demonstration, etc.

Advantages and Disadvantages

Advantages.

very low false alarm rate

passive and undetectable

well defined for human intrusion; will detect metal shoe eyelets, steel-toe caps or shanks in boots, etc.

not sensitive to the environment

can be used in some forms on fences, walls, etc.

very simple to install and commission

cheaper than many forms of buried cable sensors

reasonably easy to repair

can be run under most building surfaces, asphalt, concrete, etc.

Disadvantages.

is upset by variations in the earth's magnetic field, which is particularly a problem in areas of high thunderstorm activity

intruders must be carrying or wearing ferrous metals

a soil permeability test is needed to accurately determine the required burial depth around the site or trenching costs might prove unexpectedly high

cables should never be buried next or parallel to power cables

cables should not run near electric arc-welding or discharge-lighting transformers, generators, radio receivers, radar, or other powerful sources of radiated EMI

it is possible to enter an area and cut the cables if sufficiently skilled and knowledgeable (although this is a remote risk)

MOVABLE POINT DETECTION (MPD) SYSTEMS

Principles of Operation

MPD systems avoid the costs of trenching and/or cable supplies by coupling the sensing device to a data-processing unit and VHF radio transmitter. The whole assembly is a battery-driven, stand-alone system and individual units may be moved at will.

MPD systems normally use geophone sensors that are buried slightly remote from the processor/transmitter, which is concealed in vegetation, etc. The systems are specifically designed to detect persons running, walking, crawling, etc., across open areas or through vegetation and also to detect the movement of vehicles.

Equipment Types and Selection

These systems were developed for the armed forces to detect the movement of troops and, hence, should always be checked for armed forces certification of compliance with US or NATO standards.

Systems should always use civilian VHF wave bands permitted in the location of operation.

Problems with mutual interference can arise if sensors are not discriminating. This should be checked with the manufacturer.

A range of different sensing heads is available for different ground conditions and situations. No commitment to a final purchase should be made until the heads have been successfully field tested at the establishment.

The length of response signal and level of skill required to differentiate humans from animals varies from system to system.

Unit costs are normally high, so it is important to use

the optimum sensing range and radio transmitter range without compromising coverage.

The reliability of the system depends on its ability to transmit, which rests with the battery reliability. Batteries should be very high quality and processors should give a warning of low-voltage output from the unit.

The VHF signal should always be an encrypted tone burst to prevent sensors from disclosing their locations and allowing message substitution.

Uses

protection of temporary goods, vehicles, buildings, etc., on manned, large, open perimeter sites

rapid deployment system for fast response to a threat or potential threat

Application

Systems are simple to apply, provided ground condition and radio transmission path surveys are properly carried out before deployment (or during commissioning checks).

Systems often claim signal radio transmission ranges under "test environmental conditions." While ideal conditions rarely coincide with intrusion, normal bad weather–condition transmission units should be considered.

The technology of the sensor is the same as that used on any other system, so the same warnings regarding snowfall, icing, etc., apply.

Advantages and Disadvantages

Advantages.

transportable and reusable

rapidly deployed

requires no trenching or cabling

completely concealed (provides time for response force to arrive)

Disadvantages.

manned site control needed

costly

not as reliable as buried line sensors

no visible deterrent effect

may require local radio license to operate, and frequency and interference checks

covert, so no deterrent effect

4
Sensors for Internal Use

INTRODUCTION

Internal use sensors are generally simpler, smaller, less robust, and more numerous than external use sensors. Internal use sensors do not normally face the harsh environmental conditions of external sensors and their signal processors do not need to cope with the variety of signal noise that can beset external systems. All this does not necessarily mean selection is simple! In order to ensure that *any* internal use sensor will operate reliably throughout the space in which it is to be used, the following checklist should be considered before focusing on a particular device.

Environmental Considerations

range of temperatures within space
exposure to direct solar radiation through windows
exposure to natural drafts or artificial ventilation
range of humidities within space
dust- and dirt-laden air
vibration of structure or material to which sensor is affixed, induced by normal conditions (e.g., wind, passing traffic, pumps, etc.)
types of room surfaces, finishes, absorptivity, reflectivity

Physical Considerations

position in relation to furniture, partitions, etc., to ensure maximum coverage and no blinding of sensor
position in relation to ease of testing and maintenance, and ease of attack and tampering
physical protection from accidental damage

Operational Considerations

presence of radiant-heating appliances
presence of convective-heating appliances

proximity of air supply diffuser or air extract grills
proximity to source of EMI (e.g., transformers, rectifiers, radio transmitters)
penetration of glass, partitions, and gaps by active beam sensor emissions

Security Considerations

maximum reliable sensitivity
visibility of sensor to passersby or visitors to the area
accessibility of the device during unarmed hours
security of power supply, routing of cables, and position of control panel

PRESSURE MAT/PAD

Introduction

Table 4-1 summarizes the types, uses, typical dimensions, other characteristics, and advantages/disadvantages of pressure mats or pads like the one shown in Figure 4-1.

Principles of Operation

The pressure mat (or pad) is a simple switch. In its basic form, it consists of two foiled metal or tinned copper plates or strips. These are separated by a compressible insulating medium, usually foam rubber, or an air gap with nylon springs. The foam rubber type is most common. The rubber is perforated with holes, usually 0.4–3.1 in (10–80 mm) across. The foil elements are usually profiled so that when pressure is placed on the topmost element, the foam compresses and the profile area presses through the hole in the insulating material, making contact with the lower element and completing an electrical alarm circuit. When the pressure is released, the foam expands and the circuit is open once more. Elements and insulation are enclosed in a sealed, insulated, watertight,

Table 4–1   Pressure Mat/Pad

Types	Pressure mat or pressure pad, open or closed loop.
Activated by	Direct pressure. Pressure mats to BS 4737, Part 3, Section 3.9 must operate when a force of not less than 22.5 lbF is applied to any portion of its surface by a solid test disk of 2.4 in (60 mm) ±0.2 in (5 mm) diameter, and shall *not* generate an alarm for a force of 4.5 lbF if applied by the same means.
Standard	BS 4737, Part 3, Section 3.9; UL 63.
Uses	Point detection by direct contact with body weight of intruder at door, windows, etc.
Not to be used	As sole means of detection or in areas of high door traffic.
Typical dimensions	Mats: to any length (usually in sections up to 16.4 ft (5 m). Thickness 0.6–0.8 in (15–20 mm). Pads: 37 in (950 mm) wide × 31.5 in (800 mm) long × 1.6 in (40 mm) deep.
Associated work	Concealment: rebates, additional underlay for carpet, etc.
Advantages	Cheap, simple, reliable.
Disadvantage	Very easy to circumvent.

FIGURE 4–1   Pressure mat

airtight plastic bag. It is important to remember that the device is a trap in the true sense of the word and it cannot detect intruders if they do not stand on it long enough to compress the foam.

Uses

The pressure mat is used principally as a good, domestic or low-risk retail outlet intruder detection device. It can be extremely useful in very secure buildings as a cheap, secondary, totally separate alarm system for emergency or last-ditch use. When properly concealed, these devices can be difficult to spot and more than one intruder, who has taken time to identify and beat a conventional sophisticated system, has been caught by the last-ditch

pressure mat. The most common uses are in positions where the intruder is likely to step—door openings, under windows, stair treads, near high-value items, etc.

Application

The overriding concern in the application of this device is concealment. The pad or mat must be totally hidden or the element of surprise will be lost and the burglar will find a simple way to circumvent it. The mat should never be placed directly under a carpet, as the bulge will show. However thin the device, given time, carpet wear will emphasize its outline. Pressure mats should be recessed into thick felt underlays. Pressure pads should always be placed in a smooth-finished, lined, structural rebate (groove chase, etc.). Neither should be laid directly on bare concrete, wood, etc., as chafing, condensation, etc., can cause their premature failure. Pressure pads or mats should never be used for door access control and alarm devices (dual function—day/night) or in areas where pedestrian traffic is high, as the device will be compressed continually, which will dramatically reduce its life.

Pressure mats and pads should never be used in buildings with pets; loose, patrolling, guard dogs; rolling vehicles; etc. Also, they should never be placed in positions where furniture could be placed on them. This makes fitting them under windows difficult, as this space is often occupied by desks, chairs, etc. Furniture weight overcomes the natural resilience of the mat and eventually the contacts fail, causing false alarms.

Equipment Types

There are three basic types of this device.

- pressure mat
- pressure pad
- piezo mat

Pressure Mat.   The pressure mat is the cheapest of the three devices and is much thinner than the others, typically 0.6–0.8 in (15–20 mm) or less. The mat can be cut to or supplied in virtually any shape or size and is easily concealed. It is usually only available in the open circuit type; it is not fail-safe and seldom has any antitamper loops.

Pressure Pad.   The pressure pad is more substantial and is designed specifically for commercial applications, as opposed to domestic use. The pad is usually supplied in preformed sections of approximately 3.1–3.3 ft (950–1000 mm) wide, 2.6 ft (800 mm) long, and 1.2–1.6

in (30–40 mm) thick; hence, it always requires a structural rebate. The pressure pad is the most reliable of the devices and is available in the closed loop (normally energized) or open loop form, with antitamper loops.

Pressure mats are often supplied with special, ultraflat, armored cable for connection. Ensure that this is *never* visible and that it is suitably protected from wear. With pressure pads that use normal wiring, connections should be made by concealed, *in situ* conduits.

Piezo Mats. A piezo mat is created when a very thin, piezo crystal film is placed inside a semirigid mat. The mat is laid down on a flat surface; any person standing on the mat compresses the piezo crystal and, as a result, the mat generates a small voltage. The voltage is in millivolt levels and its resulting current flow is amplified by an integral amplifier. The sensitivity, number of contacts, pulse counting, etc., can be adjusted in a processor to which the mat is connected. The piezo mat is an extremely effective device when compared to all other piezoelectric sensors (see "Piezoelectric Sensors" in this chapter). Add to this the freedom from false alarms due to animals, additional robustness, longer life, many shapes, and thinness for additional concealment, and it is likely that piezo mats will replace normal pressure mats and pads, and find a wide variety of applications.

Selection

Look for devices that comply with UL or British standards.
Choose a closed loop, rather than an open loop, if available.
Ensure that guarantee and product description clearly defines the action of the device and cites examples of what will and will not activate the mat or pad (e.g., weight, duration on the mat or pad). The more expensive the product, the more durable the materials used—this is a reasonably safe maxim with pressure switches.

Limitations, Advantages, and Disadvantages

Limitations.

will only activate after intrusion is made
fairly easily avoided if located
cannot define or differentiate direction of travel
daytime switching out is required to prevent false alarm at doors in normal use

Advantages.

cheap (not the piezo mat)
simple
reliable

can be easily replaced, removed, or reused by nonspecialist

Disadvantages.

unless properly sited it may give false alarms because of pets, children, etc., in area at night during alarm-armed hours
moving parts wear (except piezo mat)
furniture movement can give false alarms (e.g., cleaners could leave a chair resting on a pad or mat; control panels should prevent alarm setting when this occurs)
most types are open circuit and, hence, not fail-safe during a power interruption
easy to bridge or bypass electrically

MAGNETIC REED SWITCH (MRS)

Introduction

Table 4–2 summarizes the types, uses, typical dimensions, other characteristics, and advantages/disadvantages of the MRS.

Principles of Operation

The MRS is comprised of two parts—the permanent magnet, which is fixed to the door or window, and the reed switch, which is fixed to the frame. The magnet is a simple two-pole, hard-core permanent magnet that has very little magnetic strength, but an exceedingly long magnetic life span. The reed switch is comprised of a

TABLE 4–2 Magnetic Reed Switch

Types	Surface or flush; single, double, or triple reed; balanced or coded.
Range	Point of contact only.
Activated by	Direct displacement between door and frame or window and frame of *not more* than 4 in (100 mm); 1–2 in (25–50 mm) normal.
Standard	BS 4737, Part 3, Section 3.3 and UL 634.
Uses	Primarily door and window openings, but also cabinets, drawers, etc.
Not to be used	Alone for protection of high-value goods; in any area with easily penetrated surfaces (partitions) or where vibration could give rise to false alarms.
Typical dimensions	Flush circular, typically 0.4 in (11 mm) diameter × 1.3 in (32 mm) deep.
Associated work	Best installations have conduit to door or window frame fully concealed.
Advantages	Extensively used, easily understood, reliable.
Disadvantage	Will not detect cutting through door or window material to affect entry.

glass or plastic envelope that is hermetically sealed at both ends. Within this envelope, there is a simple switch that consists of a flexible ferrous metal foil strip (or *reed*) and at least two contacts. The ends of the reed and the contacts are brought through the envelope and connected to circuit conductors, which carry a very low electrical current. The envelope is either evacuated or filled with an inert gas (often dry nitrogen) to reduce corrosion of contact surfaces by dust, moisture, etc.

On the simplest switches, the reed is deflected and held against one of the contacts (one side of the envelope) by the influence of the magnet affixed to the door. Hence, when the door is closed, the magnet repels (or "throws") the switch to its "on" (or closed) position, a circuit is maintained, and no alarm is generated. When the door is opened, the magnet is drawn away from the switch and its reed, no longer compelled by magnetism, springs back to a straight position. In doing this, it breaks free of the "on" contact and hits the "off" (or open) contact. The breaking of the circuit activates an alarm; hence, the device is fail-safe. In either an open door or no electrical supply condition (e.g., wires deliberately cut), the alarm will activate.

Magnetic reed switches are by far the most common alarm activation device.

Uses

> doors, windows, skylights, cabinets
> very useful to protect doors that might be left unlocked accidentally and, hence, unprotected (since alarm system cannot be set until door is shut)
> protection of individual items (silverware, paintings, etc.) in a special thin-surface, fixed form

Application

The MRS is so simple and reliable that false alarms are invariably caused by incorrect installation, application, or maintenance, and seldom by part failure. The following should be noted:

> The MRS should only be used on fairly heavy, well-fitting doors or windows that will not vibrate excessively in the wind. Although vibration might not actually cause a false alarm immediately, it varies the magnetic pull on the reed, causing it to flex, and ensures premature fatigue and subsequent failure.
> If fitting the MRS to the top, leading edge of the door, ensure that nothing can be slid between the magnet and switch from the unsecured side, and ensure that the alarm will activate if the door is opened a maximum of 4 in (100 mm) (the usual setting is 0.4–2 in, or 10–50 mm, maximum).

On steel doors, successful installation can be more difficult and special arrangements are required. This usually means providing a nonmagnetic collar or shield around the actual magnet contact area, usually made of brass or bronze. Steel doors do give additional security, as they are more resistant to attack. This forces the intruder to try and open the door and any attempted magnetic bridging will fail due to short-circuiting the magnetic flux through the door material.

Ensure that all wires to the unit are concealed (*in situ* or chased in) within a steel conduit. Wherever possible, use factory-fitted stainless steel, flex conduits with prewired MRS connections for final connection of the MRS device.

Ensure that all doors, windows, etc., to which the MRS is fitted have good, firm, security dead bolts or latches that will stop any rattle. Security latches force the intruder to use a jimmy (if time prevents cutting) and the shock movement as the door or window opens will invariably set off the alarm.

Do not use excessive amounts of putty or fillers over the surface of the recessed magnet or MRS when they are installed in wooden frames. These will reduce the sensitivity, reliability, and life span of the MRS. If properly installed, a single coat of finishing paint should make the sensor parts completely flush and nearly invisible.

MRSs that are installed so that the axis of the reeds is parallel to the face of the magnet, will generally operate more successfully than those with reeds at 180° to the face of the magnet (i.e., in-line).

In most applications, the recommended mounting positions are on the top of the door's front leading edge (for maximum sensitivity); on top of the door, but 12 in (300 mm) toward the hinged edge (for lower sensitivity, but fewer occurrences of false alarms caused by rattles, etc.).

On doors or windows with several moving parts, a separate switch should be used on each part.

Remember to consider the type of room or door in which you place the switch, particularly if the switch is a surface type. Rust-encrusted switches on doors do not often please the client's eye! Stainless steel or *anodized* aluminum are normally the best materials for MRSs in sensitive situations.

Flush (linear) mortise types are more secure than surface-mounted types but less tolerant of alignment mistakes by fitters.

Units that have antitamper and factory-set adjustments with an alignment template add security while ensuring easy installation.

Magnets can have their life span severely damaged by being dropped or knocked against other objects. Never accept MRSs unless they are delivered sealed in the manufacturer's packs. Always ensure their installation recommendations are followed to the letter. The manufacturer knows best, *not* the installer.

Equipment Types

Classification systems for MRSs differ between countries, but, in general, a simple, single-line switch diagram similar to that shown in Figure 4–2 will get the customers and salespersons of any country talking on the same wave-length. Figure 4–2 uses the common type/class description that specifications might contain (if any are provided). It is very common for security salespeople to fail to specify or explain contact switch types properly with a view to providing only what they stock. Clients should check that devices are specified *and* fitted.

The most common means of ensuring a good quality product is to specify that it is UL 634 listed, tested, approved (or stamped with the mark BS 4737) and quotes the class required.

		Type/Form	Class		General
1	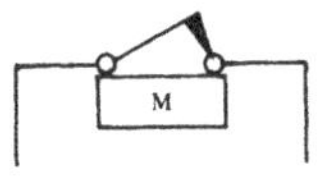	1 Door closed, switch closed	Single reed, closed loop Form A	I	Low-risk residential
2	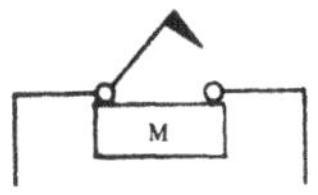	2 Door closed, switch open	Single reed, open loop Form A	I	Low-risk residential
3	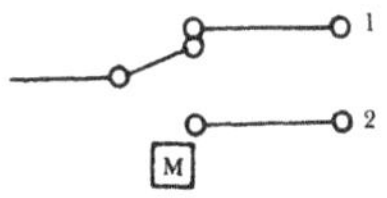	3 Door closed one position, door open opposite position	Single reed, open and closed loop; single pole double throw (SPDT) Form C	I	Low-risk residential or small commercial
4	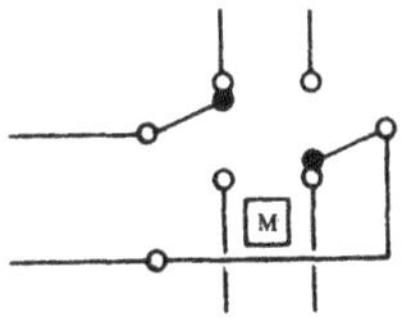	4 Door closed one pair one position or the opposite	Single reed, open loop, double pole double throw (DPDT) Form C	I	Low-risk residential or small commercial
5	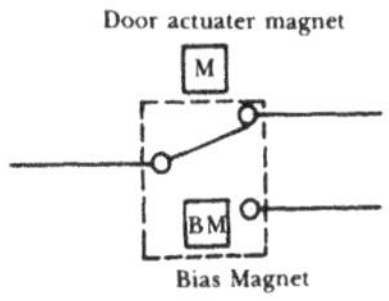	5 Door closed, door actuator magnet overcomes bias magnet	Single reed, closed loop, single pole double throw (SPDT) Form C based	II	Where bias magnet is nonadjustable field strength	Medium-risk residential, commercial, or industrial
6	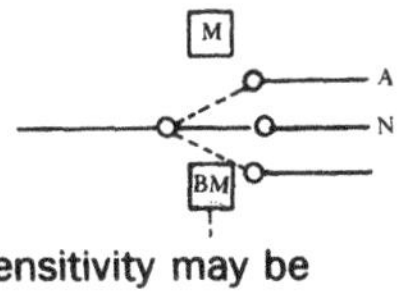 Sensitivity may be adjusted	6 Door closed, reed held in control neutral state in balanced magnetic field	Open or closed loop, balanced bias single pole double throw (SPDT)	II	Where bias magnet is nonadjustable field strength	High-risk residential; medium-risk commercial or industrial
				III	where bias magnet is adjustable field strength	As above but for greater values
7	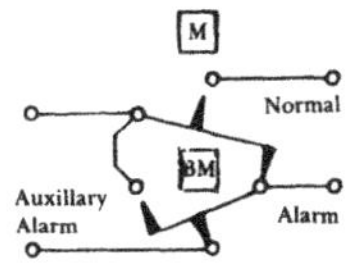	7	Double reed, closed loop, balanced bias double pole double throw (DPDT)	III		As above

M = Door or window frame mounted permanent magnet

BM = Bias magnet within the body of the MRS

FIGURE 4–2 Classifications of magnetic reed switches

There are several main types of magnetic reed switch. These are listed in order of increasing security. These types may or may not align with their class.

- surface-mounted, single reed (see Figure 4–3)
- flush, concealed, single reed (see Figure 4–4)
- coded, balanced, double reed
- balanced single- or double-reed (surface or concealed) (see Figures 4–5 and 4–6)

Surface-Mounted, Single Reed.   This unit (Figure 4–3) was the first unit produced and is the cheapest alarm switch available. The magnet is surface-fixed to the top, leading, rear edge of the door. The switch is placed directly above it. While perfectly adequate for domestic systems, the cheapness of alternative, more secure types of MRSs are gradually making this magnet obsolete. Surface-mounted, single-reed switches should not be used on any commercial system. These switches are normally referred to as *closed loop, Form A, Class I.*

Flush, Concealed, Single Reed.   The flush, concealed, single-reed magnet (Figure 4–4) is cylindrical in shape and is placed in a hole that has been cut into the door's top leading edge. The switch is of similar shape and is concealed within the frame directly opposite the magnet. By concealing both parts of the switch, the

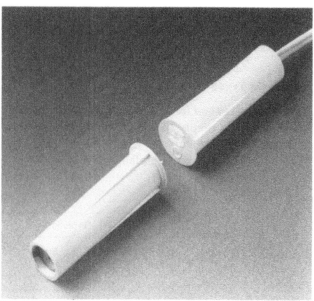

FIGURE 4–4   Flush-mounted, single-, double-, or triple-reed magnetic switch

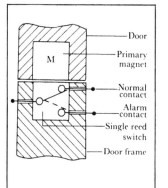

FIGURE 4–5   Flush-mounted, single-reed switch arrangement (M = primary magnet)

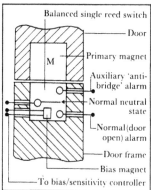

FIGURE 4–6   Flush-mounted, balanced, single-reed switch arrangement

device is far more difficult to locate and circumvent. Switches of this type are normally recommended for domestic systems, but should never be used for commercial systems. These MRSs are the same classification as the surface units.

Biased, Single Reed.   In the simplest form of *biasing,* a switch is used that has a common, central reed and two output reeds (Figure 4–6). However, unlike Figure 4–6, the *biased switch* is factory-assembled so that *one* of the output contacts touches the reed that holds the biasing magnet. When the door is closed, the stronger magnet field of the actuator magnet pulls the reed off the contact and across to the second contact. By biasing in this way, good, solid contact is made in both the alarm and normal states, thereby reducing wear, etc. Biased, single reed switches are normally referred to by their form type and class.

FIGURE 4–3   Surface-mounted, single-, double-, or triple-reed magnetic switch

Biased and Balanced-Biased, Single or Double Reed.
The balanced-biased, single-reed switch consists of the normal door-mounted permanent magnet and door frame–recessed switch. The door switch has a three-contact-position switch—status alarm, neutral (no alarm), and auxiliary alarm (tamper). The switch also contains a small biasing magnet that may be either a permanent or adjustable DC, field electromagnet. With the door closed, the magnetic fields of the normal door and bias magnets interlace and cancel out each other's force of attraction over the reed switch. Hence, the switch remains in its no-alarm, neutral position between both magnets. If the door is moved, the reed switch is attracted to the bias magnet and the alarm contact. If an external magnet is introduced in an attempt to bridge the device, the external magnet distorts the field and the reed moves to the auxiliary alarm contact. As the bias magnet is an electromagnet, its field strength may be altered, using a reostat, to adjust its sensitivity to door movement and reduce false alarms. This device is normally open circuit–unmonitored only.

The balanced-biased, double-reed switch is a similar device, but greater security is achieved with the addition of a second reed to provide closed circuit monitoring. The switch consists of two reeds that are just long enough to overlap. A bias magnet placed between them repels both reeds and causes them to maintain an open circuit. This bias magnet also causes one reed to be pushed toward and onto another contact to form a closed circuit. With the door closed, the door magnet helps to keep the circuit closed. If the door is opened, the bias magnet, alone, cannot keep it in position and the circuit opens and raises an alarm. If a rogue bridging magnet is introduced, the second reed is pulled onto an auxiliary (or tamper alarm) contact and an alarm is raised. Because the second reed is dedicated to antitamper protection, and the device is closed circuit–monitored (fail-safe), this device offers more security than the previous devices at very little extra cost. It *should* be the minimum standard adopted for all commercial systems.

Coded, Balanced, Double Reed (Class IV or V). The coded, balanced, double-reed switch is identical, in almost every respect, to the normal, balanced, or double-reed switches. The coded switch differs only by having the biasing magnet or magnets well shielded from any magnetic field not emanating directly from the door magnet, and by coding the magnets—having their poles arranged in a particular sequence (e.g., NS, NNSS). Unless the bridging magnet matches this sequence, it cannot equalize the field as the door is forced and an alarm results.

Balanced or Double Reed with Antimagnetic Ring or Security Ring (Class III, IV, or V). This switch is sim-ilar to the coded switch, but an antimagnetic device surrounds the switch.

Triple Reed (Class VI). The triple-reed switch is a double-reed switch that has a third reed dedicated solely to raising an alarm should a magnetic bridging attempt take place. The third reed is a dedicated contact.

Selection

UL 634 will normally ensure sufficient MRS product quality. Products *not* developed to UL 634 standards are usually inferior or very high security, state-of-the-art electronic contacts. UL 634 lists at least 18 tests that switches must pass to get UL recognition. Some of these are very simple and effective, and can be fun to carry out on suspect samples supplied to you!

Reed switch quality is generally indicated by the number of poles and the number of reeds. In general,

- select balanced, double-reed switches (or Class II switches) as the minimum standard
- ensure that the switch is in a hermetically sealed and molded case
- ensure that the switch is not oversensitive to door vibration
- on more secure projects, ensure that the switch offers the capabilities to prevent tampering and bridging (electrical and magnetic), and that the switch itself is surrounded by a magnetic shield. These switches are often referred to as either *double pole magnetic with security ring* or *triple reed.*
- select switches that are supplied with integral wiring long enough for them to pass through the door and jamb before connection or jointing is required
- use units with balancing (biasing) magnetic sensitivity adjustments for high-security projects
- the words *wide gap type* in catalogs usually indicates a more reliable switch that requires stronger magnets
- never use switches that have body materials that might corrode. Always use stainless steel, bronze, etc.
- never use MRSs in which the reed have been cropped to reduce size or sensitivity, as this can cause premature failure
- use MRSs that have silver or rodium contact surfaces. Rodium is particularly hard-wearing and indicative of a well-manufactured switch. Gold contacts should *not* be used.
- for high reliability, sensitivity, and stability, select switches that have low-power magnets, low-pull in switching (rated in ampere turns, AT), little contact bounce, and fast operation (3–6 ft/sec, or 1–2 m/sec, maximum)
- the switch sensitivity is rated in AT. The greater the number of ampere turns, the less sensitive the switch

is to positions of the door actuator magnet, and the more reliable it is on low-risk buildings.

higher-class MRSs (Class IV, V, and VI) often have separate customer specification requirements, which may be available for general use or closed user groups only. These include military, bank, and Defense Intelligence Agency models. If the risk is high, try to obtain and use the higher-class MRSs on a compatible system.

higher-class MRSs should always include both anti-tamper and magnetic tamper features that are connected to 24-hour, supervised alarm loops with remote testing capabilities

switches should always have a life expectancy of 10,000,000 cycles (opening/closing) and should always activate if the relative movement of the magnet to the switch exceeds 0.25 in (0.64 cm)

all reed switches should be individually wired to overcurrent protection

Limitations, Advantages, and Disadvantages

Limitations.

position not too difficult to locate

door can be cut through without breaking the switch and activating the alarm

not very suitable for outer doors or heavily trafficked doors

while encapsulated, they are not inherently safe under all conditions. In potentially explosive environments (e.g., petroleum gas, dust) special Ex-rated MRSs are essential.

Advantages.

better than most mechanical switches

very reliable

very cheap

effective

not affected by heat, moving objects, etc., which affect volumetric devices

balanced double- and triple-reed switches will trigger an alarm if the reeds fail

Disadvantages.

can be bridged

only protects individual item to which it is affixed

PERSONAL ATTACK SWITCH

Introduction

Table 4–3 summarizes the types, uses, typical dimensions, other characteristics, and advantages/disadvantages of the personal attack switch.

Table 4–3   Personal Attack Switch

Types	Kick bar, duress button, or hand-held radio transmitter, all latching or non-latching. Hold-up alarm (special). Types X, Y, Z.
Activated by	Force applied to switch by hand or foot at right angles. Force for foot-operated devices not less than 0.11 lbF (0.5 N) and not more than 0.45 lbF (2.0 N) pressure; for hand-operated devices not less than 0.11 lbF (0.5 N) and not more than 0.22 lbF (1.0 N).
Standard	UL 634 or BS 4737, Part 3, Section 3.14 (deliberately operated devices) and Part 2 system.
Uses	To allow any authorized user to deliberately initiate an alarm under duress or emergency condition. Cashiers, store assistants, senior citizens, etc.
Not to be used	*Where activation would put person operating the device at a level of personal risk he or she is not prepared to accept.*
Typical dimensions	Kick bar: 4.7 in (120 mm) high × 20 in (500 mm) long × 6.3 in (160 mm) deep. Duress button: 2.4 × 2.4 × 1.4 in (60 × 60 × 35 mm). Radio transmitter: 4 × 2.5 × 1.2 in (100 × 65 × 30 mm).
Associated work	Kick bar: Fully recessed, concealed conduit to it. Duress button: position dictates work (often on desks, etc., so joinery required to conceal).
Advantages	Kick bar: Hands free. Duress button: Small. Radio: Portable
Disadvantage	Risk in use: any device used for raising an alarm during a robbery must be totally silent in operation, and not raise any local audible alarm. These specialized devices are commonly called "holdup" alarms.

Principles of Operation

Most personal attack switches are simple, electrical switches that when physically opened by the person under duress, stop the current from passing through a circuit, causing an alarm to be generated. Some of these switches operate on radio frequency sound generators, which transmit an alarm signal to a receiver/communicator when a switch is opened. It is important to remember that these devices are not sensors; they must be physically operated. Correctly used, they can be very effective, as they can summon help covertly or overtly, before or after an intruder has actually broken into a secured area.

Unlike many alarm devices, personal attack switches *must* be wired to a circuit of the control panel, which is continuously operational even when the majority of sensors are off (i.e., daylight hours). For this reason, these switches are usually wired to a separate zone that cannot be isolated accidentally. Current wisdom dictates that all devices of this type are connected to silent alarms (i.e., no local bells, sirens, etc.), as noise can provoke the intruder into reacting violently.

Wiring the device to a separate circuit allows the control panel to have particular processing functions suited to panic buttons. On normal alarm circuits, minor voltage/current variations are common and can be caused by a variety of things. Unless checked, they will cause constant false alarms. Normal circuit field sensor wiring connected to the control panel delivers signals that are desensitized, counted, checked for encryption, etc. In other words, time passes while a signal's validity is checked. With a duress or panic signal, this time delay could literally be fatal for the user. Always check alarm circuits for duress buttons and panic attack switches to ensure they are correctly connected and they raise an alarm, without delay, with a minimum force application to the button or switch, for the minimum time, in the least likely scenario.

BS 4737 requires

a device that will operate after temperatures of 30°F to 131°F (−1°C to 55°C) have been applied for 16 hours and with the correct force applied

a device capable of withstanding impact blows of 1.9J ±0.1J and still operate

a life of up to 10 years

the device to operate correctly after not being used for 1 year

Devices are classified by BS 4737 as either

Type X—single force operation

Type Y—two simultaneous forces to separate sides of the operating mechanism

Type Z—two sequential forces to separate sides of the operating mechanism; the first force should be held until the second is applied

BS 4737 does not recommend the use of Type X devices.

Devices may be *latching* plus *reset* with an *operated* indicator or *nonlatching, automatically resetting* with an *operated* indicator.

Devices are further classified by the noise they generate during normal [60dB(A)] or quiet [30dB(A)] operations.

Equipment Types and Selection

Kick Bar.

Operation. Kick bars are floor-mounted, steel arch enclosures containing a switching device that operates when a person's foot is slid along the floor under the arch, making contact with a pivoting bar (see Figure 4–7). (The word *kick* refers to a now obsolete type of switching mechanism.)

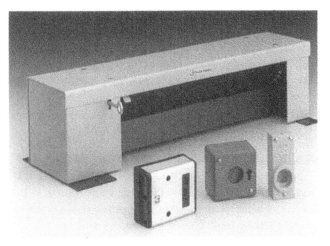

FIGURE 4–7 Kick bar with key reset and concealed panic duress buttons

Uses. Kick bars are primarily used as floor-mounted, concealed-from-view, silent alarm actuators. They are commonly used at guard and reception desks, cash registers, bank tellers' windows, etc., where the person who will operate the bar is permanently seated and the person's feet are not visible.

These criteria are important, as the kick bar is a hands-free device that can only be used if the person is actually seated at her position when the attack occurs, and can use the device without being seen operating it, and, hence, does not put herself at risk. The kick bar should *never* be used when it is exposed to view.

Selection.

Always select UL, BS 4737, or better kick bars.

Ensure the device has a robust steel enclosure.

Ensure that it cannot be accidentally operated by resting feet, handbags, etc., on it.

Ensure that it is completely silent during use. Noise could endanger the life of the operator. Must generate much less than 30dB(A) at 3 ft (1 m) in total ambient silence.

Ensure that it has a latching lock feature. Make sure that once operated, it continues to generate an alarm until a key is used to stop the alarm.

To comply with BS 4737, Part 3, Section 3.14, the device must be closed circuit—monitored to prevent disconnection and tampering.

Ensure that it has a minimum life span of 50,000 cycles.

Advantages.

• hands-free
• covert

Disadvantages.

• fixed position
• puts operator at some risk

Installation. The enclosure is usually bolted to the floor, at each end, through carpet finishes. Most devices require a conduit outlet box that is recessed into the floor, fixed to the side of the desk, etc., at the left-hand side (plan view) and has a conduit feed into it, for connection of concealed cables.

Size. The device is typically 4.7 in (120 mm) high × 19.5 in (500 mm) long × 6.3 in (160 mm) deep.

Hand-Operated Electrical Switch (Panic Button).

Operation. A standard microswitch or proximity switch is used and a latching/locking/reset feature is included in most panic buttons. The switch is operated by a single-finger push button (see Figure 4–7).

Types.

surface or recessed
positive push or proximity push

Uses. Panic buttons can be used in any situation where the user has a good chance of using the button before he is actually attacked or seen. These buttons are commonly used in computer suites, bank managers' offices, and next to beds in domestic premises. Like the kick bar, they are stationary devices and must always be within easy reach. These devices are sometimes referred to as duress buttons or personal attack buttons (PABs). Their primary function is to call local assistance to the area if an emergency or unusual situation begins to develop. In banks, they are often placed in the manager's or chief cashier's office. This allows them to call assistance if an interviewee or customer becomes violent and, more importantly, acts as a secondary holdup alarm system. Seeing a bank robber hold up a cashier, the chief cashier can operate the device in comparative safety, from her office, whereas the cashier might be in grave danger if he tried to use a kick bar. It is essential that panic buttons, like kick bars, are completely silent during use, and do not raise a local, audible alarm.

Selection.

Always select a UL, BS 4737, or better panic button.
Ensure the button is easily reached and large enough to operate with confidence — 0.6 in (15 mm) diameter or greater. Buttons less than 0.2 in (5 mm) in diameter are not easily operated.
Positive spring-action buttons are more reassuring for users than proximity buttons, as users can feel that they have actually operated the switch successfully.
Ensure they include a latching/lock/key reset feature.
The best models have two buttons, which must be pressed simultaneously, to raise the alarm. This avoids false alarms.
Ensure the button has at the minimum, a 50,000-cycle life span.

Advantage.

small

Disadvantage.

hand-operated, so is more conspicuous than the kick bar and, hence, puts operator at more risk than the kick bar, unless the button is remote from but in sight of the action

Installation. The panic button should always be connected by wiring that is protected in a steel conduit (rigid or flexible).

Size. The device containing the button is 2.4 × 2.4 × 1.4 in (60 × 60 × 35 mm). Buttons are usually 0.4 in (10 mm) in diameter.

Radio Frequency, Hand-Held Switch.

Operation. A miniature, battery-operated, hand-held radio transmitter is activated by pressing an electrical switch. The transmitter transmits a coded radio frequency signal in the VHF band (173.225MHz or similar) to a remote, secure, radio antenna/receiver, which initiates the alarm (see Figure 4–8).

Types. Types of hand-held radio switches vary only in external appearance and method of switching. Some devices require two switches, one on each side of the unit, to be pressed simultaneously (avoids false alarms). Some have a single, recessed button. Most are housed in a GRP case that can be hooked to the inside of a desk or counter, carried on the belt or breast pocket, etc.

Uses. Hand-held radio switches are used in any situation where reaching a fixed device, such as a kick bar, would take too long. This device is commonly used by shop assistants (in stores specializing in furs, jewelry, etc.), patrolling guards, etc.

Selection.

Ensure that the device has sufficient range for the actual environment in which it must work and that the signal will not be attenuated.

FIGURE 4–8   Hand-held, radio frequency switches

Ensure that it will not interfere with, or be interfered with by, other equipment transmitting electromagnetic energy.

The device should be FCC-approved (it might require a license).

Electronics within the device should be of high quality. Look for devices that include crystal controllers to prevent frequency drift.

Where multiple units are used, it is essential that one does not interfere with another. It is also important that the frequency used is not close to those used for other purposes. To this end, devices that include the ability to be tuned and encoded after installation are very useful.

Before purchase, ask prospective sellers to carry out a site demonstration to ensure that no blind spots exist in your building, range is adequate, and interference levels are minimal.

Limitation.

The quoted range is 328–656 ft (100–200 m) *free space*, 164 ft (50 m) or less in actual buildings.

Advantages.

no wiring
completely portable
can be used in a pocket, desk drawer, secretly, etc.

Disadvantages.

not hands-free
not suitable for some buildings
range is limited
battery-operated device (can run out at a critical time)

Installation. The radio receive antenna must be situated in an unscreened, secure position where it will not suffer accidental damage and where it is within range of the receiver. The range is normally 164–328 ft (50–100 m), depending on the building and transmitter power.

Size.

hand-operated transmitters: 4 × 2.5 × 1.2 in (100 × 65 × 30 mm)
radio receiver: 3 × 7 × 2 in (80 × 180 × 50 mm)
antenna: 15.7 in (400 mm)

Magnetic Lever Switches. Magnetic lever switches are magnetic reed switches as seen in Figure 4–9. In this application, the reed switch is contained within a large hinged case with the switch on one lid and its balancing magnet on the other lid. Opening the case by approximately 1 in (2.5 cm) with the fingers, raises the alarm as the reed switch opens. The switch is normally armed in the closed position.

These switches have the advantage of almost totally

FIGURE 4–9   Magnetic lever switch

silent operation, but their severe disadvantage is that they require someone to operate the lid. Thus, they are only suitable for applications like those of the hand-operated, electrical switch.

Piezo Pressure Strips. A piezo film is cut into thin strips and concealed in a place that, although easy to press with a foot, side of your head, etc., would not normally be subject to operational pressure. Any slight pressure causes a latching relay, remote from the scene, to operate, which raises the alarm. This device is totally concealed, totally silent, and easy to operate without calling attention to yourself. Its main drawback is the very weak signal strength and electronic amplification needed. Its complexity reduces its reliability.

ELECTROMECHANICAL SWITCH

Introduction

Table 4–4 summarizes the types, uses, typical dimensions, other characteristics, and advantages/disadvantages of the electromechanical switch.

Principles of Operation

The electromechanical switch consists of a microswitch located behind a spring-loaded plunger and recessed into a door or window frame. When the door or window is opened, the spring forces the plunger toward the open space, allowing the microswitch to open. The break in the current causes the alarm to sound. When the door or window is shut, the plunger is forced onto the switch

Table 4–4   Electromechanical Switch

Types	Lever or push rod/ball action. Closed or open circuit.
Activated by	Removal of pressure applied to contact due to presence of door or window frame. Alarm must be sounded prior to a gap of 4 in (100 mm) appearing between door and frame/window and frame, etc., when opened in the normal manner.
Standard	BS 4737, Part 3, Section 3.3; UL 634.
Uses	Low-security, low-value protection against unskilled intruder. Mainly on doors, windows, drawers, etc.
Not to be used	On poorly fitted doors or windows.
Typical dimensions	All sizes to suit thickness of materials.
Associated work	Recessing of unit.
Advantages	Simple, cheap.
Disadvantages	Very low security rating, almost obsolete.

(against the spring), causing the switch to close and allowing the current to flow through the alarm circuit once more (see Figure 4–10).

Uses

The electromechanical switch is still used on many window and door frames, particularly those made of wood, where recessing is easily carried out. Although still popular for simple, low-security systems, double-reed, recessed, magnetic switches are far better devices.

These switches can be very useful on roller shutter doors (up and over) fitted to the bottom of a recessed door well, which protects the switch and prevents levering of the door. In this application, special magnetic reed switches are a very comparable alternative.

Application

The sole objectives when installing electromechanical switches are to put the switch in a position where it will operate correctly and to put it in a position where it will not be subject to tampering. On doors, the best positions

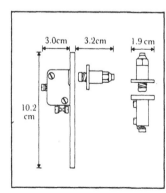

FIGURE   4–10 Concealed, push rod, protective switch and housing

3.0cm   3.2cm   1.9 cm

10.2 cm

are on top of leading edge or top back edge. The former will require the door to open the least distance before activation; the latter ensures that tampering is made slightly more difficult to carry out unnoticed. These switches should only be used on openings where the gap between the door and frame or window and frame is extremely small (less than 0.04 in, or 1 mm) or where a protective *gap cover* is securely fixed to the frame. They should not be used in public access areas that are unattended. Taping over a switch only takes seconds to execute. Also, they should not be used where door traffic is heavy.

On heavy steel roller doors, the *wheel box* type of device should be used, as normal microswitches and subminiature switches are simply not robust enough for these situations.

Equipment Types and Selection

The most common electromechanical switch type is the simple plunger/microswitch. There are many variants of this type, but all are basically the same in that a circuit contact is opened by physical movement of some part of the protected door or window. Wired door hinges that have contacts in which one leaf of the hinge forms one contact and the other leaf forms the second contact are one common variant. Other devices use lever-activated switches to amplify any door or window movement.

Being a "moving part" device, the most important factor in selection is the quality of the engineering used and whether the switches were manufactured specifically for security purposes. These switches are low-cost, low-security devices for detection of the unskilled intruder. They often form part of low-cost security packages and, to keep costs low, industrial *limit switches* used to control machine processes are sometimes used. These switches should be avoided, as they do not have the required precision. When selecting switches, always ensure that they are

hermetically sealed, watertight, and airtight
fully recessed
complete with electrical leads radiating from the molded, hermetically sealed unit
made from noncorroding metals (stainless steel, bronze, etc.) or tough nylon

Coiled springs are generally stronger, more robust, and likely to last longer and prove more reliable than *leaf springs*.

Manufacturers often list criteria for their switches, which grade the switches' actual performance. These performance levels are seldom quoted by the alarm companies that use them. In general, a well-manufactured switch will have a large contact surface, and a minimum hysteresis effect, contact bounce, and over travel.

Advantages and Disadvantages

Advantages.

simple, cheap, easily installed
completely hidden, does not invite vandalism
fail-safe
will detect frame-spreading attempts to overcome
locks

Disadvantages.

can be held in a closed position by plastic cards,
knives, etc., and can be cut from door or window
frames
cutting through door or window is still possible
without detection
can be troublesome due to dirt, moisture, etc.
can be held in a closed position using sticky tape
cannot be used on new wood doors that might distort
or warp during seasoning/drying out

FOIL-ON-GLASS AND FILM-IN-GLASS SENSORS

Introduction

Table 4-5 summarizes the types, uses, typical dimensions, other characteristics, and advantages/disadvantages of the foil-on-glass sensor.

Foil-on-Glass Sensor

Principles of Operation. Foil-on-glass sensors operate only when the foil from which they are made ruptures. The material used, although it appears very much like a baking foil, is actually an alloy of lead and tin or a special aluminum, and it has two very important properties—it conducts electricity very easily and has a very low tensile strength. The first property enables it to conduct very small amounts of current generated at the alarm control. This current continuously passes through the foil and is constantly monitored. The low tensile strength is required so that when the glass to which it is rigidly affixed breaks or is removed, the foil will fracture with it. This fracturing stops the current from flowing in the circuit and an alarm is generated. In order to prevent damage during normal use, cleaning, and tracking, or short-circuiting due to condensation, the foil is coated with a thin layer of very hard varnish, which acts as both mechanical protection and insulation.

Foil-on-glass sensors can only work on glass types that exhibit both of the following properties when struck— they crack *and* the cracks that develop always migrate toward and fracture at a supported edge of the pane. For this reason, these sensors should not be used on lami-nated glass or toughened glass (which sometimes shatters only in the area of impact).

Uses. Foil on glass is used primarily for the protection of glazed doors and larger windows. It is essentially a very cheap, low-security device, which is often part of a do-it-yourself kit purchased by many shopkeepers, householders, etc. Its use is gradually diminishing, even in the do-it-yourself market, and its place is being taken by break glass detectors.

Foil on glass should never be used to protect goods, materials, or property of medium to high value, as it is too easily defeated.

Foil-on-glass detectors are not recommended for lami-nated glass or toughened or wired Georgian glass (despite the allowance in UL 681 of the latter).

Although usually associated with glass, foil tapes can be used to protect against penetration through other surfaces such as wood panel doors, partition walls, etc. The main drawbacks in this application are that

most materials, other than glass, do not fracture clearly
toward the perimeter. Hence, a single loop of tape
is not sufficient. To ensure tape rupture, it must be
laid in parallel strips 6–8 in (150–200 mm) apart
across the whole area. This increases cost considerably, due to the labor involved.
walls are generally used for affixing objects to or
placing objects against, and unless covered with a
heavy lining paper or plywood sheet, damage will
soon result, along with many false alarms.

Application. It is essential that the manufacturer's fixing instructions be followed exactly. This often means thorough cleaning of the window, proper application

Table 4-5 Foil-on-Glass Sensor

Type	Sheet or plate glass application, continuously or "alarm time only" monitored. Single or double pole.
Activated by	Tape breakage causing circuit current to stop, or electrical bridging attempts causing resistance change.
Standard	UL 681; BS 3737, Part 3, Section 3.2.
Uses	Small retail outlets on windows or doors likely to be attacked by amateur criminals, also domestic doors.
Not to be used	For medium or high security, or on laminated glass.
Typical dimensions	Foil tape (0.4 × 0.0016 in, or 10 × 0.04 mm) circuit (loop) length to suit area protected.
Associated work	Glass cleaning, varnishing.
Advantages	Simple, cheap, reliable.
Disadvantages	Poor security rating, no protection against glass cutting if tape not cut. Can only be used on normal plate and sheet glass.

of correct adhesives, etc. The foil must never be allowed to kink, blister, or have any sharp bends, or it will fail prematurely. On plate glass, the foil is often fixed 2–4 in (50–100 mm) in from the edges and on sheet glass the grid formed should never leave uncovered areas of greater than 8 × 8 in (200 × 200 mm) (see Figure 4–11).

Never allow connections to window foil to be made in unprotected, normal electrical cables; always insist on proper security flexible electric cord and junction boxes.

Always consult UL 681 or BS 4737, Part 3, Section 3.2 and Part 4.1, Section 4.2.3 for layouts that comply with the relevant code. Failure to install in accordance with these codes could result in an unsuccessful insurance claim!

Consider these particular points in the application of foil:

Its use should be limited to glass on which condensation forms very infrequently.

On all glass (and with British Standard layouts in mind), the terminal connectors should be as high as possible to prevent any chance of condensation running into them.

Foil should not travel beyond the glass pane. Each pane should be treated separately. Foil should not

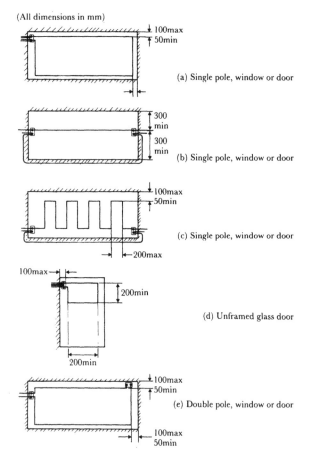

(All dimensions in mm)

(a) Single pole, window or door
100max
50min

(b) Single pole, window or door
300 min
300 min

(c) Single pole, window or door
100max
50min
200max

(d) Unframed glass door
100max
200min
200min

(e) Double pole, window or door
100max
50min
100max
50min

FIGURE 4–11 Examples of permissible locations for foils per British Standards. Consult UL 681 for similar arrangements.

be taken across glazing bars, cemented to glass joints, leaded joints, etc.

Foil should be fixed on a bed of semidry polyurethane varnish 0.24 in (6 mm) wider than the foil. The foil should then get a protective coat of varnish, with a second coat added at least 12 hours later. This second coat is often omitted due to increased labor cost.

All tapes, once applied, should be covered with a thick coat of polyurethane varnish to at least twice the width of the tape. This will protect them against reasonable wear.

Comparison of foil application methods between the UL and British Standards shows that British Standards rely far more on monitoring for bridging attempts, as less emphasis is placed on having tape connectors at different ends or corners of windows.

In general, the expensive element of foil use is labor time and yearly maintenance for damage repair.

Never accept quotations that are foil material costs plus labor for time charge rates. Get a fixed price for all window cleaning, fitting, varnishing, and testing, inclusive of all blocks, terminations, etc.

Always insist on good, solid soldered connections. Snap-fit terminal blocks are of very variable quality and reliability.

Never accept a maintenance agreement that leaves the method of foil repair to the discretion of the installer. The installer may try to splice a repair over the damaged section. This alters the sensitivity, durability, etc., of the whole system. Insist that the maintenance cost includes foil replacement as a whole.

Equipment Types and Selection. There are two basic types of foil-on-glass sensors; each is specific to the type of glass to which it is affixed.

Sheet Glass Foils (Ordinary Glass). Sheet glass is easily broken, but, unfortunately, it is also easy to cut accurately with glass cutters without shattering. In order to protect this type of glass, a continuous grid or labyrinth of foil must be fixed to the pane on the secure side, which forces the intruder to cut at least one part of the foil in an attempt to gain entry.

Plate Glass Foils. Plate glass is tougher to cut than sheet glass and often cracks when an attempt at cutting is made. When it breaks, it produces many fracture lines and these always travel toward the edge of the pane (which is less tough). This tendency reduces the need for the grid pattern used on sheet glass. Instead of the grid, a single length of foil can be run around the outer perimeter of the protected sheet, located slightly in from the four edges.

Whichever type is required, always ensure that the system includes continuous 24-hour monitoring of the foil

monitoring of the foil using a double pole, DC biasing current or balanced-bridge, end-of-line impedance. This will prevent both bridging of the circuit and deliberate short-circuiting from being carried out, even during daylight hours.

foil complying with UL 681 or BS 4737, Part 3, Section 3.2 – foil that is no more than 0.5 in (12.7 mm) wide and 0.0015 in (0.038 mm) thick

In general, lead-tin alloy foil tapes offer better security than aluminum tapes, but are more expensive. Some lead tapes do not comply with the British Standard or UL codes, being up to 0.03 in (0.76 mm) thick. Self-adhesive tapes are quicker to install, but more thorough glass cleaning is required and they damage more easily than varnished tapes.

Limitation, Advantages, and Disadvantages.
Limitation.

low security rating

Advantages.

cheap (first cost)
very stable and unlikely to give too many false alarms
fail-safe
visible deterrent

Disadvantages.

it can be bridged if a section of glass between lengths of foil is removed

to work, the glass must be broken. Hence, damage is caused before an alarm is initiated.

as it is obtrusive, it can be both a deterrent and, at the same time, an invitation to vandals

can be unsightly

difficult to apply if proper care not taken; is labor-intensive

window cleaning on the secure side can be awkward and accidental; damage is common (particularly on sheet glass doors in shops)

no protection against accurate glass cutting with cheaper, less obtrusive types

break glass or *glass monitor* sensors usually more acceptable

Film-in-Glass Sensor

Principles of Operation. The film-in-glass sensor is often called *alarm glass* in the security industry. This can, however, lead to confusion with foil on glass, wired

alarm glass, vacuum or differential pressure double glazing, etc. *Film in glass* is basically a normal laminated glass to which an extremely thin coat of a metal oxide has been added during manufacture. The result is a pane of laminated glass comprised of glass, metal oxide, polyvinyl butyral (PVB), glass, PVB, and glass. The overall glazing thickness depends on the physical attack resistance required. A cross-sectional view of alarm glass is shown in Figure 4–12. The metal oxide turns the inner glass surface of the outer layer (that on the unsecure side) into a conductor of electricity, as glass is, normally, an insulating material. The film of oxide is so thin, however, that it is partially absorbed into the crystalline structure of the glass during manufacture. This means that it transmits light and, unlike a solid metal film exposed to air, its resistance remains extremely stable.

This resistance stability allows for its use as an alarm sensor. If a small current is passed through the glass via electrodes running along top and bottom beaded edges, the current can be used to register changes in resistance, which occur when the window is attacked, cut, moved, etc.

In practice, the film only forms part of a network of resistances. Other resistors are placed near the window in the wiring/termination box. This network can be used to increase or reduce the overall contribution of the film to the alarm circuit resistance, hence giving the window a fully adjustable sensitivity.

Uses. A very high security device, laminated glass provides excellent protection against violent assault, but not much resistance to normal cutting or melting. The addition of a surface sensor layer makes it a very formidable material. The glass, however, is very expensive and, hence, its use is normally restricted to showcases, jewel-

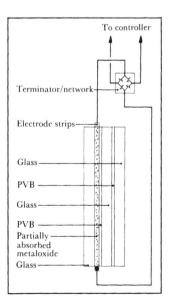

FIGURE 4–12 Film-in-glass sensor (alarm glass)

ers' windows, etc., where risks and values of protected goods are high.

Advantages and Disadvantages.   Both the selection of the glass and its application require expert technical advice that only a combination of reputable alarm companies and manufacturers can give.
   Advantages.

will raise alarm before damage caused or glass broken when set on maximum sensitivity
where accidental contact is likely (e.g., display windows/showcases), sensitivity can be reduced to ensure alarm rings only if the outer pane is broken
can be reset even when severely damaged, by resetting resistor (provided no substantial part of the glass oxide layer is missing)
gives physical and audible alarm protection

Disadvantages.

cost
delivery time
made to measure
weight

Polyfilms

Principles of Operation.   Polyfilms are two layers of transparent PVB between which is laminated, tinned copper or silver film conductors with widths of approximately 0.04 in (1 mm) spaced at distances of 1–4 in (2.5–10 cm). The laminated film is applied to existing, normally glazed windows or surfaces. Its application provides both increased resistance to physical attack (shattering, etc.) and whole surface alarm detection against glass cutting or windowpane removal.

Uses.   Polyfilms can be used when increased security is required. This method is a lower-cost alternative to film-in-glass sensors. Polyfilms can be used for higher-specification replacements for foil on glass used on display cases, display windows, etc.

Advantages and Disadvantages.
   Advantages.

reliable with a very low false alarm rate
very, very robust
much more secure than foils
looks better than foil on glass
adds strength to window and delays smash-and-grab raids

Disadvantages.

higher cost than foil
requires careful application by trained personnel

WIRING SENSOR

Introduction

Table 4–6 summarizes the types, uses, typical dimensions, other characteristics, and advantages/disadvantages of the wiring sensor.

Principles of Operation

A wire is drawn taut across an opening. If broken, cut, bridged by other wiring, or sufficiently displaced, it triggers an alarm. The wire is concealed in a number of ways, depending on the application. In order to promote alarm triggering, the wire used is hard drawn copper wire (usually 0.01 in, or 0.26 mm, in diameter), which has a low electrical resistance (allowing it to carry a small current) and is very brittle.

Uses

doors, screens, skylights, cabinets
single (taut) trip wire—door protection, low security
wire in tube—window protection (in removable frames)
wired glass—window and door panes
decorative screens—windows (small)
integrally wired—walls, partitions, doors, cases, cabinets

Existing walls and partitions can be lined with block-board wiring and block-board panels to provide surfaces with an alarm sensor.

Application

Trip Wire.

Ensure that the wire reel and contact ends are rigidly fixed to the structure.
Never install on heavily used door.
Always use a doorstop to prevent damage to unit if door is left to swing during the day.
Always position to ensure that smallest possible movement of door or window will stretch wire (particularly important on center-pivoting types).

Wire in Tube.

Ensure an antitamper switch is fitted to removable frame types.

Table 4–6 Wiring Sensor

Types	1. Trip wire. 2. Wire in tube. 3. Continuous wiring (screens or integral). 4. Wired glass.
Activated by	Barrier opening, causing distortion or breakage of wire, cutting or bridging of wire.
Standard	BS 4737, Part 3, Section 3.1 or UL 681.
Uses	Protection of entry through door, window, or wall materials.
Not to be used	In conditions where moisture in any form might be present.
Typical dimensions	Retractable trip wire: wire up to 10 ft (3 m); unit 4 × 2.4 × 1.2 in (100 × 60 × 30 mm). Wire in tube: tubes up to 3 ft (1 m) long, spaced at 4 in (100 mm) maximum intervals. Continuous wiring: limited only by circuit electrical resistance. Wired glass: normal pane sizes.
Associated work	Wiring only.
Advantages	Very stable and reliable.
Disadvantage	All but trip wire (which is for low-security areas only) are expensive to make and install (labor-intensive installation).

Do not install in positions where accidental or malicious damage can be caused by customers, vandals, etc.

Never install on windows that are inward-opening.

Integral Wiring.

Ensure that wiring and central panel arrangement is a no-gap type. If the wiring is not bonded directly to a surface and covered by another with which it is in continuous contact, the wires can move in the gap or worse, the surface can be drilled and wires tapped to bridge and loop the circuit.

Do not install on any surface that will suffer from rising or penetrating dampness, or surface or interstitial condensation.

Do not install on any surface that is likely to be used for fixing shelving (screws will penetrate and break wiring).

Whenever possible, make connections to doors through specialized prewired hinges or connectors.

Equipment Types

Single Wire. Single wire is usually referred to as *taut wire* or *trip wire*. A single wire is strung across the opening, usually a door; if the wire is broken by any movement of the door or is attached to a weak contact that is broken, an alarm will be raised.

Grooved Stripping. Grooved stripping consists of wooden battens with a fine groove formed in the rear that runs the full length of the batten. The fine wire is made taut and fixed in the groove at 12 in (30 cm) intervals. The battens are placed grooved side down on doors, across openings, etc., to give a barrier and sound an alarm. Grooved stripping is essentially a cheap, inferior form of the wire-in-tube or integrally wired (lace and board) format.

Wire in Tube. Wire in tube consists of vertically placed metal cylinders or hollow wooden dowels made to resemble window bars. They are set at a maximum distance of 4 in (100 mm) apart. The wire is passed through the dowels or cylinders and connected at each end to a metal frame, which holds the dowels. Anyone trying to force an entry will try to remove or bend the bars, not realizing that he has triggered the alarm by breaking the wire.

Wired Glass. In wired glass, the wire is cast into ordinary annealed glass to give it the outward appearance of Georgian or fire-resistant glass. If the wire is broken, the alarm is activated. Wired glass should not be confused with film in glass, which operates on the same principle but uses a different material.

Decorative or Filigree Screen. Metal mesh screens fitted to windows, doors, etc., resemble low-security, intruder-resistant grills, but, in fact, have a very fine wire either within the screen mesh and set into rebates or simply woven through the mesh. If the screen is cut, bent, or removed, an alarm is generated.

Integrally Wired (Continuously Wired) Doors. In integrally wired doors (sometimes known as *lace and board*), the wire is stapled to the inner surface of a normal, laminated panel door and is invisible from either side. Wiring is taken from the door to the frame via special, wired-through hinges. Any intruder attempting to cut through the door (in order to avoid door alarm contacts, locks, etc.) will activate the alarm. Similarly, partition-type walls can have wiring fixed to block-board with hard-board protection, and normal finishes.

Selection

General.

Select to Ul 681 or BS 4737 standards.

Ensure that the device will be activated by cutting, displacement (2 in, or 50 mm maximum), and bridging attempts.

Wired glass and integral, continuously wired doors are by far the most effective and most secure of all wiring sensors. Always have a specialist select and install them.

Single Wire.

Never select trip or taut wire devices that have the free end of the wire affixed by self-adhesive pads.

Wire in Tube.

With removable, wire-in-tube frames or decorative meshes, always specify that the frame has an adjustable sensitivity, vibration (reed) switch, which will detect attempts to tamper with or remove the entire frame from the opening.

Ensure circuit wiring is closed circuit.

Always ensure that tubes are good quality material that will not corrode or become unsightly (e.g., copper, aluminum, chrome plating, etc.).

Ensure that a vibration alarm is fitted to the wire.

Never use systems with tubes of more than 3.3 ft (1 m) in length and spacings between tubes greater than 4 in (100 mm) between centers.

Wired Glass.

Only allow specialized manufacturer and installer to install sensor—never general alarm company. This work is often subcontracted.

Decorative or Filigree Screen.

Always use custom-made screens in preference to site-wired devices.

Integrally Wired.

For additional security on walls or doors, specify traverse or grid pattern, continuous, integral wiring, which runs in both horizontal and vertical directions.

Never use open span wall or hollow panel door, continuous wiring. Wires should be rigidly fixed, continuously supported, and covered by surfaces.

Integral (concealed) wiring is much more reliable than surface-fixed type.

The best types of these sensors have the wires pressed into the wall or door panel plywood laminates during construction at the factory.

Ensure that wire is hard-drawn copper 0.01–0.015 in (0.3–0.4 mm) in diameter with continuous PVC insulation manufactured to BS 6746, of radial thickness 0.008–0.01 in (0.2–0.3 mm).

Ensure that wiring is fixed at intervals of not more than 4 in (100 mm). To comply with BS 4737, wiring must be fixed at maximum intervals of 23.6 in (600 mm), but this is generally too great.

Look for systems with low resistance.

Limitations, Advantages, and Disadvantages

Single Wire.
Limitation.
requires resetting each night

Advantage.
very low false alarm rate, as a positive contact is required

Disadvantages.
does not detect cutting through door or window and crawling under wires, only opening
is a trap for the unwary

Wire in Tube.
Limitations.
should be fixed on secure side of door or window; hence, damage will always be caused to structure before alarm is raised
fixed types can be very unsightly

Advantages.
cheap
reliable
simple
effective
relatively free from false alarms
frame type demountable

Disadvantages.
exposed wiring types are easily bridged
can only be used on small openings due to creep in strained wires and the high cost of installation
unsightly
obstructs glass if not demountable
invites attention

Wired Glass.
Limitation.
only detects breakthroughs

Advantages.
will only cause alarm if glass is broken and removed
very stable
more secure than break glass ultrasonic detectors or monitors, as with those devices, glass cutting to circumvent the alarm system is possible

Disadvantage.
expensive

Decorative or Filigree Screen.
Advantages.
stable
unobtrusive

Disadvantages.

accidental damage is common
repair is labor-intensive

Integrally Wired.
Limitations.

wall boards added after construction reduce room
dimensions
doors require expensive, specialized hinges for
through wiring

Advantages.

completely concealed
very stable and reliable

Disadvantages.

doors must still have door-open alarm
if left unlocked or unauthorized person in possession
of key, the device will not work
cost
damage results before alarm raised

PIEZOELECTRIC SENSOR

Introduction

Table 4–7 summarizes the types, uses, typical dimensions, other characteristics, and advantages/disadvantages of the piezoelectric sensor.

Table 4–7 Piezoelectric Sensor (Five Types)

1. Break glass piezoelectric *(sometimes called "acoustic" or "microphone" break glass sensor) (see Figures 4–13, 4–14, and 4–18)*

Types	All the same basic sensor. Types differ in range/sensitivity.
Range	Placed up to 6.5–13 ft (2–4 m) away from center of windowpane will cover 12–18 yd² (10–15 m²) per sensor, but depends on glass type.
Activated by	Breaking glass only, not pane removal, melting, etc.
Standards	None.
Uses	Window/door panes. Only to cover glass breakage as method of gaining entry. Plate or sheet glass only.
Not to be used	On any surface that vibrates excessively or aimed at any very large window or glass that is likely to vibrate. Not for medium- or high-security applications.
Typical dimensions	3 × 3 × 0.8 in (80 × 80 × 20 mm).
Associated work	Wiring to window head or other sensor position. Wiring preferably in conduit.
Advantages	More reliable than most other detectors used to cover glass. Not visible. Better than foil on glass.
Disadvantages	Larger than break glass monitor. Only detects breaking glass.

2. Break glass "monitor" piezoelectric sensor *(sometimes called "window guard") (see Figure 4–17)*

Types	All the same basic sensor. Types differ in physical size, whether latching, LED included, etc.
Range	6.5–9.8 ft (2–3 m) radius (maximum) of sensitivity when *fixed* to pane.
Activated by	Breaking glass only.
Standards	None.
Uses	Any plate or sheet glass.
Not to be used	Toughened, wired, or laminated glass.
Typical dimensions	1 × 1 × 0.8 in (25 × 25 × 20 mm).
Associated work	Wiring to window frame head conduit box.
Advantages	Small, unobtrusive, not line-of-sight.
Disadvantages	At least one to cover every pane. Not economical on windows of many small panes.

3. Fence-mounted piezoelectric sensor

Types	Post mounted or directly attached to fence material.
Range	Fixed at 4.6 ft (1.4 m) above ground level and intervals of 15.7 ft (4.8 m) on a taut/stiff fence material. Zones usually under 164 ft (50 m).
Activated by	High-frequency vibrations caused by cutting of fence (or high acceleration displacement) being transmitted to device through fence material to sensor position.
Standards	BS 4737 and UL.
Uses	External perimeter fences or internal precious metal or high-value goods cages.
Not to be used	On any wide mesh or loose chain-link fence or any fence not securely and continuously anchored to ground. Better on weld mesh fence material.
Typical dimensions	4 × 2.4 × 1.4 in (100 × 60 × 35 mm).
Advantage	Very sensitive with site adjustable ranges of sensitivity.
Disadvantage	Not a continuous cover device (unlike Electret cable), hence easier to circumvent.

4. Structural attack piezoelectric sensor *(sometimes called "vibration detector")*

Types	1. Wall, floor, or soffit mounting to homogeneous heavyweight material (e.g., concrete). 2. For protection of safes, strong-room doors, etc.
Range	1. Fixed at 4.6 ft (1.4 m) above floor level (AFL) and 19.7 ft (6 m) centers (five per zone). 2. Safes/doors up to 4.6 ft (1.4 m), one number on door.
Activated by	High-frequency vibrations caused by hammering, drilling.
Standards	Both to UL and BS 4737 and to various international standards for

(continued)

Table 4–7 (continued)

	protection of safes and strong rooms, i.e., DIN, VDE, etc.
Uses	1. Protection of massive heavyweight structural materials against penetration; mainly bookstores, bonded stores, strong rooms, etc. 2. Special protection of safes.
Not to be used	On any surface that might vibrate randomly (i.e., intermittently) under normal (ambient) conditions.
Typical dimensions	8 × 4 × 2 in (200 × 110 × 60 mm).
Associated work	Concealed secure conduits to point of connection.
Advantages	Very sensitive, very difficult to defeat.
Disadvantage	Very limited application.

5. Buried piezoelectric sensor

Types	1. For external ground burial. 2. For internal use buried in screeds.
Range	Depends on nature of material covering device, but typically 19.6–26.2 ft (6–8 m) radius on materials that will transmit vibration easily without great attenuation.
Activated by	Medium-frequency, ground-transmitted vibrations.
Standards	UL; defense standards.
Uses	External grounds and approaches. Internal entrances, corridors, etc., leading to high-security areas.
Not to be used	In any external ground that might freeze solid, or on any surface that has not been properly compacted under the sensor, before it is placed in position.
Advantage	No moving parts (unlike geophone).
Disadvantage	Less stable than geophone.

Principles of Operation

Piezoelectric means the ability to generate electricity from molecular vibration. Piezoelectric sensors operate like ultrasonic motion detectors in reverse. With the ultrasonic motion detector a voltage is applied to a piezoelectric material, which responds by vibrating and generating an exact frequency that is, in turn, beamed across a space to detect movement. In a piezoelectric sensor, the material generates an electrical potential when its molecules are vibrated by some external influence. This may be done by physical *squeezing* of the material, vibration caused within it due to the force of its being hit by ultrasonic waves, mechanical vibration of surfaces to which it is affixed, etc.

The magnitude of the voltage generated is dependent on and proportional to the acceleration of particles within the material. The acceleration is dependent on the force vibrating the structure to which the sensor is affixed. The frequency of the voltage generated will be

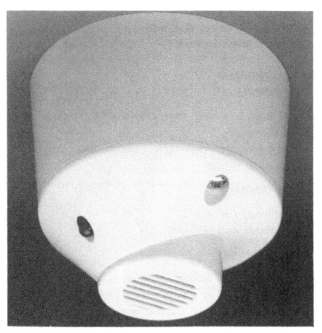

FIGURE 4–13   Break glass sensor, acoustic type

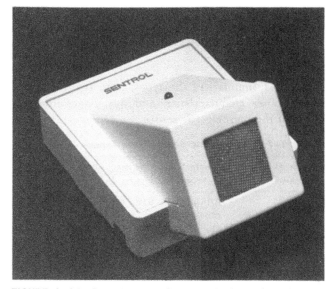

FIGURE 4–14   Remote-mounting, break glass piezoelectric sensor

directly related to the frequency of the structural vibration.

In trying to understand the operation of these devices, it is essential to remember that the voltage is not generated by the lateral or vertical movement of the device, but by relative movement *between the surfaces of the piezoelectric crystal*. It must deform in order to work; pressure or stress must be exerted on it (see Figure 4–16). Most piezoelectric detectors consist of

an outer envelope of PVC to seal the unit from any harmful environmental conditions

FIGURE 4-15 Shatterbox switch and monitor

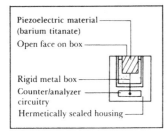

Piezoelectric material
(barium titanate)
Open face on box

Rigid metal box
Counter/analyzer
circuitry
Hermetically sealed housing

**FIGURE 4-16 General
arrangement of a
piezoelectric sensor**

an electronic frequency analyzer/filter circuit and a
    counter/accumulator
the sensing element

The sensing element is comprised of a very stiff, rigid
box with one open face, made from a relatively high-
mass metal or other material. Placed in the box is a chip
of barium titanite (consisting of many hundreds of individual piezoelectric crystals). The open box exposes one
face of the chip, while protecting the others. This face is
attached to or aimed at the source of any likely vibration.

The frequency analyzer/filter/counter/accumulator unit
takes all the sensed frequencies, filters out those regarded as normal, and counts the remaining frequencies
that are characteristic of attack. The *count* consists of
storing all significant voltage and current levels, at a particular frequency band and duration, occurring in a
given sampling period. This might be all voltages of over
2mV that last for more than 2 seconds, with a frequency
between *x* and *y* Hz. At the end of the sampling period,
the accumulator integrates all the voltages, and if the
level is above a trigger level, it actuates the alarm. On
break glass and glass monitor sensors, the sampling
period is extremely short and the characteristic frequency band quite wide, as glass breaks in a single crack,
generating a very complex frequency pattern. On structural attack sensors, the count or sampling period will be
much longer, as it takes longer to penetrate these
materials. So the longer the sampling period, the more
information that can be gathered, the more reliable the
sensor. In addition, structural attacks generate frequencies with a very narrow frequency band, typically
1500-6000Hz, and so most other frequencies can be

attenuated or completely filtered out, making the storage
capacity required far less than would otherwise be the
case. There are six basic types of piezoelectric sensing
devices:

those that are physically attached to glazing materials
    requiring protection (e.g. a glass pane)—commonly
    called *glass monitors* or *guards*. They are attached to
    the glass pane by a strong adhesive. When the glass
    breaks, it affects the device by physically (directly)
    distorting the piezoelectric material and by causing
    shock waves to travel through it, vibrating the crys-
    tal. Hence, the device generates a very high level of
    voltage and is subjected to a very wide range of fre-
    quencies of vibration. The monitor is easily distin-
    guished from break glass devices by being far
    smaller, usually 1 × 1 in (25 × 25 mm) or less (see
    Figures 4-15 and 4-17).
break glass sensors. These are not fixed directly to the
    glass, but to the window frame or remote by dis-
    tances of up to 9.8-13 ft (3-4 m). These devices rely
    purely on the ultrasound generated when the glass
    breaks and are sometimes referred to as *ultrasound
    sensors*. The "sounds" are transmitted to the detector
    surface through air (see Figures 4-13, 4-14, and
    4-18).
fence guards. These are like glass monitor sensors, but
    are designed to detect cutting of fence materials or
    climbing of a fence. The vibration caused by the
    attack is transmitted through the fence material to
    the sensor (see Figure 4-20).
structural attack sensors. These are basically high-
    security devices for the protection of safes, strong-
    room walls and doors, party walls in secure areas,
    bonded warehouse walls, etc. They are attached
    directly to the protected structure and sense directly
    transmitted vibrations (see Figure 4-21).
Buried sensors. These are far less common than the
    previous devices, but are intended as alternatives
    to geophones and fluid pressure sensors. They will
    detect the vibration caused by a person or vehicle
    moving across an area of ground under which
    they are placed. Unlike fluid pressure sensors, the
greound cover does not need to compress, as its

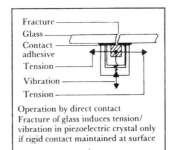

Fracture
Glass
Contact
adhesive
Tension
Vibration
Tension

Operation by direct contact
Fracture of glass induces tension/
vibration in piezoelectric crystal only
if rigid contact maintained at surface

**FIGURE 4-17 General
arrangement of a break
glass monitor piezoelectric
sensor**

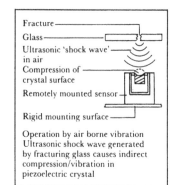

Figure 4–18 General arrangement of a break glass acoustic piezoelectric sensor

Operation by air borne vibration Ultrasonic shock wave generated by fracturing glass causes indirect compression/vibration in piezoelectric crystal

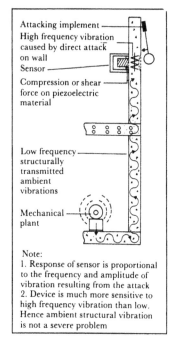

FIGURE 4–21 General arrangement of a wall-mounted (structural attack) piezoelectric sensor

Note:
1. Response of sensor is proportional to the frequency and amplitude of vibration resulting from the attack
2. Device is much more sensitive to high frequency vibration than low. Hence ambient structural vibration is not a severe problem

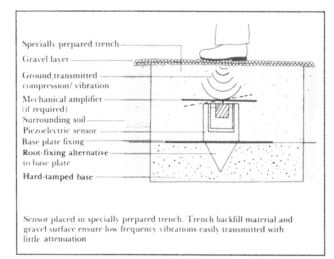

Sensor placed in specially prepared trench. Trench backfill material and gravel surface ensure low frequency vibrations easily transmitted with little attenuation

Figure 4–19 General arrangement of a buried-in-ground piezoelectric sensor

vibration detectors. These are cheaper selective versions of the structural attack sensor that fits on any door, window frame, etc.

Uses

The primary uses for piezoelectric sensors are in preventing entry by glass cutting or smashing, fence penetration, and ground cover. For glass protection, they are probably the most reliable and most commonly used sensor. The monitor is a very economical security solution in office blocks with many small, ground-floor cellular offices with single windows that are permanently shut. Remote-mounting break glass sensors are usually used on openable windows that require covert, nonintrusive protection, large display windows, or multipane windows. In this role, these detectors are normally used in conjunction with passive infrared sensors covering the corridor off which the offices lead.

In fence protection, they can overcome many of the problems associated with geophones, but do not provide the all-round protection offered by Electret cables. The wall and safe structural attack detection sensors are well established and commonly used. Care should be taken to ensure that the correct device has been specified for the particular application. Sensors designed for safes may not be suitable for walls and wall sensors come in a variety of types depending on the groups of attack to which they are more sensitive—hammer blows, high-speed drilling, explosives, nibbling devices, core drills, etc. Some devices are specifically manufactured to be insensitive to vibration caused by automatic machinery commonly used in banking (e.g., automatic teller machines and cash dispensers).

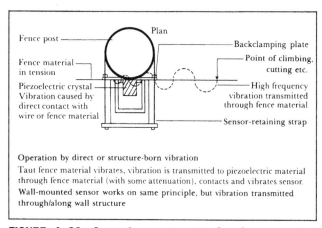

Operation by direct or structure-born vibration

Taut fence material vibrates, vibration is transmitted to piezoelectric material through fence material (with some attenuation), contacts and vibrates sensor. Wall-mounted sensor works on same principle, but vibration transmitted through/along wall structure

FIGURE 4–20 General arrangement of a fence-mounted piezoelectric sensor

they are placed. Unlike fluid pressure sensors, the ground cover does not need to compress, as its vibration activates the device, not compression (see Figure 4–19).

Buried sensors are not commonly used and if selected they should have characteristics that are matched to the surface materials under which they are buried. Their most common use is as an external sensor in areas where structural attack on a wall might commence. They are sometimes used in "hard" areas (internal or external) where normal traffic is high and ground conditions would not suit geophone sensors. As they are sensitive to vibration or shear movement, the cover surface does not need to compress when a vehicle or person passes over the area.

Application

Break Glass Sensors.

Break glass sensors should be installed as close as possible to the windowpane and in direct line of sight; preferably on the window frame.

They must be fixed to a solid structure so as not to give false Doppler shift amplification (see "Ultrasonic Sensor").

They should operate on first impact on glass, which results in breaking, but not on glass rattle caused by wind.

They normally cover 12–19 yd^2 (10–16 m^2) per sensor, but coverage depends on glass type and particular sensor model.

Break glass detectors are *not* universally applicable to all glasses. Consult the sensor manufacturer once the specific type is known.

Consider the labor costs before selecting break glass, a passive infrared (PIR) device or dual technology sensor might be more economical.

Break Glass Monitor Sensors.

Break glass monitors should be securely fixed to the window at either a single corner or two diagonally opposite corners.

The manufacturers should specify and supply the adhesive. The correct window precleaning and adhesive types are critical to successful installation and continued trouble-free use.

Will work on almost any glass type other than laminated.

Only devices wired as balanced bridge, with end-of-line resistor or constant current circuit, actually comply with UL or British Standards. Some devices offer cheaper forms of circuitry, which do not comply.

Fence-Mounted Sensors.
As the sensor relies entirely on vibration being transmitted to it through the fence material, ensure that

the device cannot be circumvented. The fence should be a minimum of 9.8 ft (3 m) high with a cranked top. To prevent tunneling, a 6 ft (2 m) (3 ft, 1 m, secure side; 3 ft, or 1 m, unsecure side) asphalt strip should be laid. To prevent lifting, the fence should be continuously and securely anchored or buried. The latter is preferable and should be encased in a strip of concrete 8 in (200 mm) wide × 12 in (300 mm) deep.

the fence material is taut, close mesh. The more taut the fence, the more faithful the transmission of high-frequency vibration caused by fence cutting. Most devices include a counting device that will only allow an alarm signal to be generated after a certain number of attacks (cuts) in the fence material have been made. Obviously, the closer the fence mesh, the greater the number of cuts required to gain entry, and hence the more likely the detection by the sensor.

Structural Attack Sensors.

Devices are rigidly fixed in a permanent manner to structural concrete or masonry walls. They should not be used on partitions or metal-skinned/clad sandwich panels or directly fixed to roofs, doors that might be subject to vandalism or child's play, or surfaces that might expand or contract significantly with temperature or solar radiation.

The coverage of each sensor will depend on the density and rigidity of the structure to which it is affixed, and the model. Do not exceed the manufacturer's guidelines.

Can be used on floors or walls.

The shielded supply cable should always be encased in conduit that is recessed and hidden within the structure, if possible. This will reduce any EMI, tampering, or accidental damage. When used, these devices often protect high-value areas against professionals, so installation standards are critical.

Sensors are often encased in steel, and covered and interlinked with metal conduits to provide a continuous vibration path. The conduit is firmly fixed to the structure every 3 ft (1 m).

Never affix the sensor to any structure that could legitimately vibrate due to lift machinery, rotating motors, pipe-work vibration, passing traffic, etc., unless a thorough field test has established that these signals will be correctly attenuated.

Never install the device on brick or block work without establishing that it will transmit structure vibrations both through the wall and across the surface of the wall. These sensors are normally fixed 4 ft (1.4 m) above finished floor level (AFFL) at 19 ft (6 m) intervals. If any attack takes place at the midpoint of spacing (i.e., 10 ft, or 3 m, away from the nearest sensor), the vibration must travel this distance to the sensor. If it has to pass many mortar joints, expansion joints, sealing materials, etc., or if it is mounted on an inner-applied finish that absorbs vibration, its amplitude will be greatly attenuated and the sensor might fail to "see" the attack. This problem is often overcome, in practice, by mounting the sensors on a continuous steel channel fixed every 4 ft (1.4 m) AFFL to all protected surfaces. This provides a continuous path for vibration detected at any point. It is, however, really a remedial measure and, if required, is either being used as insurance or because the sensor selected was really inappropriate. Obviously, if the steel channel is used, it must be fixed very firmly and very frequently to the structure or its fixings will attenuate the vibrations themselves.

Buried Sensors (External).

Usually buried in a trench 18–20 in (450–500 mm) deep or in a bed of asphalt 3 in (75 mm) deep. The trench base is prepared with 4 in (100 mm) of compacted soft sand. The sensors and wiring are then placed in the trench, each sensor resting on a metal-based plate. This forms the unmoving side of the piezoelectric crystal. Sand is then used to backfill the trench and is compacted. This sand transmits the intruder pressure onto the "moving" side of the piezoelectric crystal.

Devices are connected in chains or radial circuits that connect each sensor to the next using an electromagnetically shielded cable.

Devices are sometimes buried directly into road substructures 3 in (75 mm) below the topmost surface of asphalt.

Vibration Sensors. Like its higher-quality, higher-security brother the monitor, the vibration sensor *must* be rigidly fixed to the window or door frame. This device consists of a piezo crystal to which a cantilevered arm has been affixed. The arm amplifies any movement it feels and stresses the crystal. It is essential that the arm feels the vibration consistently, hence this device should never be fitted to uneven surfaces, unseasoned or wet timber, etc.

Equipment Types and Selection

Break Glass and Monitor Sensors. Look for devices that

are site sensitivity adjustable
are closed circuit with latching LED indicators and test capabilities
are hermetically sealed, prewired, and comply with UL or BS 4737

include low-voltage alarm, antitamper feature with an automatic reset after 3 seconds (for continued protection)
for break glass detectors, ensure that devices include analyzer and filtration circuitry and respond only to frequencies in excess of 40Hz (below this, radiator "hiss," glass scraping, etc., that can set off alarms)
are suitable for the type of glass being used

Fence-Mounted Sensors. (See Chapter 3, "Geophone Sensors.") Look for

highly adjustable sensitivity
sensitivity to cutting (mechanical or oxyacetylene), lifting, or climbing
a selective frequency counter (e.g., only above 4000Hz).

Structural Attack Sensors. Look for

a latching sensor
an alarm processor that includes a counting/accumulating device that counts the number of impacts of a certain frequency and amplitude, and, above a critical level, initiates the alarm
a device that includes an antilift or antitamper switch fitted to the piezoelectric surface, which butts up against the protected surface. This will provide protection against the device being lifted off the surface during an inactive alarm period.
superior methods of screening cable and affixing the device to the structure

Buried Sensors. Look for custom-made devices suitable only for burial under asphalt, grassy areas, etc.

Vibration Sensors. Vibration sensors are in direct competition with inertia sensors. The inertia sensor is basically an electromechanical *switch* that acts under the force of gravity. It, therefore, has moving parts, contacts, etc. Vibration sensor manufacturers claim that these moving parts are a disadvantage and their products have none. Vibration sensors are currently taking over the European market, once solely dominated by inertia sensors. Hence, reliability is probably now proven beyond any doubt. (This product was introduced in 1984.) When choosing a vibration sensor, be sure it has the following features:

double-knock alarm — first vibration will only cause an alarm if vibrating at a preset frequency and amplitude for more than 0.25 seconds. The second valid vibration will cause an instant or time-delayed alarm.
automatic, manual, and remote reset capability
alarm memory: first device to alarm, and subsequent device to alarm functions
walk-test LED

site sensitivity adjustment with calibration testing

ability to be mounted in any plane

Advantages and Disadvantages

Break Glass and Monitor Sensors.
Advantages.

can distinguish between high-frequency vibration caused by breaking glass and low-frequency vibration caused by traffic or wind

can be tuned to individual glass types by use of filters

extremely reliable, relatively inexpensive, small and compact

can be remote-mounted (break glass)

no moving parts, purely electrical device

large coverage for each sensor 12–19 yd^2 (10–16 m^2)

unobtrusive and inconspicuous

Disadvantages.

more costly than inertia sensor, monitors have high labor cost

cannot detect glass pane removal and some forms of cutting (e.g., thermal knife)

monitor type visible and could be detected

remote type must be in direct line of sight, if possible

cannot be used on certain glass types

requires more on-site testing and commissioning than inertia sensor, in most instances

destruction must take place before alarm is generated

unobtrusive, hence of little deterrent value against opportunist criminals, unlike foil-on-glass sensors, which advertise their presence

Fence-Mounted Sensors.
Advantages.

better than geophone at discriminating between vertical and horizontal fence movement; horizontal damping not needed

no mechanical moving parts to fail

low current draw on system

more sensors per "chain" than geophone (typically 100–150)

Disadvantages.

not a continuous sensor like electromagnetic cable

no listen-in capability available to give audio confirmation of attack

must be mounted at close intervals (typically 15–19.5 ft, or 4.8–6 m)

not suitable for low, unanchored, block or large mesh (chain-link) fences

Structural Attack Sensors.
Advantages.

much wider and finer sensitivity adjustment possible

than with geophone; mechanical or inertia switch used for similar protection

not as subject to false alarms caused by lower-frequency vibrations transmitted by lifts, generators, transformers, trains, cars, etc., as with most other sensors

will protect against cutting as well as drilling and hammering

Disadvantage.

more costly than geophone or inertia sensors

Buried Sensors.
Advantages.

no moving parts

can be used under hard surfaces

does not rely on compression, alone, to activate an alarm

limited coverage; does not pick up vibration from neighboring public roads, railways, etc.

Disadvantage.

high cost associated with trenching and trench preparation

Vibration Sensors.
Advantages.

small

low cost

reliable

available for doors, windows, pictures, and display cases with remote or stand-alone power

Disadvantages.

vibration signal needs major amplification and processing before use as an alarm signal so that reliability and sensitivity rest, not on the sensor, but on the amplifier and processor electronics; interior switches use large direct vibration sensors

can be affected by EMI and RFI, unlike most inertia switches

most current types are surface-fixed only and, hence, unarmed time tampering can occur

higher cost than inertia sensors

INERTIA SWITCH

Introduction

Table 4–8 summarizes the types, uses, typical dimensions, other characteristics, and advantages/disadvantages of the inertia switch.

Table 4–8   Intertia Switch

Types	Ball or cylinder. Ball usually better—more secure device for commercial applications—although cylinder has some unique features. Rasp rolling pod latest development.
Range	Depends entirely on type of structure to which it is affixed, and model.
Activated by	Actions inducing *acceleration*. Manufacturer's literature should state what it will detect, and what it will not.
Standard	BS 4737, Part 3, Section 3.10 or UL.
Uses	Protection of walls, doors, floor, etc., against drilling, cutting, etc., but should only be used on relatively rigid structures. Special devices for protection of safes and strong rooms. Window frame–mounted types to protect against forcing or breaking of glass.
Not to be used	Where accessible to those who could cause nuisance alarms (i.e., first external perimeter fence in built-up area, on exposed-to-public lightweight external cladding, etc.).
Typical dimensions	1.9 × 2.4 × 0.8 in (50 × 60 × 20 mm) for units with integral analyses; 1.9 × 0.8 × 0.8 in (50 × 20 × 20 mm) without.
Associated work	Wiring only.
Advantages	Small, reliable, better than continuous wiring.
Disadvantage	Reliability dependent on matching device to structure on which it is placed and environment in which it operates.

Principles of Operation

If a milk bottle is balanced *neck down* on a playing card resting on a level surface, it is relatively unstable. If the playing card is slowly pulled away from under the bottle, the bottle is likely to tumble down in a direction away from that in which the card is pulled. The pulling force of the card is opposed by the lesser forces of friction on the bottle and air resistance. In this case, the neck of the bottle travels in the direction of the card and the base travels in the opposite direction. However, if the card is pulled very quickly, the bottle will remain upright. The card and bottle neck friction will induce the bottle to topple away, but a force supports the bottle, hence it stays upright and in position. The force is known as *inertia* and it is a resistance to the rate of change of velocity (*acceleration*).

Inertia-sensing devices are used to detect extraordinary acceleration patterns caused by various forms of attack, which are transmitted to the device through the structure in the form of vibration. Inertia devices are usually so called in order to differentiate them from devices that recognize vibration transmitted through *air*, such as ultrasonic detectors, or those that simply detect physical movement on a larger scale, such as reed

switches. Inertia switches are essentially mechanically operated, electrical make-and-break switches. They are not geophones—a sensor with which they are often confused—and geophones are not switches; they are electrical generators.

Uses

Inertia switches are primarily used on monolithic structural elements, such as brick walls, concrete roofs, window frames, walls, floors, etc., to detect forcing, cutting, hammering, or drilling before penetration actually occurs. High-quality ball types with integral processing are often used on strong-room walls as an alternative to piezoelectric vibration sensors (see Figure 4–22). The cheap cylinder or ball types are sometimes used as alternatives to magnetic reed switches on doors, window frames, etc., but only where they are not visible to the potential intruder.

Specialized subtypes of the ball-type sensor are available for fence or buried-in-ground applications, but unless exceptional circumstances prevail, electromagnetic cables or geophones are more suitable alternatives (see Figures 4–23, 4–24, and 4–25).

Good quality inertia sensors are often used as alternatives to window piezoelectric monitors. In this role, they are affixed to the window frame and will detect violent forcing or breaking of the frame or glass.

Application

The installation will depend almost entirely on the area, object, or surface to be protected. These devices are almost always surface-mounted and connected by normal, mechanically protected, concealed wiring or screened cables.

FIGURE 4–22   Master-slave inertia vibration switch (ball type)

control panel. Normally, up to ten sensors can be wired from a single analyzer using a 4-core cable. On larger installations with either large areas of wall or many areas covered by sensors, such as larger safe stores or book rooms, considerable savings in wiring costs, reserve battery capacity, and capital costs can be made using net systems. In these systems, sensors (up to ten) are linked to detectors, which carry out some of the analyzer functions. The sensors are then linked linearly.

In very general terms, one inertia sensor will cover up to 20 × 8 ft (6× 2.4 m) of a monolithic structure.

When used on safes and strong rooms, specialized advice should always be sought from two alternative sources, unless the client has her own expert or adviser. Generally, one switch will be required per safe or strong-room door (but this is dependent on the height and mass of the door) and in addition, switches will be needed for the body of the safe or vault.

When used on nonmonolithic, homogeneous structures, such as weld mesh cages, grills, etc., the inertia path between sensors must be improved by either connection by steel conduits or provision of a steel band welded onto the frame at the appropriate height. When used on window frames, the orientation of the switch will change its sensitivity.

Equipment Types

The pure inertia switches fall into three categories:

- ball devices (single, double, etc.)
- cylinder devices

Ball Devices. The ball devices (see Figures 4–22 and 4–26) were the first to be developed. They have been tried and tested over many years, and their numbers greatly exceed those of the more recently developed cylinder types. The switch usually consists of a gold-plated ball held within a shallow "cup" formed by the placement of several gold-plated electrical contacts around its circumference. The contacts are wired to an

FIGURE 4–23   Frame-mounted inertia switch

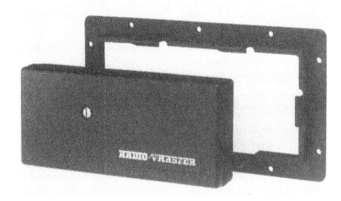

FIGURE 4–24   Radio transmission model

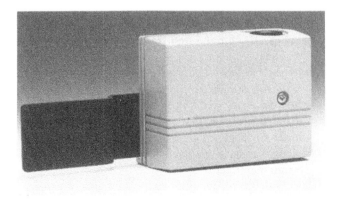

FIGURE 4–25   Strong-room, wall-mounted inertia switch

Systems usually consist of a number of sensors (inertia switches) linked to a remote analyzing unit. This unit might be one of several such analyzers fed from a zone

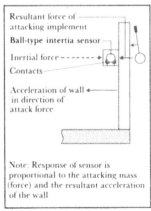

FIGURE 4–26   Wall-mounted inertia sensor

electronic analyzer (a processing and selection circuit) through the sides of a sealed, plastic, insulating enclosure, which prevents the ball from ever rolling off all the contacts completely and reduces its exposure to contaminating atmospheres. When fixed to a structure, the ball within the switch will respond to any vibration by rolling away from some contacts and up onto others, hence making and breaking electrical circuits. The level of vibration and the frequency at which the alarm state will be activated, can be adjusted by many methods, such as increasing the mass of the ball, holding it in position with the assistance of a tiny magnet or spring, or increasing the friction of the contact surfaces.

Ball-type devices are found in many specialized variations to cover particular applications and manufacturers must always be approached for advice on the final selection from a range to protect a particular make of safe, weight of door, or type of structure.

Cylinder Devices. Cylinder devices are a modified form of the vibrating reed switch (or *vibroswitch*). They consist of two contacts—one fixed and the other placed at the end of a springy metal strip. Under normal conditions, the contacts are closed and a circuit is maintained. Under vibration, the spring strip vibrates, lifting one contact off the other, opening the circuit, and triggering an alarm. In order for the device to differentiate between low-frequency vibration caused by traffic and high-frequency vibration caused by drilling, a mass is introduced to dampen the spring metal arm movement. The mass is provided by a hollow, stainless steel cylinder that is threaded over the metal arm, which oscillates around the arm as it vibrates, altering its center of gravity.

Most devices of this type also include a magnetic reed element, making it a hybrid switch for the detection of breaking through or opening a door, etc.

Although cheaper than the ball devices, cylinder devices are not as reliable, do not have true frequency selection and analysis, and should not be used to protect medium- to high-security areas. Essentially, these devices are for light commercial or domestic applications only.

"Rolling pool" sensors are claimed to have higher reliability due to a "self-cleaning" action.

Selection

When selecting a device for protection of a safe or strong room, be sure you know exactly what you are purchasing. Devices that look almost identical in case shape can be

- acoustic sensors
- structural attack, piezoelectric sensors
- inertia switches

The second two sensors are often referred to as *vibration detectors*. Sensors for the protection of safes are often

called *limpets* (they stick to the safe!), but this gives no indication of the principle of operation. Other common names are *buroguard, wallguard,* and *vaultphone*.

When selecting devices in general, compile a brief for the manufacturer and

be specific on the forms of attack against which protection is needed. Most devices will detect sawing, drilling, etc.; some will not detect scraping of mortar, light blows, etc.

always specify the materials on which the particular device will be used. Describe the type (i.e., wall: brick-cavity-brick, brick type, mass per m², area of wall, etc.).

if the client's organization gives a brief, always ensure devices comply with their brief [i.e., client may specify low- and high-frequency limits, lower gravity (g) unit, min/max displacement within which the device is required to act, etc.]. Do not take the word of a sales representative if this is the case; obtain written confirmation from the manufacturers if necessary.

Look for capabilities that will reduce the likelihood of false alarms, increase the likelihood of detection, and ease testing and maintenance, such as

units with sensitivity selection on each device (usually two to three levels)

test buzzers

latching neon indicator to show which device is activated and remote indicator capability

devices that include antitamper circuits

devices that are normally closed circuit (i.e., fail-safe)

devices that include an identified alarm frequency on a system should any device in a series covering the same structural element fail to operate

devices that include time sampling and counting capabilities (either in the device or by means of a remote analyzer)

switches that have low RFI

switches that respond and transmit an alarm within 3 seconds

switches that have separate, instantaneous response to explosive attack

switch cases that are PVC, hermetically sealed, with a metal inner case that has toughened or hardened inner surfaces

devices that have matched screen cable connections (same manufacturer)

devices that offer separate analysis circuits or replaceable, printed circuit modules

devices with a high operating temperature range—at least 14°F to 104°F (−10°C to 40°C); on external works, 14°F (−10°C) might not be sufficient

devices with gold-plated ball-and-contact sets, if possible (longer life, more reliable). Rhodium is an even better material, but it is seldom used on these devices, as it has a tendency to fuse together under pressure.

contact arms that have a crossbar to increase contact pressure. Rubidium contacts should be avoided, as these have a tendency to cold weld under pressure after a few months, disabling the switch to a "no alarm" situation and giving a false air of normality.

contacts bathed in an antioxidizing solution prior to assembly

contacts that may be swiveled to allow sensitivity plane selection

Limitations, Advantages, and Disadvantages

Limitations.

must be affixed to a fairly rigid structure

best on monolithic materials (e.g., concrete, glass, etc., not brick, mortar, or masonry)

environmental test period and presite inspection always recommended before final installation (additional cost)

Advantages.

can discriminate between levels of vibration, frequency, etc. This can be extremely important in preventing false alarms due to structural vibration caused by road traffic, wind flutter, someone knocking at a door when it is closed, children playing football against a wall, etc.

will give advanced warning of attack before entry is made, hence intruder warned off before she has a chance to see contents of building, attack internal security personnel, etc.

reliable in alarm response; can be used internally or externally

cheap to install, simple to service, and efficiency can be checked

cheaper than continuous wiring (can be used as a backup)

very versatile in application

small; some 0.5 in (13 mm) in diameter × 0.5 in (13 mm) deep

established device, well understood and trusted for use in many high-security applications

fits flush

slave strings possible

Disadvantages.

careful siting required to ensure switch is suitable for material and the environmental conditions; it can never be taken for granted that a device will work in a particular application

should only be used in areas not normally subject to continuous vibration or banging (partitions or high-trafficked doors)

not as many closed circuit devices from which to choose as there are for open circuit

many devices not as sensitive to low-speed drilling, mortar removal, etc., as they should be

usually surface-mounted

can fail due to contact welding, wear, or contamination, just like the points in a car

site sensitivity cannot be as finely tuned as a piezoelectric sensor of same type

prone to false alarms caused by low-frequency vibrations being transmitted through the structure from some remote source

PASSIVE INFRARED SENSOR

Introduction

Table 4–9 summarizes the types, uses, typical dimensions, other characteristics, and advantages/disadvantages of the passive infrared sensor.

Principles of Operation

All objects at a temperature above absolute zero radiate energy. The amount of energy they radiate is dependent on

their surface temperature level in *relation to surrounding temperatures*

the availability of energy within the object

the surface emissivity of the object (i.e., matte or reflective)

the physical size of the object

the internal energy generation (metabolism, activity)

the emissivity, absorption, transmission coefficient, etc., of surface *surrounding* the object to which it is trying to radiate

A significant proportion of the energy radiated by objects is in the infrared energy (IRE) part of the electromagnetic spectrum and is invisible (see Figure 4–27). In a typical office area, the surfaces of all materials, walls, floors, fixtures, etc., will radiate IRE. As both persons and surface are at approximately the same surface temperature, they radiate IRE of approximately the same wavelength. If all the surfaces in the room are stationary and no people are present (or they do not move), a pattern of radiating energy is set up that is fairly constant in magnitude and direction. If a person now enters the room, he will disturb that pattern in two ways:

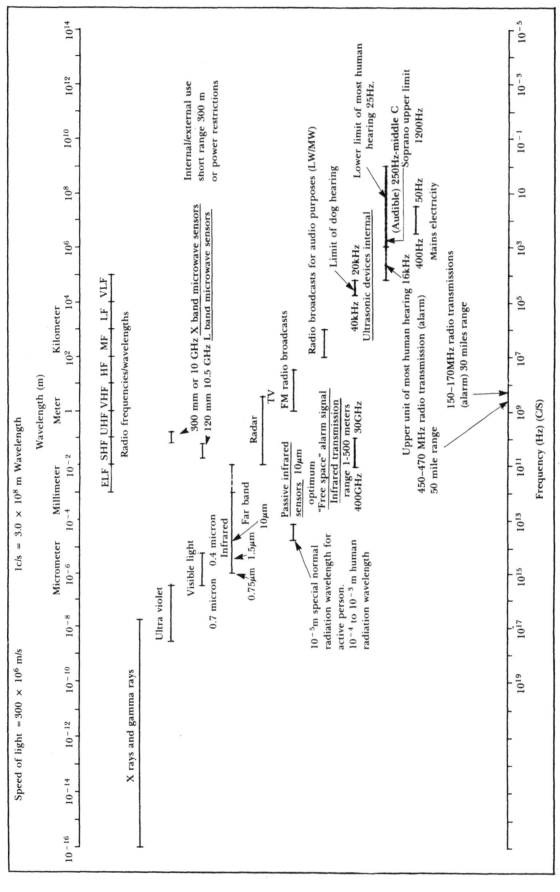

FIGURE 4-27 The electromagnetic spectrum and intruder detection sensors

Table 4–9 Passive Infrared Sensor

Types	Wall mounting, surface or flush. Ceiling mounting, surface. Long-range narrow beam for corridors. Wide-angle for rooms. Ceiling mounted for "curtain" coverage of doors/windows.
Range	Long-range narrow beam, up to 164 × 6.6 ft (50 × 2 m). Wide-angle or ceiling mounting, commonly 49 × 49 × 13 ft (15 × 15 × 4 m). Normal wall types: 33–98 ft (10–30 m), 60–120° spread "fan."
Activated by	Radiation of infrared energy from the human body reaching an infrared heat-sensitive sensing element. More specifically, a person of 88–176 lbs (40–80 kg) walking at speeds between 12 and 24 in/sec (0.3 and 0.6 m/sec) over a distance of 6.5 ft (2 m) causing: (a) a difference in radiation received by the sensor. (b) the rate of rise of temperature at the sensor being greater than 0.1°C/sec.
Standard	BS 4737, Part 3, Section 3.7 and UL.
Uses	Low- to medium-risk areas where environment unsuitable for either microwaves or ultrasonic, and/or very stable reliable detector required. Commonly used in office complexes, in corridors, cellular offices, etc.
Not to be used	In any area with a heating appliance, or luminaire, that will cause a rapid rise in temperature by the emission of large amounts of radiant infrared energy (e.g., electric fires, streamline night storage heaters, open fires).
Typical dimensions	Wall mounted: recessed—6 in (150 mm) high × 5 in (125 mm) wide × 4 in (100 mm) deep. Surface—6.3 in (160 mm) high × 4.5 in (115 mm) wide × 2 in (50 mm) deep. Ceiling mounted: 6 in (150 mm) diameter.
Associated work	Walk test: 14-day "in use" commissioning period required.
Advantages	Stable and reliable if correctly sited.
Disadvantages	Easy to overcome, less sensitive than ultrasonic or microwave.

he will interrupt the lines of radiant flux passing between some objects and others

he will generate his own IRE, causing some lines of radiant flux to increase in magnitude

If he walks across the room, he will effectively cause an IRE shadow to fall on some surfaces while increasing the level of others. The areas of increased IRE will rise in temperature. Passive infrared sensors are built to recognize the resultant consequences of a person entering a stable IRE field. They do this by

introducing an element or elements that will respond to minute changes in received IRE (generating a voltage of microvolt level)

ensuring that the device gathers IRE from a number of discrete, individual, sample zones each divided from the next by a blind area or dead zone

recognizing changes in IRE that occur in zone sections, elements, individual zones, or between different zones, within a given time period

only responding to IRE in the wavelength range characterized by the normal activities of a human intruder, typically 8–14 μm)

filtering out infrared energy from other sources such as lights, sunlight, heating appliances, etc., in order to reduce the background noise or infrared haze

amplifying any changes of IRE seen as being characteristic of the presence of an intruder

The passive infrared (PIR) sensor consists primarily of a chemical element or ceramic material, enclosed in a special envelope. This element is ultrasensitive to infrared energy and acts as a *sink* or sensor of that energy. Any infrared energy reaching the sensor causes a change in the electrical properties of the element. This change can be monitored, amplified, and made to raise an alarm.

The development of PIR devices has concentrated on taking advantage of or compensating for the following characteristics of situations in which they would normally be used.

Background Characteristics.

temperature of typical surface of no importance

emissivities likely to be seen

rate of change across given areas of emissivity/wavelength

duration of changes

uniformity of a change in the field of view

Target Characteristics.

wavelength of IRE generated (i.e., at 70°F, or 21°C, room surfaces have wavelength of approximately 1.0 μm, sunlight 2.7 μm, intruders 10–14 μm or more)

true size and perspective (i.e., a small animal that is close covers the same viewing sector as a man a long way away!)

temperature difference of man to background (non-normal surface)

temperature is between 96°F and 98°F (32°C and 33°C) characteristic of a human

speed of movement and unformity of blockage across similar, viewed zones

speed of travel through zones

All PIR sensors use the principle of differentiation (presence/nonpresence) to check if an intruder or object is really there or whether it is an internal anomaly of their electronics, such as EMI/RFI. They do this by creating radiating fingers of sensitive areas referred to as *zones*. These zones may be in one or more layers. Between zones and layers there are *areas of no sensitivity* or dead zones for comparison. These insensitive zones are one of the few weak spots of PIR detection.

Sensor Zone Types

There are various types of zones used in current PIR sensors. They are, in order increasing complexity,

1. single element per zone sensing
2. twin element per zone sensing
3. dual edge per zone sensing
4. quadruple element per zone sensing
5. octal element sensing (360° sensing)

Types 1 and 2 are now virtually obsolete and will normally only be encountered on cheaper range sensors aimed primarily at the domestic or very low value/low risk commercial sector. (See Figures 4-28 through 4-41 for examples of different PIR sensors.)

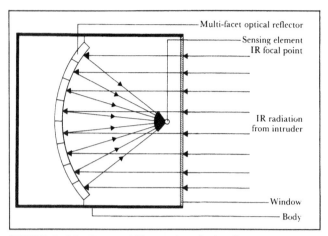

FIGURE 4-29   The multifaceted optical reflector

FIGURE 4-30   Long-range, dual zone PIR sensor

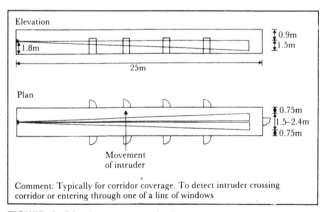

FIGURE 4-31   Long-range, dual zone, narrow coverage (single beam) PIR sensor

Dual Edge Sensing. Dual edge sensors (see Figures 4-42 and 4-43) take each radiating zone and effectively divide it down the middle to give a positive section (or edge) and a negative one. In normal use, an infrared source must

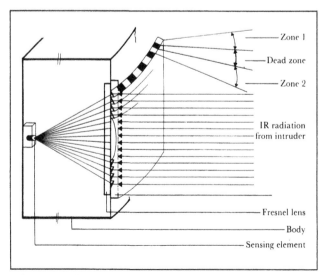

FIGURE 4-28   The Fresnel lens

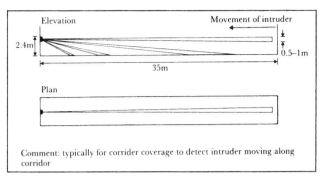

FIGURE 4–32 Long-range, multizone, narrow coverage PIR sensor

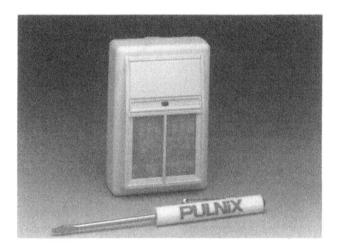

FIGURE 4–33 Long-range, multizone, narrow coverage, dual element sensor

FIGURE 4–34 Multizone, 360° curtain PIR sensor

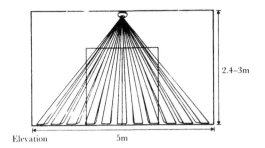

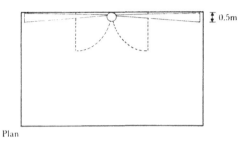

FIGURE 4–35 Multizone, vertical curtain PIR sensor

FIGURE 4–36 Wide-angle, multizone, antiblinding PIR sensor

FIGURE 4–37 Wide-angle or 360°, quadruple element PIR sensor

appear in an active zone

move through the zone from the positive-leading to the negative-leading edge or vice versa

move between the leading edge and through both edges *within a certain time period*

The sensor processor constantly monitors the edges of each zone and looks for a characteristic level of energy passing through the zone in, for example, 0.06 seconds, and obscuring a certain zonal area. In some sensors, sequential path-tracking is used to check the presence and

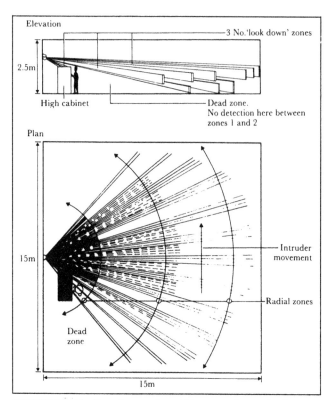

FIGURE 4-38   Broadspread, multizone, volumetric (for wide-range) coverage

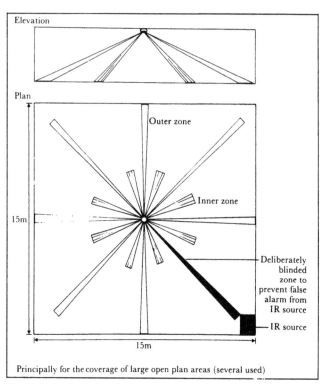

FIGURE 4-40   Ceiling-mounted, wide-range, dual zone PIR sensor

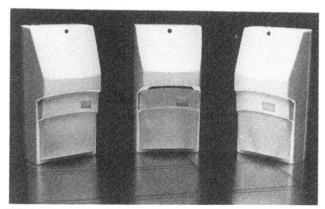

FIGURE 4-39   Single, twin, and quadruple elements

movement of an IRE source through more than one zone. The benefits of this sensing type are

   it has a high resistance to false alarms caused by *stationary*, but intense, IRE sources that might have energy levels similar to a person (e.g., radiant fires, etc.)

   it has good resistance to false alarms caused by moving IRE sources that are not characteristic of a moving person (e.g., tracking sunlight, moving hot plates, etc.)

   many systems incorporate *trouble logs* to record all sources of IRE that appeared in one zone, through one edge, or moved, but not characteristically. This allows maintenance engineers to look at records for

FIGURE 4-41   Steerable PIR sensor

each detector (if any one keeps giving trouble) and to make educated guesses as to what the problem might be.

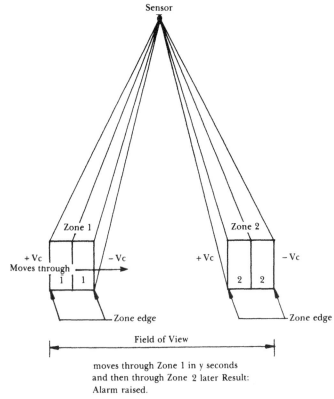

moves through Zone 1 in y seconds
and then through Zone 2 later Result:
Alarm raised.

Intruder passing through 2 edges on Zone and 2 zones

FIGURE 4-42 Dual edge sensing

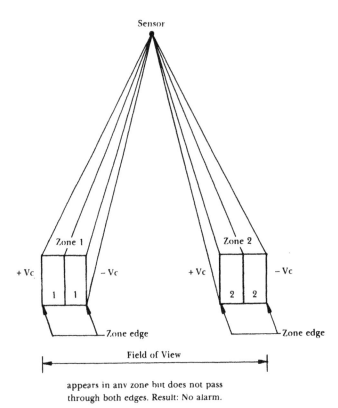

appears in any zone but does not pass
through both edges. Result: No alarm.

Intruder passing through 1 edge and zone

FIGURE 4-43 Dual edge sensing

Quadruple and Octal Sensing. Quadruple (*quad*) and octal sensing (see Figures 4-44 through 4-49) take the principle of splitting any zone into parts or elements, but do not use the pass-through-a-line principle. They use the "fill" principle. As the name implies, quad sensing splits each zone into four elements—two top and two bottom—normally referred to as *A* and *B*. The *A*s are matched halves— $+A$ and $-A$; similarly, the Bs are $+B$ and $-B$. Figure 4-44 shows this.

A processor attached to the sensor takes the radiant energy level value (over a certain threshold) from each of the four areas and performs the simple equation:

$$|A + B| - |A - B| = 0$$

If 0 is the result, no alarm will be raised.

Note that the enclosures are modulus signs, not brackets. This is important as it does not matter what value appears inside the enclosing lines; the algebraic law says it shall always be treated as a *positive* value.

Look at Figure 4-45 and see that a small rodent appears in zone 1 and presents a value of $2\mu J$ to the $B_1 +$ element. (This could be above and below background; it only matters that it is there "filling" the element.) None of the other elements "sees" the rodent.

The processor computes this as

$$|0 + 2| - |0 - 2| = |+2| - |-2| = 2 - 2 = 0$$

So, no threshold is reached and no alarm is given.

Now look at Figure 4-46. A person walks into the space and fills $+A$ and $+B$ with a value of $5\mu J$:

$$|5 + 5| - |5 - 5| = |10| - |0| = 10 - 0 =$$
$$10 = \text{alarm threshold}$$

Now look at Figure 4-47, the person has covered his body to reduce his surface radiation. He crawls to try and avoid the sensor, but he has been seen.

$$|1 + 3| - |1 - 3| = |4| - |2| = 4 - 2 = 2 =$$
$$\text{alarm}$$

Figure 4-48 depicts a diagonally moving intruder:

$$|2 + 2| - |2 - 2| = |4| - |0| = 4 = \text{alarm}$$

So every second, the processor computes the equation and determines

Was "it" present in any pair of four elements?
Was "its" value of significance?
Is "it" still there or has "it" moved to zone 2?

One loophole you may have detected is *perspective* or *apparent size*. What if the rodent is sitting *on top of the sensor*? It would appear to fill all elements and all zones (see Figure 4-49).

$$|2 + 2| - |2 - 2| = |4| - |0| = 4 - 0 = 4 =$$
$$\text{alarm}$$

Well, yes and no! If the device has a good antimasking device to recognize a deliberate blockage attempt, it would sound an alarm anyway, *assuming* a masking at-

$$|A + B| - |A - B| > \text{ or } < 0$$

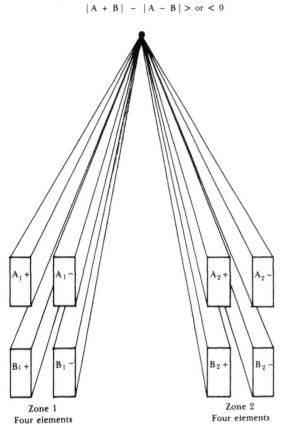

Zone 1
Four elements

Zone 2
Four elements

FIGURE 4–44   Quadruple sensing

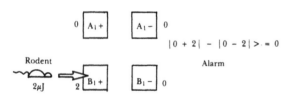

Rodent

$2\mu J$

$$|0 + 2| - |0 - 2| > = 0$$
Alarm

FIGURE 4–45   Quadruple sensing

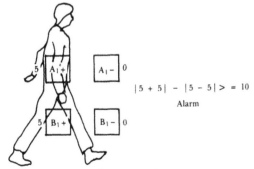

$$|5 + 5| - |5 - 5| > = 10$$
Alarm

FIGURE 4–46   Quadruple sensing

tempt (particularly if it was the active infrared integral beam type). If the device has a good IRE filter to prevent nonhuman wavelengths, sunlight, etc., from causing false alarms, it might not sound an alarm, as rodents have different metabolic rates and surface temperatures, and, hence, radiant energy wavelengths.

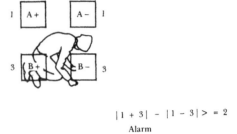

$$|1 + 3| - |1 - 3| > = 2$$
Alarm

FIGURE 4–47   Quadruple sensing

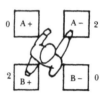

$$|2 + 2| - |2 - 2| = 4 \text{ Alarm}$$

FIGURE 4–48   Quadruple sensing

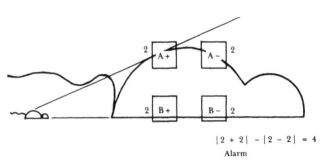

$$|2 + 2| - |2 - 2| = 4$$
Alarm

FIGURE 4–49   Quadruple sensing

Octal element sensing simply splits each zone into eight elements (four pairs). Each pair presents a slightly different area face-on to the intruder and the same effective area when viewed from any angle. This feature allows the device to be used for 360° coverage without complex processing and uses a similar equation.

The advantages of quad-type sensing are

very low false alarm rate to all human targets
will not recognize small, stationary IRE sources
will detect very weak levels of difference between background and target in the right wavelengths, thus reducing risk of radiant shielding attempts by intruders

The equation effectively amplifies the value. For example,

$$|0.5 + 0.5| - |0.5 - 0.5| = |1.0| - |0| = 1.0$$

The enhanced computation speed allows checking and rechecking of accuracy before any significant time has elapsed.

Analogue Sensing

All PIR sensors receive energy in an analogue form—as a varying, normally nonlinear quantity that one would expect from radiant energy with a *waveform*. Some sensors use this characteristic waveform to their advantage by recognizing the waveform's frequency, wavelength, amplitude, and cyclic nature. If the characteristic form appears in two zones (or layers of zones) *simultaneously*, a human is present. If it does not, it is probably a rodent or false alarm source. Future sensors will probably combine analogue sensing with human zone element mapping and digital processing to provide a sensor that will *never* recognize anything other than a human and never miss a human!

Analogue Digital-Sensing Processors

Much hype is being made in current literature about the benefits of converting an analogue signal received to a digital-sensing processor. Digital signals simply allow simpler processors to be used to perform calculations faster, at less cost. Digital signals *themselves* are not of particular benefit in detecting intruders.

Component Parts

The PIR sensor consists primarily of

- a case
- a window in the case with a filter
- a focusing device
- a sensing element (or sink)
- signaling and amplifying electronics
- a signal processor

The Case. The case should protect the contents from any adverse environmental conditions. As most PIRs are used internally, most are produced with very lightweight PVC or GPR enclosures. However, whatever the enclosure, its mounting and construction should protect the contents from

- dust
- water (accidental or otherwise)
- radio or other forms of EMI
- tampering

The Window Filter. At some point in the case, a window must be provided that will protect the internal electronics to at least the same degree as the case and filter out unwanted radiation of other types (e.g., visible light, radio waves, etc.), but gather the maximum amount of radiation characteristic of a man (i.e., 0.1–100Hz and more than 8–14 μm wavelength).

These windows are usually made of plastic or special glass that will transmit a high proportion of the IRE falling onto it. Some devices have only an opening in the case; others use the window as a site for the focusing device. Neither of these is recommended. In quality devices, the window is actually a filter that only allows IRE with wavelengths shorter than 3 μm to pass. This ensures that only IRE typical of that generated by the normal surface temperatures of the room and by man falls on the focusing device and the sensing element. The filters of most modern sensors are made from germanium.

The Focusing Device. The focusing device actually performs two functions. Being composed of many small facets or lenses, it effectively creates the discrete zones of incoming IRE and dead zones of no IRE. This allows recognition of changes of IRE from one zone to another as an intruder moves through the field of view of the sensor. It also allows the time taken for movement from one zone change to another (causing changes on the sensor element) to be accurately measured. Secondly, the focusing device concentrates all the energy of the viewed area onto a single point (the sensor element). This ensures that even small changes in IRE between zones will be effectively amplified, causing the sensor element to respond more quickly than would otherwise be the case.

Infrared energy is part of the light spectrum, hence it behaves just like ordinary visible light in that it travels at the speed of light in straight lines, and can be focused and reflected easily. The focusing device is the most important part of the PIR. Without it, the device would not "see" the IRE changes. The arrangement of the focusing device determines the coverage of a detector and the range and shape of that coverage. There are two common types of focusing devices—the Fresnel lens (FL) and the multifaceted optical reflector.

The *Fresnel lens* replaces the window in the housing and focuses IRE directly onto the sensing element. Being a true lens, it has very accurate optical properties, giving very clear zones of detection. However, being made of glass, it suffers from the fact that glass absorbs IRE and, hence, there is a reduction in efficiency (see Figure 4–28).

The *multifaceted optical reflector,* commonly called a *mirror,* is a concave, metallic, or silvered reflector with many small facets making up the whole (rather like an inverted ballroom dancing dome!). IRE is taken through the filter, past the sensing element, and reflected and focused back onto it (see Figure 4–29). The mirror facets are separated by small amounts and this produces the zoning. Being concave in shape, the mirror arrangement has an advantage, as it amplifies and concentrates the beam of incoming energy.

The Sensing Element.

The sensing element must have the property that its electrical nature changes when heated by infrared energy of very low levels. Several types of sensors have been used in the past, each responding in a particular way, including

- triglycerine sulfate in a germanium envelope: direct current production
- thermocouple: direct current production due to temperature differential
- thermistor (usually a metal oxide): resistance change with temperature
- semiconductors: lead telluride, lead sulfide, indium antimonide, etc.

Most of these materials have drawbacks and have generally been superseded by the use of *pyroelectric ceramic* or *plastic sensing elements*. Most of production now revolves around the use of pyroelectric ceramic materials, such as lithium tantalate, lead zirconium titanate, etc.

Pyroelectric ceramic elements change polarity from *plus* at one junction to *minus* at the other, and vice versa, with a change in temperature. They will exhibit this property only if they "see" a temperature change. This change in polarity is effectively a current flow in the device and its magnitude can be monitored and used to raise an alarm. Under normal conditions, a stable level of IRE exists in all zones of detection and no change of temperature at the sensing element occurs. When an intruder enters the space, this quiescent condition is disturbed. IRE levels in the first zone entered will rise and the temperature at the sensing element will rise. Hence, its polarity might be positive. As the intruder moves from this first zone to the second, the temperature of the sensor falls and the polarity becomes negative. If sufficient change occurs in a given time and this results in a significant sensing element temperature rise, an alarm will be triggered. Several manufacturers are now producing plastic sensing elements that are much cheaper to produce than pyroelectric ceramic types. Plastic elements are, however, in their infancy and, other than a small cost reduction, offer no major advantages, particularly as pyroelectric ceramic elements are tried, tested, and extremely reliable.

Signaling and Amplifying Electronics.

These processing electronics are used to

set threshold limits on the sensitivity
perform algebraic equations
amplify the received energy and use it to drive the communication down a transmission cable
provide protection against voltage reduction, current or voltage spikes on the mains, etc.

Being passive devices that react to changes in actual physical zones, there is no need for the PIR to have any of the sophisticated signal processing associated with active devices, such as ultrasonic sensors and microwave sensors.

Uses

PIRs are used principally for the detection of infrared radiation emissions at wavelengths and energy levels characteristic of humans moving through a space.

PIRs are used primarily for indoor detection.

PIRs are used extensively for trouble-free volumetric detection, particularly in long, uninterrupted corridors.

PIRs are used where noise, vibration, shifting room furniture, etc., make ultrasonic or microwave detectors unsuitable.

PIRs can be used to protect very small areas and/or objects, or to "see" cutting of steel, etc., by thermal lance.

PIR systems are totally covert, with no emission that can be sensed or measured by an intruder.

PIRs can be individually tailored or set up to avoid false alarms being caused by radiant IRE sources.

Very low power consumption makes PIRs particularly useful as stand-alone units with self-contained batteries and radio transmission of alarm signals. PIRs are useful in areas where wiring concealment is difficult and needs change—art galleries, museums, etc.

Units are small, typically $7 \times 4 \times 8$ in ($175 \times 100 \times 200$ mm) (max.) and can be used fully recessed for concealment and aesthetics.

Sensitive PIRs with a latching mechanism can be used not only to raise the alarm, but to "trip out" as the criminal passes through the protective zone of each device. In this way, even if the intruder is not caught, the sequence of tripping or latching can map the entry or exit route from the building, giving valuable information.

PIRs are particularly useful where large areas of vertical glazing are present. As they are passive, no energy can escape from the protected space, as is the case with microwave and ultrasonic sensors. As infrared energy of the critical wavelengths to which PIRs are sensitive ($3 \mu m$) is highly attenuated by normal glass, passersby are unlikely to trigger the sensor.

Application

In the early days of their use PIRs received a lot of bad press within the industry. This was partly due to delayed development of a stable and reliable sensing element (which had not been achieved), but mainly due to the lack of thought that went into their installation. Before considering their use, it is essential that

there is no infrared-generating equipment in the room that would be left on (even accidentally). This includes infrared heaters, electric fires or natural convectors, night storage heaters, heat lamps, and normal tungsten lamps near the sensor (9.8–16.4 ft, or 3–5 m). The most fundamental requirement for the successful application of the PIR sensor is a stable IRE pattern and temperature gradient within the space.

they are not placed in positions where unfiltered sunlight could directly affect them. This applies where security barred windows are left open during the day with the alarm system armed. PIRs should not be used in areas where very rapid temperature increases occur on surfaces due to direct solar radiation (e.g., very highly glazed foyers), particularly if they have glass roofs.

the room furniture layout is known before the device is finally positioned. Large cabinets, bookcases, etc., could shield the infrared energy from the intruder and prevent it from reaching the sensor. As these objects will only absorb the energy extremely slowly, the energy would not "appear" on the opposite side, facing the sensor, for some considerable time.

they are not to be placed in rooms with flapping signs, curtains, etc., or where insects, birds, etc., might be present. Any of these could either directly cover the sensor zones or mask an object normally contributing to the radiant energy in a zone. Either would give rise to a false alarm.

While many manufacturers will claim their devices have features that will cure these ills, why take the risk in the first place if they can be avoided?

Devices should always present the maximum number of zones at right angles to the most likely path of an intruder. As most devices are more sensitive to changes in IRE between zones than within an individual zone, the more zones the intruder passes through in a given time, the more likely his detection.

Passive infrared sensors are now available with *blinds* or *masks* that will remove individual zones, making them insensitive to IRE from a particular source. This should be looked upon as a latent capability should the room use or equipment change, and is not a way of installing devices in rooms for which they would otherwise be unsuitable.

The sensor must be placed in a position where it would be extremely difficult for an intruder to reach it and mask it, making it totally insensitive.

Equipment Types and Selection

The basic types are defined by the zonal pattern of sensitive areas or facets:

Long-range, dual zone, narrow coverage — Used primarily on long corridors with many crossing paths or along lines of windows (see Figures 4–30 and 4–31).

Long-range, multizone, narrow coverage — Used on long corridors to detect intruder walking toward or away from detector. Generally better than previous sensor (see Figures 4–32 and 4–33).

Multizone, barrier coverage (or curtain) — Usually mounted on the ceiling above large entrance doors, glazed areas, etc., to act as an invisible fence or curtain of detection (see Figures 4–34 and 4–35).

Broadspread, multizone, volumetric (or wide-angle) — These are the most common types of PIRs and can be used to give detection of an intruder moving about in any small or medium-sized office or similar space. In general, they are usually more sensitive to movement across the space (at right angles to the zones), because not as many look-down zones as radial zones exist (see Figures 4–36, 4–37, and 4–38).

Ceiling-mounted, 360° coverage, two-zone — The device is ceiling-mounted and produces a 360° conical arrangement of zones of two different circumference lengths. It is extremely useful in offices with medium-height workstation partitioning or high cabinets that might block the field of view of other detectors. Several devices can be used to give overlapping pools of detection and when radiant sources exist in a space, the zone in that area can be deliberately blinded (see Figures 4–39 and 4–40). Wall-mounted, swivel bracket PIRs are illustrated in Figure 4–41.

If possible, always choose devices that

have dual edge or quad sensing elements to reduce false alarms

have the maximum number of discrete zones, lookdown zones, and several sensing ranges in the same device

have surge and low-voltage protection, and RFI suppression

use pyroelectric ceramic sensing elements

have a proper window (not just an opening)

are UL-listed and comply with British Standard 4737, Part 3, Section 3.7

have zone adjustment switches that allow the range to be changed or individual zones to be disregarded. This will allow for changing room layouts.

are latching, with an LED indicator

have walk-test capabilities that incorporate a plug-in test meter connector can be plugged into a mounting base plate or ceiling rose, allowing the sensor head to be rapidly changed, should the need arise

have antimasking devices. These are normally a very low power, *active* infrared transceiver that will sound an alarm if a masking attempt is made. Whether the detector is sleeping or armed, it should generate its own audible warning.

have high RFI/EMI rejection using tantallum capacitor circuits

have a high signal-to-noise ratio

have a trouble log capability

have an automatic self-test capability that injects IRE into the device every few hours and, if it fails to respond correctly, gives a trouble or fault signal

have a temperature use range of 14°F to 122°F (−10°C to 50°C) at the minimum

have a silent alarm relay

have temperature gain compensation. The processor must be able to compensate for reducing signal strength radiated as IRE when the uniform room temperature rises. As the background temperature more closely approaches the target's surface temperature, the IRE difference becomes less and is harder to detect.

have enhanced processing (i.e., simple — one threshold level or value; better — rate of rise and threshold; better still — has preceding features plus duration in zone and presence in both elements)

have a tamper switch

have site adjustable sensitivity control and pulse count before triggering an alarm

have first-to-alarm memory

have a "low voltage" signal

have multifaceted reflectors — first choice, aspheric FL; second choice, FL; third choice, mirror reflector

Advantages and Disadvantages

Advantages.

lower false alarm rate than ultrasonic or microwave

small, light, can be recessed; simple to install

very low power consumption; cheap to operate

does not penetrate glass (emits no energy)

no moving parts

simple electronics

no detectable beam that can be found and avoided

cheaper than most other volumetric detectors

longer range than ultrasonics

range of detection or pattern easily changed *in situ* by changing the type of FL and/or sensor head

no chance of mutual interference between sensors

Disadvantages.

will only detect IRE if it is present in sufficient quantities for them to "see." It is a relatively simple matter for an intruder to wear a suit that reflects the body heat inward to beat the detector. This reduces the level of security afforded.

does not give the density or uniformity of coverage offered by ultrasonic or microwave detectors. It has dead zones and furniture, etc., can cause its field of view to be blocked.

range is limited to wide zone coverage at up to, say, 49 × 49 × 164 ft (15 × 15 × 50 m), and narrow beam coverage of 164 ft (50 m), although a ceiling-mounted disk pattern will give a 49 ft (15 m) diameter circular coverage

not fail-safe; will not activate an alarm on failure of the sensing element to sense. Most volumetric detectors produce a beam that, if it is distorted or fails, causes an alarm to be raised. This is the primary reason for its better false alarm tolerance. Some insurance companies will not approve the use of PIRs unless used in pairs or with dual sensing elements — one element effectively being a failure backup.

other IRE sources can trigger an alarm

some devices use piezoelectric sensing elements, which can mistake RFI for IRE. The best devices using these elements use "twinned" elements to cancel these effects.

ambient temperature can affect the performance. Manufacturers quote a range of 32°F to 122°F (0°C to +50°C); relative humidity 10–90%. However, after 80.6°F (27°C) the performance decreases, as this is the surface temperature of the human body, and hence differentials between background and intruder energy reduce and the intruder blends into the background. This can be a problem for radiator heating systems.

a maintenance factor is associated with the device, as the element cannot be cleaned and the mirror may get dusty, unless properly protected. This is particularly noticeable during dry conditions when dust electrostatically attaches to the device and clouds its "vision."

can be masked if an intruder reaches it without detection; better devices have an antitamper/mask alarm

if the device is wired directly to a police station, a 14-day probation is required before connection

can be set off by strong direct sunlight or by animals in the field of view; set the units high to avoid this

Conclusion

The PIR device is probably the most commonly used, least troublesome of all volumetric detectors. It is used primarily for low- to medium-security risks.

Model Specification

In order to give the manufacturer/installer a chance, you will need to specify several types of information.

Performance Specification.

area of coverage required (give plan—side and end elevations)

possible routes of intruders to secure area by normal circulation (so that detector can be arranged across the path)

type of target, size of target, and infrared sensitivity requirement (see British Standard). You might require a high sensitivity to "see" someone wearing insulated clothing or someone moving extremely slowly.

what the device is to protect; PIR devices are essentially first-line-of-defense, low-security, low-value, target devices with high reliability and are not suitable for high-security, high-value targets

drawings to scale, preferably 1:50 or 1:100, showing final furniture layouts, position of plant, etc.

Environmental and Operating Conditions.

type of space, office or factory, and dust/dirt likelihood

heat- or IRE-producing equipment within space and position

ambient temperature, relative humidity

any unsuppressed electrical equipment that might cause interference

operating voltage to be supplied and tolerance and regulation; most are 12–20V DC and incorporate low-voltage, "no function" alarm

Installation Requirements.

room height, positions where detectors cannot be placed (i.e., in direct sun or where other signs, lights, etc., might clash)

positions of furniture and fixtures, particularly racks or shelves that might block view. Always indicate windows on which curtains or loose blinds are to be used.

whether surface or concealed wiring, or radio signal transmission will be present

whether surface or flush-recessed units are required

A typical example of a US manufacturer's standard architectural specification is shown on the next page.

Model Information

At a minimum, a PIR manufacturer should provide you with

a perspective or artist's impression of the pattern of coverage

a plan (end and side elevations), drawn to scale, of the zone of detection, indicating not only the length, depth, and breadth of each discrete zone of protection, but more importantly, the length, depth, and breadth of each dead zone

a statement of the range

whether the PIR uses an FL or mirror focus, and whether the lens can be changed on-site, using a replacement or pattern changing lens

the type of window

the material of the housing

the maximum operating temperature

the optimum operating temperature

the weight and dimensions

whether the PIR can be recessed

whether the PIR includes integral mains failure battery backup (wireless type)

whether the PIR includes "latching" LED

whether the PIR includes a walk-test capability and indicator

typical applications for the pattern of coverage

the vertical and horizontal coverage pattern adjustment; is it possible once installed?

whether mounting or tilting brackets are included in the package

the operating voltage, current consumption, relay operation and rating

whether the electronics are modular, on-site replaceable (plug-in)

whether the PIR can be used with a radio transmitter

mounting height (preferred)

whether it includes antitamper and antimask devices

whether its *pattress* (base) will fit directly onto standard conduit back boxes

whether the PIR is "approved" (US or British)

to what factory testing it was subjected (e.g., element screening for failure, environmental and soak tests, sensitivity minimum and maximum, noise resistance test)

ULTRASONIC SENSOR

Introduction

Table 4–10 summarizes the types, uses, typical dimensions, other characteristics, and advantages/disadvantages of the ultrasonic sensor.

Principles of Operation

Sound is caused by vibration of an object in a transmitting medium. The medium in which most of us are used

ARCHITECTURAL SPECIFICATION FOR 360° RADIAL PATTERN PASSIVE INFRARED INTRUSION DETECTOR

1.0 Specification Requirements

1.1 Intent
The intent of this specification is to provide a detailed description of a passive infrared (PIR) intrusion detector having a 360° radial detection pattern so that this item may be accurately specified for use in appropriate applications.

1.2 Scope
The passive infrared intrusion sensor described in this specification shall be self-contained, serviceable in the field, and designed for ease of installation and adjustment.

2.0 General Requirements

2.1 Principle of Operation
The PIR shall detect the movement of a body within the given detection pattern by sensing the temperature variance between that of the intruder's body and the ambient temperature within the detection pattern. Infrared energy emitted by the intruder shall be detected by reflection through a mirror system to a single-element pyro-electric sensing unit within the PIR.

2.2 Detection Pattern
The PIR shall be capable of providing a full 360° detection pattern from a single unit which is ceiling mounted. The detection pattern shall consist of an outer detection pattern, and a second inner pattern having a radius approximately one-half that of the outer pattern; both inner and outer detection patterns shall be capable of 360° coverage. An additional single detection segment shall be located at the exact center of the radial patterns and shall extend vertically from the center of the PIR.

The outer radial detection pattern shall be thirty-five to forty (35 to 40) feet in diameter when the PIR is mounted at a ceiling height of eight to ten (8 to 10) feet. The outer detection pattern shall be approximately sixty (60) feet in diameter when the PIR is mounted at a maximum height of sixteen (16) feet.

2.2.1 Zone Shielding Disk
A zone shielding disk shall be provided which, when mounted within the PIR, allows selected portions of the outer radial pattern to be masked. Masked portions shall not detect motion.

2.3 Automatic Temperature Compensation
The PIR shall have special circuitry which allows the motion detection sensitivity of the PIR to automatically adjust relative to the environmental temperature. The sensitivity shall become maximum at body temperature range (approx. 95° F) and decrease when the temperature differential increases either above or below that of the body temperature range.

2.4 Sensitivity Adjustment
The PIR shall have a manually adjustable sensitivity control located within the unit. This shall provide for adjusting the PIR detection sensitivity relative to the ambient environment.

2.5 Tamper Warning
The PIR shall be equipped with Tamper outputs to provide an alarm signal if the body of the PIR is detached from the base, or if the cover is removed from the body.

2.6 Underwriters Laboratories Listing
The PIR shall be listed by Underwriters Laboratories, Inc.

2.7 Warranty
The PIR shall be covered by a three (3) year warranty from the manufacturer commencing on the date of shipment from the manufacturer.

3.0 Electrical Requirements

3.1 Power Requirements

3.1.1 The PIR shall operate on a supply voltage of 12V DC to 30V DC (non-polarity).

3.1.2 Power consumption shall be 30 mA @ 12V DC standby.

3.2 Electrical Requirements

3.2.1 The alarm signal shall utilize a dry contact relay output, 1C, (N/C, Common, N/O); 2 to 5 second reset time; contact capacity of 30V DC, 0.5A.

3.2.2 The PIR shall have one set of normally-closed (NC) contacts for Tamper output.

4.0 Mechanical Requirements

4.1 Case
The PIR case shall consist of an ABS plastic base with a rigid polymer cap. The case shall include a detachable base to mount the unit.

4.2 Walktest LEDs
The PIR shall have two walktest LEDs mounted 180° apart for walktesting the unit. These LEDs shall have covers available should it be desirable to cover them.

4.3 External Sensitivity Check
The PIR shall have an external jack through which an optional level checker may be attached to display sensitivity of the unit for purposes of installation and adjustment.

4.4 Weight
The PIR shall not exceed thirteen (13) ounces in weight. The PIR, when mounted in the mating flushmount kit, shall not exceed one pound, three ounces (1 lb., 3 oz.) in weight.

4.5 Dimensions/Mounting

4.5.1 The PIR shall be 3.875″ in height and 5.750″ in diameter.

4.5.2 The PIR base shall be capable of mounting on a standard electrical beck box. The distance between mounting holes shall be either 2.675″ or 3.250″.

4.5.3 The PIR, when mounted in mating flushmount kit, shall be 4.250″ in height and 7.50″ in diameter.

4.5.4 The cut-out hole required for mounting the PIR in a flushmount kit shall be 6.750″ in diameter.

5.0 Environmental Requirements

5.1 The PIR shall be capable of operation in an ambient temperature range of 14°F to 123°F (−10°C to +50°C).

Table 4-10 Ultrasonic Sensor

Types	1. Wall mounting. 2. Corner mounting. 3. Ceiling mounting. With transmitter and receiver in one housing; or separate units. Former "transceiver" is self-contained; latter either "master-slave" or "multihead."
Range	Linear ranges: usually 33-49 ft (10-15 m) (effectively) × 10-13 ft (3-4 m) across. Ceiling mounted: sphere or cone 16-52 ft (5-16 m) across at base of cone. Master-slave units: coverage approximately 78 yd^2 (65 m^2) each. Slave multihead pair: 191-215 yd^2 (160-180 m^2) with systems having 20-100 pairs.
Activated by	Any solid object, or person of mass 88-176 lbs (40-80 kg) moving at any speed between 1 and 2 ft/sec (0.3 and 0.6 m/sec) through a distance of 6.5 ft (2 m) toward or away from the detector, or by 20% of his radial separation from the detector, whose reflected ultrasonic energy saturates the detector for greater than 200 msec.
Standard	BS 4737, Part 3, Section 3.5. UL listed.
Uses	Low- to medium-risk areas. Best used in very stable environments. Very useful in foyers, lobbies, display window areas, all of which have a highly glazed facade (ultrasonic energy is almost totally attenuated by glass).
Not to be used	In any area with moving or swinging objects, machinery, etc. In any area where fans, quick response heating, plant, etc., could affect air movement or density severely.
Typical dimensions	Ceiling mounted transceivers: 6 × 6 × 4 in (150 × 150 × 100 mm) (smallest, least reliable) to 6 × 31 × 4 in (150 × 800 × 100 mm) (more reliable). Wall mounted: 23 × 4 × 1.6 in (250 × 100 × 40 mm). Corner mounted: 4 × 4 × 8 in (100 × 100 × 200 mm).
Associated work	Walk test: 14-day "in use" commissioning period required.
Advantage	Less prone to false alarm than microwave detector.
Disadvantages	Large and visible. Far less coverage per sensor than microwave.

to hearing sound transmitted is air. In a perfect vacuum, no sound is transmitted. This is because there is no surrounding medium against which the vibration of the object can exert force or pressure. Sound can be defined as a physical disturbance caused by the vibration of a solid object in a transmitting medium, resulting in a pulsation of pressure that can be detected by the human ear.

Sound travels as a series of waves with crests and troughs, just like the ripples from a pebble thrown into a pond. The distance from one crest to the next is called the *wavelength*. The number of wavelengths passing a point in one second is called the *frequency* and is measured in Hertz (i.e., 1 wavelength/second, or 1 cycle/second, = 1 Hz). In air of constant density, the wavelength (X) of sound is fixed and the speed (S) of sound is fixed; hence, the fundamental relationship *frequency × wavelength = speed of sound* shows that, in air, the only variable in sound transmission is the frequency (Hz).

The human ear can hear sounds in the frequency range of 18-16,000 Hz. The midband frequency of audibility is normally 1000 Hz (1 kHz). Sounds of a frequency lower than 18 Hz are called infrasonic and those above 16000 Hz range are *ultrasonic*. Ultrasonic detectors work based on *standing waves* or *Doppler shift* effects.

Standing Wave. If a sound transmitter is placed on one side of a room and a receiver on the other, and the transmitted sound is given a standing wave frequency of 40,000 Hz, the receiver will receive a sound with frequency 40,000 Hz, because S is constant and X is constant. This relationship will only occur if nothing is placed in the path of the sound wave to topple the wave. If an object is placed in the path of the wave, some sound energy will be bounced back toward the transmitter or take a more indirect path to the receiver (see Figure 4-50).

This indirect path will result in a change to the net frequency and energy of the sound received over a given period. The change may be detected and used to produce an alarm. The device will detect any new obstruction in the space. The use of the standing wave principle device has now been mostly superseded by the use of the Doppler shift effect.

Doppler Shift. If an object *moves* within a room filled with sound being transmitted at a particular frequency (pressure waves per second), that object will, itself, set up pressure waves in the air. These might work *with* those being transmitted by the device, giving an apparent over-

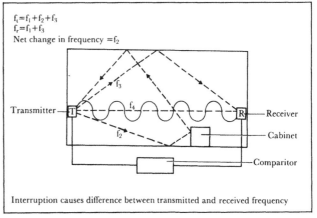

$f_t = f_1 + f_2 + f_3$
$f_r = f_1 + f_3$
Net change in frequency $= f_2$

Transmitter — T   R — Receiver
— Cabinet
f_2
— Comparitor

Interruption causes difference between transmitted and received frequency

FIGURE 4-50 Standing wave

all frequency increase, or *against* them, causing an apparent overall frequency reduction from the source (see Figure 4–51).

Hence, with the Doppler shift device, the moving object is "seen" as a frequency change. *Stationary* objects will not be "seen." The essential thing to remember about ultrasonic detectors (and all beam detectors) is that to recognize an intruder, emissions from the device must actually hit the intruder. A person walking behind a large bookcase or partition would cause no Doppler shift, if all the energy transmitted in that direction were reflected or absorbed by the bookcase.

As the ultrasonic sensor is arranged to sense a Doppler or frequency shift of a certain magnitude caused by an object entering a space (or movement within the space), the object and its actions determine whether it will be "seen." The Doppler shift produced depends on

the mass and density of the object and the reflectivity of surfaces around it

the *speed* at which it moves through the space under observation

the transmissivity of the moving material or object

the direction in which the object is moving; with the sound wave produced against it or across it

the size of the object and the number of waves it cuts through

the density of the air within the room, which is a function of temperature and humidity

To ensure that the detector can differentiate between a man (a thing of high mass and density) and any ethereal object, such as a draft of air or flapping curtains, requires careful design.

Luckily, human beings fall within easily defined limits in terms of mass. They also move within fairly close speed limits (0.7–33 ft/sec, or 0.2–10 m/sec, maximum) and, as sound waves are curved, it is difficult to move forward without cutting through a wave. Hence, humans will generally produce a characteristic frequency change across a wide frequency spectrum when moving toward a target (normally 10–800Hz). The receiver will be tuned to recognize this characteristic shift from its set frequency. If the shift is any less or any greater than expected, the device will ignore it; herein lies the greatest weakness of the device. If a device is tuned to a normal Doppler shift

recognition for a human in a room, it is possible that a person moving extremely slowly (i.e., less than 8 in/sec, or 0.2 m/sec) or a piece of apparatus (e.g., mechanical grappler, etc.) will not be detected, as it will not produce the expected response. Obviously, in areas where ambient conditions and equipment are very fixed and stable, such as a strong room, a more finely tuned device may be used, but in general areas this is not possible. The ability of the signal processor to receive and analyze frequency changes is of paramount importance for all devices.

All ultrasonic detectors are comprised of the following items. In some devices, the various parts may be contained in a single cover, while in others the parts may be physically separated.

The Transmitting Transducer. A *transducer* is basically a device that transforms one form of energy into another. In the transmitting transducer, electrical energy is provided to the device, which it transforms into ultrasonic sound energy by vibrating a material using a primary source (usually a piezoelectric crystal). This crystal is used to vibrate a disk of metal that then radiates ultrasound energy as a series of waves of fixed frequency, amplitude, and wavelength. The frequency of the sound will depend on the application and sensitivity required and the country of origin of the device.

BS 4737 calls for all ultrasonic devices to have a frequency of at least 22,000Hz. American and Japanese manufacturers offer devices with 19,000, 21,000, 25,000, 26,000, and 40,000Hz. The frequency chosen is a matter for the expert and manufacturer. They should base their recommendations on

the range required. The higher the frequency of a noise, the greater the energy loss over a fixed distance through air and, hence, the less energy available at a receiving transducer.

the physical objects and finishes within the room. The harder and more dense the surfaces, the greater the reflection of sound. The softer the finishes (i.e., carpets, curtains, etc.), the greater the attenuation of high frequencies.

the possibility of penetration to other areas. Low-frequency sound will be transmitted through walls more easily than high-frequency sound and could cause nuisance tripping of the devices in other areas by appearing as Doppler motion within that space. The higher the frequency chosen, the lower the effective range of the device, but the greater the sensitivity.

the amount of air movement envisioned in normal circumstances. The lower the frequency, the less effect air distribution and disturbance has on signal fidelity.

The transmitting transducer can be shaped to produce various types of radiated sound patterns (similar to the beam shaping of infrared detectors). The most common

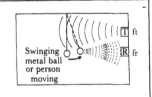

Swinging metal ball or person moving

Moving object 'compresses' waves, hence reflection frequency to receiver is different to transmitted frequency fr > ft

FIGURE 4–51 Doppler shift (R = Receiver; T = Transmitter; Ft = Frequency transmitter; Fr = Frequency receiver)

forms are disk, dome, ring, and tube. The typical
arrangement of these and their coverage patterns are
shown in Figures 4–52 through 4–54.

Some devices on the market offer *magnetocoil generators*,
which are basically high-frequency electromagnetic
speakers. These are usually of lower and greater frequency range and are more sensitive to intruder movement,
but they are less reliable.

The Receiving Transducer. The receiving transducer is matched in characteristics to the transmitting
transducer and works in the reverse manner—the metal

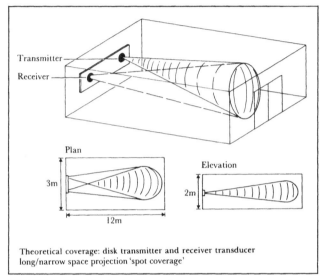

Theoretical coverage: disk transmitter and receiver transducer
long/narrow space projection 'spot coverage'

**FIGURE 4–52   Typical ultrasonic sensor coverage patterns:
disk**

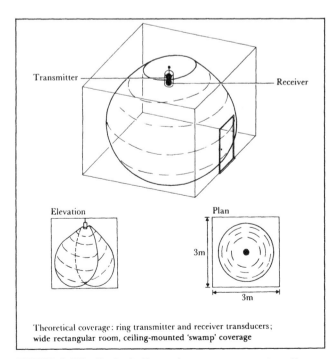

Theoretical coverage: ring transmitter and receiver transducers;
wide rectangular room, ceiling-mounted 'swamp' coverage

**FIGURE 4–53   Typical ultrasonic sensor coverage patterns:
ring**

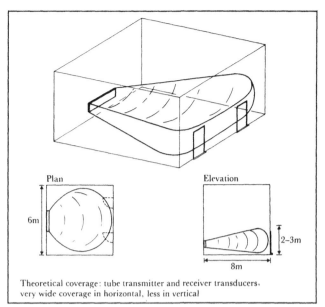

Theoretical coverage: tube transmitter and receiver transducers,
very wide coverage in horizontal, less in vertical

**FIGURE 4–54   Typical ultrasonic sensor coverage patterns:
tube**

disk is vibrated by the sound energy received in turn
compresses the piezoelectric crystal, which then generates electricity. Its functions are threefold—to gather as
much as possible of the available sound energy being
transmitted, to transform this energy into electric energy
with a maximum efficiency, and to use this energy to activate an alarm signal only under strictly defined conditions determined by the signal processor.

The first two functions are dependent on the physical
characteristics of the materials used in the device. Within
the latter function, the materials can differ, even in
devices of the same physical makeup.

The Signal Processor. The signal processor is the
intelligence of the detector. The processor must decide
whether the Doppler shift was produced by a bird or a
man! It does this by being programmed only to recognize
and react to a very specific range of frequencies and
other characteristics that have the human signature.

A series of mixers, filters, and end pulse counters is used to separate the characteristic frequencies and these
are passed to a *sampler* (or discriminator), which actually
analyzes the Doppler shift and makes the decision
whether or not to raise the alarm.

Most good, modern sensors achieve this signature
recognition discrimination by simultaneously monitoring specific parameters. If all the parameters are present,
the processor raises an alarm. The parameters are

> *overall frequency limit.* The noise environment in which
> sensors are placed ranges from 1Hz to more than
> 100kHz. Air movement, flapping signs, etc., usually
> have frequencies from 20Hz to 12kHz; electrical
> equipment vibrates at 50–60Hz; noises from tele-
> phones, printers, forced ventilation, etc., have a
> bandwidth of 20–70 kHz. Ultrasonic detectors

generate 19–25kHz and are looking for Doppler shift around the selected frequency of 10–800Hz. They are, therefore, right in the middle of the noise environment! The detectors must remove the unwanted noise. The first step is to introduce upper- and lower-frequency filters to remove all frequencies below 19kHz and above 26kHz. This clarifies the scene, but still leaves vibrations from fast-moving surfaces and noises from bells within the bandwidth of detection. These cannot be physically removed without losing some or all of the desired signal from the intruder. Hence, a fundamental signal-to-noise ratio for the room in which the sensor sits is established. This ratio will differ for every single installation.

net forward or backward displacement (Doppler shift). The net displacement reduces the likelihood of false alarms due to swinging objects.

amplitude of the frequency change. The amplitude of the frequency change is dependent on the speed, size, mass, and direction of the object moving in the covered zone. By having an amplitude filter or window, a fast-flying bird of low mass or a slowly moving rat can be differentiated from a man.

duration of disturbance. The detector times the duration of the net forward or backward movement and if it continues beyond a set period (e.g., 656 ft/sec or 200 m/sec) or repeats within a given time (shuffling), the detector raises an alarm. This is often called *sampling* or *counting.*

characteristic sample frequencies within a range or spectrum

It is the signal processor quality that will determine the reliability of the device. The signal processor should be matched to the particular application. Some devices can be tailored for particular applications in order to reduce the likelihood of false alarms (notably in retail outlets).

Uses

primarily indoor, enclosed-environment devices

widely used for the protection of small-volume spaces requiring a good, reliable level of security

commonly used in highly glazed areas, as glass will attenuate the high frequencies to a great degree and reduce any chance of spillage from the space or from outside interference

often used in very large-volume spaces or irregular plan areas in the multifrequency or multihead form to give the most reliable and economical solution to a volumetric detection security problem

can be used in very hot or cold *stable* temperature/humidity environments without loss of sensitivity, if the normal ambient temperature limits are specified

to the manufacturer (−0.4°F to 138°F, or −18°C to 59°C).

Application

The device should be set high within the room in order to give the maximum space or volumetric coverage for the radiating sound waves.

The mounting position should be carefully chosen to take account of beam shape, beam range, etc. Generally, the best position is ceiling-mounted, looking down into the space and reflecting back from the floor. With this mounting, there is little chance of shadowing caused by shelves, etc., commonly encountered with wall-mounted devices (see Figure 4–55).

It should only be used in rooms with fairly hard reflective surfaces and finishes (e.g., modern offices, factory spaces, etc.), *not* furniture stores, retail warehouses, etc.

Most devices have a linear range of 23–49 ft (7–15 m) and a coverage of 12–60 yd² (10–50 m²) depending on the mounting and transmitter transducer shape. Multihead units offer 191–215 yd² (160–180 m²) per pair of transducers and master-slave units approximately 78 yd² (65 m²) per unit.

As sound is being radiated from a point source, the nearer the object is to that source, the greater its apparent size (it will cut more waves of energy). This increase in apparent size of an object can generate spurious alarms, as the device responds to size, mass, density, and speed. Potential sources of these spurious alarms are pets, birds, insects, etc.

The operation of the device relies on a fixed range recognition of Doppler or frequency shift. To do this, the speed of sound through air and the wavelength of that sound must remain constant for a given application. If the physical nature of the air changes, the device will become less sensitive or more sensitive and false alarms, or even worse, no

FIGURE 4–55 Wall- or corner-mounted ultrasonic receiver

alarm, could result. To avoid any likelihood of this, use the device with care

where strong air currents caused by either natural or mechanical ventilation are present when the alarm is armed

where large air temperature changes cause air density changes

where there are large relative humidity changes (e.g., hothouses, bath processes in factories, near air-conditioning units that only operate intermittently)

where steam is produced

above radiators, or near or in the line of sight of infrared radiant heaters

commission the device when the humidity in the space is at its lowest. Ultrasonic sensors are at their most sensitive at low and high humidities, with least sensitivity at 20–30% relative humidity.

near ventilation systems; the sensor should be at least 16–20 ft (5–6 m) from any fan or ventilation grill

The ultrasonic sensor can be activated by other equipment within the space that will produce a Doppler shift due to intermittent or unforeseen running.

Problems are sometimes encountered with

telephones

rotating machinery noise

high ambient noise

fans or noise introduced via air ducts

printing machinery

any machinery causing vibration in the 50–100Hz frequency range, which produces ultrasonic sound

fire alarm bells; whenever possible, replace bells with bleepers, buzzers, sirens, or klaxons to avoid false alarms

Ultrasonic devices should only be used in areas that are completely unoccupied. Do not alarm one area of a room during the day when there are staff in another area. Ultrasonic sound can cause headaches and other side effects, and the devices have not yet received a clean bill of health from international health research agencies; the higher the frequency, the greater the power output of the device, the greater the risk.

Devices should not be wall-mounted above doors or windows. An intruder could easily gain entry under the device's protective radar screen, without even entering the covered space, and disable or mask it.

Ultrasonics should not be used in communication areas where there is any chance of mutual interference or feedback.

Wall-mounted devices are directional and should be aimed at the most likely intruder entry point in order to maximize their sensitivity.

Equipment Types and Selection

These devices are divided into three categories:

- transceivers
- multihead devices
- master-slave devices

Transceivers. Transceivers are devices that have both the transmitting and receiving transducers *and* the signal processor in the same enclosure. They are sometimes referred to as *monostatic devices.* The transceiver is very widely used and has the advantages of

single-point power supply and low wiring cost

integral signaling and processing

low sensitivity to swinging motions from objects within the space

beam-shape focusing

no vulnerable wiring to the signal processor

Its disadvantages are

physical size

if more than one unit is used in a space, the operational frequency limits of each detector must be mutually exclusive. That is, if one device operates on 39kHz and detects a +50Hz Doppler shift, the other device should be on, say, 29.9kHz + 50Hz or 30.1kHz + 50Hz. If they are not, the operational frequency drift of one might activate the other.

high cost in multiroom or multiunit installation

Multihead Devices. Multihead devices have the transmitting transducer and the receiving transducer separated and enclosed in separate cases. With these devices, the transmitter and receiver both contain identically tuned piezoelectric crystals, but neither contains any signaling, processing, or alarm-generating power electronics. This function is carried out by a third separate component—the *controller* or *signal processor.* The controller receives signals from the transmitter and receiver, carries out the comparison, identifies any Doppler shift, and investigates an alarm.

Multihead devices offer the following advantages:

the transducers are smaller, more easily concealed, and consume less power

they are most commonly used by large office complexes requiring high security to multiple cellular office spaces. In this application, there is a considerable cost savings over transceivers, due to the single, central, signal-processing device.

they have an effective area coverage of at least three to four times that of transceiver types

They have the following disadvantages:

they are more difficult to commission than transceiver devices
because of their size, latching electronics and antitamper switches are less common features
wiring is more vulnerable

Master-Slave Devices. These systems are a hybrid of the preceding systems. A master unit contains all required processing, signaling, antitamper, and standby power capabilities and a transmitter/receiver pair. Slave units are simple transceiver extensions, or *outstations*.

Master-slave systems offer the following advantages:

they are more economical on small-scale, multicellular office accommodations, than self-contained transceivers
they are simple to extend at a future date
they offer slightly greater coverage per transceiver than simple, self-contained transceivers

Their disadvantages are:

interwiring can be labor-intensive
mutual interference must be avoided

With transceiver-type devices, never be tempted to select compact units, which have the transmitter and receiver very close together. This will increase the likelihood of false alarms. With all types of device, those that can detect a net forward or backward general movement will prove more reliable and less troublesome than those that react to any movement.

Additional advantages of all devices are:

they require no license (unlike microwave sensors)
they can be continuously monitored and interrogated as to their condition
they are self-limiting in range

MICROWAVE SENSOR

Introduction

Table 4-11 summarizes the types, uses, typical dimensions, other characteristics, and advantages/disadvantages of the microwave sensor.

Principles of Operation

The internal use microwave detector is similar to the ultrasonic detector in that they both rely on a frequency change or Doppler shift to detect an intruder. The ultra-

Table 4-11   Microwave Sensor

Types	1. Side by side. Single transceiver. 2. In-line. Single transceiver. 3. Separate transmitter–receiver units.
Range	49–328 ft (15–100 m) long × 13-33 ft (4–10 m) wide × 10-13 ft (3-4 m) high.
Activated by	Person (or object) having a mass of 132 ± 44 lbs (60 ± 20 kg) moving a distance of 6.6 ft (2 m) toward or away from the detector at any speed between 1 and 2 ft/sec (0.3 and 0.6 m/sec) (approximately 0.6-1.2 m/hr, or 1-2 km/hr, walking speed).
Standard	BS 4737, Part 3, Section 3.4 and UL.
Uses	Primarily large-volume offices or long internal corridors with block-work walls. Can cause problems if detector too powerful for room.
Not to be used	When it might be aimed at windows, light partitions, or metal objects. Never use in temporary buildings and use with care in buildings with lightweight steel portable frames and metal cladding, skins/roofs, etc., which might vibrate.
Typical dimensions	Recessed models: 4 in (100 mm) deep × 1 in (25 mm) high × 4 in (100 mm) wide. Surface bracket mounted models: 3.7 in (95 mm) high × 8 in (200 mm) wide × 4 in (100 mm) deep.
Associated work	1. Walk test, 14-day "in use" commissioning period required. 2. Chasing into structure with recessed models. 3. Check whether FCC approved.
Advantages	Very sensitive, very controllable, can give early warning (prior to actual penetration).
Disadvantage	Can penetrate most partition walls, hence causes frequent false alarms if overpowered and used in multiuse or multiuse tenancy office blocks.

sonic detector does this by emitting waves of energy at frequencies in the 20–40kHz range, just above the maximum human hearing level. Microwave detectors also emit waves of electromagnetic energy, but with frequencies of 9.3–10.7GHz (i.e., 1,000,000 times greater than ultrasonic). As the fundamental equation *wavelength × frequency = constant* holds true, the wavelengths of energy with this frequency must be far shorter, hence the name *microwave*.

The actual wavelengths of energy produced vary between 0.4 and 9.8 in (10 and 250 mm), depending on the device. Figure 4-27 shows the various types of electromagnetic energy and their ranges of wavelengths. From the figure, it can be seen that microwaves have wavelengths longer than light and heat, but shorter than subsonic sound. This puts microwave energy in the enviable position of traveling at the speed of light (not sound) and of having a greater penetration effect on objects than either heat or light. Figure 4-56 illustrates typical microwave coverage.

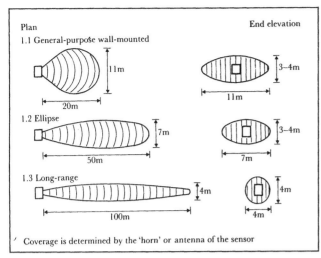

FIGURE 4–56 Typical microwave coverage

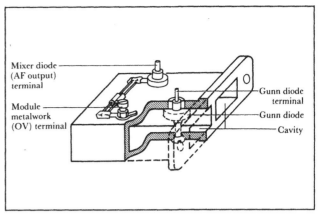

FIGURE 4–57 Typical arrangement of a side-by-side microwave detection device

If we consider the effects of these properties, we begin to see the supreme advantages that microwaves can have over ultrasound. The greatest advantage is that, traveling at the speed of light, its energy level is so high that it passes through air as though it does not exist. Indeed, microwaves can travel through a vacuum just as easily as air and do not rely on air pressure to transmit their energy (unlike ultrasonic energy). If microwaves do not recognize the existence of air and do not rely on it for their action, air currents, temperature changes, etc., do not trigger false alarms.

Traveling at the speed of light, a particle of microwave energy (with a very short wavelength) can penetrate many common materials while being only moderately attenuated. The result of this is that microwave detection devices can be used to detect intruders *before* they actually penetrate a secure space. Microwaves are only attenuated significantly by very dense material such as metal, very dense concrete, water, etc. Just like visible light, microwave energy can be focused, refracted, reflected, etc. This makes for a very controllable, precise form of energy that can be aimed very accurately.

Internal microwave detectors respond to Doppler shift or frequency changes. In order to produce a Doppler shift, some of the transmitted energy must be reflected and some of it must be compressed by the object moving toward the detector. As microwave energy can penetrate many materials, it is fortunate that humans are not one of these materials. This is primarily because we are composed mainly of water, which reflects and attenuates microwaves. If intruders are to be detected by an internal microwave sensor using Doppler shift, they must move *within boundaries* that will not only contain some of the transmitted microwave energy, but also reflect some of it.

Microwave detectors are comprised of the following (see Figure 4–57):

the transmitting transducer or *Gunn diode.* Just as with ultrasonic energy, a transducer is used, but here it is

used to convert electrical energy into electromagnetic energy traveling at the speed of light. This very special transducer was called a *Gunn diode oscillator (GDO)* and was named after its inventor, John Gunn. It has *not* been superseded by a disk of semiconductor material (usually gallium arsenide – GaAS) in a field effect transistor (FET). Microwave literature usually refers to *GaAsFET microwaves* to give the impression that we've moved on from the inventor's time, but we really haven't! When supplied with a voltage of approximately 7–16V DC this material will have its electrons excited, causing it to emit energy as a series of microwaves traveling at the speed of light and with a frequency in the "X" band range (1–100GHz).

the resonant or microwave cavity. The GaAsFET is enclosed in a solid metal case with thick walls and polished internal sides. One side of the case is open and the opening is of precise dimensions. This opening is called the *microwave* or *resonant cavity.* As most metals will reflect 100% of incident microwave energy, the three, closed walls of the cavity act as a reflecting and focusing device for transmitting the energy to the space to be protected. The microwave cavity is sometimes referred to as the *wave guide.*

the receiving transducer and mixer diode. The function of this device is to absorb the reflected microwave energy and compare it with that radiated by the GaAsFET. To do this, it is supplied directly with a small sample of the transmitted signal, either by being placed in the field of radiation of the GaAsFET or by electronic means.

the mains filtration or ripple rejection unit. Microwave energy is being used in the device to recognize a Doppler shift of 10–100Hz. Unfortunately, the main electrical supply to the unit can supply this shift, as well as an intruder, in the form of a *mains ripple.* Hence, with microwave sensors, complex ripple rejection electronics are needed. The better the ripple rejection, the more reliable the device.

the antennae and horn. The microwave cavities on most devices are of standard size and will result in a coverage of approximately 328 × 1.6 × 1 ft (100 × 0.5 × 0.3 m). However, most applications within buildings do not require a range of 328 ft (100 m) with such narrow width and height. As microwave energy is easily focused into different beam shapes, this can be achieved by the use of steel caps with a variety of different opening shapes. These steel caps are called *horns. Antenna* is the more technical term for a horn and some manufacturers use this term in their literature. When this term is used, the coverage pattern is usually described in *decibel gain.* Generally, the higher the gain, the longer the range and/or sensitivity; the lower the gain, the broader the coverage and the less the sensitivity.

the amplifier. A person walking in a microwave field will only generate a frequency change of approximately 10–100Hz or 0.1–1.0 × 10⁶ percent from the radiated frequency. For the sensor to use this difference, it must collect and amplify it with high fidelity. The larger the receiving transducer microwave cavity, the greater the chance of capturing a signal. Once the signal is captured, the amplifier must perform two distinct tasks:

1. It must only amplify signals resulting from characteristic Doppler frequencies (3–300Hz).
2. It must provide a large increase in signal strength (amplitude) without amplifying noise. It must, therefore, have a high gain and low signal-to-noise ratio.

signal pulse counter/accumulator integrator. This takes each characteristic Doppler shift occurring for a given set duration and stores it. If, at the end of a predetermined period, the sum content of the store is over a given level, it raises an alarm. Hence, a person moving a short distance very quickly and producing a single, large Doppler shift is just as likely to trigger an alarm as a person moving a short distance very slowly.

the voltage regulator. This ensures that all parts of the device receive a stable, even voltage with very few spikes, which cause malfunctions of the sensor.

Uses

Primary use is for large-volume spaces that require protection of high-quality or high-value goods. The very long range and large volumetric coverage make it a very economical solution for protection of the following (when the necessary precautions listed in the next section are taken):

warehouses—long-range, narrow pattern aimed down from high level to cover the passageway between high stacked goods

factories—covers areas of particular production importance or materials and goods of high value

school, community, and sports halls

exhibition halls during closed hours

any very long corridor that has solid walls

any structure of value that requires protection before penetration in order to prevent fabric damage (e.g., a church)

museums and art galleries

Other uses include the spot protection of items of very high value. The device has excellent beam focusing, shaping, and pinpoint coverage of the target area. It is also used in any area where radiant heating, forced ventilation, or air property changes make ultrasonic sensors impractical or unsuitable.

Application

The small size of the microwave Gunn diode mechanism enables it to be recessed into the building fabric for complete concealment. Typical recessed units measure 3 in (80 mm) high × 4 in (100 mm) wide × 6 (150 mm) deep; some in-line units are 50% smaller!

Surface-mounted modules are still very small, 4 × 4 × 7 in (100 × 100 × 180 mm) or less, and are mounted on wall or ceiling pivoting brackets for on-site adjustment or future coverage alteration.

With many units, it is possible to have remote control and signal processing electronics.

Devices that incorporate net movement discrimination will generally prove more reliable and more sensitive than normal, forward-motion-only sensors.

Devices are available that will give only a 33 ft (10 m) range. These are marketed as direct competitors to ultrasonic devices. Many of these short-range devices are, however, simply throttled-back versions using the same source as their long-range, large-volume brothers. If these devices are used, there is obviously a payment made in reduction of operating efficiency, as the device is not performing in its true role. Some of these short-range devices can have short life spans, because the incorrectly designed microwave cavity and antenna cause too much internal rebounding of energy, and, hence, overheating. Short-range microwave devices should always have low power consumption (25–40mA).

The maximum power consumption, microwave power generation, and bandwidth of operating frequencies for these sensors is covered by the FCC. If foreign or cheaper equipment is considered, a check should be made to see whether it is unlicensed or approved. These devices must be firmly fixed in or on a surface that is free from vibration. Recessing into solid concrete or masonry is probably best. Sites such as partitions, false ceilings, or positions near heavy doors, should be avoided, as false alarms will be more likely. Positions near small transformers, equipment within buildings, or radio frequency equipment can also increase the potential for false alarm.

Devices should not be aimed at fans or louvers of duct-work systems, as both can cause false alarms. Fans can be effectively screened using metal mesh over the discharge grill. Louvers with thin metal vanes should be stiffened.

Fluorescent lights can be another major source of false alarms, due to their electron excitation frequency. Never aim or allow microwave energy to be reflected onto lamps if possible. A newer hazard is the bounce/vibration effect caused by luminaires with metal blade louvers (VDU luminaires). Micro-wave devices should not be aimed at these.

Unless prepenetration protection is required, never aim the sensor at glass windows or lightweight par-titions. Penetration of these surfaces by the micro-waves can cause false alarms due to passersby, equip-ment movement in other rooms, etc.

Never aim the sensors at any metal surface, partic-ularly metal foils or skins that vibrate. This will be mistaken for an intruder movement. A "rule of thumb" is that if any metal surface can move by more than 0.6 in (15 mm) it may be potentially troublesome.

Remember that vibration or movement across a beam field is less critical to false alarm prevention than movement toward or away from the sensor.

The manufacturer's ranges shown in their advertising literature will be expressed to impress the customer and not to tell the whole truth and nothing but the whole truth! The advertised coverage limits are generally those resulting from a test of the device in an extremely large space or free space that has very hard surfaces that are matched to maximize the surface reflectance for the transmitted frequency of the device. The range is generally the limit at which a person, moving at a prescribed speed, will trigger the device successfully. If the test is carried out to BS 4737, the person will weigh 132 ± 44 lbs (60 ± 20 kg) and will walk at a speed of 1–2 ft/sec (0.3–0.6 m/sec). What this information does not tell the reader is that if an object with greater mass and speed, or greater reflectivity to microwave energy and less speed, were to move outside this range, the device would probably detect that object! Hence the "range" is *object dependent*. The actual range to very reflective fast ob-jects can often be at least 30% greater than the stated range and when microwaves penetrate glass or light-weight structures, they often go into alarm conditions.

Also, the "range" will be dramatically affected by the absorbability of the surfaces of the room and the room geometry. If the surfaces are highly absorbent, most of the generated and transmitted energy will be absorbed and not reflected. Hence, the device cannot gather microwaves from the area concerned in sufficient strength (amplitude) to make effective decisions on whether an intruder is present.

A room with many alcoves, lines of cabinets, etc., will reduce the effective coverage considerably. Although microwaves can penetrate most surfaces, they cannot penetrate metal filing cabinets or similar objects, and these will produce blind spots and false alarm–generat-ing reflections. As microwaves are much more narrow beam and have shorter wavelengths than ultrasonic sen-sors, they do not tend to flow into all recesses and alcoves as well as ultrasonic sensors do. Hence, 100% coverage of a space is rarely, if ever, possible and increasing sensor numbers in a space simply increases problems of mutual interference. It is, therefore, extremely important to position the sensors so that the maximum coverage is given to those areas in which the intruder is most likely to be present.

Microwave fields do not have sharply defined edges and this makes aiming and range finding difficult. In principle, the higher the frequency and shorter the wave-length, the sharper the field edge. However, above 10.52GHz, only 22.125GHz is allowed without a license and this limits the choice of sharp-edge devices.

Equipment Types and Selection

There are three main categories of this device:

transceivers with side-by-side (SBS) assembly of transmitter (oscillator) and mixed diode (receiver) (see Figure 4–57). As can be seen from the illustration, this unit has a transmission cavity and a separate receiver cavity. Both cavities are surrounded by a common antenna.

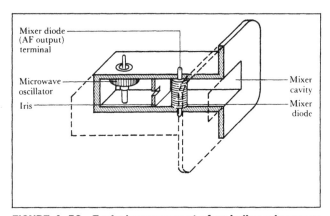

FIGURE 4–58 Typical arrangement of an in-line microwave detection device

This antenna will syphon a sample of the transmitted frequency and provide this as a reference to the receiver (with some electronic boosting bias given by a current supplied from a DC source). The receiver accepts the signal bouncing back from the object and compares the frequency to the reference frequency. The main disadvantage of this form of device is the inefficiency of having to provide power for a biasing current.

transceivers in which the oscillator is placed in a separate chamber behind the receiver; usually called an *in-line microwave* (see Figure 4–58). Here, the oscillator is placed directly behind the receiver. A small aperture (or *iris*) focuses a proportion of the beam onto the mixer, for reference, before it is shaped by the receiver cavity and antenna for transmission and detection purposes. The advantages of this type of device are that it does not require a bias current and hence is more efficient and it is less susceptible to the production of harmonic frequencies, of the reference frequency that could cause false alarms within the device, or in others in the area operating at different frequencies.

devices that separate the transmitting oscillator and receiver; sometimes referred to as *unitary devices.* These offer the advantage of reduction of false alarms arising due to objects moving in close proximity to the detector. This also means, of course, that unless properly used, the devices can have increased blind areas. These devices operate on the line-of-sight/ loss-of-signal-strength principle, not true Doppler shift, and are more suited to outside use. (See Chapter 3, "External Microwave Sensor.")

Exterior sensors should not be used indoors. These sensors normally operate on a different principle—a *bistatic transmitter/receiver* that looks for loss of signal, out-of-phase reception or deviation from pulse, or coded pulse signals on a carrier frequency—*not Doppler shift.* These bistatic devices are really beam breakers. Some outdoor detectors use monostable/monostatic transceivers, which look like large versions of their interior casings. These transceivers do use Doppler shift principles, but only in a limited way, as nontarget objects of considerable size may be legitimately present and moving in the field, but should still not cause false alarms!

Ripple Suppression. All microwave devices rely upon the production of a very fixed constant frequency by electrical means. Any variation in the produced frequency can be seen by the mixer diode as a Doppler shift and the alarm may be activated. For this reason, it is fundamental to good design that any ripples or frequency anomalies on the mains supply be suppressed before reaching the oscillator, otherwise these anomalies may activate the alarm. *Ripple rejection* is given in the form of

decibels and indicates, to some extent, the quality of electronics manufacture. Any device that quotes 85dB or above, will not suffer unduly from ripples. It is common for good devices to have in excess of 100dB ripple rejection.

Radiated Electromagnetic Suppression. Fans, fluorescent lights (particularly high-frequency types), and computers can radiate electromagnetic energy or reflected microwave energy, which could be seen as a Doppler shift, if not blocked. The supplier should be told of any likely sources and asked what type of narrow band filter the device contains to block or suppress this unwanted energy.

Pulse Counting and Integration. Many manufacturers advertise sophisticated pulse counting and integration features as an advantage of their device over competitors. This feature might reduce false alarms, but it will also significantly reduce overall security and reliability, and for this reason no "bonus points" should be awarded! For maximum security, devices should be sensitive, accurate in response, and *fast in response.* Pulse count and integration generally reduce these criteria.

Filters and Container Construction. False alarms caused by insects are reduced by twin transducer cavity construction. False alarms from fluorescent lighting is reduced by frequency tailoring of band pass filters.

Net Forward and Backward Movement Discrimination. Although slightly more expensive than devices without this capability, sensors with this function should be viewed as superior, because not only will they reduce false alarms caused by the flapping curtain type of hazard, but they will also reduce false alarms caused by electrical interference and reduce susceptibility to false alarms caused by internal signal noise amplification. This is a result of dual, simultaneous response, signal channeling processing techniques used within this device.

Mounting. Devices are usually either concealed (i.e., recessed into the structure) or surface-mounted on a swiveling bracket. The former is the more secure device and, if properly installed and sited, far less prone to false alarms. It is, however, more expensive. The surface-mounted model is obvious to the skilled intruder, but has the advantage that its aim can be changed easily as items such as cabinets, etc., are moved around the room, thus reducing the likelihood of blind spots (see Figures 4–59 and 4–60).

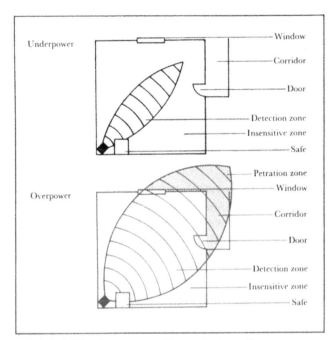

FIGURE 4-59 Sensor mounting and strength

FIGURE 4-60 Surface-mounted microwave for single- or multiple-head application.

Sensitivity. The setting up and sensitivity adjusting of microwave detectors to prevent false alarms can be difficult. In order to reduce these problems, ensure that detectors include test meter and self-checking capabilities, and full, site adjustable sensitivity. Microwave sensors are extremely sensitive devices, due to their very high frequency stability and short wavelengths. Hence, any reduction in the sensitivity level of a microwave sensor to overcome some small application problem is far less worrisome from a security viewpoint than a similar reduction in an ultrasonic sensor would be.

Features to Look For.

- alarm memory
- antimasking
- day release and disable
- mutual interference rejection

FIGURE 4-61 Microwave sensor in explosion-proof enclosure for use in potentially explosive environments

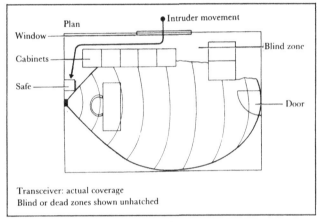

FIGURE 4-62 Microwave blinding

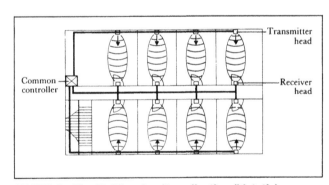

FIGURE 4-63 Multihead unit application (bistatic)

- noise monitor circuitry
- automatic noise compensation

For other types of sensors, see Figures 4-61 through 4-63.

Frequency. The available frequencies include 9.15, 2.45, 5.85, 10.52, 10.7, and 22.12GHz. In the USA, 10.52GHz is the common standard for small to medium volumes with full containment of the wave (10.7GHz on European models). For larger volumes or prepenetration protection through windows, partitions, etc., 2.45GHz units are sometimes recommended.

Advantages and Disadvantages

Advantages.

smaller size than PIR or ultrasonic sensors

cheaper for larger volume than PIR or ultrasonic sensors (less labor, as larger coverage per head)

longer range than PIR or ultrasonic sensors

can provide prepenetration protection

not affected by air movement, humidity, temperature, and most other environmental changes (unlike ultrasonic sensors)

very efficient

can be focused and aimed very accurately

very sensitive, but with easily adjustable sensitivity ranges

beam pattern can be changed easily at later date by changing the horn fitted to the front of the microwave cavity

more difficult to mask than PIR or ultrasonic sensors

transmitter and receiver can be separate from each other, thereby reducing size and providing tailor-made cover patterns

a medium- to high-security-risk device

Disadvantages.

not as stable as PIR or ultrasonic sensors; more prone to false alarms if not installed and commissioned properly

should not be used in excessively humid atmospheres, as water droplets reflect microwave energy

penetration of plasterboard walls, wooden floors, etc., can cause false alarms due to movement in unprotected areas, such as offices of other tenants on the same floor or office block

can be affected by small vibrations or mains ripple of electric supply

can be masked (with difficulty)

false alarms can be caused by microwave penetration of windows, hail on windows, rainwater in internal plastic rainwater downspouts behind stud work, etc. (metal foil papers or fine mesh screens will usually overcome problem)

metal objects can create blind spots

requires a 14-day probation period before direct connection to the police station

will not saturate an area like an ultrasonic detector; microwaves create well-defined pools of cover

bounce amplification off metal objects introduced into the space after commissioning; can increase the overall sensitivity and promote false alarms

unsuitable for small spaces due to penetration properties and inherent long-range and large-coverage output

Model Information

Manufacturers should provide the following information on the products.

General Layout.

drawings to scale; plan; side elevation and end elevation of the pattern of coverage provided and the dead spots

numbers and references to all standards to which the equipment conforms and key to devices shown on the drawings to enable the client to find out electrical and physical information about a device from specifications and schedules

Electrical Information.

operating (supply) voltage, frequency, current, and power consumption

level of ripple suppression provided

type of oscillator used (e.g., GaAsFET)

transmitting frequency

whether the device is of in-line or side-by-side construction

whether an LED test lamp is included to aid walk testing

whether the device will require separate frequency ranges if more than one is used per room or area

the level of radio interference protection provided. This should be given as a practical example (i.e., CB radio at power x watts transmitting at distance y meters from device)

whether a low or no voltage alarm is included

whether a latching device is included

whether test lamp connections are provided

the rating of the alarm contacts (amps and watts)

site sensitivity, range, and levels of adjustment possible

Physical Information.

design range

operating range in this application

dimensions

weight

finishes available

whether it is suitable for recessing, wall-mounting, ceiling-mounting, etc.

whether it is protected from dirt, water, moisture, etc., in any way, and the level of that protection (i.e., IP rating)

all materials of construction, casing, etc.

the level of structural vibration (in terms of frequency and deflection) that will cause false alarms

ACOUSTIC SENSOR

Introduction

Table 4–12 summarizes the types, uses, typical dimensions, other characteristics, and advantages/disadvantages of the acoustic sensor.

Principles of Operation

Acoustic sensors act in the same way as the human ear and brain, only with much less sophistication and discrimination. Any intruder trying to gain entry to a space or walking within a space will generate a noise. That is to say, the person will cause the air to vibrate at frequencies audible to humans. The type, frequency, amplitude duration, repetition, and pattern of these noises can be recognized as being characteristic of intrusion and not being the ambient noises normally encountered. The acoustic sensor is generally comprised of a very accurate, stable (hi-fi) microphone, a band pass filter, and an electronic integrator/summation unit. The microphone transduces the acoustic energy into an electrical voltage. The band pass filter ensures that only characteristics of particular sound frequencies are passed into the integrator/summator for processing. In very simple terms, the noises generated by structural attack, walking or moving within a space, etc., are at a relatively high frequency (1500–6000Hz), while those characteristic of structural or ground-born vibrations that generate air-born noise are low frequency (5–1500Hz). Very high frequency noises (10,000Hz and more) are usually caused by whistling or whining of equipment for short durations, or things such as breaking glass, which are likely to occur only once and are not repeated. The band pass filter progressively attenuates these high- and low-frequency noises and allows the integrator/summation processing unit to concentrate

on discriminating noises likely to have been caused by an intruder, and hence provides a device that is both very sensitive and very stable.

UL and other standards require that devices shall not be placed in rooms with ambient noise levels greater than 55dB. This is very misleading as no weighting, frequency spectrum, etc., are quoted. In practice, 45dB(A), with a mid-range of 1000Hz, is the maximum level recommended. Similarly, standards call for devices to raise an alarm if a level of noise 30dB *over* the ambient noise or a maximum of 70dB (whichever is less) is reached. However, we know that noise levels are frequency-dependent and, quite rightly, sensors should be decibel-weighted for known attack noise characteristics.

Uses

All noises are of relative loudness and while a pin dropped in a hushed cathedral might be heard, it certainly would not be heard at a rock concert! For acoustic sensors to operate under maximum sensitivity, they must be sited in areas that are normally quiet when the intruder system is armed (i.e., in areas where the sounds the devices are programmed to recognize will appear relatively loud). For this reason, their use is generally restricted to bank vaults, strong rooms, book safes, basement stores, or rooms with very heavy structural elements and very little plant and equipment in them. They are generally regarded as a very high security device. There are, however, some very good new products that are specifically designed to detect very noisy intruders, such as vandals attacking schools or other premises, and these are proving to be very effective.

Application

A manufacturer's advice should always be sought for the particular sensor. In general,

the harder, more reflective the room surfaces, the better

the more dense and heavy the room's structural materials, the better; soft furnishings should not be present

the more stable the room temperature, the better; rooms should never be allowed to fall to below 41°F (5°C)

flush-mounted units are preferable and more secure than surface-mounted units; structurally mounted units are preferable to those mounted on false ceilings, floors, etc.

never use where ambient noise levels exceed 45dB(A)

the sensor should always be placed at the center of the volume it is to cover, if possible

Table 4–12   Acoustic sensor

Types	1. Vault protection.
	2. Vandal protection.
Activated by	Audible sound energy resulting from the presence of an intruder, or attack on the protected structure.
Standards	BS 4737, Part 3, Section 3.6. IS 199.
Uses	1. Protection of vaults, strong rooms, safes, etc.
	2. Schools, galleries, museums, etc.
Not to be used	In areas of high ambient noise level (NC 30 or above).
Typical dimensions.	3 in (80 mm) diameter × 2.8 in (70 mm) deep.
Associated work	Recessed conduits.
Advantages	Reliable, stable.
Disadvantage	Limited application.

Equipment Types and Selection

There are four basic types of audio sensor:

the external or internal "listen-in" microphone. This is very seldom used, as it requires a guard to listen to the sound activity being generated. The guard acts as the processor/discriminator.

the office or commercial user sensor (sometimes called a *vandal alarm*). This is a medium-grade security device for the protection of relatively large spaces that are quiet when the alarm is on. Examples are schools, churches, theaters, art galleries, etc. This can be very useful as a less-sophisticated alternative to the ultrasonic, microwave, or passive infrared sensors mentioned in earlier chapters. This type of acoustic sensor has a much wider band pass filter and more sophisticated processing than the vault sensor described next, as it must operate in environments that have (in relative terms) a much higher ambient noise level. The sensors used are usually of the master-slave type, with the master having the processing power and the slaves acting as outstation collectors of sound in order to give greater and more economical coverage.

the single-source vault sensor (see Figure 4–64). This sensor is specifically designed for vault or other high-security area protection. It takes all its sound samples from the space in which it is sited and is very frequency selective. Devices of this kind usually recognize frequencies in the 2500–3500Hz frequency spectrum and greatly attenuate all others. This sensor can discriminate shallow breathing!

the dual or comparator sensor. This sensor samples noise from within the protected space and compares it with ambient noise outside the protected space. This gives the sensor one additional, discriminating feature over the single-source sensor.

FIGURE 4–64   Single-source vault sensor

Regardless of the type of sensor appropriate to your client's needs, all good sensors incorporate the following functions that increase reliability, stability, and sensitivity, and hence reduce false alarms:

characteristic frequency selection.   Usually the band pass filter will attenuate all frequencies over 6000Hz and under 1000Hz as a minimum standard.

pulse count. This should be two types—exceptionally high amplitude, single-pulse alarm, which will allow for the detection of a single explosion, heavy hammer blow, etc.; or accumulation, which takes sounds at the same frequency occurring at random intervals of time, or sounds of slightly different frequencies occurring at regular intervals of time, and stores them and accumulates them for a given time period (usually adjustable between 1 second and 1 hour). If, at the end of the accumulation period, the resultant accumulated sound is above a predetermined level, an alarm is generated.

high gain amplification. This allows the sensor to pick up very quiet sounds at frequencies that are extremely characteristic of particular forms of attack, and to amplify them for use in the normal way. This capability will increase the ability of the sensor to raise the alarm before any significant attack has been made on the structure that might seriously weaken it and be costly to remedy (scraping of mortar, etc.).

test bleeper. This is a device that emits a range of sound frequencies within the sensor in order to periodically check its operation for accuracy, sensitivity, etc.

latching indicator. This is important in multisensor or master-slave sensor systems, as it indicates which sensor on the alarm loop raised the alarm and, hence, leads to quicker discovery of where the structural attack took place or where the intruder went.

integral antitamper switch

loop monitor. This increases the security of the wiring from the sensor to the control panel and reduces the likelihood of bridging.

sensitivity control/adjustable trigger level. To comply with Section 3.6 of BS 4737, all sensors must

   have a stated or adjustable nominal ambient sound level [say 20dB(A)]

   only trigger an alarm when the characteristic frequencies give sound levels of greater than 15dB(A) above this nominal [e.g., greater than 35dB(A) on nominal of 20dB(A)]

   have a maximum trigger level in normal circumstances that does not exceed 85dB(A)

   trigger an alarm for any sound in excess of 120dB(A) that lasts longer than 100msec

   trigger an alarm if any trigger level sound is present for more than 5 seconds in any counting period of 30 seconds

Limitation, Advantages, and Disadvantages

Limitation.

only consistently reliable in quiet environments

Advantages.

adjustable sensitivity
alarm rings before penetration
simple, reliable principle of operation
very established use
can have local or remote listen-in capability
passive device and therefore hard to detect presence
  before penetration
alarm verification reduces false alarms and waste of
  manpower

Disadvantages.

cannot be used in areas containing machinery
very easy to upset commissioned sensitivity

AIR PRESSURE SENSOR

Principles of Operation

A sealed area of a building has a constant air pressure. Opening a door or window, or breaking through a wall rapidly changes this pressure, which is measured by a sensitive barometric sensor that generates an alarm signal. The sensed rate of pressure change is normally around 20Hz. This low frequency is inaudible and of such a low frequency that it will travel a considerable distance without attenuation.

Uses

Air pressure sensors are only used in small, very well sealed, nonventilated spaces, cases, or vaults. Some devices are sold for small offices or large homes, but I do not recommend them for these applications. This sensor is considered an alternative or backup to structural vibration protection in specialized situations, extremely hazardous environments, etc.

Application

The area must be well sealed and pressure tested to determine application. Sensors should be site commissioned for a particular duration, *not* factory set with site sensitivity adjustment.

Equipment Types and Selection

There are two basic types—high-security sensors and low-security sensors. High-security sensors cannot be fully described here, but seek specialized advice. Low-security sensors should not be considered unless your house or office is sealed to Scandinavian building regulation levels of airtightness and infiltration resistance and you are prepared to close internal *and* external doors at night.

Advantages and Disadvantages

Advantages.

extremely difficult to circumvent
prepenetration alarm
quite a low cost for high security

Disadvantages.

sealing of space is critical
volume limit
not for use in large, open spaces
not possible to use in rooms with doors ajar, ventilation systems, etc.

DUAL TECHNOLOGY SENSOR

Principles of Operation

False alarms are the largest single problem for the security industry, the police, and the customer. They annoy everybody and waste enormous amounts of money, manpower, and resources. False alarms are rarely caused by device failure or internal electronic faults. They are most often caused by changes in the environment around the sensor that the sensor cannot tolerate, distinguish, and differentiate from deliberate attempts to overcome, mask, or avoid it. These environmental problems may be due to

poor surveying and device selection, such as a passive
  infrared sensor placed opposite an intermittently
  operating gas-radiant fire
unforeseen, intermittent, and difficult to detect
  changes in the environment (e.g., very windy days
  that cause excessive internal air movement within a
  building due to a draft seal problem on particular
  doors—the fault only occurs when the wind is +20°
  off NNW!)
normally hostile environments for any sensor (e.g.,
  very humid, very dusty, misty, very large spaces, etc.)

Each particular type of sensor (PIR, microwave, ultrasonic, microphonic, etc.) is prone to certain types of environmental problems to which its brother sensors may be more or less (and sometimes totally) immune. Dual technology sensors (as seen in Figures 4–65 and 4–66) aim to reduce false alarms by combining two different sensor types, normally in a single head, both of which are aimed at the same area. The sensors act on *double-knock* principles, that is

if one sensor type sees an intruder and the other does not, further verification by one or later confirmation by the other is needed before triggering an alarm

if one sensor type sees an environmental change, it will not automatically sound an alarm, as it assumes the other sensor will still provide adequate intruder detection coverage while it is temporarily disabled. If both sensors report the same "environmental" problem, it is highly unlikely that the problem is

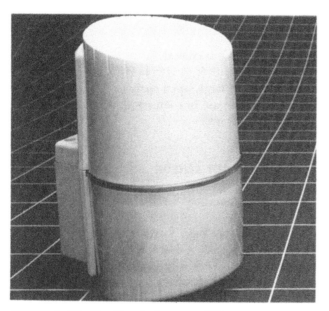

FIGURE 4–65   Combined microwave/PIR sensor

FIGURE 4–66   Combined ultrasonic/PIR sensor

environmental, but either an intruder or a catastrophic event.

The most striking differences between the common, single-technology sensing devices' principles of action are those between PIRs and active microwave and ultrasonic sensors. PIRs measure the level of infrared energy change between zones in a defined period at a defined rate. This gives it a high immunity to a wide range of environmental effects, most, if not all, of which can be overcome quite simply. PIR devices receive energy in very small amounts and must perform very complex amplification and filtration processes to provide alarm signals. This has led to a very high quality of production. Its major drawback is that what it gains in stability and reliability, it loses in depth of protection, speed of response, and coverage of area. This restricts its use primarily to low or average risks for residential and commercial buildings.

Microwave and ultrasonic devices, which actively "see" an intruder due to minor movements causing Doppler shift, give very, very good detection probabilities, as they swamp an area with a signal. They are suitable for both above-average risks and high risks in residential and commercial properties. Their major drawback is that they are very sensitive to environmental effects—microwave sensors being particularly problematic with minor vibrations and movements and ultrasonic sensors with noises due to water, aircraft, phones, cars, etc. These differences have led to the common combinations of

- ultrasonic/PIR
- microwave/PIR
- microphonic/PIR (occasionally)

to maximize detection sensitivity while retaining the range and cover to reduce environmental false alarms.

Another major difference between PIR and microwave or ultrasonic sensors is their directional sensitivity. PIR devices tend to be far more sensitive to movement across the field of detection than toward the device. Microwave and ultrasonic devices tend to be more sensitive to Doppler shift toward or away from the sensor field than across it.

This combination of technologies allows

the sensitivity on each device to be greater than normal with little increase in instability

equal sensitivity across or toward/away from the sensor

backup protection of a single element on any detector should one technology fail (although the backup will have temporarily increased instability)

Types of Confirmation

Dual *technology* devices should never be confused with dual or multichannel *devices,* or multihead devices. Dual channel or multichannel devices split the incoming

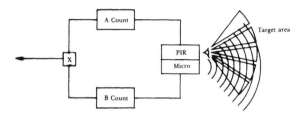

FIGURE 4–67 Dual technology—parallel count ratio

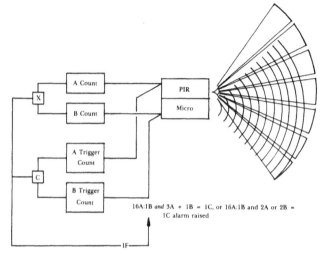

FIGURE 4–68 Dual technology—simultaneous count ratio

signal from a *single* sensor type and process the division separately to confirm internal processing. Multihead devices simply use the same sensing head types and spread them around to increase cover and provide verification of intrusion. While they may verify intrusion and improve cover, they often simply confirm the environmental problems in an area and trigger an alarm anyway!

Figures 4–67 and 4–68 show, in very simplistic terms, the type of signal confirmation used in systems and those that are particular to dual technology devices.

Parallel Count Ratio. (Count first, trigger later; see Figure 4–67.) A ratio of A and B is set by a user. For example, X = the alarm threshold. X is reached at a ratio of $16A$ counts on the PIR sensor to $1B$ count on the microwave sensor. When the dual device senses any change in the area it is covering, it starts counting the number of occurrences of the change. If the PIR count reaches 16 *before* the microwave sees one count, then the device will assume that a fault *might* exist on the PIR *or* the microwave, perhaps due to an environmental change in the detection area or a fault in either device. It raises a *fault* signal on the device or at the control panel. If the $16A:1B$ count ratio is still not reached after a certain time period, the device concludes that it was extremely likely that the counts were caused by a real fault or environmental change and not an intruder. This time period confirms the fault and the counters on the sensors reset

to begin counting again; either device may initiate (trigger) an alarm condition if the count ratio of $16A:1B$ is confirmed.

Simultaneous Count and Trigger Ratio. With this system (Figure 4–68) each sensor contributes a count to counter X. This is the environmental problem counter, and it operates in the same manner as the parallel count ratio, i.e., it looks for a ratio of $16A:1B$ within a certain time period, to assess the likelihood of faults of the sensors. At the same time this count starts, each device also contributes counts to a counter, C. This is the *trigger counter,* and it *immediately* arms the circuit ready to emit an alarm. Alarms will be raised if certain ratios or summations of counts exist on one or both counters, for example, if

- $16A:1B$ on X occurs in under 30 seconds, alarm is raised
- $16A:1B$ on C occurs in under 30 seconds, alarm is raised
- $3A + 1B$ on C occurs anytime, alarm is raised
- $16A:1B$ occurs on X in under 30 seconds *and* $2A$ or $2B$ is raised on C, alarm is raised

Multipoint Dual Technology. Multipoint dual technology (Figure 4–69) uses a processor that is remote from the sensing head to carry out either simultaneous count and trigger or parallel count processing. This technology is used where *very* large area coverage is required and a single dual technology sensing head could not provide the coverage. The system is arranged so that a single long-range microwave or ultrasonic sensor is covering the area. It reports to the same processor as multiple PIR sensors of par (each of less range ability) arranged around the protected space. This removes the need for the multiple use of dual technology sensors, which would include microwave or ultrasonic sensors in *every* device, that cost considerably more to purchase. It also improves overall sensitivity as the heads are positioned individually to their best detection direction (e.g., microwave looking toward the intruder's entry path; PIRs looking across it). This is not possible with dual technology sensors, which must aim in the same direction, as they are within the same physical container.

Single versus Dual Edge Technology. Dual technology manufacturers make many claims for the advantages of single or dual edge technology in dual technology devices. These claims are really of a very esoteric nature. It is claimed, for instance, that single edge detection (see PIR sensors) actually enhances the detection performance *toward* the PIR device. If the device had multiple look-down zones and quad or dual edge sensing, this would not be true.

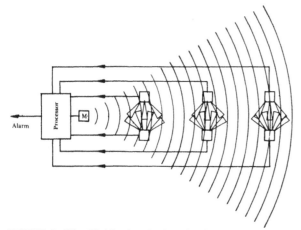

FIGURE 4–69   Multipoint dual technology

Nondual Technology. Nondual technology (Figures 4–70A through 4–70D) has the same sensor types of verification (PIR shown, but applies to any sensor).

   "And" Verification.   See Figure 4–70A.

* twice coverage or area
* provides redundancy in emergency
* no improvement in environmental problems

   "Or" Verification.   See Figure 4–70B.

* twice the coverage and sensitivity
* detection probability increased
* mutual interference probability increased
* no improvement in environmental problems

   "If" Double-Knock or Multichannel Processing. See Figure 4–70C.

* no increased coverage
* no improvement in environmental problems
* better self-diagnostics

   Low-Energy Field Changeover. See Figure 4–70D. If a low PIR (or low microwave or ultrasound) sensor environmental field exists, a more sensitive version is brought up automatically. Used where high internal temperature might blind a normal PIR to an intruder's presence. This system is not good. PIR, microwave, and ultrasound devices should have their *own* low-energy-field compensation.

Uses

Dual technology devices were, until quite recently, only used in areas where hostile environments for individual detectors precluded the use of the natural selection, single-sensing devices. This has now changed and dual technology units are beginning to replace single-sensing devices in even average-risk environmental situations. The primary reasons for this are that a few years of use have now *proved* former claims of reduced false alarms. In

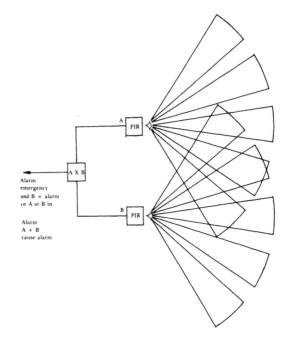

FIGURE 4–70A, FIGURE 4–70B   Nondual technology "and" or "or"

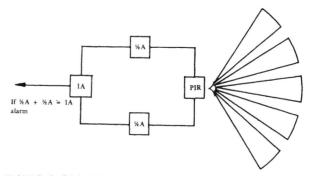

FIGURE 4–70C   Nondual technology

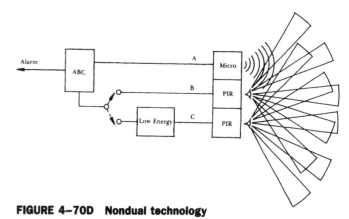

FIGURE 4–70D   Nondual technology

addition, ever-rising labor costs for repairs and reduced price differentials made dual technology devices an economically sensible solution.

   When considering whether to purchase dual technology devices in preference to single-sensing devices, determine the repair costs you've paid for the last few years (or someone else's if you are buying for the first time). It is often found that reducing repair calls by a fraction of the reduced false alarm rate will more than offset the additional cost of the device over a 3-year period.

On systems that are direct purchase without repair agreements, or on domestic systems, it is often argued that these devices do not pay back their initial investment. What is often left out of the equation in these situations is the owner's reduced confidence and peace of mind, and the ever-increasing bad feelings generated in the local police force if your system keeps generating false alarms.

Application

It is essential that when considering dual technology devices they are not seen as the panacea for all ills. They *will* reduce false alarms significantly due to environmental causes; they will *not* remove them all. They will not stop false alarms caused by poor selection, installation, owner abuse, etc. No reduction in the surveying effort to list and identify all sources of potential problems for *each* sensing element should ever be considered or allowed. Customers will not be too happy if the microwave/PIR solution is used, only to later find that false alarms are being caused due to massive microwave field strengths within the police transmitter station in which they are sited!

Aiming, field testing, and performance evaluations should be more thorough than with any other volumetric-type sensor.

Equipment Types and Selection

The four basic types of sensor combinations are

- PIR/ultrasonic
- PIR/microwave
- PIR/acoustic
- PIR/vibration

At first consideration, PIR/ultrasonic would seem the natural selection, as they are complementary in type of application and have similar range sizes. However, as out-of-hours environmental problems within buildings are more often associated with changes in air temperature, humidity, air movement, and noise (phones, water, etc.) from which ultrasonic sensors suffer, than with fluorescent lights, penetration of glass, etc., from which microwaves suffer, PIR/microwave is generally the preferred combination. PIR/microwave devices are particularly useful for large warehouses, factories, and large, open-plan office spaces. PIR/ultrasonic devices come to the fore where multiple use of microwaves would result

in mutual interference problems or where microwaves would give too great a coverage for the area (small cellular offices, for example). While ultrasonic/microwave sensors can be purchased, the list of environmental and building services systems that can give rise to false alarms makes them difficult to apply.

PIR/acoustic and PIR/vibration devices are normally used in situations where penetration through a structure (normally sensed acoustically or vibrationally) requires very early confirmation to prevent further advance into a building before an alarm is given. The PIR sensor would be armed as soon as a possible intrusion count was started on the vibration/acoustic processor and would only shut down if no dual sensing was actually found. These devices can be of use where a multisensor chain of break glass sensors covers windows that may be subject to intermittent wind buffeting, knocks from passersby, etc.

It is generally considered that claims that a single-sensor, false alarm ratio of 500–400:1 for PIR/microwave, 400:1 for PIR/acoustic, and 300:1 for PIR/ultrasonic are fairly accurate but this is highly dependent on the actual installation.

Weighed against the lower false alarm rates of PIR/microwave devices, PIR/ultrasonic detectors tend to cost less or be smaller, require no license, and be easier to commission.

Things to look for include the following:

equipment from manufacturers specializing in the field of dual technology. Reject any suspiciously low unit costs for the dual sensors from smaller companies, as it will invariably mean a compromise of quality and reliability.

intelligent reporting and continuous monitoring of microwave *and* PIR elements separately to control panel using a built-in or base-mounting addressable transponder. This will inform the panel of any faults, failures, masking attempts, false alarms, alarms, etc.

sensors that incorporate

- self-testing of individual elements, (e.g., three hourly doses of IRE injected into the PIR for up to 1.5 seconds to test for masking)
- at least 25 PIR zones (in three levels) and backup and normal range microwave fields
- site selectable field-counting ratios (e.g., 16:1, 32:1, 64:1, 128:1, 256:1)

dual signal path processing

2-year, comprehensive warranty

6-month environmental false alarm money-back guarantee

separate sensor fail LED lamps for diagnostic purposes

unlocking capabilities to allow either the PIR/microwave or PIR/ultrasonic elements to be unlocked and the device reset to operate only on the remaining sensing element should the other fail. This gives emergency coverage until the unit can be replaced.

disable capabilities to allow only one element to work during the day or night and automatically bring in the other at predetermined times. This is useful where daytime fluorescent lighting, passage paste windows, or similar problems might exist that could upset one element unnecessarily.

built-in zone finder or test port for portable meter

PIR elements that have

- a temperature control feature
- mask-out zones for small animals and problem areas
- double-edge zones (i.e., positive and negative signal across zone within specific time—intruder only needs to move across one zone before an alarm-raising sequence starts)
- an internal generator and testing
- optical filter suppression of direct sunlight and heat changes
- low temperature differential capability to detect small changes of temperature
- a Fresnel lens
- 10–50Hz cross-field detection on the microwave sensor element

Advantages and Disadvantages

Advantages.

lower false alarm rate than single technology sensor when false alarms affected by changes in the envi-

ronment in which the sensor was placed (normally quoted at 500:1)

sensitive to movement toward, across, vertically through, or tangential to the device (unlike single-technology sensors that usually have reduced sensitivity in one direction)

sensitive to all types of motion (e.g., crawling, shuffling, jumping). The second sensor usually detects the avoidance techniques used against the first and vice versa.

normally includes features that are only found in the best of each single, sensor head in auto-test, auto-reporting temperature gain compensation

can be used when problems would exist for any single sensor head (e.g., a windy area of a large size, with high temperature gradients over short time periods)

useful for higher-risk, poor environments or remote sites where false alarm charges would be very high

Disadvantages.

more expensive than single-sensor heads

often physically larger than single heads

slightly reduces *overall* probability of detection, when compared to single-sensor heads in perfect environments

increased internal complexity requires even better quality assurance testing than normal to ensure very long periods of time between failure of internal components (2–3 years minimum). Only the devices offered by established manufacturers of dual technology devices are normally able to provide these features at prices that customers will pay. Beware of cheaper devices claiming similar ranges; it normally means that quality has been sacrificed.

5
Closed Circuit Television Systems

INTRODUCTION

The last 5 years have seen an explosion in the use and variety of applications for closed circuit television (CCTV) and closed circuit video recording (CCVR) systems in the security industry. Why? Taking the cynic's view, one could say it is because people seem to find it easier to manage machines than other people. Hiring, managing, and checking the performance of guards takes time and effort. Good, reliable guards or guarding companies are difficult to find. Pay rates are low and risks are high. Hence, staff turnover and lack of real commitment to protect goods or property are hardly surprising. Add to this the real cost of providing wages and other payroll costs for three people working 8-hour shifts to cover a building for 24 hours, and the cost to a building owner can be very high.

A well-designed CCTV installation will significantly reduce manpower requirements. Many such systems pay for their installation in saved payroll costs within 3 years. Never forget, however, that while CCTV systems can perform many tasks that guards cannot and can do them with total vigilance, the converse is also true — guards can provide levels of protection that CCTV cannot. The two are complementary. A CCTV system is only as good as the guards using it and interpreting its findings. Primarily, CCTV systems are extensions of a guard's own eyes.

CCTV SPECIFICATION, SELECTION, PURCHASE, AND INSTALLATION

The key to successful CCTV acquisition is to recognize the simple things:

1. You are the person who knows the risks and where they are located. You must help specify and review designs.
2. You are not the expert in CCTV design. Do not pretend that you are. Seek professional advice from a CCTV expert. It will save you money.

Either hire an independent security consultant who specializes in this area or select several, good, *specialized* CCTV design and installation companies, and invite them to submit proposals based on your outline brief. How do you determine the good companies? Well, use the same basic rules that apply to all purchasing:

Find out whether they do only CCTV work. If so, it is their bread and butter and they should know more about CCTV than a general security company.

Take references of past installations. Ask around.

Ensure that you know enough to be able to identify a knowledgeable salesperson from someone who was selling hamburgers last week.

Insist on a turnkey installation.

Get each company to consider at least two options/ proposals.

Good companies will show you exactly what they will supply and will give you camera specifications, etc. Very good companies provide easily understood price quotations in plain English that detail their proposals and describe their products so that *you* understand the system's capabilities, functions, limitations, maintenance costs, etc. One such company is Focus Ltd. of Bristol, England, to which I am indebted for expanding my own knowledge and for permitting me to pass some of their knowledge on to you in this chapter.

SYSTEM COMPOSITION AND FUNDAMENTALS

Systems are generally comprised of the following parts that are linked in various configurations:

- the scene and its ambient lighting level
- the camera and lens viewing the scene
- the camera housing and a drive or mount
- the picture transmission system
- the picture monitor
- peripheral devices for camera switching and activation, picture recording, picture production, etc.

Systems may be simple or complex, but the fundamental principles are the same:

> The viewed scene must have sufficient light reflected back from it, toward the camera, for the camera to generate a good, clear picture.
>
> The picture signal transmission must be fast, efficient, and without loss of strength or quality. The signal should arrive at the monitor in as good a condition as when it left the camera.
>
> The monitor should faithfully convert the signal into a high-quality picture of the viewed scene and, when necessary, enhance what was actually "seen" to improve guard recognition of any potential trouble.

CCTV picture production is a simple chain of events and components. However, like all chains, they are only as good as their weakest link. *Every* part needs careful consideration for eventual success and satisfaction.

Availability of Light

One of the very first questions that should be asked in the design of CCTV systems is, "What is the minimum lighting level that will *arrive* at the camera lens under any reasonable situation?"

To function, *all* cameras, no matter how sensitive, must receive some reflected light from the scene being viewed. The lighting level for which your camera is selected may be so low or at such a high wavelength that, to the human eye, there is no light, such as the case with infrared imaging or starlight cameras. This apparent lack of visible light does not alter the fact that there *is* reflected or generated light present and cameras will not work without it.

Proper consideration of light is needed for two reasons. First, the lower the available minimum ambient lighting level, the more expensive the camera. As levels drop, CCTV costs rise exponentially. Second, provision of lighting to reduce camera cost can provide secondary benefits to guarding, but is expensive to install, run, and maintain.

There is obviously a crossover point between these two levels of provision and cost, and, hence, an optimum solution. With interior security, CCTV camera selection is normally fairly straightforward. As lighting is normally present for other functions at very high levels, relatively cheap cameras with low sensitivities can be used. On exterior CCTV systems, this is often not the case and many company proposals do not consider the optimum crossover point sufficiently. They want to sell CCTV cameras, not light fittings! "Camera Sensitivity and Light Level" discussed in this chapter *and* Chapter 8 give further guidance on this subject. Let us now consider picture production, sensitivity, and light levels.

Electronic Picture Production

Today's cameras fall into two categories—*tube cameras* and *solid state cameras*. To show how light actually generates a picture, let's look at the tube camera process. The camera lens focuses light from the viewed scene onto a glass faceplate and photoconductive layer. This faceplate is at the lens end of the camera housing and forms the front of a sealed vacuum tube. At the far end of this tube, an electrical (cathode) element releases electrons. These form a beam that travels down the tube and collides with the photosensitive faceplate. Wherever the beam hits the photosensitive layer, energy is given off. The amount of energy given off at that point is dependent on the amount of light falling on the other side of the layer, as reflected from the viewed scene. A signal electrode arranged around the plate, gathers this energy and transmits it down the transmission line to the monitor. In order to form the picture viewed at the monitor, the energy is released into a similar tube where it forms a second beam. This beam collides with a photosensitive screen and one minute element of the remotely viewed scene is presented to the observer. In order to form a complete picture, the electron beam sweeps the photosensitive plate (and hence the viewed scene) very rapidly and transmits that collected energy to the monitor. The entire scene is scanned 30 or more times per second, and hence a viewed scene is scanned before it has time to change radically. Figure 5–1 illustrates a "tube" camera.

From the preceding description of the camera tube, it can be seen that the process of producing a picture on a CCTV monitor is little different from any normal television. Essentially, a scanning electron beam rapidly travels, to and fro, across a small area of a scene, as presented on the photosensitive layer by the lens. As the pattern of light arriving at the camera affects the beam, it simultaneously transmits these electrical variations to a monitor where the process is reversed. That is, the beam now "paints" the viewed scene on the monitor by scanning across a photosensitive layer and activating phosphors on the screen to produce the visible light patterns, shades, contrasts, etc., that our eyes perceive as

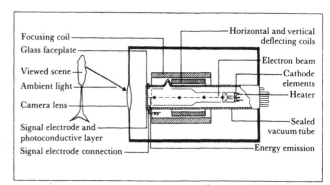

FIGURE 5–1   Vidicon tube—cross section

that which the camera saw. To ensure that both camera and monitor scanning are synchronized, a special "sync" signal is passed with the scene information. When selecting systems, better resolution will result from "full" or 2:1 sync signals and not the more normal, less expensive "random" type.

"Solid state" cameras, or, as they are often called, *charge coupled device (CCD) cameras,* modify the normal picture reproduction in several ways. They present a direct array of electrical signal generators to the viewed scene and do not have a scanning beam within the camera assembly, and they store the picture received momentarily, frame by frame, and then pulse this picture as electronic signals to the monitor receiver where a conventional scanning beam reproduces it on the screen.

Whatever the camera type, the crucial factor in picture reproduction is the amount of light that actually hits the photosensitive faceplate. This is known as *faceplate illumination.*

Camera Sensitivity and Light Level

Light is a form of energy and cameras use this source of energy to excite phosphors on the photosensitive layer of the faceplate.

The amount of light received on the faceplate of a camera is crucial, because the camera is designed for a specific range of ambient lighting levels, with the minimum level being the threshold of good picture production. Cameras will often give pictures below this level, but not *good* pictures.

The magnitude of the signal produced (in volts), its rate of change from 0V to, say, 1V, and the overall duration of the signal produced will be dependent on the light reflected and received from every tiny part of the viewed scene (i.e., every photon of light that *actually arrives* at the very small part of the faceplate at exactly the same instant the scanning electron beam hits the opposite side).

Camera manufacturers are often very shy in their *specifications* of lighting levels required for their cameras to operate correctly. This is deliberate. It is nice to have something like the vagaries of light to blame when camera pictures are not good. To properly specify a camera's sensitivity, the following information should always be given:

- the faceplate illumination level required (minimum) in lux (lumen/m², or footcandle)
- the percentage of the viewed scene reflecting light toward the camera that is needed (e.g., 89% scene reflectance). This is a measure of the uniformity of the light.
- the level of illumination reflected off the scene as measured on-site (e.g., 10 lux on the vertical plane)
- the f-stop number. This is a measure of the size of lens aperture at which the available light enters the camera. The higher the f-stop number, the less light that will be let through. An f-stop number of 1.4 is the normal bench mark for measuring sensitivity. If other f-stops are used, it is often to hide inadequacies.
- the maximum f-stop number for the "usual" lens associated with the camera when under test
- the color temperature of the light source if artificial light is used (e.g., 2856°k, 3500°k)
- for low-light cameras, the minimum full-video illumination level (e.g., 10 lux), *not* the usable picture level. The latter is a usable level if you do not mind not being able to tell a person from a dog! The *usable* level may be 20 times lower than the *full video* or *recommended* level.

Faceplate illumination level, while the ideal method of indicating a camera tube's *sensitivity,* is a totally impracticable method for actually specifying an illumination level for a camera and ensuring it works. To do this, the CCTV designer must calculate whether (a) the minimum illumination hitting the viewed scene, under all reasonable scenarios, will result in the minimum faceplate illumination to produce a good picture or (b) the automatic iris and lens combination will be able to regulate the level of illumination entering the camera in such a way that the optimum faceplate illumination is always present while still giving a wide and high enough viewed scene area.

Essentially, to start with, the designer assumes that (a) the former calculation will be easier than the latter determination and therefore (b) concentrates on calculating the required minimum level of ambient illumination.

Before considering this process in more detail, it is useful to look at the problem globally. The ambient light falling from a subject will go through a series of contortions, reflections, travel distances, transmissions through lenses, etc., before reaching the faceplate. Each of these interactions represents a loss of available light. These losses are considered, in sequence, to arrive at the final, remaining figure in order to compare that figure to the actual needs of the faceplate (or, in the reverse direction, to arrive at the average illumination of the scene required for a particular camera selection). Consider the following losses.

Average Values. The *average* versus *minimum* level of illumination on the *subject plane considered* — i.e., what is the minimum level of illumination falling onto the actual viewed scene? Let's say that a light meter is placed horizontally on the ground beside a wall some distance from the luminaire being used to light the area and the meter produces a reading of 100 lux. This reading is the level at that point only. If a further 20 readings are taken in a regular grid pattern, it is soon evident that lighting levels vary greatly over small distances. By adding together all the values and dividing by the number of points we get the *average illumination.*

This *average value* is only applicable in that plane and at that angle, that is, on the horizontal. If we wish to view the wall that is on the vertical plane, the *minimum* horizontal illumination may only be 50% or so of the average lux. The actual light reflected *onto* the viewed scene may be 50% of this (i.e., 25 lux). The reflectivity of the wall may only be 50% at best, hence the illumination directed *onward* toward the camera may be only 12.5% of that on the ground and the minimum illumination on the wall 12.5 lux.

Distance to the Camera Loss (The Inverse Square Law of Illumination).   As light travels from the wall to the camera, it diverges and its intensity diminishes, with the result that less light falls on the lens than came off the wall. The reduction is distance dependent; each doubling of distance results in one-quarter of the illumination. Let's say we put the camera close to the subject and we get a 50% further reduction (i.e., 6.25 lux).

F-stop Value and Lens Losses.   Cameras usually operate under a variety of lighting levels. Camera imagers have particular sensitivity at a particular level of light and this level is the optimum for good picture production. At lighting levels above and below the optimum, the picture quality deteriorates. To prevent this, CCTV camera lenses are provided with an iris like that of the human eye, but with fixed settings. The CCTV iris is normally automatic and it "throttles back" or "stops" the light from entering the camera as the ambient level increases and lets more light enter the camera as the ambient level decreases, thereby maintaining the optimum level actually reaching the lens. This "stopping" by the iris/lens combination is referred to as the *iris f-stop number.*

$$\text{f-stop} = \frac{\text{focal length}}{\text{lens effective diameter}}$$

The f-stops are set so that each reduction in f-stop number is equivalent to 50% of the previous level of illumination that would have been passed. Similarly, each increase in number would double the previous illumination entering the lens. So, the iris of the lens will only admit a certain amount of the light falling on the lens and the lens will reflect, absorb, and eventually transmit, to the faceplate, a percentage of that light, which is determined by the formula

$$C_I = \frac{1}{4f^2}$$

where C_I = the internal light level passed with 100% transmittance and f = the given f-stop value. Thus, using this formula, if our camera has an f-stop setting of 2.0 for the view, C_I = 1/16 (or one-sixteenth of the 6.25 lux available). (Note that we began with 100 lux "available" at

the subject and now have 0.39 lux *at the faceplate.*) In practice, the lens transmission loss and other factors between the lens and faceplate give higher losses than just the formulas would suggest, and, as a result, only one-twentieth of the 6.25 lux will actually arrive at the faceplate (i.e., 0.3125 lux). Table 5–1 gives the typical value by which light falling *onto the lens* would have to be *divided* to determine the actual faceplate illumination level. Table 5–2 shows how this would change for alternative sensitivity/ f-stop combinations quoted. Table 5–2 is important for when a lens may be f-stopped *above* the level of f-stop quoted for faceplate illumination. For different camera lens formats see Table 5–3.

In conclusion, it can be seen that the various losses dramatically reduce the level of illumination reaching the faceplate. Hence, in general, CCTV "rules of thumb" are often used to approximate a calculation. For example, if the *faceplate illumination* is quoted as *x* lux, the actual average illumination falling onto the horizontal should be 200 × lux or more to receive *good* pictures (e.g., 0.1 lux faceplate = 20 lux average horizontal requirement).

If the *camera illumination level* 1 × lux is quoted, then it will need 10 × lux average horizontal for a good picture and 50 × lux for full video recording quality pictures.

Table 5–1   Faceplate Illumination Reduction with Increasing F-stop

For ½ CCD cameras. Divide the illumination value falling on the lens by the following figures to determine faceplate illumination for given f-stop values.

f1.0	5
f1.4	10
f2.0	20
f2.8	40
f4.0	80
f5.6	160
f8.0	320
f11.0	620
f16.0	1280
f22.0	2560

Table 5–2   Faceplate Illuminations for Alternative F-stop Bench Marks

If ½ CCD camera quotes faceplate illumination required = 10 lux at f1.4, the following would apply to different f-stop values for the same camera.

f1.2	7 lux
f1.4	10 lux
f1.7	15 lux
f1.9	18 lux
f2.4	30 lux
f2.8	40 lux
f3.4	60 lux

Table 5-3 F-stop Variation with Camera Format

To allow the same light through the camera for different camera formats, put lens on:

1-inch camera   f2.8
2/3-inch camera f1.4
1/2-inch camera f1.0

Example 5-1: Typical CCTV Calculation
Data from Field Survey.

Area to be viewed: building wall 24 m (72 ft) wide ×8 m (24 ft) high

Distance subject to camera: 10 m (30 ft)

Average *horizontal* illumination: 100 lux average (minimum 50%)

Plane of subject to be viewed: vertical (i.e., person upright in front of wall)

Reflectivity of subject: wall or person average 50% intensity of lighting off wall 312.5 candelas

Reflectivity of ground in front of wall: average 50%

Assume f-stop value of camera lens and iris: f1.4

Camera Data.

Minimum faceplate illumination of chosen camera: 0.1 lux for good pictures

Minimum scene illumination of chosen camera: 0.6 lux, f1.4, 50% reflection

1. *Available illumination at camera lens*
   a. Light falling in front of wall: 100 lux average
   b. Minimum light falling in front of wall: 50% of average = 50 lux
   c. Minimum light reflected from ground onto wall: 50% of minimum at ground = 25 lux
   d. Minimum light from wall toward camera: 50% of that arriving on wall from ground—i.e., 25 lux × 0.5 = 12.5 lux. Intensity at this point is 312.5 candelas.
   e. Loss of light due to distance to camera (assuming light reflected directly toward camera): $E = I/d^2$, where E = lux level at camera, I = intensity in candelas at wall, and d = distance from wall to camera

$$\therefore E = \frac{312.5}{10^2} = 3.125 \text{ lux}$$

2. *Theoretical illumination at faceplate of camera with lens* $f = stop = 1.4$

$$C_I = \frac{1}{4f^2}$$

where C_I = illumination level at faceplate with 100% transmittance and f = f-stop number used on iris

$$C_I = \frac{1}{4 \times 1.4^2} = \frac{1}{7.84} = 0.127, \text{ or } 12.7\% \text{ of that}$$

arriving at camera

Thus, faceplate illumination available = 3.125 lux × 0.127 = 0.39 lux

Check calculation using rules of thumb:

3. *Camera minimum faceplate illumination = 0.1 lux*
   Rule of thumb: required faceplate illumination × 200 = average horizontal illumination required at scene: 0.1 × 200 = 20 lux. Actual horizontal average (100 lux) and minimum (50 lux) both well above 20 lux required by rule of thumb calculation.

4. *Camera scene illumination level*
   Rule of thumb: 10 × camera illumination level required for good pictures: 0.6 × 10 = 6 lux. Both 100 lux average and 50 lux minimum well over this.

Lighting level calculations should always be provided by the salespeople or system designer to show *how* a chosen camera would be suitable. Many manufacturers make it difficult for the true camera needs or performance to be established by publishing either partial information, introducing unseen assumptions, which enhance apparent performance, or deliberately inflate claims. Typical things to watch for include

faceplate illumination: 1 lux with 89% scene illumination (see Table 5-4)—no f-stop quoted; *scene illumination is irrelevant; reflection *assumed* is what matters

10 lux camera—useless; no definition of what or where

10 lux f1.4—This implies 10 lux with 100% reflection toward the camera, although it does not say so! In practice, the reflection could be 20%, hence the lighting level needed would be much higher.

infrared reflective assumed—this will be presented as a light level of *x* lux at 85% reflectance. Only *invisible* infrared (IR) illumination wavelengths can achieve such high reflectance levels. Visible light reflectance rarely exceeds 70% and is normally 20–40%. If quoted 85%, this indicates economy with the truth!

light source assumed at color temperature 2856° Kelvin—this, again, indicates that IR is being used to boost performance; 2856° Kelvin indicates a tungsten light source. Only tungsten and high-pressure sodium light sources have any significant infrared content. Tungsten sources are uneconomical and should only be used in special circumstances in CCTV work. Tungsten halogen, low-voltage tungsten metal halide, and high-frequency instant-strike fluorescent lamps are often better for CCTV work.

filter removed—this assumes that IR filters provided to *protect* the camera during normal operation have been taken out. This would be impractical and assuming it boosts sensitivity figures by 400%. If actually practiced, it could result in an unusable camera!

If any of these "conditions" is used in the specification, either increase the level of scene illumination actually required by a factor of 10 in order to compare that camera

Table 5-4 Light Reflectivity for Common Building Materials

Percentage of scene illumination reflected toward camera (for white light color temperature 3500k)

	Average	*Minimum Likely*
Asphalt/tarmac	5%	3%
Parkland, trees, grass	20%	15%
Red brick/blue brick, dark mortar	20%	15%
Yellow brick	25%	15%
Dark stone	30%	25%
Red brick, light mortar	35%	30%
Middle-colored stone	35%	30%
Plain concrete (white)	40%	30%
Cars parked	40%	30%
Painted woods, light colors	50%	35%
Ceiling tiles (white), floors of light color	55%	37%
Portland stone, bath stone, or other smooth white/cream stone	60%	40%
Stone or profiled aluminum cladding	60%	40%
Smooth aluminum building cladding	65%	40%
Curtain glazing/windows	70%	60%
White brick or polished white stone	80%	65%
Large areas of snow	85%	70%

For any of the above horizontal surfaces with a vertical chain-link fence in the field:

Green-covered PVC, subtract 1%
Black-covered PVC, subtract 3%

If minimum levels of light reflected are required, use second column.

Source: Data courtesy of The Electricity Council, Great Britain.

with other cameras, or find yourself a more honest manufacturer or salesperson!

A good specification should say at least "camera sensitivity 10 lux of horizontal scene illuminance at f1.4 maximum aperture with 60% reflectance."

LENSES AND FUNCTIONS

Function

Lenses are the eyes of the CCTV system and the remote eyes of the guard watching the monitor screen. What must they do? They must provide a picture on the monitor that is commensurate with the objectives and aims of the CCTV system. These are normally defined as either

simple detection—"something moved somewhere"; fixed camera or scanning wide-angle camera

recognition—"a person moved across the screen, look!"; scanning camera, medium lens

identification—"that's not one of our guards"; scan, tilt, and zoom lens

automatic detection—motion detection by camera brings picture on at the monitor

The second stage in selection and positioning of CCTV systems (after determining the illumination level and camera type) is to determine which lens should be used for the camera and for the optimum numbers of cameras and the positions to cover the required area and give either detection, recognition, or identification capabilities. Figure 5-2 shows how CCTV can be linked to various sensors, paging systems, motion detectors, etc., to provide an invaluable tool to the security guard. Figure 5-3 shows a simple package, single unit, CCTV system.

Area Viewed

To determine the key areas to be viewed, a site survey is essential. Initial plans to establish outline numbers, costs, etc., are fine, but final lens selection should always be by surveying existing or new premises.

Surveyors normally look at an area with a customer and establish how long it would take for him to disappear from view behind cover (or out of the field of view of the lens). Running fast, most intruders cover 20–26 ft/sec (6–8 m/sec) across open ground or floor space. The normal time required for eye contact and action at a monitor is 3–4 seconds. Hence, the camera should be able to cover a width of viewed scene of at least 79–105 ft (24–32 m) in open space work or the space between areas of possible concealment elsewhere. The height of the viewed area is generally less important than the width. On most designs, a height of 26 ft (8 m) is normally recommended to cover fence climbing, ladders to first-floor windows, etc.

Lens Definition

A lens is only fully defined if the following are quoted:

- format: ½", ⅗", 1"
- focal length: say, 50 mm
- fixed focal length or zoom
- f-stop number: 1.4, etc.
- function and view: split optic, pinhole, etc.
- light transmission ratio
- optical distortion
- whether an IR filter is included or not
- mount type (i.e., C or CS)
- if a zoom lens, motorized or manual
- iris type: manual, automatic, or fixed

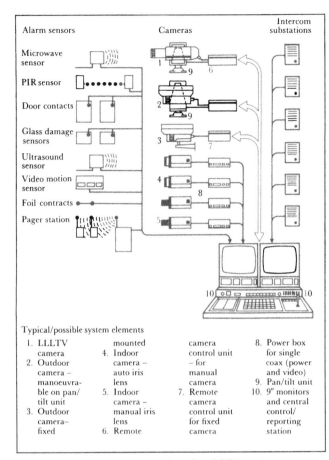

FIGURE 5–2 Typical comprehensive CCTV system configuration

FIGURE 5–3 Simple package, single unit, CCTV system

There are very few lenses of the following types *specifically* made for ⅔ ″ format; most are ½ ″ design. Common descriptions used for CCTV lenses are:

super wide (focal length (FL) = 4.5 mm) — this gives a very wide field of view, usually 70°6′ angular field

of view horizontal. This is sometimes called a *fish eye,* as the field of view is actually wider than that of the human eye.

wide (FL = 6.0 mm) — 56° angular field of view horizontal, short distance to subject, large view scene

normal or standard (FL = 12.5 mm) — 30° angular field of view horizontal; approximate to human eye and probably most commonly used lens

telephoto or long-range (FL = 36 mm) — narrow field, 10° horizontal, but long distance to subject possible

zoom — variable focal lengths typically 10–100 mm (1:10)

Various lenses are displayed in Figure 5–4. So let's say we wish a lens to cover 79 ft (24 m) with a minimum height of 26 ft (8 m). What other criteria apply? Well, if you have determined the area, but the camera is so far away that on the monitor the intruder is a tiny dot, your guard will not see him! The general "rules of thumb" for recognizable size are

For an intruder, standing

5.0% of scene height for detection
10% of scene height for recognition
40–60% of scene height for identification

For an intruder, crawling

0.75% of horizontal field of view or 10% of scene height for detection
3.0–5.0% of horizontal field of view or 20% of scene height for recognition
10% of horizontal field of view or 60% of scene height for identification

Do not forget that these are simple rules of first approximation. The actual *quality* of the picture for detection, recognition, or identification will still be very dependent on the scene illuminance, the scene uniformity of illuminance, transmission losses, etc., which only field trials can fully determine. See Example 5–1.

FIGURE 5–4 Various lenses and lens drive assemblies

Lens Focal Length Selection

Having established the width and height of the area you wish to see and the minimum height a person should appear on the screen, and having looked at possible camera mounting positions, you may select a lens. The normal planning process would be to consider the likely distance the camera will be from the viewed scene and then undertake an iterative process, using lens specification charts, until a suitable lens is determined. Before this is carried out, the lens size or *format* should be determined. Lens formats are normally ½ ″, ⅔ ″, and 1 ″. The golden rules are that the lens format *must* be larger than or equal to the camera tube/imager format (½ ″, ⅔ ″, 1 ″) and that the larger the format, the bigger the field of view, and the crisper the picture production.

Tables 5–5 and 5–6 give examples of typical lens selection data. Using these tables for an example of lens selection, review Example 5–2.

Example 5–2:   Lens and Focal Length Selection.

1. Use the same data as in Example 5–1.
2. Area of scene: 24 m (72 ft) wide × 8 m (24 ft) high
3. Distance to scene: 10 m
4. Detection conditions: Assume intruder is *standing* and about to scale wall. We wish to detect intruder's presence (not recognize or facially identify).
5. Use Table 5–5 and rules of thumb for recognizing intruder. Intruder (say, 2.0 m tall) would need to be at least 5% of scene height displayed. Taking height of scene first on Table 5–5:
   a. A height of 8.3 m at a distance of 10 m results in selection of a ⅔ ″ format, 8.0 mm focal length lens, but the width of the viewed scene would only be 11.0 m, so half the area would not be covered by the CCTV (22% person-to-scene height is over that required).
   b. Using a ⅔ ″ format, 4.0 mm focal length would give a field of view 22 m wide and 16.5 m high, with the intruder representing 11% of the viewed scene height (5% being the minimum for detection, 10% for recognition).

Selection *b* would at first sight seem better than *a*. However, we can see that the camera is very close to its subject when compared with the subject's width. We also know from Example 5–1 that the illumination level is good, and hence we could consider a third option, which would be to move the camera to 20 m away from the wall. Using Table 5–5 again, we get:

⅔ ″ format, 8.0 mm focal length lens, scene width 22 m, scene height 16.5 m with 11% person-to-scene height

The advantage of this would be a slightly better choice of focal length.

Using the Rule of Thumb Check for Focal Length. (See Figure 5.5.)

$$\text{Distance to camera from viewed scene} = \frac{\text{scene width} \times \text{focal length}}{\text{format width}}$$

Using ⅔ ″ format: format width = 8.8 mm

$$10000 \text{ mm} = \frac{24000 \text{ mm} \times \text{FL}}{8.8 \text{ mm}}$$

$$\text{FL} = \frac{10000 \times 8.8}{24000} = 3.6 \text{ mm}$$

$$\text{Distance to camera from viewed scene} = \frac{\text{scene height} \times \text{focal length}}{\text{format height}}$$

$$10000 \text{ mm} = \frac{8000 \text{ mm} \times \text{FL}}{6.8 \text{ mm}}$$

$$\text{FL} = \frac{10000 \times 6.8}{8000} = 8.5 \text{ mm}$$

We are choosing either the 8 mm or 4 mm focal length.

Using Tables 5–5 and 5–6, it is possible for a designer to consider many alternatives to a particular problem, which could include increasing distance and illumination level, fewer cameras with wider fields of view, lower resolution at same viewed area but lower cost, etc.

Focal Length Formulas

CCTV companies often produce simple, ready reckoners for focal length (FL) selection. These are based on simple geometric formulas (Figure 5–5) simplified to give a common constant. Examples are

1 ″ format lens: format width (FW) = 12.7 mm
                    format height (FH) = 9.5 mm
⅔ ″ format lens: FW = 8.8 mm
                    FH = 6.6 mm

$$\text{Distance to camera from viewed scene} = \frac{\text{scene width}}{\text{FW}} \times \text{FL}$$

or

$$\text{Distance to camera from viewed scene} = \frac{\text{scene height}}{\text{FH}} \times \text{FL}$$

Some examples will show how focal length requirements can be checked.

Example: Using a 1 ″ format lens with a viewed scene width requirement of 24 m, at 10 m distance would require

Table 5–5 Lens Selection Data for ⅔″ and ½″ Format Imaging Devices

Focal Length ⅔″	½″	Angle of View Horizontal	Vertical	Width of Scene at Distance—Meters 5	10	15	20	30	50	Height of Scene at Distance—Meters 5	10	15	20	30	50	2m Person as Percent of Scene Height at Distance—Meters 5	10	15	20	30	50
25	16	20°	15°	1.8	3.5	5.3	7.0	10.6	17.3	1.3	2.6	4.0	5.3	7.9	13.3	138	69	45	34	23	14
16	12	30°	23°	2.8	5.5	8.3	11.0	16.5	27.5	2.1	4.1	6.2	8.3	12.4	20.6	86	44	29	22	15	9
12.5	8.0	41°	32°	3.5	7.0	10.6	14.1	21.1	35.3	2.6	5.3	7.9	10.6	15.8	26.4	69	34	23	17	11	7
8.0	5.6	58°	45°	5.5	11.0	16.5	22.0	33.0	—	4.1	8.3	12.4	16.5	24.8	—	44	22	15	11	7	—
6.5	4.8	68°	54°	6.8	13.5	20.3	27.1	—	—	5.1	10.3	15.3	20.3	—	—	35	17	12	9	—	—
4.8	3.5	84°	68°	9.3	18.2	27.5	—	—	—	6.9	13.8	20.6	—	—	—	26	13	9	—	—	—
4.0	2.8	95°	80°	11.0	22.0	—	—	—	—	8.3	16.5	—	—	—	—	22	11	—	—	—	—

Telephoto Lens mm to mm	Horizontal	Vertical	Width of Scene at Distance—Meters 50	100	150	200	300	500	Height of Scene at Distance—Meters 50	100	150	200	300	500	2m Person as Percent of Scene Height at Distance—Meters 50	100	150	200	300	500
50	10°	7.5°	8.8	17.6	26.4	35.0	53	88	6.6	13.2	19.8	26.4	40	66	27	14	9	7	—	—
100	5°	3.8°	4.4	8.8	13.3	17.6	26	44	3.3	6.6	9.9	13.3	19.8	33	55	27	18	14	9	5
150	4°	3.0°	2.9	5.9	8.8	11.7	17.6	29	2.2	4.4	6.6	8.6	13.2	22	82	41	27	21	14	8
200	2.5°	1.7°	2.2	4.4	6.6	8.8	13.3	32	1.7	3.3	5.0	6.6	9.9	17	106	55	36	27	18	11
250	2.0°	1.5°	1.8	3.5	5.3	7.0	10.6	18	1.3	2.6	4.0	5.3	7.9	13	138	69	45	34	23	14
300	1.3°	1.0°	1.5	2.9	4.4	5.9	8.8	15	1.1	2.2	3.3	4.4	6.6	11	164	82	55	41	27	16

To determine the correct zoom lens for an application it is necessary to check the field of view at "wide angle," that is, low focal length (e.g., 12.5 mm), and telephoto setting (e.g., 7.5 mm).

Standard zoom lens: 6:1 (12.5–75 mm)   10:1 (10–100 mm)
                    8:1 (12.5–100 mm)   15:1 (10–150 mm)

Special zoom lens: 10:1 (16–160 mm)   15:1 (15–225 mm)
                   13:1 (16–208 mm)

Data courtesy of Focus (CCTV) Ltd.

⅔-in (8.8mm) format. Focal length (mm)	Horiz. field of view	(12.7mm) 1-in format on ⅔-in Vidicon. Focal length (mm)	Horiz. field of view	1-in format on 1-in Vidicon. Focal length (mm)	Horiz. field of view
4.8	100°	6.5	69°	6.5	103°
8.5	56°	12.5	36°	12.5	54°
11.5	35°	14	27°	14	50°
12.5	33°	18	21°	18	39°
16	28°	25	19°	25	28°
50	9°	50	9°	50	14°
75	5°	75	6°	75	9°
90	4°	90	5°	*90	8°
100	4°	180	4°	108	7°
		135	4°	135	5°
		140	3°	140	5°
		144	3°	144	5°
		150	2°	150	5°
		160	2°	160	5°
		350	1°	250	2°

1-in format lens : format width (FW) = 12.7mm.
                 : format height (FH) = 9.5mm.
⅔-in format lens : format width (FW) = 8.8mm.
                 : format height (FH) = 6.6mm.

1. Distance camera to scene = $\dfrac{\text{scene width}}{(FW)}$ × focal length.

2.   or = $\dfrac{\text{scene height}}{(FH)}$ × focal length.

e.g. using a 1-in format lens, scene viewed 1m × 1m, with focal length of lens 90mm:

distance = $\dfrac{1000\text{mm}}{12.7\text{mm}}$ × *90mm = 7.085m (or 8° angle, see above).

Formula may be transposed to give any missing value.

FIGURE 5–5   Common, fixed focal length lenses

$$10000 \text{ mm} = \frac{24000 \text{ mm}}{12.7 \text{ mm}} \times FL$$

or

$$FL = \frac{12.7 \times 10000}{24000}$$

$$FL = 5.23 \text{ mm}$$

Say, 6.0 mm focal length selected.

Example: If height is critical, say 3 m

$$10000 \text{ mm} = \frac{3000 \text{ mm}}{9.5 \text{ mm}} \times FL$$

$$FL = \frac{9.5 \times 10000}{3000} = 31 \text{ mm}$$

Say, 25 mm focal length selected.

Fixed Lens or Zoom Lens?

In general, for CCTV surveillance, fixed focal length lenses are generally preferred to zoom lenses for the following reasons:

> To zoom in on a scene requires operator action at a critical time. Seeing something more closely does not necessarily assist the guard in gathering vital information and acting *quickly*. At the critical time of intrusion, you can guarantee that the intruder will be at the edge of the scene. People love fiddling around with zoom lenses and hence they are often zoomed in too close for the broad picture to be appreciated or the intruder presented center stage.

Table 5–6   Lens Data

1″ Camera FL	⅔″ Camera FL	½″ Camera FL	Approximate Angle of View	
			Horizontal	Vertical
160	112	75	4.5°	3°
140	100	70	5°	4°
100	75	50	8°	4.5°
75	50	35	10°	6.1°
50	35	25	14°	9°
35	25	18	20°	12°
25	16	12	30°	18°
16	11.5	8.5	41°	27°
12.5	8.0	5.6	58°	36°
9.5	6.5	4.8	68°	50°
8.0	5.5	4.0	80°	55°
7.0	4.8	3.5	84°	63°
6.0	4.0	2.8	95°	80°

This table provides approximate comparisons mainly as a quick guide to selection and has been devised using the most popular lenses. For accurate details reference should be made to the manufacturers' specifications.

Examples using the table:

1. A 160 mm lens designed for a 1″ camera when fitted to a ½″ camera will provide the same field of view as a 75 mm lens designed for a ½″ camera.
2. A 5.6 mm lens designed for and fitted to a ½″ camera will provide the same field of view as an 8 mm lens designed for and fitted to a ⅔″ camera.
3. Note from example 2 that a 12.5 mm × 1″ lens fitted to a ⅔″ camera will equate to an 8 mm × ⅔″ lens and fitted to a ½″ camera will equate to a 5.6 mm × ½″ lens.
4. Note that using a larger lens on a smaller camera has the effect of widening the field of view.

A word of caution for engineers specifying lenses, or customers interpreting competitors' specifications: A 140 mm zoom lens is not necessarily more powerful than a 100 mm zoom lens. If the 140 mm lens was designed (as it probably was) for a 1″ camera, it becomes a 100 mm lens when fitted to a ⅔″ camera.

> Zoom lenses are complex, motor-driven devices that are costly and fail.
> A fixed lens with a 75 mm focal length (a common size) on a $^{2/3}$″ camera sees as much as a 1″ zoom lens on a 1″ camera (the camera becomes a 100 mm lens!) at a much lower cost if the fixed lens is properly situated.
> Zoom lenses have less resolution than other lenses at an equivalent cost. Zoom lenses should only be considered for interior identification, facial recognition, specialized, covert surveillance work (e.g., casinos), etc.
> Fixed focal length lenses are simpler, cheaper, more reliable, and of known, predictable performance.

Table 5–5 shows the effect of different focal length lenses by considering a zoom lens at its maximum and minimum fields of view, respectively. A rule of thumb often used in CCTV work is that an intruder must occupy 40–60% of the viewed height to be distinguished by general appearance, as friend or foe. Table 5–5 shows that to achieve this, the width of the field of view must be severely restricted. So the zoom lens is useful *if used wisely.*

Lens Application

Format.

Use a common format throughout the system whenever possible (either ½″, ⅔″, or 1″) (i.e., match lens and camera).

Use 1″ formats whenever cost will allow, to reduce geometric distortion.

Use 1″ format on all low- or very–light-level cameras.

The ½″ format is *only* suitable for ½″ CCD cameras. It is not suitable for tube cameras.

Never, never use a lower format for the lens than for the camera (e.g., ½″ lens with ⅔″ camera). If you do, the edge of the picture viewed will be underfilled, with severe picture loss at the edge of the screen.

Always specify fully (e.g., ⅔″ = 8 mm = f1.4).

Quality. Quality is determined by transmission and distortion of light. The lower the level of barrel or pincushion distortion, the better. Always choose lenses that have a resolution of 1000–1500 lines. To ensure quality of the overall picture, never choose a lens resolution that is smaller than the monitor resolution (i.e., 625 lines). The higher the number of optical elements (4 to 12), the less the distortion but the higher the cost.

Focal Length and Width of Field.

Put cameras as close as possible (without compromising security) to the viewed scene. Use wide focal length lenses (i.e., less than 16 mm) to maximize the general surveillance area while reducing the number of cameras installed.

Only use greater than 50 mm focal length or zoom lenses when distant objects need to be brought into close view with sharp focus.

To maximize the width of field for a particular lens, match its format to that of the camera. For example, a 1″ lens on a 1″ camera. A 1″ lens on a ⅔″ camera overfills the imager.

In selection, use actual focal length (i.e., FL = 36 mm), not common terms for type (i.e., not "telephoto," "medium," etc.).

Intensified, low-light cameras should have the lowest possible f-stop values, and these values should never be larger than 1.8.

Iris and Filters.

Use automatic iris, not manual.

The larger the number of f-stops on the automatic iris, the greater the camera's ability to function under a variety of light levels, focal lengths, etc.

Values of an automatic iris from f1.2 to f1500 can be found. Try to avoid these and use f1.2 to f16—the most common range.

Low-light cameras [intensified CCD (ICCD), etc.] are primarily used for night viewing, but are also available for use during daylight hours. To achieve daylight use an automatic iris *and* light attenuation device will be required. Light attenuation devices are either neutral density filter (NDF) or spot filter (SF) types. Spot filters are generally thought to be better and more reliable.

In general, Newvicon, Silicon Diode, Ultricon, and CCD solid state cameras will require SFs or NDFs, giving an equivalent of f-stop 360 for daylight operation.

Low-light silicon-intensified target (SIT), intensified SIT (ISIT), and ICCD cameras will require SFs, giving an f-stop equivalent of 1500 for daylight operation.

Lenses and Mounts.

Standard ⅔″ and 1″ cameras use C mount lenses. One-half-inch cameras use CS mounts.

Use CS mounts whenever possible. Never use CS on a C-type camera.

It is possible to use an adapter ring to allow for the use of a C-mount lens on a CS-mounted camera, but this is not recommended.

Do not use *extenders* (also called *range converters*); these are an unnecessary way to extend focal length.

CAMERA TYPES

Cameras are normally defined or grouped by their sensitivity to light, the method of picture generation, and the format of the picture-generating device.

Sensitivity to light is often the first consideration in CCTV camera selection and the area where greatest care must be taken in preparing specifications. The following section on camera sensitivity explains this importance.

The types of picture-generating devices used with camera tubes are defined by the date of their development. New is not always best; the type used depends entirely on the application. In chronological order, we have the

- electron tube—30 years ago (1960)
- silicon diode—20 years ago (1970)
- charge coupled device—8 years ago (1984)

Camera diameters, or *formats,* are usually expressed in inches: 1″, ⅔″, and ½″. Generally, the larger the diameter, the better the resolution and the lower the limit of usable light.

Standard Vidicon (10 Lux Min Face Plate Illumination)

Standard Vidicon cameras are the oldest and most widely used CCTV cameras. They are now being challenged by

CCD cameras in all areas of camera work, but they remain the cheapest camera type for general indoor work on low-security premises, video access control, etc. They have the advantages of

> low capital cost
> established, well-understood performance, pitfalls, and repair types
> many producers, therefore good competition
> good resolution sensitive to most daylight conditions
> camera response closely approximates the human eye and gives low guard eye and viewing fatigue at monitor
> longer life than all but CCD cameras

Vidicon cameras should not be used

> in static, fixed-scene applications where high light levels might occasionally occur (sunlight or artificial), as this causes *burn in* or *bleaching* of the tube that considerably reduces tube life
> when low or no light is present. Lighting levels of 5 lux or more require some lighting to be present at night; this costs money.
> when targets are expected to move fast (e.g., in vehicles, etc.), as a *lag* or *smear* will appear on the camera as the object moves, making viewing difficult
> outdoors at night unless the installed lighting is very well designed

Vidicon cameras are generally suited for indoor surveillance work with good, 24-hour lighting levels. These cameras should be placed at fixed distances and have scan, pan, and tilt mechanisms. Generally, it is necessary to change a tube on cameras at least once every 1 to 2 years; the cameras also require a 6-month maintenance check. These costs should be included in the initial cost package.

Newvicon

Newvicon cameras, a trademark of Matsushita Electric Company, Incorporated, are an enhanced version of the Vidicon cameras and aim to provide greater sensitivity. These cameras give good quality pictures during daylight hours, indoors or out, and still take usable quality pictures at dawn, dusk, and most outdoor security lighting levels. The Newvicon is the workhorse camera for most indoor and outdoor perimeter security work where a minimum faceplate illumination of 0.005 lux, f1.4, 100% reflectivity or 1.0 lux minimum, 10 lux average horizontal illumination is provided. In this role, the Newvicon camera is steadily being superseded, in number of installations per year, by silicon diode cameras and CCD cameras.

Advantages.

> better sensitivity than Vidicon cameras
> spectral response (wavelength of light) good—particularly near infrared level
> cheaper than silicon diode, CCD, or Ultricon cameras
> lower maintenance and slightly longer tube life than Vidicon cameras
> generally a better resolution than Ultricon cameras, but a poorer resolution than Vidicon cameras

Disadvantages.

> lag and smear still poor at low levels of illumination
> burn-in potential and *flare* problems still exist
> most include automatic iris with automatic control (AGC) NDF or optical filter for control of faceplate illumination as ambient levels rise and fall. This adds unneeded complexity.

Ultricon III, a Registered Trademark of the Radio Corporation of America

Ultricon III cameras were developed after Newvicon cameras. Ultricons incorporate a silicon-target photosensitive plate, instead of the Vidicon-style antimony plate, and have improved electron gain to make them both more sensitive to available light and almost immune to bright light burn-in problems. In addition, they are *very* sensitive to infrared light. This combination makes them a normal first-choice camera for lower-cost, 24-hour, outdoor or indoor covert surveillance, using infrared light sources to illuminate particular areas of a building or perimeter. These cameras are the most sensitive, nonspecialist, nonamplified-intensity cameras available and they are particularly cost- and performance-effective for 1″ diameter formats. Lighting levels of 0.01 lux minimum horizontal, f1.4, 75% reflectivity or 0.0065 lux faceplate illumination result in average illumination of 0.5–1.0 lux actual minimum light requirement for good picture quality (full video).

Advantages.

> better sensitivity to available light than either Vidicon or Newvicon
> very wide spectral response bandwidth (from near ultraviolet to well into infrared)
> more sensitive to infrared light than Newvicon or Vidicon
> better resistance to burn in than Newvicon or Vidicon
> longer tube life than Newvicon or Vidicon
> no ghosting or lag of picture

Disadvantages.

> poorer resolution than either Newvicon or Vidicon

still produces a smear at low- and high-level light ranges

more expensive than Vidicon (3:1); comparable to Newvicon in price

not as robust or as long a life span as CCD cameras

Superchalnicon

Superchalnicon cameras were developed from the Chalnicon cameras used in the medical field. They have a very good sensitivity to visible and infrared light, but are more sensitive to infrared sources than the Ultricon III cameras to which they are normally compared. Superchalnicons are used when totally covert infrared source surveillance is needed. For this application, they are better, but slightly more expensive than Ultricon III types. Superchalnicons are not widely used or manufactured.

FIGURE 5-6   CCD camera

CCD Cameras

As the quest for more sensitive camera technology continued, it was realized that different branches of research were providing similar answers—silicon diode and charge coupled device technologies. It soon became apparent that while both these technologies could accommodate amplification and intensification to reduce their minimum sensitivity way below normal electron tube technology, CCD devices would produce more immediate and far-reaching solutions to CCTV problems. As a result, CCD research took off and produced a range of imaging types:

• frame transfer (FT)
• interline transfer (IT or IL)
• x-y coordinate, metal oxide silicon (MOS-xy)
• charge injection device (CID)

MOS-xy and IT/CID devices all gave resolutions comparable with Vidicon or Newvicon cameras, but with only a light sensitivity of Vidicon. CCD frame transfer (CCDFT) devices proved the most popular line by having a light sensitivity and resolution better than the then available Newvicons and Vidicons and comparable with silicon diode (SD) cameras.

SD cameras held their ground for a while until it was proved, through use, that CCDFT cameras overcame the biggest problems associated with all electron tube cameras and some SD cameras—

smearing or lag
burn in and premature tube failure
blooming
very short tube life

CCD cameras (see Figures 5–6 and 5–7) are now flooding the market and replacing all former forms of Newvicons,

FIGURE 5-7   CCD color camera

Vidicons, Ultricons, etc. In practice, the improved CCD's performance is very well worth the extra cost incurred in buying one.

Advantages.

no lag or burn in
can be "no flare"
not affected by EMI
more compact, lighter, and consume less power
longer tube life (probably 10 years versus 1 ½ years)
no geometric picture distortion
freeze-frame capability possible
lower voltages used; safer product
very robust and not affected by vibration

Disadvantages.

higher cost

very infrared-sensitive with special precautions needed

currently has poorer resolution than some camera types

not normally recommended for covert infrared lighting applications

Costs are falling dramatically, but infrared flaring is more difficult to overcome and infrared filters are essential. The specifications for minimum illumination should always be assessed *after* the filters are fitted, as a 4:1 increase in level required usually results. Resolution is steadily being improved by using ½ " lenses with CCD cameras (which differ from all others, as they use $2/3''$ or $1''$ formats). This has produced two effects—fewer lens types available for use and cameras require special CS lens mounts (normally C mounts) to incorporate the infrared filters, the automatic iris, etc.

CCD cameras that have infrared filters in place normally require a minimum of 2.0 lux at f1.4, 75% reflectivity scene illumination for a *usable picture*, and 8.0 lux for full video output. Higher Newvicon equivalent specification cameras require 0.6 lux scene illumination, f1.4, and 73% reflectivity with an infrared filter present.

Very Low Light Level Cameras (VLLLCs)

All the preceding camera types use tubes or *plates* with sensitivity levels at which the human eye can function quite well. However, once the 0.1 to 0.01 lux level (or moonlit sky) is gone, the human eye becomes quite poor at discerning objects, cannot adapt quickly, etc. (see Chapter 8). Below 0.005 lux (which corresponds to a moonless, partly cloudy night with some starlight) the eye performs extremely poorly. It is at these lower levels of illumination that CCTV cameras provide invaluable vision assistance.

VLLLCs provide this assistance by employing illumination intensifiers, which gather the available light and electronically modify it and the resulting signal output. These intensifiers are currently applied in the following ways:

Single-stage intensification applies to conventional imagers (e.g., Newvicon, Vidicon, etc.). These devices are now obsolete.

Single-stage SIT (SS-SIT) cameras use silicon diode-type imagers similar to those found in Ultricon IIIs and couple the intensifier in the same housing. This combination provides low ghosting, low blooming, and one-quarter to one-eighth moonlight (i.e., works under a cloudless sky when a quarter-phase moon is present). This illumination level equates to 0.0004

lux faceplate, 0.08 lux or less scene. SS-SIT cameras are commonly used for covert surveillance or overt, poorly lit surveillance. These cameras were developed in the 1970s and are now being superseded by ICCD cameras.

ICCD Cameras

In ICCD cameras, an image intensifier is coupled to a standard CCD camera. The result is a camera with the same sensitivity as the SS-SIT camera, but with the added benefits of

no lag at extremely low levels (picture fades away fast, but until this point, the target is clear)

crisper picture quality

Second-generation ICCD cameras came onto the market in the late 1980s. They extend the sensitivity level of CCDs still further—down to one-quarter starlight, 0.00002 lux faceplate illumination, and 0.0004 lux scene illumination—with the same benefits as the SS-SIT and the ISIT cameras. ICCD cameras are now replacing ISIT and SS-SIT cameras due to lower production costs.

ISIT Cameras

For many years, ISIT cameras (see Figure 5–8) were the standard, one-quarter starlight covert surveillance cameras. While still widely available, they should be carefully considered against second-generation ICCD cameras on cost and performance grounds.

Third-Generation ICCD Cameras

Third-generation ICCD cameras are beginning to appear on the nonmilitary market and extend the range of CCD cameras to total darkness or very little starlight. These cameras are very expensive and a second-generation ICCD camera with infrared illumination assistance will provide similar performance at a much lower cost.

FIGURE 5–8   ISIT camera

Infrared and Thermal-imaging Cameras

There are three camera alternatives when CCTV is required for interior or exterior, covert, night surveillance in totally dark environments:

- infrared-sensitive cameras with infrared illumination
- laser spot illumination cameras
- thermal-imaging cameras

Only the first option is usually affordable, as both laser and thermal-imaging cameras are extremely expensive and should only be selected based on the advice of a specialized consultant or as required by a military specification.

Infrared-sensitive cameras now come in a variety of types, but probably the best infrared camera available is the Ultricon III. The Ultricon III would normally be used in conjunction with solid state, 6–50W infrared-generating luminaires, which produce invisible wavelengths that are very reflective off normal scene materials. This configuration is often far less expensive than either illumination to 1.0 lux of a scene (1.0 lux being too low to be of much use for eye vision to guards) or starlight cameras with no or very little artificial illumination.

Infrared-sensitive cameras have the following advantages over visible light cameras:

the illumination level does not vary, so picture quality is very stable

no lag or burn in

no shadows on picture

infrared penetrates fog, mist, etc., more readily than visible light from a normal luminaire, thus providing a better all-weather capability

Color CCTV Cameras

The following points should be noted about color CCTV cameras:

Color CCTV cameras are usually of the CCD type. They require a faceplate illumination of 0.2 lux minimum, at f1.4 with 100% reflectivity. In practice, this translates into an actual measurable horizontal illumination average of 40 lux *minimum*. However, as the good quality threshold is normally 100 lux and above, their 40 lux minimum level restricts their normal use to indoor applications, as exterior lighting rarely exceeds 10–20 lux.

Color tube imagers for low light levels are very expensive.

More expensive in initial capital cost (twice the cost of black-and-white CCD).

Require more frequent maintenance and color adjustments.

The more expensive, the truer the color.

Uniformity of scene lighting levels over time is very important. If they vary, color hues become dull.

Good quality color monitors are required.

The primary advantages of color CCTV cameras are

color brings more realism to the scene

color improves recognition of faces, dress, etc.

movement is easier to detect in color than in gray-level contrasts

color results in lower guard fatigue at monitor positions, improves guard reaction time, and increases the likelihood of recognizing and arresting the correct person in a crowd

Camera Costs

The initial capital cost of any scheme of CCTV protection is largely determined by the cost of cameras and the lighting installation required. (The labor costs are normally comparable for a given number of cameras and lamps.) By comparing the relative costs and illumination levels, an earlier appreciation of the economic sense of various alternative camera installations can be determined. For example, 1.0 lux illumination would be relatively cheap to provide. Newvicon and CCD cameras cost the same. Would paying ten times more for an SS-SIT camera *just* to avoid providing 1 to 10 lux be worthwhile?

CAMERA AND CCTV SYSTEMS APPLICATION

Introduction

Successful CCTV application requires consideration of many factors, including:

Camera	Relative Cost Ratios	Camera Average Scene Illumination (lux)
Color	5:1	40–100
Vidicon	1:1 (base (cost)	5–20
Newvicon SD	4:1	1–10
Ultricon SD	5:1	0.5–1.0 (or infrared)
Superchalnicon	6:1	Infrared
CCD	4:1	2–8 or 0.6–5.0
SS-SIT	40:1	0.002–0.008
ICCD	30:1	0.0004–0.0025
Third-generation CCD	70:1	0.0001–0.001

Decide whether night viewing is required and whether viewing is to be passive or active. The passive system uses a *tube* and intensifiers to amplify the available ambient lighting (natural or artificial). Simple passive systems cannot "see" in total darkness (i.e., cloudy, moonless night in open country). Active systems employ infrared light sources to illuminate the target area. Infrared light is invisible to the naked eye and hence the camera can see the criminal but not vice versa. Good active systems using infrared tubes will see a man at 1312 ft (400 m), and intensified cameras will see a man at 1968 ft (600 m) in conditions of starlight or less. Although passive systems are often not as good as active systems in low-light conditions, they generally give much better quality pictures with more definition.

Remember, it is not enough to simply specify a passive camera; the available ambient reflected light level must be given.

Remember, if lighting fails or is vandalized, a passive system that requires artificial lighting will not be able to "see" anything.

In terms of camera economics, decide whether improved artificial lighting of the scene will allow a less sensitive, passive camera to be used, resulting in a lower overall cost. Ask the manufacturer to give system cost alternatives with improved lighting costs treated as a single element.

In terms of camera performance, decide the performance required from a system and tell the manufacturer [e.g., "I want to read a car license plate at 400 yd (366 m) at an angle of 20° from center horizontal and 60° from vertical, in conditions of starlight, in a rural location"].

In terms of camera/monitor resolution, just as the lens sets a limit on resolution and definition, so does the camera/monitor combination. Choose the system that gives the highest possible *overall resolution* for an affordable cost under exactly the same environmental conditions for monitors. Overall resolution is a function of vertical and horizontal resolutions. Vertical resolution is fixed by the number of "lines" quoted (e.g., 500, 525, 550, 625, etc.). Horizontal resolution is seldom quoted and, without it, *resolution* is meaningless. Horizontal resolution is a function of speed or frequency of the scanning electron beam in the picture tube. The combination of horizontal and vertical resolution is known as overall resolution. Overall resolution may be quantified and graded by the number of *dots* or *picture elements*. A good overall system resolution is in excess of 165,000 picture elements. Ask the manufacturer for the system's overall resolution.

Cameras and monitors should use the same electrical phase.

The following section will look more closely at some of these application factors.

Camera Housings and Accessories

A CCTV camera is a sensitive piece of electronic equipment. A good housing will turn it into a robust piece of equipment. An outdoor camera is shown in Figure 5–9. What environment will the camera operate in? Many systems have failed because the camera has simply blown over, or the lens has frosted or become covered with dirt. Even an indoor camera will need dust protection, so consider whether the camera will need to be protected from

- extremes of cold
- extremes of heat
- ice formation
- rain, mist, spray, and other water sources
- corrosive elements
- vandalism or malicious damage
- dust
- theft
- radio interference or other electromagnetic waves (including lightning)
- snow
- high winds

Always choose a camera with a housing that has been matched to the camera it contains and is fit for its purpose. With all camera housings ensure that

wiring through the housing is by means of a watertight, sealed grommet

the enclosure is highly resistant to dust and water; cleaners have been known to wash them!

the housing has an ingress protection (i.e., IP or NEMA rating suitable for protection from dust and moisture)

the camera is secure within its housing and the housing is bolted to a strong, secure mounting. All housing access should be padlocked or antitamper screwed to prevent covert tampering.

the enclosure is of noncorroding material

the case is metallic and it shields the camera effectively if electromagnetic interference could be a problem

For external camera housings,

try to select a camera assembly that requires only a 110V supply or lower

ensure all electrical connections are completely weatherproof and internal

ensure cable materials and insulation are suitable for extremes of cold and heat and will not deteriorate when moved in these conditions; ensure outer sheaths will not fail in strong sunlight

pick a housing with a reflective "umbrella" or cover to reflect the sun in the summer and to deflect snow and rain in the winter

If high temperature and heat generation from the camera might occur, ensure that the camera has a housing with either a chimney ventilator or a fan for cooling.

FIGURE 5-9 Outdoor camera and accessories

If frost is a problem, use a screen heating element or integrated fan-heater unit. Ensure heaters are automatically temperature controlled and not manual only. If the camera is switched on to view an intruder, the intruder will not wait for you to defrost the screen! To disperse rain, a wiper will be required. Always use the appropriate washer in conjuction with the wiper.

Remember that the low temperature limit of the camera may be well below the level at which either the automatic iris or powered zoom lens might fail; look for the weakest part of the system. Alarm the camera via its own transmission cable to prevent removal.

Camera Mountings and Drives

Pan-and-tilt camera drives allow the camera to be remotely trained onto a specific area of a viewed scene in order to improve the direction of viewing or to follow an intruder. Unless the camera is of the covert, infrared type, the criminal will usually spot the camera and will move on to try and dodge it. *Pan* refers to horizontal angular rotation; *tilt* refers to vertical angular rotation. A camera that has either or both of these mechanisms is known as a *scanning camera* to distinguish it from the simple, fixed-position camera.

Autopan and autoscan cameras continuously move over the target area in order to reduce the cost and number of static cameras required to cover the same area. However, unless at least two, interlaced, camera sweep patterns are used (i.e., with overlapping fields of vision), the intruder will probably dodge the overt camera.

As all motors, gears, and brakes are complex, expensive, and liable to break down unless well maintained, scanning cameras should be avoided unless the situation or economics demand their use. Never use pan-and-tilt or autopan and autoscan cameras when one or more simple, fixed, or manual scanning cameras will do an adequate job. When choosing a motorized pan-and-tilt drive, look for

sealed bearing races and sealed, pregreased charged gear trains
a drive that brakes effectively, does not overshoot, and

will not bind. Eddy current braking is usually the best, as it is frictionless.
drives that are as highly weatherproof as the camera enclosure
drives that are classed by the manufacturer as internal, external, light duty [below 22 lbs (10 kg) camera and ancillaries], medium duty [above 22 lbs (10 kg) camera and ancillaries], or heavy duty [above 44 lbs (20 kg)]
a drive that is compact with a very low center of gravity
camera leads that have movement loops that cannot foul the drive

Pan-and-tilt drives (as seen in Figure 5-10) are available that are intruder/sensor actuated. There are two types—a drive system (i.e., a fence detection device tells the guard where the intrusion is occurring and automatically drives it to the specific coordinates, then allows for manual control) and video motion detection (see later section). Drive systems are not recommended, as intruders may deliberately decoy cameras to a position and be out of the range of the viewed scene by the time a second sensor or camera responds.

Interior cameras should not be suspended from false ceiling grids or affixed to wall partitions, as they vibrate. Use a solid, structural element as the mount.

Exterior cameras should be mounted on galvanized and synthapulvin-painted lattice masts with built-in ladder eyes to allow for proper maintenance. A steel standing platform should be cantilevered from the tower at approximately 4 ft (1.2 m) down from the camera. The platform should have an upstanding edge to allow tools to lie safely on it without crashing to the ground. The mast manufacturer should give structural calculation guarantees for likely maximum wind, ice, and snow load combinations and a service warranty. Masts should always be supplied and installed by the camera manufacturers so that they are responsible for workmanship, alignment, vibration, resistance, etc. Wooden telephone poles, straight scaffolds, etc., should not be used, as they

FIGURE 5-10 Pan-and-tilt drive unit

normally vibrate too much in high winds. On higher masts, cameras can vibrate too much in high winds. On higher-security work, masts can be fitted with movement detection sensors on alarm loops to prevent unauthorized climbing of a mast. This is particularly important on remote sites, without camera overlap (i.e., when one camera can see the next).

Camera Positioning

Positioning is primarily determined by the available, secure, mounting positions and the area that is to be viewed with the selected lens combination. General pitfalls to avoid include

   aiming at bright light sources, such as directly into oncoming traffic (headlights) or directly into the rising or setting sun

   aiming straight at or near trees or grass, which can become covered in snow

   allowing too great of a dead zone under the camera (i.e., the area the camera cannot tilt down to see)

   aiming toward a fence (weld mesh, pallin mesh, etc.) without sufficient angular offset (the fence appears solid!)

   insufficient mast height to look down on ground hollows, over the tops of cars, etc.

SIGNAL TRANSMISSION

Transmission Medium

CCTV, as the name implies, is normally comprised of a television camera that is linked, by cable circuit, to its monitor (unlike public network *broadcast* TV). While this is usually the case, CCTV signals may be transmitted via

- cables (twisted pair, coaxial, triaxial, fine wire, or optic)
- microwave link
- infrared line-of-sight beam
- telephone line (real time or slow scan time)
- radio frequency multiplexing over the cables

Some of these methods of signal transmission are discussed in detail in Chapter 2.

Signal Composition

When the camera for the application has been selected, it will have a defined resolution quoted in *lines* (e.g., 450 lines, 500 lines, etc.), in *pixels* (e.g., 510 horizontal by 492 vertical), or as a bandwidth (e.g., 10 MHz). These measures define the potential, *maximum clarity* of the picture.

The picture at the monitor cannot be any clearer than that at which the camera sees. *Lines* are the most common units of resolution quoted and, in most cameras, resolution will be quoted at 400–600 lines; 600-line resolution is the normal maximum and it equates to a normal bandwidth of 7.5MHz picture signal transmission. Therefore, all transmission system parts on most CCTV systems should be matched to provide the minimum loss, attenuation, distortion, etc., in this frequency bandwidth.

Unfortunately, attenuation, distortion, etc., are *particularly* frequency dependent and hence in various parts of the bandwidth, the effects will be greater. In general, the attenuation losses at higher frequencies are greater than attenuation losses at lower frequencies (e.g., at 5MHz a quarter signal loss may result, while at 1MHz an eighth loss may result). As the actual amplitude of the signals also varies, it is easy to see that transmission systems and components *must* be made to optimize the resultant signal. A general rule of thumb for systems is that a loss of no more than 4dB should occur if *noticeable* resolution loss is to be prevented.

The principal requirements of any transmission system are that it

   be available at all times (particularly in bad weather!)

   faithfully transmit the signal with high fidelity and little signal loss, distortion, etc.

   be economical to install and run

   be secure from tampering or signal injection

Transmission will form a significant proportion of the total system cost and, hence, its type and logistics should not be taken lightly.

Losses

All methods of transmission suffer from various forms of interference and loss. The essence of good design is to minimize their impact. Some transmission systems are immune to particular forms of loss and interference (e.g., optic fiber link cables and RFI/EMI devices). These systems may, however, have their own particular, less understood problems, which add to unforeseen costs.

Cable Transmission Systems

Twisted Pair Systems.   Unshielded, twisted pair systems should not be used.

Coaxial Cable Systems.   Coaxial cables were designed specifically to assist with the reduction of many forms of interference found in TV work. They are the most common, most well understood form of signal transmission and should always be a first consideration.

While many salespeople will insist that coaxial cable transmission is good for a distance of 2–2.5 mi (3.2–4 km), on most systems this is not true due to bends in the cable, etc. Outdoor limiting distance is normally 1 mi (2 km) and indoor is limited to 0.5 mi (1 km).

Coaxial cables are graded using *RG numbers,* which define the attenuation loss in decibels per meter. They are also more commonly classified by their standardized per unit impedances as

- 75 ohm, unbalanced, indoor
- 124 ohm, balanced, indoor
- 75 ohm, unbalanced, outdoor
- 124 ohm, unbalanced, outdoor

Indoor cables are smaller, more flexible, and easier to hide, as they have a braided outer conductor. However, they are easily damaged and have high attenuation at high frequencies. Outdoor cables are stronger, have larger diameter cores, and have less attenuation, but are more costly. Uses are as follows:

- overhead, ducted, indoor: 75 ohm UB RG-59/U
- overhead, ducted, outdoor: 75 ohm UB RG-11/U
- buried direct: 75 ohm UB RG-13A/U

Similar use 124 ohm cables would be 16AEVL, 754E, and V-1-DSAL types. The most economical solution for many situations is to use RG-59/U with amplifiers when necessary. (See Figure 5–11 for maximum distances.)

Always ensure that the manufacturer of your system has matched the cable selection to your site requirement and not used a "universal" cable. If a "universal" cable is used, you will not get the best from the system.

The method of termination at both ends of the coaxial cable is very important. If the connection is not made correctly, ghosting will appear on the monitor. To eliminate this, the end terminations should have a characteristic impedance equal to the cable impedance at a given frequency. When a system has many lines of resolution, is of high frequency, or has cable lengths in excess of 400 ft (122 m), the terminations used should always be complex terminations.

The best method of reducing any hum, ghosting, etc., on a system is to use a *balanced line system,* which has self-correcting, matched signal transformers at each end of the transmission line.

To summarize, to ensure a good quality picture with contrast, little loss of definition, and no ghosting, try to

keep cable routes straight, short, well protected, well isolated, and segregated

use the best quality coaxial cable and end terminations, and the lowest frequency transmission that money will allow

ensure that the coaxial cable is *matched* to the termination

When cable runs exceed the limits given in Figure 5–11 there are three alternatives that can be selected to still use cable transmission:

Video signal amplifier on the output from the camera to the coaxial cable with intermediate signal boosting when attenuation becomes too great. This normally increases the distances quoted in Figure 5–11 by a factor of six.

Transmission on powered coaxial cables. This is normally referred to as *vidpower, videoplexing,* or *powered signal.* The actual installed distance limits are normally two times greater than those quoted in Figure 5–11. This system uses the same cable to power the camera and send the signal. Special output port camera types or converters are required, but savings in power wiring can be considerable.

Twisted pair or *telephone grade cables.* These cables use transponders at each end of the cable to send and receive signals. The system is direct, dedicated, and should be electrically screened in conduit and mechanically protected. Cables are normally limited to 2000–5000 ft (610–1524 m), depending on the quality and size of the wire cores used.

When multiple camera installations are required to transmit to a remote building, a coaxial trunk cable is often used, which transmits by using modems at either end of the coaxial cables to simultaneously frequency multiplex the various channels of the camera signals down the single cable.

Problems

The most common problems encountered during signal transmission are interference suppression and fidelity control on cable systems. These problems are often encountered during *commissioning* usually reflect directly onto the monitor and, thus being identified, can be corrected. Some problems are less easy to determine due to their intermittent nature or particular harmonics. Typical problems include

loss of picture resolution or *blurred screen details.* This is usually a result of loss of part of the signal ampli-

Cable	Maximum Transmission Distance in ft/m (no bends or connectors)	Normal Limit in ft/m
RG-59/U	750 (229)	500 (152)
RG-6/U	1500 (457)	750 (229)
RG-11/U	1800 (549)	1000 (305)
RG-15/U	2400 (732)	1500 (457)

FIGURE 5–11   Cable transmission distances (unamplified)

tude at the higher end of the frequency spectrum and is generally caused by too low a specification cable for the distance covered; too many bends, kinks, joints, etc., in the cable; damage to the braided outer or inner stranded cores.

smearing. The picture content has unclear, poorly defined edges. This is generally caused by signal phase addition, particularly in the lower part of the frequency spectrum.

hum. The picture has horizontal black or white, moving or semistationary bars on the screen, which block out part of the picture. This is caused by radiated energy (radio frequencies) from adjacent power lines (normally 60Hz). The transmission cable cuts the radiation and a current is induced onto the cable.

snow. Snow appears as lots of black and white dots on the screen occurring in a very fast, random pattern. Its source is normally a high-frequency form of radiated energy. It is heavy random noise.

noise. Noise is a less frequent random form of snow and is usually associated with directly injected interference or radiated energy from switching operations, welding, motors, etc.

jumps. Jumps appear as a partial or full loss of picture quality. They are always associated with a bad connection or joint, damage, etc., which infrequently affects the picture.

These effects may be mitigated by taking the advice offered in the following paragraphs.

Resolution Loss. Use 75 ohm or 124 ohm cables as the minimum standard. Use flexible core, coaxial cable and ensure the minimum bending radius is exceeded by at least a factor of two. On cable runs more than 900 ft (274 m), including changes of direction, etc., employ signal amplifiers and equalizers at the monitor end of the cable. On cable lengths more than 500 ft (152 m), use amplifiers and equalizers at both ends. When cable runs exceed 2000 ft (610 m), fit either preamplification devices at the camera end and convert to a balanced line, twisted, shielded cable or a 124 ohm, balanced, coaxial cable with intermediate amplification.

Hum and Smearing. Use either *hum dampers* at the camera end or combined, signal-isolating/impedance-balancing transformers at the camera and monitor ends, and transmit the signal on balanced, 124 ohm transmission cable.

Snow. High-frequency RFI/EMI snow is most readily removed by a balancing transformer and balanced, 124 ohm cables.

Noise. Noise can be removed by physical separation of the cable from its source, metal conduit or additional foil screening, or correction of the offending device noise output by suppression devices. Suppression devices are always the best alternative.

Jumps. Insist on unjointed cable drum and run lengths, on-site inspection of installation quality, etc. Use of the correct connectors will help avoid signal transmission jumps.

Long-Distance Cabling Using Telephone Lines

When distance from the camera to the monitor exceeds 1 mi (2 km) outdoors and 0.5 mi (1 km) indoors, and if picture quality is to be maintained, coaxial cables should not be used. In these situations, the public or private telephone network should be considered. For these transmission media, there are two alternative methods of picture transfer—slow-scan TV (SSTV) and real-time video coder/decoder (videocodex).

SSTV. Television signals operate at frequency bandwidths of MHz, while telephone lines operate at audio frequency bandwidths of kHz. To avoid using 1000 lines of telephone cables and splitting the signal to transmit it, data compression packet-switching modems are used to gather, send, and decode the signal. In this way, it is possible to send up to 16 camera signals simultaneously from a modem down a single voice grade or data grade telephone line to a monitor station at any distance, national or international. See Figures 5–12 and 5–13.

Points to note when selecing systems using this transmission method are

these systems are general surveillance or alarm-activated video-recording systems primarily suited to fixed target viewing of vaults, doors, utility installations, etc. They are often marketed as an alternative to manned patrol of remote sites.

these systems are *not* real-time picture transmissions, but are a series of picture *stills*, where each picture is separately scanned, stored, sent, and displayed in sequence. Most systems operate on selectable picture display and send rates from 2.5 to 35 seconds, where 8 seconds is a normal compromise for reasonable definition without too great a delay between picture frames produced.

the resolution of the picture is determined by the scanning rate. Hence, a picture that takes 2 seconds to receive will be of far poorer resolution and quality than a 20- or 30-second picture. The scanning is normally selectable.

a separate line through the same system is needed for telemetry control and if a voice alarm (i.e., "intruder, intruder!") is used, a third line is required.

line rental is often very expensive and should always be a key element when comparing cost to other

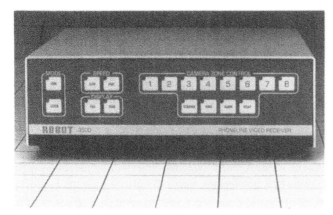

FIGURE 5-12 Slow scan TV system receiver

FIGURE 5-13 SSTV transmitter

transmission systems. In its initial cost, SSTV may be cheaper than infrared or microwave links, but these systems do not have line rent charges!

the common standard SSTV is 256 pixels × 256 lines × 64 gray shades. This is *below* most television monitor line resolutions. A better standard for selection is 512 pixels × 256 lines × 128 gray shades. Most systems only provide a resolution half as good as real-time systems, even at the highest scan times, so don't use SSTV for identification purposes.

Digital transmission is better but more expensive than analog.

Videocodec. Videocodec is the current method of transmitting pictures at very fast rates to approximate *real-time* CCTV and is rapidly being developed to a point where videocodec will compete directly on a cost

basis with SSTV within 5 years. Videocodec differs from SSTV in major ways. Videocodec systems

transmit and display pictures at 5, 10, or 20 pictures per *second*

transmit digital signals, not analogue, using fast telephone system rates (typically 64 kilobytes/sec), similar to computer data transmission

generally use fiber optic trunk lines capable of transmitting at 500 megabytes/sec.

Currently, good videocodec systems use units with selectable transmission rates of 64–1920 kilobytes/sec, such as British Telecom's VC2100. These systems allow for future expansion of picture rates and selectable speeds of 10, 20, or 30 pictures per second.

With international standardization on data transfer, and falling fiber optic cable costs, it is likely that real-time transmission down data lines used for other purposes will gradually replace SSTV and dedicated coaxial transmission.

Fiber Optic Transmission

Currently, fiber optic transmission cable is expensive and specialized. It does, however, provide an almost ideal method of transmitting picture data. It is virtually interference-free, is immune to EMI and RFI, will transmit farther than coaxial cables without amplification, and has a data capacity vastly greater than twin or coaxial cables. On larger systems (say, 30 cameras or over) or where reasonably long distances are involved, harsh environments are present, or SSTV or videocodec systems are required, fiber optic transmission should be seriously considered. There is no doubt that fiber optic cables will, during the next few years, become the preferred method of transmitting all forms of data signals. It is worth remembering that on many existing computer networks or telephone networks for which fiber optic cables are used, the huge data-carrying capacity of the fiber optic cable is rarely fully used and there would probably be space for CCTV transmission. See Chapter 2, "Fiber Optic Transmission," for a detailed discussion of fiber optic cables.

Direct Beam Transmission

Under circumstances where no telephone lines exist or it would be extremely difficult to run a new cable, direct beam transmission is an alternative. The most common transmission media are infrared beams and 10-milliwatt microwave beams (23GHz).

Chapter 2 covers the technical description of these media; see "Microwave Transmission," "Free Space Infrared Transmission," and "Radio Transmission." In prac-

FIGURE 5–14   Monitor

tice, the choice of these systems is always based on economic grounds (e.g., would it be cheaper to beam the signal across the Grand Canyon or to put up pylons and run a cable between them?). Cost for particular systems depends on system complexity, but if we look at relative costs, the initial decision may be easier. A direct beam, infrared signal transmission system, operating over its normal distance range of 2000–3000 ft (610–914 m) would cost approximately three times as much as a simple coaxial transmission system.

Microwaves are considered when the operating range is normally 5–500 mi (8–80 km). However, beam line restrictions, FCC power limits, etc., normally mean range is commercially economical in 10-mile jumps. Equipment for this range will cost more than $20,000. Microwaves do, however, allow for real-time CCTV at a cost similar to videocodecs.

MONITORS, RECORDERS, AND DETECTORS

Monitors

The monitor (shown in Figure 5–14) is where the buck stops! Everything in camera selection, transmission line selection, and protection has been done with the single purpose of producing a good, clear picture on a monitor. The monitor must take the signal it receives, with its defined bandwidth, horizontal resolution, etc., and reproduce a picture of the viewed scene with maximum efficiency and clarity. A good signal and good monitor give a good picture, but a poor signal cannot be improved significantly by even the best monitor.

Knowing the system bandwidth, you can begin the search for a suitable monitor. The first point of consideration is the amount of definition and resolution possible and desirable. In general, this is related to the CCTV application and its function. If the system is a simple, single-camera, low-cost, low-risk system (e.g., for preventing retail shoplifting), then a lower resolution

smaller monitor may suffice. If the system is high security, a high-resolution, large monitor may be needed. Monitors are manufactured in standard tube sizes and resolutions. The *tube size* is the diagonal measurement across the screen—normally 5, 7, 9, 12, 19, 20, or 21 inches, with either 625 or 525 lines scanning and 600, 700, or 800 horizontal line resolution. Five-inch screens with 600-line resolution are often used on very low-cost systems, 12-inch screens with 700-line resolution are used on medium-cost systems, and 19-inch screens with 800-line resolution are used on better or higher-security systems.

Having made an initial selection of monitor, you must now choose between monitors of equal size and quality. Good monitors in any price range should

be robust (particularly controls)
be reliable (should have mean time between failure, warranty, etc., provided)
be tested to standards (UL, CSA, etc.)
have a high resolution
have a bright, sharp picture
have a wide amount of contrast
be suitable for video time-lapse or normal, continuous viewing
be easy to operate and set, with tamper protection

The picture quality will be defined by a number of parameters:

resolution—defined by general (center) horizontal line resolution (normally 600 lines/525, 700 lines/525 or 625, or 800 lines 525/625 lines); the higher the number of lines for a given diagonal picture size, the better the picture. Corner resolution (which is often not specified) should be as close to the horizontal center resolution as possible

brightness or sharpness—is defined by the kilovolt (kV) level at which the internal picture tube operates. The higher the volt level, the better the quality. Common levels are: 15.5kV equals high quality and high resolution, 14.0kV equals medium quality and medium resolution, and 8.5kV equals minimum quality.

grayness levels or gray scale—determines the amount of contrast between white and black in a picture. It is not often quoted in literature, but should always be in excess of ten discrete levels.

picture width—domestic televisions overscan the screen to ensure a full picture is presented. This results in some loss, on the edges, of what the camera actually saw. CCTV monitors should underscan the screen so that the whole field of view is presented. While this looks odd, as the picture does not now fill the screen, at least you know you are seeing all the camera sees.

Other points for consideration include

Total system performance should always be evaluated and compared to the perfect system by commissioning trials using television test cards.

Monitors should be mounted at horizontal eye level with an optimum distance of five times the diagonal screen dimension from the eye to the screen. This reduces eye fatigue.

Monitors should be capable of split-screen division to allow multiple cameras to display simultaneous images on a single monitor.

Use a monitor that gives good definition even if transmission attenuation rises to 10dB. Transmission systems age, but the monitor should give a good output despite this.

Choose a system and monitor that give uniform resolution across the entire screen.

Match the manufacturer of the monitor to the camera.

Use a monitor that has a good focusing capability and a large, adjustable contrast range.

Always choose a monitor with the largest *diagonal* "tube" or "screen" dimension that money will allow.

Always choose a monitor that is capable of being rack-mounted, as your client might require extensions to the system.

Monitors should have back level damping, constant variable videotape recording time, and line locking as standard features.

If it can be avoided, never use more than three monitors per guard station if the monitors are continuously watched and not alarm activated.

Never use a monitor smaller than 12 in in size if facial recognition is a requirement.

Remember, monitors are not televisions; you will have no sound output!

Video Image Recording

In certain situations, a monitor will not be continuously watched by security personnel. In others, although guards may be present, independent and automatic reproducible confirmation of what occurred in front of the camera at a particular moment will be required. This is when CCTV should be linked to video recording devices (see Figure 5–15) to give an added dimension to security.

The main criteria for video selection are that the device should be off the shelf and capable of carrying at least two different manufacturers' tapes, and the video recorder should have an instant replay capability while still recording on other channels (monitors). This recording should be channeled to a spare monitor so that what just happened and what is happening (as viewed by the same camera) may be viewed simultaneously. This will allow confirmation of crimes that are hard to spot, such as shoplifting in a large, crowded

FIGURE 5–15 Commercial CCTV video recorder

department store. The act is seen and recorded. The instant replay confirms the action, if doubt exists, and the subsequent actions of the criminal as she heads for the door.

Other video recorder selection criteria include these: The system should be able to record events on several monitors, on the same tape, in sequence, by using a manual monitor selection switch. The recorder should include the ability to run off the same emergency power supply as all the other security equipment.

Video recording may be intruder activated, real time (intruder alarms), continuously set on real time (e.g., banks, etc.), or continuously on time lapse (long-term surveillance work). Time-lapse video recording offers the benefits of continuous recording, but at a fraction of the cost. Care must be exercised in selecting when and for how long the tape is off or the tape could miss the action.

There are three main types of video camera recorders (VCRs) — 4/8-hour, time-lapse, and event recorders.

4/8-Hour VCRs. Most of the 4/8-hour VCRs are produced for the domestic market and are not really designed for continuous, industrial use. In fact, the majority of manufacturers specify this in their warranty terms and cover machines for DOMESTIC USE ONLY. Use of this machine is a low-cost solution to continuous recording.

Some companies modify the standard machine so that at the end of the tape, the VCR automatically rewinds and commences recording again. This is not recommended, because it causes far too much wear and tear on the mechanism, recording heads, and tape.

The main application of 4/8-hour VCRs is in situations when 8 hours or less recording is required each day (e.g., nightclubs, offices, etc.). These VCRs should not be used for continuous, 24-hour recording unless the customer is aware of the reduced VCR life span and additional servicing.

If the time and date are required on the tape, then a separate time and date generator must be added.

Time-Lapse VCRs.   Time-lapse VCRs are tape machines designed for industrial use and continuous recording. Recording periods can be set for 3, 12, 24, 72, 120, 240, 360, and 480 hours of recording. All time-lapse video recorders include a built-in time and date generator; other features include

an alarm input that, in the event of an alarm, will switch the machine from any time-lapse mode into 3-hour standard play recording for a set period

one-shot mode, which will take a single frame on the input of an alarm and then wait for the next alarm

automatic rewind and record (on or off)

recording the time of the first and final alarms

recording the number of alarms received

a switch that can deactivate all other controls when the machine is set to record, to prevent unauthorized interference.

recording which camera or alarm sensor activated the system

CAUTION: Some machines state that audio recording is possible in 3-, 12-, and 24-hour modes. This is misleading, because although audio can be recorded, it cannot be played back! Therefore audio recording is only practical in the 3-hour mode.

Event-Recording VCRs.   Event-recording VCRs are 3-hour disk or tape recorders designed for industrial use and designed to remain in a standby mode until activated by an alarm. With other tape recorders it can take up to 15 seconds from the time the record button is pressed for the tape to spool up and start recording. With an event recorder, the tape is already spooled up and the heads are rotating. Therefore, when an alarm triggers the mechanism, the VCR starts recording almost immediately. The time the VCR remains in the record mode is adjustable or it can be determined by an external timer.

(Video disk event recorders act on the same principle, but store their information on a compact disk. While even faster than tape event recorders, their storage capacity of 100–200 frames is so limited and the choice of equipment so restricted that their use is not recommended.)

Event-recording VCRs also have a built-in time and date generator, which is used when recording is only necessary to cover specific incidents. This feature allows for very easy reviewing of the tape.

The machine is designed to protect the mechanisms and the tape. If there are no alarms, about every 5 minutes the tape will automatically advance a couple of frames. Therefore, when reviewed, the picture will appear to flash rapidly in 1-second steps, which can be seen on the clock. This does not mean that recordings have been missed, only that the VCR is operating correctly. In fact, this feature can be a useful check to satisfy the customer that the machine has been correctly set up

and is working normally. If the power has been cut or the machine has been switched off, this feature can be used to quickly establish the time of this occurrence.

S-VHS.   There are new video recorders coming onto the market that can provide improved resolution of video pictures. The current VHS video recorders produce a resolution of 280–300 lines; the new generation of S-VHS recorders can produce a resolution of 500 lines, which is a significant improvement in quality. Two points about S-VHS VCRs must be noted:

only S-VHS tapes must be used

only a limited number of monitors are fitted with the special terminals to connect to an S-VHS recorder

VCR Tapes.   All current machines are the VHS format. The manufacturers' recommendations are to use well-known brands of standard, quality tape. Do not use professional quality tape for industrial recording purposes. (See Figure 5–16.)

Multiple Camera Simultaneous Recording (MCSR)

Many security systems have multiple cameras wired to a *sequential switcher*. The switcher looks at each camera in turn and records, on videotape, what it has seen in that scanning period. During this time, action on other cameras is lost. Hence, recordings of events on a single camera are not continuous, but may have 2–6-minute delays, depending on the number of cameras per switch. Cutting this lost time requires the use of an uneconomical number of switches or video recorders (2:1 ratio of recorders to cameras). If systems are only recording on time-lapse or motion-detection principles, this problem of lost time is even worse.

A further problem with multiple camera sequential switches is analyzing a day's tape recording. Who wants to sit through five pictures, look at the sixth (the one you are interested in), then forward to picture 12 for the next event in the sequence? By the time you have got to picture 18 you will have forgotten what was supposed to be happening and your camera may have missed it anyway!

MCSR removes all of these frustrations. An MCSR unit is capable of *simultaneously* receiving pictures from cameras that are *continuously* introduced to the system. These units normally accept up to 16 cameras each. The features of the device are

no lost picture time and continuous recording

a single tape splits into 16 channels of continuous play

selection of any channel for display (either on the whole monitor screen or any fraction from one to one-sixteenth with other fractions of the screen displaying real-time action or what was simultane-

FIGURE 5–16   Cassette-recording camera

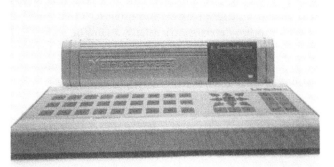

FIGURE 5–17   MCSR color or monochrome recording system and control unit

ously happening during the channel-recorded portion at which you are looking)

time, date, camera number, and area covered continuously superimposed on every continually displayed channel on the monitor, with the capability of stopping the recording while letting the other channels display their events and times, and keeping the real time going. This feature is extremely useful. It can, for example, show not only where, when, and how long the intruder took to get from point A to point B, but will also display how long your patrol took to walk past the scene of the intrusion, discover the event (hopefully!), etc. Time and date on recordings make for excellent analysis and use as criminal evidence.

ability to generate a photographic print of a single still or sequence of stills, each displaying time, date, etc. The size of print is selectable. If, for example, your monitor is displaying four quarter-sized displays, your printer will give you quarter-sized pictures of each of those events.

scanning capabilities that allow any part of the viewed scene recorded on tape to be magnified. For example, you might wish to magnify a face or license plate for easy analysis and recognition.

Most MCSR devices will allow the recorded tape channels to be played back simultaneously. Each system will allow playback to any combination of monitors with the ability to subdivide any of these monitor screens into any combination of parts from one-sixteenth of a full screen to full screen. For example, a *single* monitor might simul-

taneously show a half-screen display, two eighth-screen displays, and a quarter-screen display.

Some MCSR devices include *activity detection*, which is a very valuable feature. In some systems, activity detection is used to save tape time by only storing pictures when activity takes place (see video motion detection). On other systems, the full picture, record on, time lapse is recorded, but on playback, the person reviewing the tape may select either "all tape" or "activities only." The latter will obviously save considerable review time, particularly with multiple fraction simultaneous displays of each camera on the screen. When investing in MCSR super-resolution VHS, tapes and computer controller switching should also be considered. See Figure 5–17.

CCTV Motion Detectors

CCTV motion detectors are devices that reduce or remove the need for continuous high-concentration viewing of a monitor by a guard. Basically, the monitor is fitted with a device that has the ability to store, compare, and analyze differences in the pictures presented to the monitor. It does this by dividing the screen into its component vertical and horizontal grid lines of resolution and recording the grayness or contrast level of each point on this grid. It then compares this contrast level with another picture presented a short time later. If the device recognizes any significant difference between pictures caused by windows being opened, lights being switched on, or persons casting shadows or simply blocking light, the device will initiate an alarm. With this system, the monitor screens are usually blank and only turn on when changes are spotted. This feature reduces the boredom factor, increases the reaction time of the guard, and increases the life of the monitor. These devices can be extremely useful and cost-effective on systems requiring large numbers of monitors or guards to be away from the monitors for any length of time. CCTV motion detectors are not a replacement of guards, but a supplemental tool. Like all complex devices in security systems, the main drawback of this device is false alarms. When they initiate alarms these should only be to local guards and never direct to police stations.

CCTV motion detectors are usually analogue transmission digital storage systems. Some older systems review actual monitor screen brightness. These are outdated and not recommended.

Application. Motion detection should only be considered when

the application is either within a building, or in external applications where either the normal view is reasonably constant or the system has the ability to blank out troublesome, moving, nontarget objects (e.g, trees, etc.) External use false alarm rates can be high particularly in regions of high rainfall or fog.

the lighting level is high and constant. The level should be at least 100 lux and, for very good detection, 500 lux. These levels provide the correct contrast between the intruder and the scene. As illumination levels fall, the contrast falls and detection becomes less certain. The lighting should be constant and not able to be deliberately switched off. Standby power is also recommended.

with arrays of up to eight cameras in number field trials show an equivalent observation success compared to a permanent guard. Above eight cameras, motion detection success (compared with guard vigilance) rises exponentially. Most systems will handle 16 cameras.

Selection. The sensitivity and ability of the system to detect intruder movement are determined by

the number of detection elements or squares within the grid that are superimposed on the viewed scene (typically, 640 or more)

the number of distinguishable contrast levels (typically, 256 and never less than 150)

the frequency of scanning for change in grayness and rapidity of comparison. This should be selectable, as a balance between changes of significance and environmental changes must be struck to avoid false alarms.

the number of selectable threshold variations provided. The *threshold* is normally selectable and the percentage duration permissible before alarm.

detection zone selection. This allows the user to blot out, on any field of view, those areas where motion detection is not required or where it could result in false alarms due to headlights of passing cars, etc.

pixel storage (*PIX STORE*). This allows constant replay of the screen, which causes the motion detection system to switch on the monitor while the real-time action is displayed on a second monitor.

a "take control" feature. This interrupts the pixel storage recording being played on the first monitor if any other intrusion is detected (i.e., simultaneous intrusion at two different points).

CCTV motion detection systems should not be used as primary alarm systems. They can be used with PIRs to assist in detection. All systems should allow connections to freeze-frame features, event recorders, video tape recorders, slow-scan TVs, etc.

Video Signal Switching

When several cameras are used to survey an area, several monitors are usually required. There is a limit to the number of monitors that a single person can watch continuously and efficiently (usually three). If more cameras are needed, some form of switching is required. The simplest form would be manual switching, where each monitor with three cameras could be selected, in turn, by the use of a rotary switch. This system, although simple, can allow the viewer's interest to become fixed on one viewed scene or group of cameras to the detriment of the attention paid to the others.

A better system is the *electronic scan time switch with manual override* (sometimes called the *homing sequential switch*). With this system, each of the three cameras per monitor could be switched on to the monitor in a predetermined sequence for a given length of time. Any one camera scene could be viewed or held on the monitor by means of an override switch, should something suspicious be seen. In situations where an alarm signal is possible (e.g., an intruder after hours), the monitors may be left off until signaled, at which time the camera covering the zone in which the alarm was activated is automatically turned on.

Sequential switching of this type can, of course, be programmed to any sequence of cameras and any sequence of video time-lapse or real-time recording.

Remote sequential switches are extremely useful in a system as they can cut down transmission cable paths and hence costs. With multiplexing, it is possible to have 16 cameras at one location and 16 cameras at another, both with their own switches. The two switches can then be synchronized to send pictures at a slightly different time. Hence, what was, potentially, a 32-line system or at best two multiplexed lines, becomes one multiplexed line!

The most recent advance in sequential switching is the dedicated microcomputer switch. These switches are used, essentially, on a computer management system for larger installations (16 cameras or more, when MCSR is normally used). Each camera feeds its video signal into a central data store (like the PIX STORE of the video motion detection unit). Command or other data are also entered. A programmable keyboard is used to instruct the data store on how to manage the system. Usual capabilities include

local- and remote-monitoring screen capabilities

any number of cameras hooked up to any number of monitor combinations

simultaneously sending one camera signal to any number of monitors

manual, semiautomatic, or fully automatic sequencing of camera activity or recording and playback

video motion detection and activity detection recording

activity tracking (i.e., once a camera has seen a motion using video motion detection or the guard has spotted motion on the monitor, either a fully automatic or a single-keystroke command will bring either of the cameras along the intruder's path to the correct direction, zoom, pan, tilt, etc., and be ready to record it on the VCR)

on-screen messages and instructions that are preentered at the keyboard. These systems are very good tools, but should only be considered when fully trained, intelligent users will be in control of the system and stay with it long enough to take advantage of its full potential.

Systems currently under development bring together all the functions of video motion, intruder detection, and telemetry control of pan, tilt, zoom, sequential switching, MCSR, etc., to form an expert system. These not only detect, track, and record intrusions, but control and counter them by identifying (to a remote station) what is happening, locking doors normally open, directing response forces to projected escape routes of the intruder, etc.

LIGHTING AND LAMPS

General

Lighting plays such an important role in CCTV that its selection must be given as much thought as that of the entire CCTV system. Lighting may be divided into options, which should be analyzed for the best, most economical solution based on discounted cash flow or more simple payback methods. The common lighting options are:

1. Maximize daylight use and minimize artificial lighting time by using a camera that operates at dusk and dawn, and only requires illumination after dark.
2. Same as 1, but use a more expensive camera that will operate in moonlight and will only need artificial lighting (approximately 10–12 lux) under starlight or totally dark conditions.
3. Same as 2, but use a starlight-sensitive camera (i.e., no artificial lighting).

4. Same as 2, but use a less sensitive camera and an infrared source to illuminate the intruder.
5. Same as 2, but turn artificial lighting on only when the alarm is activated and the camera is required.

A coincident fundamental decision that must be made early in the design of a CCTV lighting system is the amount and manner of integration it will have with the building or perimeter alarm system. In general, the various options are

no direct link to the alarm system—CCTV used purely as monitoring aid for a *stationary* guard. If anything suspicious is seen, the guard operates the alarm system. With this system, CCTV and lighting are continuously on and monitored.

continuously on, with a CCTV motion detector to monitor the screen. If this system is used, the monitor should alert the guard, who should consider the picture before operating an alarm system. Automatic, directly linked CCTV motion detection and alarm systems are not recommended, as intruders can often create diversions causing the motion detector to raise alarms repeatedly. Eventually, the guard will either become completely confused as to the area under threat or decide the system is malfunctioning.

CCTV and any lighting normally off, but brought on by the direct action of any part of the alarm system. These are often called *trip lighting systems,* which are less expensive to run, but not recommended. If any sensor on the general system can activate the lighting, where is the intruder? Inside buildings the system can be of use if linked directly to a *particular* sensor in a particular area, say a strong room, but, again, the guard should view the scene before activating the alarm. On fence sensor systems, trip lighting should never be used unless the systems are zoned to cover a particular area or length of fence. In this application, low-light cameras or those using infrared sources are better than artificially lit cameras as intruders will not realize they are being watched. When the response force is in position, the lighting can be switched on manually. With artificially lit CCTV, intruders will immediately try to avoid the beam of generated light and may escape before the response force can react.

watch and patrol. Although not as economical to run as the two immediately preceding options, using low-light cameras, watch-and-patrol systems probably give the highest level of security offered by CCTV. With this system, the general building, perimeter, and area security lighting system is designed to simple camera CCTV standards (i.e., operating at a level of 10–20 lux; standard patrol lighting is normally designed for levels of 1–50 lux).

The lights are positioned so that they are at an angle of between 15° and 30° to the camera line of sight. This ensures that the camera is never dazzled by the light and that any intruders will cast a slight shadow, hence making them even more recognizable on the monitor. Guards can carry out their rounds with the system continuously running and report their positions at intervals on a pocket radio. A guard at the monitor position monitors the progress on the CCTV, switching from camera to camera. To increase the security of patrolling guards, the stationary guard can look into any room before security guards enter, in case any intruder intends to ambush them. Further sophistication is the inclusion of a CCTV motion detector to "watch" the camera not being viewed by the stationary guard!

Figure 5-18 shows common combinations and applications. *The maxim should always be "keep it simple."* It used to be the case that lamp energy consumption for a given light output (efficacy) was so low, that continuous artificial lighting during the night was very expensive and hence even a high capital cost, sensitive camera could be considered. Lamp efficacies and lifetime are now good and this is no longer the case. On a multicamera installation, the use of a good lighting system will offer the following advantages:

Lighting itself is a deterrent and makes a security guard less wary of patrolling a property.
Simpler, less sensitive cameras do not require expensive, automatic filters or attenuators to operate under daylight conditions as well as nighttime, artificial conditions
Cheapness of individual cameras will allow the purchase of additional cameras to view the blind spots that more sensitive, but fewer cameras would inevitably have.
Good lighting allows the police to work effectively.
Multiple cameras and multiple lighting source systems allow for more failures before a whole system becomes inoperative. Don't put all your eggs in one basket.
With single, sensitive cameras using no artificial lighting, if you miss the criminal as she goes through the field of view, she is gone for good. With good lighting, it is still possible you may see her with your own eyes. The camera is only there to augment your senses!

Do not assume that lighting is simply a matter of flooding the area with light. Too much light or incorrectly positioned sources can be as troublesome as too little light. It is always wise to consult a lighting engineer before deciding on your scheme requirement. The following notes give some guidance; for further details see Chapter 8, "Security Lighting Systems."

The percentage of reflected light from the lit surface should be assessed. This can be approximately assessed by placing a sensitive light meter (gradations of <5 lux) on the surface and taking a reading of the light (do not overshadow the meter). Now pick up the meter and aim it *at the surface* with the light source directly behind the meter. Take a second reading. Carry this out several times at various places on the surface. The second values, divided by the first, will give an approximate percentage reflectance. If several different surfaces are reflectants, repeat and obtain a mean percentage. Multiplying the mean horizontal illumination level by this percentage will give you an approximate value of the illumination level at the camera lens once the *inverse square law of illumination* has been used to take account of the distance to the camera from the object.
When illuminating perimeter areas use lights of equal output with overlapping fields of illumination to give a uniformity of at least minimum illumination not less than 70% of maximum anywhere in the illuminated area.
To reduce cost and improve security, use multiphase supplies and cables, and overlap them so that no two adjacent lamps are on the same phase of the same cable. Phase failure will still provide two-thirds of a full light output.
Access whether any surface is likely to cause dazzle or glare when brightly lit.

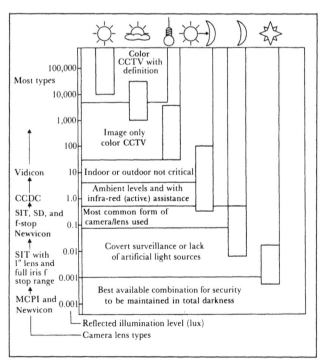

FIGURE 5-18 Levels of reflected illumination available and common CCTV camera tube and lens combinations for these levels

Tungsten general lighting service (GLS) lamps have too low an efficacy for economical use.

Tungsten halogen lamps are good for very small scale floodlight use, but annual replacement is required due to short lamp life.

Fluorescent lamps are good for internal use over large areas of offices.

Metal halide (MBI) lamps, such as General Electric Company (GEC) multivapor lamps, have output that is almost as high per watt as high-pressure sodium (SON) lamps, but with excellent color-rendering properties and camera spectral response. These lamps are often specified for use in stadia for broadcast TV recording under standard lighting.

Mercury vapor (MBF) lamps are sometimes used for area floodlighting and CCTV schemes.

MBI lamps are better than MBF lamps, but are usually more expensive. For color CCTV, these lamps are essential.

Low-pressure sodium (SOX) lamps are generally unsuitable for security use.

SON lamps are generally the best, lowest-cost, large-scheme lamps available, when color rendition is not too important.

The last four lamps cannot be used on alarm-activated trip-lighting systems due to restrike times in excess of 1 minute.

When an instant strike capability is required, manufacturers now make *Mackwell units.* These units incorporate a tungsten halogen lamp inside the MBI or SON lighting fitting that comes on during MBI or SON warm-up time and automatically cuts out when full lumen output is reached.

Infrared Illuminators

Infrared illuminators produce invisible infrared light at 715 nm or 830 nm cutoffs. They are of various outputs, usually up to 500W. The particular type of pattern of infrared (i.e., spot, narrow beam, wide beam, or flood) must be exactly matched to the camera/lens combination. Hence, these lamps should always be specified and preferably installed by the CCTV installer. The lighting assembly should be camera-mounted and twin lamps with automatic photocell on/off switches provided to prevent daylight running. See Figure 5–9.

MODEL SPECIFICATION

In order to purchase a CCTV system for yourself or on your client's behalf, you will need a brief from them on what the system is intended for and what they or you require from it. You can then add other crucial details concerning the building environment, and surroundings to

this information to form your specification to the companies bidding for the installation.

Performance Brief (Part 1)

What is the CCTV system required to do?

What is the target area or item? Is it internal or external?

Will the monitor be continuously manned or alarm activated, or will it be a movement-detection system?

What is the value of the items protected? What would be the consequence of any failure in the system?

After considering the last question, what is the greatest reasonable capital cost (and annual cost) that can be expended on the system?

Will the system need to operate in all weather, total darkness, etc.?

Is a covert or overt system required? Are dummy cameras necessary?

From the answers to questions like these, a performance specification can be formulated where Part 1 reads, for example,

The system shall be required to monitor the approaches to alarmed doorways in a bonded warehouse. The cameras shall be external to the building and the monitors shall be located within a continuously manned guardhouse at the main perimeter fence gate. The monitors shall be on continuously 24 hours per day, 365 days per year. The system shall be constructed so that a guard watching a monitor shall be able to read the license plate of any vehicle in line of sight within 164 ft (50 m) of the camera, under the worst combination of environmental conditions (see Part 2) while the vehicle is stationary in front of the doors.

Environmental/Operating Conditions (Part 2)

In order to give the manufacturer a good idea of the demands to be put on the system, Part 2 of the brief should contain the following information:

minimum lighting level at the camera

type of lighting (lamp or natural)

maximum lighting level

surface reflectance values

types of surface finishes (matte, black, etc.)

maximum and minimum temperature and humidity conditions

direct heating or cooling effects (solar, cold air exhausts, etc.) that might affect siting of cameras

vibration (important on bridges, gantries, etc.)

snow, wind, water, corrosive atmosphere, amounts of of airborne dust, etc.

highest likely wind speed (for external, mast-mounted cameras); wind rose if possible

Armed with these data, the engineer can now invite interested companies to survey and inspect the plans (or existing building), discuss these with them, and finalize the installation requirements (Part 3 of the brief).

Installation Requirements (Part 3)

The following should be made available to and discussed with the bidders:

site location plan showing surrounding buildings, trees, etc., and ambient nighttime lighting level (e.g., street lighting), if system is external

detailed site or room layout plan preferably at 1:50 or 1:100 scale, showing

- target area
- shadows cast by other buildings (in case of external system)
- furniture and goods layout, and any future arrangements
- possible cable routes and position of trees
- room or building elevations highlighting any overhangs or porches, projections, etc., that might conceal an intruder
- routes and sizes of any delivery or parked vehicles that might block the view of a badly located camera
- position of monitoring room
- possible camera position; lines of sight, or panning, scanning, tilt coverage
- points of electrical supply (most CCTV camera systems need to have all cameras connected to the same electrical supply)
- security patrol routes and any requirement for cameras to monitor these routes

Cost Limit (Part 4)

All things are possible with limitless cash. A 99.999% coverage, using very high definition equipment might, however, cost too much or be too sophisticated for a corner shop's shoplifting prevention CCTV system. Always give the prospective bidder not only a quality and performance specification, but an upper cost limit and emphasize that the bidder with the lowest-price, highest-quality system will win the contract.

The bidders are now in possession of all the necessary facts to enable them to select equipment from their standard equipment and arrange it to give you a custom-made system at an off-the-shelf price. Without these specifications, you will get an off-the-shelf system at an off-the-shelf price, with off-target performance!

Remember that costs should be turnkey and you should be aware of costs that are not included. Cables for transmission and power, trenching for cables, erection of lighting, etc., may far outweigh CCTV equipment cost even *with* installation charges!

Manufacturer/Installer

Given the specification, bidding manufacturers/installers should provide the following information with their bids:

- company profile
- system plan—coverage area and blind spots
- operational diagrams—how monitor switching operates, what a system can and, more importantly, cannot do
- system description—fixed camera, etc.
- equipment specifications

Electrical

Camera.

- camera type
- power consumption
- power supply and voltage limits
- size and type of lens
- other compatible tubes
- minimum reflected light required at particular f-stop value
- maximum resolution value (lines and dots)
- operating temperature limits
- bandwidth
- connections used and coaxial cable recommended
- type of light level adaptation device

Lighting System for Covert Systems (Infrared).

- lamp type
- power consumption
- voltage
- life span
- type of lamp housing (integrated or externally mounted)

Cable Transmission System.

- cable type used
- connection system (complex, balanced, etc.)
- transmission losses on the various planned routes

Monitors.

- resolution (in lines and dots)
- power consumption
- voltage
- voltage limits
- video output requirements
- capabilities such as contrast, black-level lock, vertical and horizontal hold

Mechanical

- types of motors, gears, etc. (if used)
- automatic pan and tilt system diagram showing coverage, clear swing zone requirement, etc.
- physical dimensions
- stacking arrangement
- weight
- heat output
- housing construction materials
- protection ratings, standards and approvals
- origin of manufactured parts

Installation Requirements

- plans and elevations showing mounting positions
- diagrams showing supports required for equipment
- plans of proposed cable routes
- work scheduled for others (e.g., power wiring, trenching, building, lighting, etc.) and estimated costs

After-Sale Service

- guarantees offered on equipment
- servicing agreement and costs, replacement costs, statement of spares held
- maximum repair response time
- statement of system capability, such as

The system, as described herein, shall be able to . . . and upon installation this shall be demonstrated to the full satisfaction of the architect. Any remedial works, additional parts, equipment, or labor required to fully satisfy the architect shall be carried out at no cost to the client.

Tender Price

As the purchaser will want to know how the total cost of the system was arrived at, you will need to ensure that a breakdown of the total cost is given in the format you require. This should be built up with prices for each camera, cable, route, etc., listed so that if the tender is over budget, realistic savings can be negotiated.

Deviations from Specification Requirements

Any deviations should be fully listed.

Programs of Works

Details of your installation process should be provided.

TENDER APPRAISAL, TESTING, AND PURCHASE

Before bids are submitted, you will probably be offered the choice of outright purchase or leasing of the system. Which option to choose is a matter for your client and your client's accountant to decide once you have given them some idea of equipment replacement cost, system life span, service costs, etc. Generally, if the system is complex, leasing is a better proposition in most circumstances.

Once the interested bidders have been invited to look at plans or survey the building, many different solutions will be offered. The specification should help sort out the wheat from the chaff, but you will still be left with the bid appraisal and final selection process. Fortunately, with CCTV systems, as long as you have some knowledge of what you are being shown and some common sense is applied, seeing *can be* believing. Once the bids have been compared, select the two systems you think are the most suitable, then

ask for a full demonstration of both systems under similar operational conditions in, if possible, similar premises

ask for two other references and obtain their impressions of the system, its limitations, advantages, etc., in person, without the bidders being present. Always consult the security guard (alone, if possible) using the system. Ask her for her opinion of the system and whether she considers it useful. These opinions might differ greatly from those of her boss, who was responsible for its purchase!

When comparing bids, look for missing information.

NEVER BUY A CCTV SYSTEM WITHOUT SEEING AT LEAST ONE SIMILAR INSTALLATION IN OPERATION.

Always use tried and tested equipment if possible. Don't be a guinea pig! Always look for simplicity in the system offered. The more complex the system, the greater the maintenance and chance of system failure. If possible, keep to systems using medium-sensitivity cameras; simple lenses; short, well-protected cable routes; and few monitors.

Look for bids that include a separate, itemized amount for commissioning and testing the system that is compatible with the sophistication of the system. Cameras, cables, monitors, etc., need a fair amount of adjustment, matching, and positioning on-site.

INSTALLATION

CCTV installation is specialized work and should be carried out by the company offering to sell the equipment.

Never buy a system and expect the local electrical contractor to install it. Some companies specialize in CCTV installations of many makes and offer specialized backup services, spares, leasing, etc., that make their offer very competitive, in terms of price, on small- to medium-sized installations. Always have the installation carried out by staff members who are permanently employed by the manufacturer and not by a subcontractor to the subcontractor!

When the system is first installed (and at regular intervals after installation) it should be fully commissioned. This involves not only looking at the target area on the monitor to see whether a good picture results, but includes a series of tests to establish whether the performance is the best that the equipment is capable of giving. The tests should include

synchronization tests using a reference set
focus tests on lenses
brightness, adaptation, and filtration tests on lenses
test card procedures to evaluate contrast, ghosting, resolution, etc. These tests involve using a modified form of the test card seen on ordinary television screens.

Proper commissioning and regular recommissioning are essential. CCTV systems should be tested, as a whole, on the overall *performance mark principle* recommended in the British Standard Code of Practice (Draft 1989), "CCTV Security Systems." This involves two fundamental tests:

a *rotating mannequin walk test.* For this test a special mannequin camouflaged on one side with a picture test card on the other are positioned along all points in the specified fields of view to confirm that coverage is achieved, to find blind spots or poor lighting areas, and to indicate the limits of image size and limit of field of coverage.

a *system-operator-monitor detection and response test.* For this test the mannequin is randomly placed in any position within the field of view and a point scale is awarded for speed of detection on any monitor and accuracy of detection (i.e., did the guard see it or mistake it for something else?).

The resulting scores are assessed to give an overall performance grade from 0 to 3. Grade 3 is easily and immediately seen, no mistakes; grade 2 is fairly easily seen, search of area or monitor refocus needed, no mistakes, fairly fast; grade 1 is difficult to find, slow, mistakes are made or the intruder is missed; grade 0 is no detection at all. Only a grade 3 system should be accepted.

Always make the responsibility of supplying suitable power supplies, or at least definition of the electrical supply requirement and tolerance, rest with the manufacturer.

If the system incorporates cameras mounted on poles or masts, or catenary coaxial cables, ensure that these are protected from lightning strike damage. This can be done by either linking (bonding) to other suitable grounded metalwork or lightning conductors, or by having the manufacturer provide and install surge arresters.

6
Locking Systems for Doors and Windows

INTRODUCTION

In the design of a building, the determination of a locking system and the correct selection of locks can be one of the most difficult and frustrating elements with which the architect or security adviser has to cope. There is often what at first appears to be an unresolvable conflict between the needs of security, fire safety, normal access, and the financial limits of the budget. Discussions between security consultants, fire officers, and architects during the design meetings can often degenerate into fruitless arguments!

The best way to avoid this situation is to obtain, in writing, the specific fire safety requirements from the fire officer and match the locking system to these. Security, in whatever form, is provided to increase safety and reduce risk, not to create it or endanger life. There are many excellent, secure, fire exit locking systems now available to the architect, which eliminate the need for 24-hour alarm sensors. The days of insecure crash bar fire exits have gone!

Once you have decided on a locking system that satisfies the fire officer, security division, and clients, you will still have to decide on lock types and quality. The best advice is to read this chapter, be aware of the various locks available (their uses, advantages, and disadvantages) and then, for smaller buildings, consult a reputable locksmith. The good locksmith will advise you on quality, manufacturer, and cost; and will prepare quotations, etc. A locksmith's opinion will usually be unbiased and based on wide experience. If you order and have your locks fitted by the locksmith, the fitting service is often free.

On larger buildings, it will often be beneficial to look at the services offered by a certified door consultant (CDC) or an architectural hardware consultant (AHC). Both of these advisers will be members of the Door and Hardware Institute. Members of this organization offer a wide range of services that the local locksmith (who should be a member of the Association of Locksmiths of America) may not.

If you would like a second opinion, contact the representative of a reputable lock manufacturer and ask him to prepare a quotation for supply only. Compare it with the materials in the locksmith's quotation. Many good lock manufacturers have *lock centers,* where their wares are displayed and free expert advice is given. Use them! Some manufacturers even have lists of approved installers for their locks. As the reputation of their locks rests as much on the precision and quality of installation as it does on inherent design strength and security, these manufacturers have a good reason to ensure that approved installers offer a first-class service. With all these services available, there is no excuse for bad lock selection or installation.

The level of security required from a locking system (structure, frame, door, lock, and key) is a direct function of

the value of the contents of a room or building and the consequences of intrusion

the surrounding criminal activity level. This is usually the immediate geographical location, but remember a criminal can cross the world in 8 hours!

the ease of unseen approach and access to a door, window, etc., and the ability to work undisturbed. Internal doors often require more resistance than exterior doors, as once entry is gained, time to "work" increases and risk of discovery decreases in buildings with poor security.

In normal circumstances, the doors and windows of the external facades of the building are those that face the greatest risks of vandalism; opportunist entry; inexpert, brute force entry; arson; lock picking; etc. The areas that are most at risk are

ground-floor windows or doors that are left partially open at any time (accidentally or deliberately). This is the classic opportunist criminal entry method and the most common type of burglary.

ground-floor windows or doors normally closed *and* locked, but out of sight (not overlooked by other buildings, etc.). These areas provide good concealment, sound muffling, etc., and may be jimmied or forced. Doors should *never* include normal glass, however small the pane size.

roof skylights or upper-floor windows that are out of sight and to which access may be gained from adjacent flat roofs, fire escapes, rain pipes, etc.

The following sections of this chapter give both detailed information on lock types and methods, and general points for consideration. NIJ Bulletins and Standards, UL Standards, and British Standards are quoted. Remember these are *not necessarily* the best or most appropriate for a given situation. Determining the most appropriate locking system requires the involvement of building and insurance representatives, police, architectural crime prevention advisers, AHCs, CDCs, locksmiths, and others with specialized experience and expertise. Use them.

Lastly, remember National Fire Protection Association (NFPA) fire standards and codes; means of escape in case of fire are preeminent to security requirements (even in prisons!).

LOCKS, FRAMES, AND OPENINGS

Principles

Before considering any lock in detail, it is vital that the fundamentals of locking are understood. Locks should never be seen as single items of equipment that are added to a door or as appendages that will improve security. Locks should always be considered as part of a system. This system will delay the entry to or exit from a space, or prohibit the removal of an item from a space by a *length of time that will significantly increase the risk of detection for the intruder.* This is the prime function of a locking system. For doors (or windows), the locking system is comprised of the

- opening and surrounding materials
- frame and fittings
- door or window and its materials
- hinging mechanism
- door handle or other hardware
- key cylinder
- locking mechanism and receiver

All of these parts of the system should be given the same consideration as the last two, which comprise the actual lock. If a lock is specified that offers significantly greater resistance to attack than the other items constituting the overall system, money will be wasted. Locks vary enormously in the level of security offered — building materials and methods of construction in modern premises do not. Hence, mismatches are commonplace.

The Opening and Surrounding Materials

The construction materials of the enclosure, boundary walls, and floors should be considered first. Generally,

these are in order of increasing attack resistance.

- plasterboards, fiberboards, etc.
- hollow panels often with wooden grids
- softwood panels and boards
- hollow panels of softwood with wooden grids
- lightweight, hardwood versions of panels and boards, and hollow panels with wooden grids
- solid, substantial, hardwood construction
- metal sheetings, glass-reinforced plastics, etc.
- block work, single layer
- precast concrete panels
- block work/cavity/block work
- single brick over 9 inches thick
- brick/cavity/block work
- brick/cavity/brick
- double brick
- *in situ* concrete and mesh
- reinforced concrete
- specially reinforced concrete

Having identified the structural class of the construction, consider the door material and method of construction and fixing. This should be as strong as the class of structural materials (within the boundaries of common sense!). For example, a hollow panel door and specialized cylinder mortise lock fitted to a reinforced concrete wall would not be cost-efficient. Cutting through the door would be far easier than defeating the lock or attacking the wall!

Next, consider whether the enclosure is secure enough to warrant protection. If good security is required to protect items of some value, it can often be cheaper to reconsider the choice of enclosure material and to upgrade it, than to provide expensive locks, and detection and alarm capabilities. The client will not thank the architect who designs paper-thin walls and is robbed on the night the alarms were left off!

Before going on to determine the next item in the system, the eventual use of the room and its integration with other areas should be considered. Are the windows through the structure as secure as the door? Are there any ducts or false floor or false ceiling entry points to the room that would allow circumvention of the door? Could a person secret himself within the room during its normal use?

If the answer to any of these questions is "yes," either block up the entry points or ensure that they are strong and secure (particularly windows).

A common problem in many multitenancy offices is the construction of secure rooms within offices or partition constructions. The building owners will rarely allow the use of block work or brick. In this case, consult your security adviser. There are many lightweight, patented security panel constructions that can be used. These usually comprise a frame of 4 × 2 in (100 × 50 mm) studding at 16 in (400 mm) centers with noggings at 16–24 in

(400–600 mm). This frame is covered on both sides with cladding-plywood sheet-plaster-wire mesh-plywood sheet-wire mesh-cladding laminated panels, which gives excellent resistance.

The Frame and Fittings

A strong door with a strong, secure lock, fitted within a strong structure by a weak frame will be obvious to an intruder and a natural point of attack. The common methods of attack are

pulling or pushing the door and frame out of the opening by means of winches, vehicles, ramming, etc.

spreading an ill-fitting frame until the lock bolt is released

cutting, bending, or chiseling the frame to remove the staple casing

Door frames usually fall into two categories—those that are attached to or form part of a wall panel (these are normally used on internal, lightweight partitions) and those that are fixed to and within brick, block, or concrete walls. The first category is the least secure and most easily attacked. They are often oversized for the door or poorly secured (with as few as six no. 8 wood screws) and are invariably of cheap softwood or aluminum. To improve the security of frames

ensure that the frame is at least 2 × 3 in (50 × 75 mm) thick if wooden

fix a wooden frame to the structure using wood screws and plugs at 12 in (300 mm) maximum intervals. Ensure the screws project 2–8 in (50–200 mm) into the structure.

increase the rigidity of any wall partition and door frame (aluminum, etc.) by providing additional bracing and fixing to the structure using steel, and by incorporating steel frame inserts

ensure that joints are not folded, sprung, or tack-welded, if frames are aluminum

fix frames directly to the structure, never to the false floor or ceiling

frames should be constructed of steel and fixed using grouted bolts

frames should be rabbeted into the wall at least 0.8 in (20 mm)

frame thickness should always be sufficient to allow the lock staple box to be recessed flush, with ample material behind it

steel frames should be fixed at a maximum of 1.8 in (450 mm) intervals

metal protection and reinforcing should be placed around the frame striking plate area

contain the frame in a masonry jamb where possible

In many multitenancy buildings, such as office blocks, insecure partitions and aluminum frames are a fact of life. The answer to this problem comes in two forms.

Form a hardwood door with laminated glass or vertical slot windows at the end of each common corridor. Install a good, registered key lock and issue limited keys (one per office) to all the tenants on that floor, and

Provide sufficient infrastructure in the form of secure conduits, trunking, etc., for individual tenants to have an intruder alarm system. A single-company service system for each office within a block can be surprisingly cheap for each tenant and is a good selling point.

As good wood becomes more expensive, door and frame manufacturers are saving timber by using frames that, on the surface, appear to be solid, but are actually veneer-faced timbers spaced with softwood dowels. This construction saves up to one-third timber thickness, but makes the whole door system useless in terms of security.

Government agencies have standards for security and buy security grade frames and doors. These are manufactured by private companies and can be purchased directly from the manufacturer.

Doors and Windows

Doors are provided to allow access, restrict access, prevent the spread of fire and smoke, and provide privacy.

The design guidance found in NIJ Standards (NIJ-STD-0305 and NIJ-STD-0318), Architects Graphic Standard, British Standard 8220, etc., should be followed with respect to door classification and overall security provision whenever possible. BS 8220 recommends that before door security measures can be detailed, a formal classification and functional assessment should be carried out for each door planned.

Table 6–1 shows the general tabular approach that can be used to complete a functional assessment. This is simply an extension of a normal hardware schedule commonly compiled for a building under construction.

Windows are provided to allow the entry of daylight while preventing the entry of rain, wind, noise, etc.; allow the entry of fresh air when required; provide a view or allow visual communication. The normal window schedule compiled for buildings in design can similarly be extended for security purposes. Windows are normally classified as

• opening and accessible
• opening and nonaccessible
• fixed and accessible
• fixed and nonaccessible
• fixed display
• other special window

Table 6–1 Typical Door and Lock Schedule

Lock system	Zeus Blockade							Notes
Project	190168	Other details	Cash handled					
Address	1289 Long Rd., Staffs.	Lock ref	1	2	3	4		
Person	F. Steel							
Firm/company	H.M.S.O.	Between rooms	2	3	4	5		
Date	10/7/90		1	2	3	4		
System no.	1046.	No. door	1	2	3	4		

LEGEND: Door Construction	Function		f				
1: Solid timber	Position		I				
>1.7 in (44 mm)	Thickness (in)		1.6				
2: Hollow timber panel	Construction		6				
3: Partly glazed	Door opening		SO				
4: Aluminum frame or	Cylinder side		IS				Insecure side
door	Door hang L/R		R				
5: Laminated glass	Lock type		101				
6: Steel or steel lined	Finish		C.				Chrome on stainless steel
LEGEND: Door Position	Accessories		R				
I: Internal	Cylinder type		102				Escape furniture
E: External	Strike plate type		104				
LEGEND: Door Opening	Lock case type		106				
D: Double-leaf inward	Backset		107				
S: Single-leaf inward	Number		01				
SO: Single-leaf outward	GMK		✔				
DO: Double-leaf outward	SGMK						
SL: Sideways sliding	SMK						
UP: Up and over	PGACK						
RS: Roller shutter	Group No.		1				
RT: Rotating							
FL: Folding sideways	Key holder names/types						
LEGEND: Door Function							
a: Final exit: security	GMK	G. manager	1				
b: Final exit: fire							
c: Access/egress staff							
d: Egress staff only							
e: Access staff only							
f: Access/egress goods							
g: Public door							
h: Special purpose							

Opening and accessible windows are the most vulnerable to attack. Both openable and fixed windows present entry opportunities. Display windows are large and materials normally offer better attack and cutting resistance than smaller windows. The principal danger is that of smash-and-grab activity.

If a vulnerable window cannot be blocked up, the next, most secure means of adaptation is to provide a grill or shutter. This can be far more effective and less costly than expensive glazing materials, break glass sensors, etc.

As soon as either a door or window is considered in terms of security, a conflict with these provisions arises. Of all illegal entries to buildings, most occur through ground-floor windows. If security were the *prime* consideration, these openings would be blocked up. Obviously, and fortunately, security is not the only consideration. However, when considering window and door selection, remember that form can flow from function and the secure door and window can be indistinguishable in appearance from the less secure.

Doors. For security purposes,

minimize the number of entrances and exits to the minimum level required for fire safety

avoid positioning doors *directly* next to traffic, parking areas, etc. Put barriers between these areas to prevent accidents, ramming, loading with the vehicle backed up to the door, etc.

don't create internal or external recesses, porches, etc., as they can be hiding places

ensure that windows next to doors allow for a clear view of a well-lit, safe area outside every exit and entrance

avoid using sliding doors, rolling shutter doors, folding or aluminum doors, and double doors whenever possible. Single, outward-opening doors are generally more secure.

avoid glazing in the door unless absolutely necessary

doors should present the same level of resistance to attack as the surrounding structure if the target for the criminal is of any value or the consequential losses of an intrusion are high

double doors should be avoided

security doors should always open outward toward the attacker (although fire escape considerations can sometimes prevent this)

rabbeted doors should be used

prefabricated doors should be avoided, and shims and makeup pieces to overcome poor workmanship should never be tolerated

doors that, when fitted to frames, have gaps of greater than 0.04–0.08 in (1–2 mm) on any edge should not be used. Doors should be oversized and planed to a good fit.

timber doors should always be of top quality, fully seasoned timber

inward-opening doors should have frames protected either by masonry doorjambs or by an angle iron that is affixed to the wood of the frame and to the masonry

external or secure area doors should be fitted with two locks—one at one-third of the door's height down from the top, and one at one-third of the door's height up from the bottom

never use flush, paneled, Georgian wired, annealed glass, aluminum-framed, lagged and braced, or sliding doors unless absolutely necessary. (If they are used, alarm sensors will normally be required.)

never allow openings in doors unless they are stronger

than the door material (e.g., steel and brass, not PVC, etc.)

strength of doors can be increased by the addition of steel plates [0.061 in (1.5 mm) sheet thickness] secured with dome-headed coach bolts through to the secure side

timber door frames, panels, and subframes should always be made in accordance with and of materials specified in NIJ Standards or BS 1186, Part 1, 1986, Appendix Species B, "Softwoods," and Species C, "Hardwoods." Workmanship should always be comparable to at least BS 1186, Part 2.

Grades of doors in descending order of resistance to attack:

- flat steel, 0.3 in (6.5 mm) or thicker (BS 6510 and BS 1245)
- security laminated doors
- flat steel, less than 0.3 in (6.5 mm) thick, but greater than 0.1 in (3.5 mm) thick
- 0.2 in (50 mm) solid hardwood
- 0.2 in (50 mm) solid hard *core* door
- 1.8 in (45 mm) hardwood or softwood with solid panels
- 2 in (50 mm) softwood with glazed panels
- 1 in (25 mm) or less close-paneled door (tongue-and-groove boards)

Timber doors should be at least 1.7 in (44 mm) of solid timber. Panels, glazing, etc., in doors should be avoided. If viewing panels are required, they should be as small as possible with their bottom edge no less than 57 in (145 cm) above the floor level. Door stiles should be at least 4.7 in (119 mm) deep.

Steel doors should at least comply with BS 6517 rolled section, BS 1245 pressed steel, and welding should be compliant with BS 5135. All fixings should be concealed and visible from the secure side of the door only.

Aluminum doors present particular security problems, but these can be overcome through careful specification, as many commercial premises now use aluminum doors (either fully or partially glazed) extensively. The British publication *National Association of Shopfitters: Security of Aluminium Framed Doors in Commercial Premises* is a good reference. Points for consideration of aluminum doors include

gaps should never exceed 0.2 in (5 mm)

swing bolt or hook bolt mortises should always be used

many good locks *will not* fit the narrow stiles of aluminum doors. Many locks that do fit are poor. Locks should have laminated and ceramic core mortises of bull nose profile, with cylinder guards, armored striking plates, etc. These locks should resist a *minimum* upward or downward blow to the mortise of 8.9 kN (2000 lb force).

PVC doors and frames are not recommended for security purposes

Windows. Windows are opened primarily to provide ventilation. All openable windows should be provided with fixed stops to limit the amount of opening and reduce the likelihood of opportunist theft. However, window stops *should not* be considered as security devices, as they often are. They are too easily overcome. A lock should be fitted. For security purposes,

ensure that numbers of windows are kept to an absolute minimum

sizes of window openings should be kept as small as possible. An opening of 8 × 11 in (200 × 280 mm) will admit a person.

give consideration to vertical or horizontal "arrow slit" windows, when a view is required (particularly on the ground floor)

pane sizes should be a maximum of 77.5 in^2 (500 cm^2) with one dimension of a maximum of 5 in (125 mm) or panes should be larger than 3 × 3 ft (1 × 1 m). The latter will be difficult for the amateur to cut or break without great risk of the pane shattering and considerable noise.

glazing should be antibandit or multiple-laminated (not annealed, toughened, wired, etc.)

glazing should be internally fixed

consider substantial, internal, structurally fixed grills (or even shutters) when ordinary, annealed glass is used

never use louvered windows

never use sash windows, unless fitted with good quality, brass sash locks or dual screws and stopping bolts

always consider the opening mode and ease of entry. Top-hung sashes, outward-opening vents, and horizontal-sliding and casement stays should be avoided if possible (or provided with good locks); side-hung, bottom-hung, opening-in, and horizontal and vertical pivots should be chosen instead.

always place openable windows high enough to require the intruder to use a ladder to gain entry to increase the likelihood of detection

avoid tilt and turn windows

when converting openable to fixed windows or fitting security devices, do not damage the integrity of the window or frame. Screws damage galvanizing; steel screws accelerate wood rot, weaken the frames, etc.

avoid external hinges

when using security glazing materials constructed to UL 972 or BS 5357, ensure that the frame is very well fixed or an assault on the pane might result in its remaining intact, while the attacker removes it in its entirety!

use external or internal security shutters or grills at night

The Hinging Mechanism

The hinge must perform two functions—it must allow the door (or window) to open to the desired angle and it must resist attacks upon it. These attacks could take the form of

removal of the pin

cutting of the knuckle to release one leaf of the hinge

horizontal distortion to increase the gap on the lock side of the door

vertical lifting to take door off the hinge pin

crashing or cutting to remove the hinge from the frame

For security purposes,

a minimum of three hinges should be provided per door

the hinges should be of hardened, thick, stainless steel, brass, or bronze—never aluminum or plastics

hinges should never be visible from or fixed to the side of the door that is vulnerable to attack

hinges should be as long as possible

hinges should have totally enclosed pins that cannot be removed using a blunt-ended instrument or by sawing through a domed end of the pin

on all outward-opening security doors, secret concealed dog bolts or mortise bolts should be used

pivots should not be considered as secure devices, but, if used, the vertical tolerances between the frame and the door must be very fine to prevent door lifting

hinges should always be butt-type and not strap-type hinges

fixings should be adequately sized; never weaken the frame by placing fixings too close together on the hinge

door handles should never form part of the locking mechanism

door handles should not be fixed too securely, as they could be used as leverage or a fixing point for a winch to pull the door open

letter slots should be avoided and never placed where hands or tools could reach the internal lock release

chains and door viewers should always be used on domestic premises

door bolts fitted at the top and bottom corners will increase security

single, through-the-door bolts will significantly increase the security of a door

Glazing and Specialized Materials

Glazing within security doors should be minimized on grounds of security, complexity, and cost. Where glazing is thought essential (viewing panels, fire discovery) the type of material selected may be required to fulfill several roles—be resistant to fire, ballistic attack, and hammer blows; and be linked to an alarm system. Whenever possible, the advice of specialists should be obtained along with competitive quotations for the same specification. It is usually better to have a door or window manufactured with its glazing rather than risk an inexperienced glazier fitting the material incorrectly. Companies such as Norshield, Sierracin/Trans Tech, Viracon Inc., etc., are all specialized in these fields of construction.

Glazing in Doors.   Glazing in doors should only be used if absolutely necessary. Side slit windows should be used instead. Glazing should be above 57 in (145 cm) from the floor, laminated to same strength as the door, and impossible to remove as a whole from the outside. Do not use plastics, acrylics, or polycarbonates when glazing. They are only impact-resistant and do not resist many common forms of attack, unless the glass is in very small panels, and is very well framed and beaded.

Security Glazing Materials.   Only antibandit glazing and bullet-resistant glazing are true security glazing materials. Double-glazing units of tempered glass will deter opportunist amateurs. Glass blocks more than 4 in (100 mm) thick, set in concrete frames, will prevent entry, but not arson. Wired; float; sheet; rough cast; lead, single-layered tempered; and thin laminated glasses are of limited or no security value. Polycarbonates, acrylics, polyester, and glass-reinforced plastics (GRPs) are poor substitutes for *any* glass used for intruder resistance, but can be used as sacrificial outer panes when vandalism by stone throwing is a problem. (All these plastics should be used with care as they could contribute to a fire and smoke production.)

LOCKS

Introduction

Having considered all other parts of the system protecting an opening, attention can now be given to those parts of the system that allow or restrict access to the space beyond the opening. Before reviewing the types of lock available, note the following points:

Good locks are no more expensive to fit than poor locks. The major cost of installing locks is the labor element, which is fairly consistent for each type of lock.

Locks are cheap when considered from the point of view of what they protect against; they are cheap when compared to overall building material costs.

Always match the security level of the lock to that of the door and enclosure.

Never be tempted to buy locks that do not conform to internationally recognized standards and do not bear the manufacturer's name.

Never select a lock on the premise that "the more levers it has, the greater the security it offers." A lock

with a plastic mortise fails easily, no matter how many levers!

Try and select locks that are proven and tested; have UL 437, 768, and 887 approval; etc. If in doubt, always ask the salesperson to supply, in writing, evidence of approvals for the *particular* lock (not the class of locks).

The following sections cover some of the types of locks in common use. It is not an exhaustive list, but it should provide a good introduction to lock technology and hopefully help in selecting a particular locking device.

Locksmithing is an ancient art with many traditional names and terms. It is not necessary to know the meaning of all of these terms in order to select a good lock. It is far more important that the security level, advantages, disadvantages, methods of attack, etc., are known for each lock in order that a judgment may be made based on fact and not by relying on manufacturers' advertising or product claims (e.g., "impregnable dreadnought lock!").

Composition.   A lock consists of three parts:

- key
- mechanism
- bolting device

Keys and mechanisms are obvious matched pairs. Bolting devices can vary, even when using the same locking mechanisms.

The following list indicates the most commonly found locking mechanisms, which are discussed in the following sections:

- warded locks
- pin tumblers—straight key, twisting pin, dimpled key, two-level keys, rounded key, and magnetic key
- disk tumblers
- lever locks
- Bramah locks
- split follower locks
- combination locks
- magnetic locks

Warded Locks

The *warded lock* works on the silhouette principle. A key and lock are manufactured as a pair. The profiles or side and end elevations of the key *bit* (the end opposite the handle) are mirrored as a silhouette of the key within the lock. The bit is usually a rectangular metal blank with slots cut through it in a particular pattern, to leave projections. These projections are known as *wards*. The slots and wards in the key exactly match fixed internal projections and slots within the lock body. The key bit is given a particular length and a thickening of the stem allows the key to stop at the correct depth of entry into the lock. If the key is inserted to the right depth and a correlation

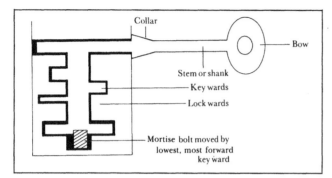

Figure 6–1   Simple warded locks

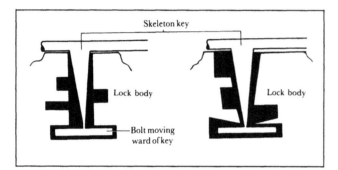

Figure 6–2   Different lock profiles; skeleton key defeats both locks

exists between bit slots and wards, and projections and depressions within the lock, it will be possible to turn the key. If there is no correlation, the key will not turn. The only part of the key that takes part in the action of physically moving the mortise is the bottommost forward ward (see Figure 6-1).

The major drawbacks with the warded lock have always been

the ease of picking
the limited number of skeleton keys required to pick many locks of different ward shapes
the limit to the permutations of ward shapes

As Figure 6-2 shows, although wards may be of many shapes and configurations, any simple, *skeleton key,* which will slip past the internal lock projections and push the mortise bolt sideways, will open the door.

Warded locks were among the earliest types of locks invented and although there were many variations, all relied upon the principle of a key of a particular shape passing through fixed obstructions within the body of the lock (i.e., the jigsaw or the round-peg-in-the-round-hole principle). These locks have been universally superseded by locks that rely on a different principle—that of the key moving obstructions to preselected positions in a predetermined sequence. All of the following sections refer to locks that rely on this more recent, more secure principle.

The term *warding* is still used throughout the lock industry to define a difference between one suite of keys and another.

Methods of Defeating.

taking a key impression and cutting a spare; a very unskilled person could do this easily as only the skeleton shape is required
forcing the mortise sideways by slotting it outside the lock or *jimmying* the lock

Security Level.   The security level is zero. Warded locks should never be used on any external or internal door to a room containing any item of value. Warded locks are not a security device and many padlocks are more effective.

Selection.   "Improved" versions are available that include tapering wards and nonflat bits (i.e., bits that have offset, cranked, wards and a corresponding keyhole so that only a key of a particular cross section will be able to enter the hole).

Uses.   Still found occasionally on garden shed doors!

Advantages.

cheap
robust

Mortise Lever Locks

The warded lock was superseded by the *lever lock.* In principle, the lever lock differs from the warded lock in only one major aspect. In the warded lock, the bottommost ward actually forces the mortise bolt sideways during the rotation of the key. In the lever lock, this action is conditional, not upon matching the fixed key wards with fixed internal lock projections, but with matching fixed key wards and *movable* lock projections. These projections form a series of flat plates known as *levers* (see Figure 6-3).

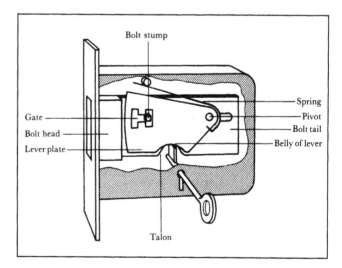

Figure 6–3   Lever lock

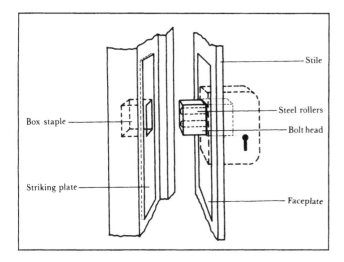

Figure 6-4 Mortise lock

The bottommost ward (or terminal stop) pushes the end of the bolt, known as the *tail*. This tail has a piece, known as the *talon*, removed from it. On the tail, there is also a side projection known as a *stump*.

For the bolt to move forward as the key turns, the stump must pass through a composite slot created by the perfect alignment of a number of segmented slots or gates cut into openings within the levers. The rotation of the key inches the bolt and stump forward while at the same time raising a lever (against the spring) to a predetermined height. If the height to which the lever is moved brings the lever gate to a position corresponding to the stump height, the stump will pass through that lever gate and the key will continue to make the bolt travel forward. The heights to which the lever gates are raised are determined by the key ward shape. Once the stump has successfully negotiated all the gates in the levers, the levers fall behind the stump, preventing its return without the reverse rotation of the key. The falling action of the levers at this point is said to *deadlock* the bolt. For an external view of a mortise lock, see Figure 6-4.

Methods of Defeating.

picking
drilling to drill out the stump and allow the mortise bolt to "drift"
jimmying the mortise from the staple
sawing through the mortise bolt
using keys of similar type (particularly on old locks with worn gates or levers)
using skeleton keys taken from an impression

Security Level. The security level offered is dependent on

the antipicking mechanism
the package offered with the actual lock (e.g., screws, metal staple box, etc.)

the number of levers (to a limited extent) and *key differs* (different key shapes)
materials, bolt strength, etc.

Generally, mortise lever locks are suitable for all applications requiring a moderate level of security (i.e., external doors to houses, apartments, small commercial premises, etc.). BS 3621, "Thief-Resistant Locks," dictates a minimum standard of five levers that would be adequate for the above applications. More importantly, it also defines the quality of the lock construction. BS 3621 requires a minimum of five levers and 1000 key differs to meet the label "thief-resistant lock."

In America, UL 437 is the most common key lock standard. In Britain, if the lock carries a British Standard number and the mark of the British Standards Institute, the lock has also been tested *in situ* as a complete package of hardware and will be more secure than a lock that merely complies with BS 3621.

Locks are available with up to ten levers. It is important to remember that levers do not necessarily increase security. A seven-lever lock with an equal (or fewer) number of differs and good materials, mark, etc., might be a far more secure lock than a five-lever model.

The number of key differs potentially available on a seven-lever lock is seven factorial (i.e., $7 \times 6 \times 5 \times 4 \times 3 \times 2 \times 1 = 5040$). If the lock and key are strong, and the key bit large enough, this potential can be approached. If, however, a lock is required that will open from both sides of a door, the key must have symmetry and the steps cut in the key actually must be symmetrical pairs. Hence, a seven-lever lock with a double entry might only have 24 differs at worst. Although an overexaggerated example, this illustrates the foolishness of specifying levers rather than key differs.

Whichever lock is chosen, always ensure that the frame into which any mortise lock bolt will fit has a minimum depth of 1 in (25 mm) behind the 0.8 in (20 mm) bolt throw, and that the frame is at least 1.8–2.0 in (45–50 mm) thick (preferably hardwood). To improve the resistance to attack of the staple, protect the frame on the visible side with nonremovable steel strips of at least 4 in (100 mm) in length.

Selection. When selecting mortise lever locks, look for

brass bolts with a minimum thickness of 0.6 in (15 mm), reinforced through the center with floating, embedded, circular, hardened steel bars. These will not corrode, and will be difficult to saw through, because bars rotate.
a mortise, bolt, stump, and tail that are one item produced from a solid block and not made from component parts and brazed, welded, or bolted together
locks with substantial bolt stumps
locks offering many differs

antipicking devices fitted as standard (e.g., false levers, notches, etc.)

a strong lever spring

closely manufactured parts that operate smoothly without sticking, but have no play in them

locks that are sold inclusive of fixing screws; pressed, hardened steel, staple box; full fixing instructions; and maintenance instructions

maker's name, British Standard number, and kite mark on lock

locks that include notched stumps

single-entry locks that can only be locked from one side of the door, when additional security is required. For a given number of levers, this will increase the number of available key differs.

minimum of 2000 key differs for additional security

locks with either a hardened steel case or a cover to prevent drilling of the stump

a *bolt throw* (the depth the bolt recesses into the metal staple case) of at least 0.8 in (20 mm)

Mortise Cylinder Locks (Pin Tumblers)

Mortise cylinder locks are often referred to by their alternative name, *pin tumbler locks*. The latter name describes the internal mechanism very well. The lock is comprised of

- an inner, or key, cylinder
- an outer cylinder
- pins
- drivers

The outer cylinder is fixed and held in place by the door into which it is recessed. The inner cylinder may rotate within the outer cylinder. A number of holes are drilled in a line in the top half of the inner cylinder (see Figure 6–5). Steel pins are placed in these holes, each pin having a slightly different length than its neighbor (see Figure 6–6). All of the pins are free to move up and down, but none of the pins is long enough for its leading edge to project out of the top of the hole when its bottom edge is level with the internal cylinder circumference (see Figure 6–7).

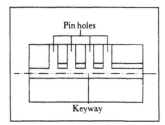

Figure 6–5   Inner cylinders

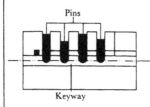

Figure 6–6   Inner cylinders with pins protruding into keyway

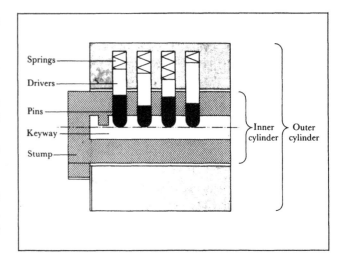

Figure 6–7   Drivers lock inner cylinder to outer cylinder, preventing rotation

This inner cylinder is placed inside the larger, fixed, outer cylinder. The outer cylinder is also drilled with holes in the same fashion as the inner cylinder and these holes align perfectly when one cylinder is fully inserted into the other. The holes in the outer cylinder have closed tops and contain a spring and another set of pins known as *drivers*. The springs force the drivers into the inner cylinder, causing them to rest on the pins. With the drivers in this position, midway between the inner and outer cylinders, the inner cylinder cannot rotate.

The key for this type of lock has V-shaped wards. As the key is inserted into the inner cylinder, it raises each pin a certain distance. Once the key is fully inserted, each pin will have been forced upward against the spring behind the driver, thereby pushing the driver out of the inner cylinder. If the key has the correct notches and wards in the right sequence and they are the correct height in each case, then the pins will raise the drivers until they are removed from the inner cylinder (without the driver being pushed into the outer cylinder). At this point, it is possible to rotate the entire inner cylinder (see Figure 6–8).

The inner cylinder is attached to a stump. The rotational movement of the inner cylinder is translated into horizontal motion by the stump, which retracts the bolt or mortise (against the force of a spring) from the staple within the door frame. Because the stump is fitted over the inner cylinder end, most locks of this type are single-sided (i.e., the key may only be used from one side). On escape doors, a manual override thumb turn will rotate the bolt with the key in or out of position to allow for exit. The pin tumbler lock can be found in the following common forms:

night latch—has a thumb wheel escape mechanism on the secure side

latch lock—has a spring-loaded unattached bolt

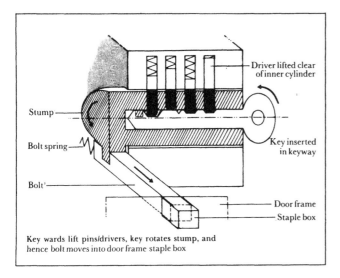

Stump

Bolt spring

Bolt'

Driver lifted clear
of inner cylinder

Key inserted
in keyway

Door frame

Staple box

Key wards lift pins/drivers, key rotates stump, and
hence bolt moves into door frame staple box

Figure 6–8 Pin tumbler action

deadlock — has a lever or catch that, when depressed,
drops a stump into the bolt making it a deadlock,
which cannot have the bolt pushed back from the
outside

automatic deadlock — offers the advantage of a latch lock
(i.e., it is self-locking, when the door is slammed
shut, with a deadlock mechanism that falls automatically without the need for a catch on the inside)

twin entry — is lockable from both sides

key-in-the-doorknob type — is commonly found in hotels
and guest houses. The locking cylinder is covered by
the doorknob and the key is inserted into the doorknob.

Most common cylinder locks are referred to as *rim
locks,* because the most common forms have a totally
enclosed, metal, staple box that surface-mounts to the
rear rim of the door frame and is not actually rabbeted
or mortised back into it.

Methods of Defeating.   The most common method to
defeat the simple locks of the thumb-turn-escape type is
to smash or cut through a door panel beside the lock,
reach through, and turn the thumb-turn manual override. *Loiding* is a term used to describe inserting a piece
of celluloid, flexible steel, or credit card between the
door and frame, and gradually edging the bolt back. This
only works on bolts that are free to move in either direction and are not fixed to the inner cylinder stump. Other
methods of defeating include

drilling to remove pins or drivers
picking
twisting out the entire outer cylinder from the door
forcing, jimmying, etc.
sawing the bolts
spreading the door frame

Security Level.   The basic mechanism of the pin tumbler and the lever locks are the same, hence the level
of security should be comparable. However, this is not
necessarily the case. In the United Kingdom, mortise cylinder locks are generally regarded as less secure than
mortise lever locks. This prejudice stems partly from the
materials to which they are fitted. On the continent and
in the United States, much greater use has been made of
integral locks within metal-framed doors. Cylinder locks
are secure in this application, as they are difficult to remove and the panels are difficult to cut or break through.
In the United Kingdom, they have been extensively marketed as locks for timber front and back doors. As the
fashion for glass and wood screen doors and lightweight
panel doors increased, the locks fitted (i.e., pin tumbler
and thumb-wheel locks) remained the same. In this application, they were totally unsuitable and offered little
security, hence the perhaps unjustified bad reputation.
Applied properly, the lock should be of equal security
rating to an equivalent cost/standard mortise lever lock.

The basic levels of security offered by pin tumbler
locks are (irrespective of key differs):

Least secure

• key in knob
• spring latch
• spring latch with night latch thumb wheel
• manual deadlock
• separate spring latch bolt and dead bolt in same lock
  (commonly referred to as a *draw back* lock)
• automatic deadlock

Most secure

• key deadlock — key on the outside throws the latch;
  slamming the door shut throws dead bolt automatically.
  A dead bolt may be locked in position from the inside
  and only operation of the key from either side will release it. This prevents a hand-through-the-hole attack.

Great care must be taken in selecting rim locks, because,
as the plethora of names indicates, many different security levels exist. Fewer locks of this type meet thief-
resistant lock standards than lever mortise locks. Rim
locks should not be considered for sole application on a
medium- or high-security-risk application, or even on
domestic and small commercial/industrial projects. Rim
locks are best used in conjunction with mortise lever
locks.

Selection.   Never select a rim lock for fixing to a door
that is less than 3.5 in (90 mm) thick. Look for locks that

have UL or BS 3621, 1980 approval (and, if possible,
are kite marked)
have a beveled outer cylinder face on the unsecured
side to prevent wrenching
have hardened steel cases and pins to prevent drilling

have antipick bars, latches, mushroom-headed drivers, side bars, etc.

have ten pins or more, telescopic pins, side-angled or twisted/twisting drivers and keys (on higher-security-level models)

quote minimum and preferred style thicknesses in sales literature and also give a graded star rating to the security level offered

have hardened steel, rotating bars in the bolt

Uses.

inner area doors of solid construction (not with normal, annealed, glazing) protecting low-value (or low-risk) items

outer doors to domestic premises when used in conjunction with a mortise deadlock

Pin tumbler locks should only be used on doors to offices or domestic flats within a *normally secure area* (i.e., a manned entrance or access-controlled entrance). They should never be used for the protection of high-value items or high-risk, classified materials, storage areas, or unobserved entrances.

Split Follower and Emergency Exit Mechanisms

Two particular problems in the locking of doors are those of nonuse of deadlocks and deadlocking mechanisms, and the insecurity of emergency exit locks. There is little point in providing a secure deadlock if the device is never used and the door is only locked by the latch bolt. Police investigators confirm that in many instances of burglary the deadlock existed, but the building owner could not be bothered to use it and its key, preferring instead to slam the door closed using the latch bolt, which is not as secure.

Normal lock cases only allow the same lock functions on either side of a door (e.g., key open/closed dead bolt, latch bolt with keyhole on both sides of the door). This considerably restricts lock capability and flexibility and results in nonuse of any function that may hinder fast exits.

Split follower locks (see Figure 6–9) allow for independent functions and security levels *on either side of the door.* The case usually contains either two latch bolts (one that is *automatically deadlocking*) or one latch bolt and a mortise dead bolt. A whole variety of key operations, handles, and thumb turns may be incorporated. A common example is the key operation of the dead bolt and the latch bolt *combined* from the exterior side, with a snib to allow operation of the latch bolt only using the exterior handle when the door is unlocked. On the interior side, a thumb turn would dead bolt the door, while a handle would open both the dead bolt and latch bolt, allowing for a fast

Figure 6–9　Split follower lock

exit. On exiting, the dead bolt automatically deadlocks on the closed door.

Emergency Exit and Access Mechanisms on Locks.
There are two principal considerations in selecting emergency exit and access mechanisms—ease of use during a *worst case* scenario emergency and security in normal circumstances. The first of these considerations is the most important. Remember that

in a smoke-filled corridor (even *with* emergency lighting) the lock operating mechanism may be very difficult to see and operate.

locks can stick even in normal circumstances. In the presence of fire or smoke, this risk increases, particularly on aluminum doors.

To reduce risks, emergency unlocking door handles should be large and easy to identify by *touch,* and should throw the locking bolt *completely* into the lock case or clear of the door staple *and* door gap. Traditionally, there have been two types of emergency exit devices:

integral or surface-fixed vertical/horizontal rods or bolts, or agastral (four-point fixing) bars linked to *crash* (or *panic*) bars (which are pushed or pulled to release them from the door staple) (see Figure 6–10).

normal locks (mortise or latch bolt) with override handles

The former device has existed for many years and modern variations of this device work particularly well. They are, however, very poor security devices, as they are easily overcome from outside the door. Solutions to this problem include *day alarm* or *door open* intruder detection, magnetic reed switches fitted to the lock. This combination can work well, but ownership and responsibility for the alarm system can be problematic, particularly in

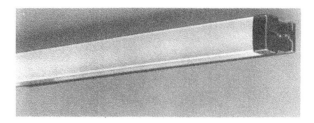

Figure 6–10   Rim-fixed emergency exit bar

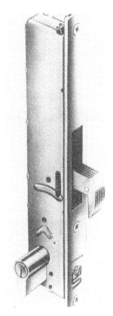

Figure 6–11   Swing bolt deadlock for aluminum doors

tain a thumb turn that overrides the lock latch or mortise bolt when turned, after breaking the dome. These devices can be used for emergency exits or in internal, secure areas requiring emergency entry (see Figure 6–12).

emergency escape deadlocks with handles—these locks are similar in action, but a bar or handle is attached to the thumb turn and passes through the dome. Turning the handle up or down breaks the dome and operates the lock (see Figure 6–13).

The thumb-turn device is more secure from outside attack, inside abuse, etc., but is less easily recognized and used. Deadlocks with emergency handles are less secure, but with their large, brightly painted fluorescent handles (separate from the everyday handle and stamped "FIRE USE ONLY") they are easier to use. When children or the

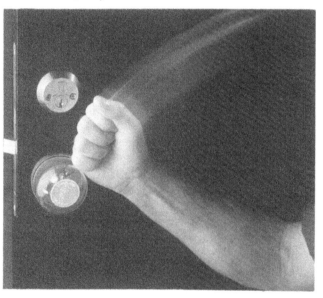

Figure 6–12   Break dome emergency night latch

Figure 6–13   Emergency escape deadlock with handle

multitenanted buildings. Crash bars or pads have the primary advantages of size and public familiarity. *Use of these devices on aluminum doors can be very insecure* unless proper care is taken during specification (see Figure 6–11). Aluminum door leading edge stiles are often bull-nosed, rounded, or beveled and not square-edged. This configuration provides an exposed panic bar, lock mortise, and hinges, which may be easily sawed through unless bolts are laminated and/or ceramic core cylinder types, or doors are fitted with rabetted hinges and/or agastral cover plates that armor and cover the doorjamb gap from top to bottom, at the hinges and the leading edge. This prevents leverage and attempts to cut through. On doors swinging in both directions, this configuration can be particularly problematic and door closer coordinators and special hinges are required. With this situation (often found on large, front entrances that are also primary final escape doors), it is often better to insist that both doors only open outward or to install large, revolving doors for normal use, with secondary, normally secured escape-only doors.

Normal locks have override escape devices that may be classified as

break dome or *break glass dome cases*—these cases are of a nonsharp, fragmenting PVC or Plexiglas and con-

Figure 6–14   Narrow profile, emergency deadlocking latch bolt

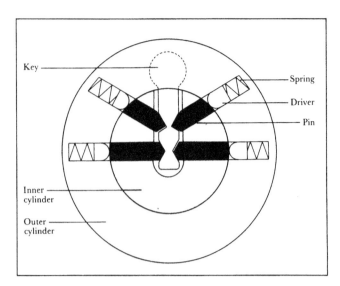

Figure 6–15   End elevation of a Kaba axially offset pin lock

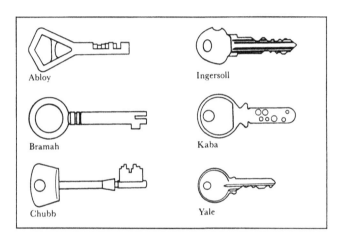

Figure 6–16   Common key shapes

elderly may need to use this type of device unassisted, neither of the two emergency escape devices are recommended above the large, traditional crash pad.

A common problem of most security and fire exit locks is that after use, the door can automatically relock on a spring and latch bolt, preventing *reentry* into the building by fire officers and thereby increasing rescue time. The split follower lock with an emergency lever overcomes this problem. The outside of the door has a normal key entry lock and dead bolt, where a latch bolt is thrown (after key use) by an exterior door handle. On the *inside,* when the door is locked, an emergency lever can be used to unlock the dead bolt without a key. The normal internal handle linked to the lock throws the latch bolt and exit is possible.

The emergency lever is single-shot one use only and the dead bolt cannot be relocked from the inside or outside without a key. The latch bolt remains operational by either inside or outside handles and hence, while the fire doors reclose to restrict the spread of fire, reentry is always possible (see Figure 6–14).

Specialist Locks

Standard locks are medium security, medium price range. For more secure areas, a specialist lock and key may be required. These include

- dimpled key pin tumblers (Kaba Ltd.) (see Figure 6–15)
- magnetic key pin tumblers (Corkey Control Systems)
- clearance, fit-fine tolerance, pin tumblers
- twisting pin tumblers
- side pin tumblers (ASSA/MEDECO)

See Figure 6–16 for common key shapes.

Advantages.   The increase in the level of security offered by these locks is not really significant and for all practical purposes is primarily due to the rarity of the keys, the need for specialized cutting equipment, and, with some locks, the registering of the issued keys. Most of these locks claim increased security by increased key differs. Although this is important, the antipicking mechanisms and the body strength are of more fundamental importance.

Most of these locks are resistant to a specialized forcing device known as a *lock gun.* As cylinder locks should not be used for the protection of high-value goods to which a professional thief, with access to a lock gun, might be attracted, this advantage is invalid.

Specialized locks have greater resistance to forcing if materials or construction are genuinely superior to less expensive locks. Quite often, the materials used and method of construction are not much better than the standard ranges, even though they appear to be superior merchandise. The main operational advantages of most of these locks are

they are often more compact
they are usually of a more prestigious finish (i.e., brass, not steel)

they often offer facilities for emergency release from the inside, auxiliary alarm contacts, double-sided locking, etc.

they usually offer extensive key mastering and suiting facilities for large installations

Most specialized locks offer various levels of manufacturer-controlled blank key issuing, cutting, etc. The usual levels are

unregistered keys — one master plus *x* number of a group of keys. A letter of authority along with the master number is required before a legitimate locksmith will cut a replacement.

registered keys — similar to unregistered keys, but a record is kept of to whom all keys and copies are issued; letters of authority and supplementary proof of identity are required to obtain a registered key

area service — the manufacturer has a local office. A lock service agreement is made with an identified person and a signature is recorded. Keys are only produced when that person appears at the local office, and her signature and that of the manufacturer (who signed the original agreement) is obtained.

card service — a special card (like a credit card) is issued with every agreement between the manufacturer and client. The card contains coded information on the client and/or his representative, which is retrievable. Keys are only issued on receipt of the card, letter of authorization, signature, and matching information held on the card (with secret, recorded information held by the manufacturer).

On all specialized lock installations, it is advisable to enter into at least a registered key type of agreement. This will aid key control, reduce unauthorized copying, and, with a service agreement, ensure that locks are kept in good order for many years.

Mortise Cylinder Locks (Disk Tumblers)

Disk tumbler locks are similar in principle to mortise lever locks; however, two cylinders are used, as in the pin tumbler lock. The inner cylinder rotates when the key is inserted; the outer cylinder is fixed and contains a number of disks. Each disk is notched and, without the key in place, the notches are not aligned. When the correct key is inserted, the notches line up allowing a talon to fall into the groove formed by the aligned notches. The talon is shaped so that at rest, its upper surface projects into the outer cylinder, preventing the inner cylinder from rotating. When it falls, the talon disengages from the outer cylinder, thereby freeing the inner cylinder. The mortise bolt is connected to the inner cylinder and is moved horizontally as the key is turned (see Figure 6–17).

Locks of this type are usually recognized by their key, which is notched on both sides (unlike most pin tumbler keys) and by the keyhole, which is usually a wave profile.

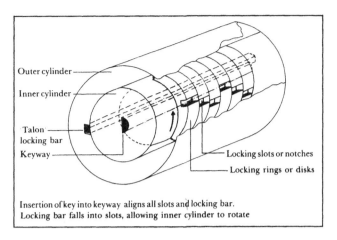

Insertion of key into keyway aligns all slots and locking bar. Locking bar falls into slots, allowing inner cylinder to rotate

Figure 6–17 Abloy disk tumbler lock

Methods of Defeating. The methods of defeating disk tumbler locks are similar to those for pin tumbler locks. Because of the increase in the number of disks and the use of better materials, they are less easily picked, drilled, etc., than a similar-looking pin tumbler.

Security Level. Disk tumbler locks are usually considered to support the middle ground, in terms of security, between pin tumbler locks and the more secure mortise locks. Disk tumbler locks are usually considered more secure than pin tumbler locks because

more disks can be included than pins for a given lock depth and, hence, a greater number of key differs is possible

keys are double-sided and often not cut from standard flat blanks. Hence, they are much harder to copy and to cut (without specialized equipment) than normal pin tumbler keys.

by having more key differs, manufacturers offer an intrinsically higher security lock. In order to enhance this, many disk tumbler locks are provided with anti-picking devices, ceramic antidrilling plates, etc.

as the locks are more expensive than the normal pin tumbler locks, fewer disk tumbler locks exist. Rarity and unfamiliarity increase security.

Selection. Selection is generally similar to that for pin tumbler locks, but is usually more straightforward, as fewer types are available and the percentage of the locks that are "approved" is higher. Most hook-bolt, mortise deadlocking disk tumbler locks offer better security than standard mortise types.

Uses. Generally, disk tumbler locks are used as pin tumbler locks of the standard type. They are often useful for double-sided, key-use variances on narrow frame, metal-stile, laminated glass doors or in internal areas where depth of frame is a problem. On external doors of this type, however, Kaba dimpled key locks offer better security.

Bramah Locks

Bramah locks have a unique key; it is shaped from a small-diameter, hollow cylinder. The end of the cylinder has notches of varying depths cut into the circumference of the cylinder and a stump behind these notches projects from the cylinder at 90°. The key enters the lock guided by a rod at the center of the keyway, which slides into the hollow cylindrical key. As the key is pushed into the lock, its end pushes a spring disk that is fixed to the rod. A circular barrel and stump-moving bolt are attached to the end of the rod that, when rotated, moves the stump horizontally, thereby locking or unlocking the device. A series of sliders (or levers) is arranged radially around this disk. These sliders have a free end to engage the key slots and another end that is notched (see Figure 6–18).

Without the key inserted, the notches in the sliders do not line up with a number of detainers that are arranged radially in an outer, fixed cylinder. When the correct key is inserted, the key pushes the spring disk into the body of the lock. The notches on the key engage the sliders, which move radially outward and into the lock body, until the notched ends align with the detainer and the central barrel is free to move. The key stump rotates the barrel as the key is turned, thereby opening the lock.

Methods of Defeating.

jimmying the mortise from the staple
sawing through the mortise

Security Level.
The Bramah lock offers a security level above that usually found with the more common pin tumbler locks. The reasons are

a practically infinite number of key differs
key rarity
difficulty of reproducing, copying, and cutting keys

resistance to picking (far superior to lever or pin tumblers)
the lock is a mortise deadlocking device, in most cases
the inner barrel is much harder to wrench out than the inner barrel in pin tumblers

Uses.
Selective use on medium-security-risk doors of solid construction is recommended. Cost of the lock and key sizes make it impractical for mass use. Bramah locks should not be used on final exits, as an escape mechanism would lower the security level provided by the lock to that which could be achieved with conventional, lever mortise deadlocks. Commonly used on high-security steel cabinets and secure stores.

Keys and Lock Suiting

To the specifier of locks and keys, the mysteries of key and lock suiting sometimes appear to be a particularly daunting form of black magic, in which the chances of having the correct locks, for the correct doors, openable with the correct group of keys, are remote! Lock manufacturers, like clock makers, are precise and pedantic people who do not deal in abstracts. You may rest assured that if the locks you receive are incorrect in any way, an honest investigation will reveal that 99 times out of 100 the specification given to the lock manufacturer was wrong! This often happens because the compilation of lock and hardware requirements is delegated (i.e., abdicated!) to inexperienced juniors.

Essentials of Specification.   The basics are:

1. *Keep it simple.* The fewer the locks, the fewer the grand master, subgrand master, submaster, pass groups, etc., the better. Key issuing and key control will be simplified. Suiting eases lock use convenience, *not security.*
2. *Do not specify until* you know *how* your building will function; you know the hierarchy of directors, managers, and staff within the building (i.e., the company's organizational structure); and you know the persons who will be permitted to enter certain groups of rooms or who will be refused entry to others.
3. *Be very precise.* Use the lock manufacturer's standard ordering data collection sheet and *fill in all columns and details for every lock,* if you do not, expect mistakes.
4. *Order last.* Do not order until you are *sure* you will not change your mind on any fundamental concept. Changes can be very expensive once orders have been submitted.
5. *Allow for expansion.* Always allow for and state any future expansion requirements for the master key suite.

Types of Master Keyed Suites.   The lock industry uses precise terms to define the role of a particular key in relation to a suite of locks:

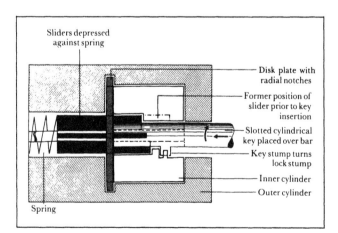

Figure 6–18   Bramah lock

Sliders depressed
against spring

Disk plate with
radial notches

Former position of
slider prior to key
insertion

Slotted cylindrical
key placed over bar

Key stump turns
lock stump

Inner cylinder

Outer cylinder

Spring

Figure 6–19   Key suites

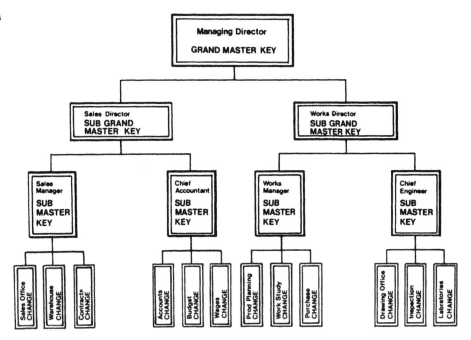

alike change keyed suite — a number of locks, *all* of which may be opened by the same key

simple master keyed suite — each lock is opened by its own key *and* the master key of the suite of those locks. *Only* the master key will open other locks in the suite.

simple master with pass group keyed suite — the master key opens *all* locks in the suite. Each lock has its own key. Keys will open locks (by "alike change") within particular subgroups, known as *pass groups*. For example, locks 1–8 all opened by the master key; locks 2–5 all opened by the same pass group key (i.e., via "alike change"); locks 4, 6, and 8 each have their own key, which will open *no other lock;* and locks 1, 3, and 7 open on another pass group key (via "alike change").

apartment or equipment keyed suite — each apartment has a key that will open the lock on the common main entrance door, the lock on the door to a floor, and the lock on a particular door to an apartment on that floor.

Types of Keys and Their Relation to the Key Suite (See Figure 6–19).

grand master key (GMK) — opens *all* locks in the building or area owned by the security manager, *chairman* of company, etc.

subgrand master key (SGMK) — first division of key control (lock suiting, usually only two or three suites). Key will only open locks in a particular suite. Owned by operation *directors* of particular areas or buildings.

submaster key (SMK) — a division or divisions of the subgrand master key area. Owned typically by a production or administrative manager.

pass group alike change key (PGACK) — the final division. Subdivision or division of a submaster suite; to allow operatives or trusted employees into their particular work space/area. Two or more locks in succession.

unique key (UK) — opens one lock only (locker key, machine key, etc.)

Groups of Locks.   Lock companies classify locks by groups. A *group* is a locksmith's term for a lock with a cylinder of a particular diameter [usually 1.3 in (32 mm) or 0.9 in (22 mm)]. Locks within a particular group may be opened by any suitably coded key cut from the correct-sized blank. The keys of one group cannot open the locks of another group, as they are either physically too large or too small for the other group. Only locks in each group are suited together for this reason. Groups *are not* subdivisions of key opening passing, etc. Cross suiting, emergency suiting, and single shot mastering are beyond the scope of this book but can be used to improve lock/key flexibility in certain applications.

Combination Locks (No Key!)

Combination locks come in two forms — disk and the more recent push button.

Traditional Disk Combination Locks.   Traditional disk combination locks were invented before the Rubik's Cube, are infinitely more complex, and are definitely more fun to try to solve! They rely on the principle that a series of objects must be placed in a unique alignment, by following a series of definite, limited actions. The "key" to the combination lock is the use of an alpha-numeric code (see Figure 6–20).

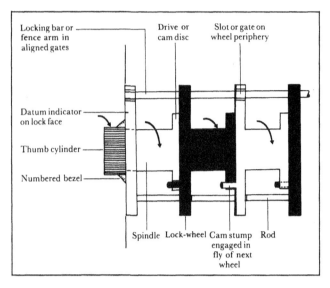

Locking bar or
fence arm in
aligned gates

Drive or
cam disc

Slot or gate on
wheel periphery

Datum indicator
on lock face

Thumb cylinder

Numbered bezel

Spindle   Lock-wheel Cam stump   Rod
engaged in
fly of next
wheel

Figure 6–20   General arrangement of traditional disk combination lock

On the face of the lock, a disk is visible on which numbers are usually engraved from 0 to 100. In the center of this disk is a thumb cylinder (or knob) that rotates freely and is engraved with a single mark. Attached to the back of this knob is a rod and spindle, which projects into the body of the lock. Fixed to the spindle is a cam disk that has a steel pin (or stump) projecting from it. If a knob is rotated, the cam disk and pin rotate with the knob. This assembly is held within a space left in the body of the door. Fixed within the door is the remainder of the lock, which consists of several wheels. The first wheel closest to the cam pin is known as the *lock wheel* and has a slot (or *fly*) cut into it. If the cam is turned to its first correct position, the pin engages the fly in the lock wheel and turns in the same direction—*if* the direction of rotation is chosen correctly! Assuming the correct direction is chosen, the lock wheel then turns and a pin on its circumference turns the wheel placed behind the first wheel. The number and series of wheels can be many. When the final number in the first sequence of the code key has been maneuvered correctly, the direction of rotation is reversed and the second sequence must be carried out. Eventually, all the wheels will be positioned correctly and the slots (or *gates*) will be in perfect alignment, allowing a *fence arm* or *locking bar* to drop into them. The lock can now be opened.

Methods of Defeating.   On doors of standard types, jimmying, cutting around the lock, frame spreading, and key copying are difficult; picking is difficult; and "discovering" the combination is possible. Photography from a remote position should also be considered a risk. Where secured goods are of high value, threats such as radiography, fingerprinting, rigging (unauthorized insertion of alternative lock), picking (manipulation), thermal lance (high-temperature) cutting, and acid attack exist.

Security Level.   From a purely engineering standpoint, the security advantages are

no keys to use, lose, or carry

no keyhole

the combination is easily changed by the users

removal of the knob by wrenching is usually ineffective

written records of combination are easily hidden and stored among other secure paperwork, unlike keys, which must be stored in large cabinets that are obvious targets

The disadvantages are

access is limited strictly to authorized code carriers

the setting must be changed regularly (every 3–6 months) to maintain security

if the code is learned by unauthorized persons, the loss of secrecy may remain undetected, resulting in a false sense of security. Hence, the need for frequent changes of the combination.

The most common causes of security failure of combination locks are undetected knowledge of the code by unauthorized persons and incomplete scrambling of the mechanism after use, allowing for easy manipulation.

If protected goods are of high value, expert advice should always be sought from a registered locksmith before final lock selection. Essentially, good four- to six-wheel combination locks are a high-security device.

Selection.   Always select locks that have at least ANSI/UL 768 status and have

at least a four- to six-component, internal wheel train

a wheel train that is remote from the knob and in a position that is not visible

been approved by the Association of Locksmiths of America

an easy capability for rapid combination changing from the rear of the lock

to be fixed in such a manner that removal (for substitution) of wheels from the rear of the lock is not easily or quickly carried out

all-metal construction with fine tolerances

resistance to acid attack or smoke excessively when attacked; some devices have ceramic shields to prevent successful thermal attack

manufacturers who offer full maintenance repair-and-replacement backup services

drill resistance

a combination lock that incorporates a conventional key lock, as a backup device on a high-security door

As these locks are essentially high-security devices, they should be selected in conjunction with the door on which they are to be used. If possible, the lock manufacturer should provide the door to which the lock is already fitted and has a better resistance to attack than the lock. Combination locks of fewer than four wheels should not be used.

Uses. As high-security devices, combination locks should only be used on structures that have considerable resistance to attack:

• strong-room doors
• secure rooms
• security closets or safes

For strong rooms, locks are only effective if fitted to solid, heavy steel doors with built-in frames. Secure rooms, book rooms, etc., should have locks fitted to steel or steel-faced doors hooked up to an alarm. Security closets should have locks fitted to steel-faced or solid, hardwood doors with hidden, strengthened hinges. The construction and siting of these enclosures is a specialized field and expert advice should be sought.

Push-Button Combination Locks

Push-button locks are either mechanical or electrical. The mechanical type is similar in nature to the disk combination lock, having all parts within the door. The electrical lock differs in that the coding/selection assembly is remote from the actual mortise bolt and the electrically generated signal that operates a lock bolt release solenoid. In both types of locks, the overall security level rests on the level of security offered by the push-button device.

Methods of Defeating. The most common methods of defeating are

overlooking, remote telephoto lens photography, etc., while someone is operating the device. This is very easily done and to counter this action only authorized personnel should be allowed in the vicinity of the lock. The lock should not be used on doors leading to public areas or on doors of rooms within public rooms.
fingerprinting, marking, and wear. The operation of any one combination of buttons, for even quite a short period of time, will show which of the buttons out of the number of available buttons are being used. Hence, if a five-button system has a three-button code, these three buttons can be recognized by dirt, wear, etc., on the buttons. This recognition dramatically reduces the security level offered.

Security Level. Mechanical devices are low-level security only. Electrical devices offer higher security, but very careful selection is required in order to ensure that the system actually provides what it promises.

Selection. In selecting push-button combination locks, the most important points for consideration are

actual available combinations versus theoretical. Manufacturers will often quote a five-button mechanical lock of cheaper construction as having 5^5 possible combinations (i.e., 3125). This is only true if all buttons may be used and each button can be used more than once (e.g., combinations such as 1-2-3-4-5, not 1-1-1-1-1). If the second condition exists, the actual available combinations are 5 factorial or $5 \times 4 \times 3 \times 2 \times 1 = 120$! In practice, many users find that staff can only readily remember four digits. In this case, the device could offer at the most, 4^4 (256), or at the worst, 4 factorial (24). There is a considerable reduction in security between 3125 and 24 combinations! Running through 24 different codes of the recognized four digits would only take minutes.
that electrical combination locks invariably offer "same-number-repeat" codes and are inherently more secure. When mechanical devices are used, they should offer same-number repeats (or integral key locks) as backup.
the device should be able to have its combination changed quickly and easily. This will encourage users to change the combination frequently, removing the recognition-of-wear risk. This occurs when some digits are repeatedly used, as they are easier to remember (e.g., 9999), and buttons become worn or dirty compared to others on the pad.
that electrical devices should have mains failure electrical backup capabilities. This should, if possible, be of at least 72-hour duration and should be provided by sealed, nickel cadmium batteries fed by an integral trickle charger, with the whole assembly concealed within a secure cabinet near the push-button unit. All wiring should be protected from deliberate interference by being encased in steel conduits. If the device does not have this feature, it should lock on mains failure, and should *never* be used on any escape route or potential escape route.
that no electrical lock assemblies, to date, carry British Standard approval or kite marks and hence information on the application of such locks should be sought from insurers
that, for good security levels, locks should have ten digits (0–9) and be able to accept or reject codes consisting of up to all ten digits, even if some are repeated (e.g., 5566213091)
that both mechanical and electrical devices suffer from abuse of the push buttons. If users cannot remember the code, it is frustrating and the lock may be vandalized by its users! These locks should only be fixed if they are extremely robust and vandal-resistant.
that all doors fitted with these devices should be fitted with strong spring-closing mechanisms. The most common cause of security failure is to find that the door used did not actually latch shut. A mortise deadlock/mechanical lock/alarm/electrical lock com-

bination offers the advantage that there is no uncertainty as to whether the door is locked or unlocked. to always select push pads that have an antioverlooking shield

Uses. Push-button combination locks are commonly used when full access control systems are too expensive and the room is already within a secure area. For example,

central computer processing rooms within access-controlled computer suites

secure offices containing classified documents locked within security containers

areas with sensitive processes or experiments (e.g., small laboratories within larger rooms that contain high-value equipment or long-term experiments)

Push-button combination locks are very useful as a cheap method of security zone control. Using several units of the same type, concentric rings of security can be formed within the building by having different security codes for each area. In this way customers could have, for example,

outer area code 128652 — code known to all employees

inner area code 1158612 — code known to all accounts staff only

inner sanctum code 1155633 — code known to manager and company secretary only

Combined with keys for each area, this concentric configuration is a very cheap and effective method of zone control.

Advantages and Disadvantages.
Advantages.

no key
inexpensive
combinations easily changed

Disadvantages.

low-level security with most units (as only five digits without repeat)

codes (like keys) are easily lost and forgotten, and, in commercial premises with users frequently using the door, often committed to paper that is then lost or stored in an obviously insecure place

temptation to give away codes to nonauthorized users is greater, as no loss of property is involved (i.e., key) and hence chance of detection is smaller

security breaches not readily detected as there is no visible loss (i.e., key)

Manufacturers and Ranges. Types of push-button locks include

mechanical plus keyway — the key, which has its own inherent security, can only be turned in the lock after the correct button sequence has been used

mechanical or electrical with timers — only allows entry during certain hours (mechanical) or time for selection default (electrical) (e.g., "you have 30 seconds to press the correct sequence")

mechanical and electrical with "freeze" or alarm-triggering feature if wrong number is selected

Magnetic Locks

Magnetic locks (see Figure 6–21) fall into three categories:

- magnetic field locks
- keyless, encoded magnetic strip locks
- magnetic card key locks

Magnetic Field Locks.
Principles of Operation.   The magnetic field lock is comprised of two parts — the *electromagnet*, which is fixed to the door frame, and the *armature*, which is fixed to the swinging door. The electromagnet is comprised of a special, steel-laminated core around which is wound fine copper wire to form *turns*. A low voltage (normally 12V or 24V DC) is applied to the ends of the wire to induce a current to flow through the wire. The current creates a magnetic field in the steel core, which becomes polarized — north and south. The armature is a simple strip of metal similar to the core. When the door is closed, with the voltage applied to the electromagnet, the armature allows for the completion of a magnetic path from the north pole to the south pole and this magnetic path attracts the armature to the electromagnet. Switching the applied voltage on or off secures or allows the door to swing open, respectively.

The power (or holding force) of the magnetic field lock is a function of the number of amperes flowing through the wire and the number of turns of the wire. As low voltages and current are used, many turns of wire are needed for strong fields, hence the lock tends to be large. The unique feature of these locks is that they have *no moving parts*.

Figure 6–21   Magnetic field lock

Uses. Magnetic field locks are used primarily for lower-security applications when they are not the first line of defense and when an internal door with heavy traffic flow requires fast, reliable operation. Most magnetic locks are commonly found on internal, access-control doors, particularly interlock doors, internal fire escape doors (hold open or hold shut), etc.

Application. The essential points to remember with the application of magnetic locks are

the door must be as strong or stronger than the lock. There is no point in providing a lock with a holding force of 500 lbs (227 kg), if the door panels may be pulled out or rammed out with a 200 lb (91 kg) force.

the door must be rigid. Most magnetic locks have only one point of contact, which is often the top, leading edge. If the door is not rigid, the gap at the bottom of the door may be exploited by simple leverage.

the door frame materials should be capable of accepting large, strong bolts and locks without being weakened

the armature and magnet must align properly every time. If they do not, then the holding force will be dramatically reduced.

it is often necessary to provide concealed wiring within the frames, heavy aluminum mounting brackets, etc. These should always be supplied by the manufacturer.

magnetic locks are always fail-safe. Hence, security is only maintained at the same level as the physical security of the wiring and the integrity of the power supply. Fluctuation in voltage or current will dramatically affect the holding force. Power supply units incorporating filters, surge arresters, battery backup, etc., are essential and often not quoted in the lock proposal.

Mortised locks and armatures are far more secure than surface versions.

Equipment Type and Selection. There are only a few manufacturers of these locks and most are manufactured in America. Trade names include Em-lock, HiTower, and Mlock. Magnetic field locks should

have a 500 lb (227 kg) holding force on internal, light-security doors

have a 200–1500 lb (91–680 kg) holding force on full security doors

have the holding force rated as "true, direct pull" (i.e., pulling the armature directly off the face of the magnet—90° from the face)

be⁻ of a stated "sheer force resistance." Magnets resist a 90° pulling force extremely well, but are far easier to overcome if the armature is sheered sideways from the face of the magnet. This indicates a probable means of defeating the lock. Shielding from insertion of steel blades to "cut" the holding force is required.

have a 24V DC power supply whenever possible

have both voltage spike and surge filters of the metal oxide–varistor type and heavy-duty, high-current tracking capacity switches

always include monitor kits to determine door status, tampering, loss of magnetic field, and loss of power

Advantages and Disadvantages.
Advantages.

no moving parts
no keyhole or key
extremely reliable
fail-safe
does not stick or bind
mortise- or surface-mounted
sliding or hinged doors
acceptable to many fire officers for use on escape routes and doors
very little maintenance
very fast action
easy, automated control, and interfacing with other systems allows for many complex functions or sequences
DC types are silent in operation

Disadvantages.

large
heavy
costly compared to traditional mechanical lock
requires strong doors and frames, and hence a specialized installer
often runs hot to the touch

Keyless, Encoded Magnetic Strip Locks. These locks consist of flat, celluloid cards into which is embedded a unique arrangement of very small, thin magnets. The arrangement of the magnets within the card creates a unique magnetic field pattern. This pattern is recognized by the *receiver*, within the body of the lock, when the card is brought close to the lock. An electronic signal is then generated, which is used to release the bolt or mortise housing, and causes the door to open.

Methods of Defeating. The most common ways to defeat keyless, encoded magnetic strip locks are by damaging the receiver with a powerful electromagnet or by simply attacking the door or lock. It is all too common to find this novel device fitted to doors that are totally unsuitable, as if the device itself offered some increased security to the door.

Security Level. Their security level is low. These magnetic locks should be used in conjunction with more conventional systems as a means of keyless access control.

Selection. As these devices have not yet been extensively tested and approved by responsible organizations, they should be selected with care. The best approach is to disregard their integration as an unoccupied or night-time security device and rely, instead, on conventional locks for this purpose. The merit of the magnetic strip lock is considered as a means of access control. The following points should be considered:

robustness of the card and receiver

effects of demagnetization (either accidental or intentional)

how unique is it *really?*

cost of replacement cards and replacement of lock if unit card and lock cannot be reprogrammed or if a card is lost or stolen

Uses.   Used across the United States and to some extent in Europe for hotel bedroom security when higher-security, magnetic key-in-lock or access systems are too expensive. The cost-effectiveness of this type of scheme needs careful consideration as it is possible that, despite disadvantages, a conventional key system might prove to be a better answer on the grounds of security and cost throughout the life of the installation. The advantages of these devices are no key, no keyhole to vandalize (receiver can be mounted behind a *lexan* or other screen for protection), and not easily copied.

Magnetic Card Key Locks.   The simplest magnetic card key locks (see Chapter 7, "Access Control Systems," and Figure 6–22) have punched, magnetized, laminated plastic sheets. The holes form gaps in the north–south array to which the receiver responds. Magnetic card key locks are only suitable for automatic barriers and authorized use of machines, and not as lock controllers. The solid, laminated sheet key types available are suitable for internal security. These operate on the same principle as the keyless type of magnetic lock, but usually offer an increased level of security due to signal encoding between a controller and the lock. This prevents tampering with the electrical supply, thereby affecting lock security.

Security Level.   Magnetic card key locks are only as secure as the key used. Loss of the key will have the same effect as the loss of a normal key. Principally, the device should only be used as a means of restricted access control within an already secure area.

Selection.   Look for a device that

includes battery backup, reprogrammable cards, a small robust key, and maximum key differs (codes) on the same lock

is solid state electronics

includes mechanical mechanisms within the card reader

incorporates an electrical push-pad coder as well as a magnetic key, for better security

incorporates antitamper, wrong-key alarms

incorporates a one-person-at-a-time restriction (e.g., the device is within a turnstile or revolving door, or is coupled to a beam detector) for higher security

Uses.   Magnetic card key locks are used as access control for bookstores, laboratories, computer rooms, etc.

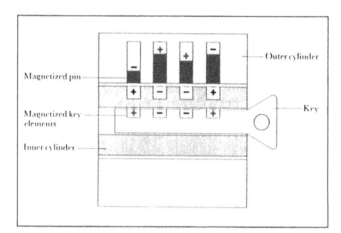

Figure 6–22   Magnetic lock

They should not be used on facades, as they are not robust enough and usually not corrosion- or weather-resistant.

Electric Locks

Electric locks fall into two broad categories—*solenoid mortise bolts* and *electric strikes* (often called *strike releases* or *electric staple box*).

Principles of Operation.

Solenoid Mortise Bolts.   Solenoid mortise bolts (see Figure 6–23) are dead bolt devices. The dead bolt is

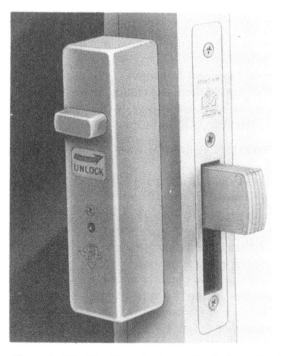

Figure 6–23   Electric actuator and snib for swing bolt

Figure 6–24 Electric strike releases for access control systems

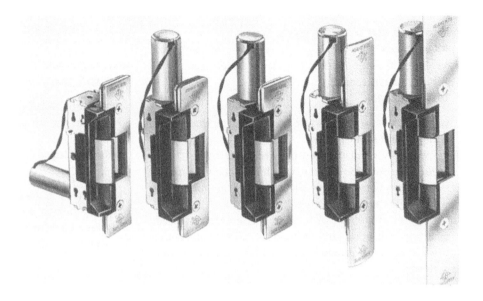

concealed within the door in a housing that contains an electromagnetic solenoid coil. When energized, the coil creates either an attraction or repulsion magnetic force. The force moves the mortise bolt backward or forward from a traditional striking plate in the door frame.

Electric Strikes. The electric strike (see Figure 6–24) is essentially a staple box for a latch bolt or a dead bolt in which one side can be hinged open to allow the bolt to be released while it is still extended from the lock. The hinged side (or *jaw*) only swings when a solenoid releases a holding pin or a magnetic field releases the strike directly.

Uses. Electric locks of both types are now used extensively in combination with traditional mechanical locks and on their own. They are commonly found as the door actuating mechanism in access control systems. Units can be of any security level required, stand-alone battery powered or mains wired, AC or DC operation, and fail-safe or fail-secure. They are used on every type of door and opening barrier. The reasons for their wide acceptance and use are that they

  reduce key use, copying control, etc.
  remove the need for extensive suiting of locks
  provide remote, automatic control
  can be reprogrammed (via controllers) to perform different functions without lock removal
  allow timed opening, multifunctioning, and interfacing with safety system
  allow added features to the normal lock
  are as reliable as mechanical locks

reduce patrol and manpower requirements by allowing central controlled closing, opening, etc.
give *positive* door status and lock status indications

Application. Units should always comply with UL 1034. *They should not be used on fire escape or safety doors,* as they can (and do) stick if not properly maintained.

All solenoid-operated bolts should be continuous duty rated (CDR) if fail-safe. All fail-safe locks should be DC powered, where possible. The CDR rating is necessary, because by being continuously energized against a spring, a high current draw and heat result within the unit. If not CDR, the unit would suffer premature failure due to the effect of heat buildup on windings and insulation. Heat also affects the magnetic holding force and slows down the action of the lock.

It is equally important that fail-secure locks be intermittent duty rated (IDR). With these locks, power is required to pull the bolt out of the lock or to open the striking plate from its normally closed, secure position. These long idle periods make it essential that when required, sufficient magnetism is generated quickly, drawing large "inrush" current. If this current is too great, fuses blow. If it is too small, the lock will not open or will open sluggishly.

Equipment Types and Selection. Units are either AC or DC and normally 12V or 24V. Units that are 24V DC should be used whenever possible as a purer magnetic field is created, operation is silent, electronic rectifiers are not needed, and interfacing with many other systems is simplified.

Unlike magnetic locks, electric locks may operate in either fail-safe (fail to allow access) or fail-secure (fail to restrict egress or entry) mode.

Advantages and Disadvantages.

Advantages. Electric locks have most of the same advantages of magnetic locks, with the added functional flexibility of fail-secure.

Disadvantages.

parts are more open to abuse and tampering than traditional mechanical locks

provide low security when used on their own

a good, strong door closer, which must be two speed, is essential to prevent the mortise from bouncing back off the waiting, open, staple bolt

very good alignment of lock and strike are required, along with good door hinges, to prevent excessive wear on both elements; particularly on aluminum doors

the resistance to ramming of electric strikes is not very high and mortise bolts are preferred in this situation

both the solenoid and the strike locks are more prone to mechanical failure than magnetic locks

Electromechanical Locks

Principles of Operation. Electromechanical locks combine the functions and operations of electric and traditional mechanical locks. Essentially, they are traditional locks (mortise, rim, or latch), but a magnetic solenoid is introduced within the lock case. This solenoid moves a rod mechanism that controls either the door handle (secure or insecure side), the mortise bolt, or the lock tumblers or levers.

The most common form of electromechanical lock is the one in which the solenoid bolt disables the doorknob in response to a signal and timed program inside the key. This disablement prevents the knob, which would normally retract the mortise or rim bolt, from doing so. Most electromechanical locks retain normal, mechanical, key features.

Uses. Electromechanical locks are widely used on hotel, motel, and commercial buildings to allow or prevent access to corridors, particularly those off common lobbies. Rim types are generally low or medium security. Mortise types are far more secure and certain specialized types exist for high-security applications.

Application. As the lock and electromagnets are on the *door,* and not the frame, it is essential that totally concealed wiring and secure, wired-through hinges are used.

Equipment Types and Selection. Mortise types are more secure, but draw more current than rim or latch types. Doors should always include monitoring of lock position and door position. These locks should always be UL, ASA, and ANSI approved.

Advantages and Disadvantages.

Advantages.

many functions (versatile)

lock *looks* traditional and hence does not attract attention

Disadvantages.

low security

monitoring of lock position is nonstandard

easily defeated

higher maintenance, as the lock has both mechanical and electrical parts

Interlocking, Sequencing, and Supervising

Many institutions require security *interlocking* or *air locking* to prevent final entry into an area until the person requiring entry is identified and is secured in the space between two locked doors. Only one door may be open at any time. With this system, warning of a deliberately propped-open door or taped-open electric staple release are needed. Doors may be time controlled, restricted open, etc. Companies specializing in lock and door control panels for these applications are recommended. Folger Adams Ltd.'s 3100 series will control up to 12 doors. Adams Rite Ltd. produces powered, open-and-shut mortise deadlock (swing bolt) actuators, which are particularly suited for aluminum-framed and glazed doors. This actuator incorporates many typical features:

- 24V AC or 24V DC operation
- momentary or remote, toggle switch activation
- local (secure side), press-button exit override
- remote door open/closed lock and open/closed indication

For buildings where the lobby space between the two doors is too great to be accommodated, revolving doors with automatic stopping at an intermediate secure position are particularly useful. The person enters the revolving door and pushes it. It moves to a position where neither entry into the building nor escape from the door is possible and stops in a locked state. The door can only be moved forward to permit entry by an authorized person entering a code or magnetic code card into a reader that signals the door lock to release to the next position. In practice, these systems tend to be more secure than air lock interlocked traditional door systems.

Equipment Locks and Secure Enclosures

Equipment Locks.   The result of most crime is the loss of high-value, transportable items for which there is a ready black market. The value of items of this type, found in schools, hospitals, and offices, can amount to £5000–£6000 (approximately $8000–$10,000) for a single item. Insurance will usually cover the primary loss, but will not cover the consequential loss of, for instance, a word processor or computer and the disk with all the confidential and administrative files. Many thefts of this type occur during the day, when a busy office is full of strangers, or during lunchtime, when many offices are left unlocked. Equipment locks are cheap, easy to use, and have a great deterrent effect against this type of crime. Equipment locks should be specified for

- VDUs
- computers
- word processors
- televisions
- cabinets
- disk or tape storing cabinets
- all filing cabinets, desks, and drawing case drawers

The various types commonly available are

bolt locks. These hold the normally transportable equipment to a wall, floor, desk, etc., using a specially designed bolt with a bolt head that needs a special friction tool to undo the bolt. Their disadvantages are that they must be installed by a skilled installer, they mar the surface to which they are attached, and they are only as strong as the surface to which they are secured. The most common method of defeat is the use of a small crowbar!

adhesive locks (or maze locks). These consist of an adhesive base plate containing a series of metal slots that form a maze. Pins are fixed to the base of the equipment to be secured and these interlock with the maze. The base plate is glued to the desk by a self-adhesive compound. The disadvantages of this device are lack of real strength, familiarity on the part of the thief with the sequence of movements required to remove the equipment from the maze, and permanent spoiling of the surface of the desk, wall, etc.

tag locks or ring seals. These come in many shapes and forms (e.g., plastic, steel wires, steel strips, etc.). They are usually one-use-only devices and offer only a token resistance, but can still deter. They are usually tagged with a special code or the name of an authorized user. If the seal is broken, the user is contacted to check whether she has removed the device for some reason. They are extremely useful for classified or restricted access to filing cabinets, files, disks, etc., as they can be threaded through drawers or door

handles. They can also be used in conjunction with normal locks to identify entry by an unauthorized person in possession of a key.

Secure Enclosures.   Secure enclosures include filing cabinets (fixed or loose), cupboards, closets, storage racks, drawing chests, etc. The range of secure enclosures is enormous, but their usefulness and value are extremely difficult for the layperson to determine. The "key" is to ask the manufacturer whether the item has been purchased and approved by government establishments. If the answer is "yes," determine its use. Selection of equipment that meets this base standard should at least be a good value for the money!

Padlocks, Hasps, and Staples

Padlocks and hasps are still among the most used security devices (see Figure 6–25). They can provide efficient and good security only if carefully selected and properly applied. With padlocks and hasps, it is essential that the material of the door be strong, sound, and resistant to drilling, cutting, etc. This usually means a minimum standard of a very thick, softwood door of solid (not core and panel) construction. Security is increased if the door is solid hardwood or steel/hardwood/steel, or, better still, solid steel. Fixings must always be hidden by the hasp and should be through-the-door bolts of hardened steel. Fixings should pass through a load-bearing, steel plate on the back of the door. Lock nuts should always be fitted.

Figure 6–25   Normal shackle and high-security, close shackle padlocks

Padlocks should

be heavy duty; minimum weight 1.6 lbs (750 g) for a
2.2 in (55 mm) body width

be hardened metal outer case

be laminated, hardened interior

be close shackled; the hasp should have a short, covered
staple

have a minimum of six levers and 5000 key differs for
low-security padlocks. Higher-security padlocks
should have ten-lever locks (minimum) with more
than 5000 key differs.

be provided with a keyway cover to prevent moisture
from entering the lock. A moisture drainage hole
should *always* be provided.

be comprised of nonferrous surface material to prevent corrosion or should be specially treated (e.g.,
stainless steel or solid bronze)

not be subjected to rivets during construction

Hasps and staples should be securely fixed with bolts
fitting flush in countersunk holes in the bar. Bolt heads
should be domed. Hasps and staples should have hinges
(knuckle joints) with strenthened hinge pins that are
fully encased and of close tolerance. The pins should be
made of hardened steel; the hasp should be of hardened,
surface-mallaeable iron to prevent sawing.

Locks should be key retaining and the disks should not
be spring-operating at low temperatures. Their cylinder
cores should be changeable.

Good devices have very short, hardened staples that
are completely enveloped by the pad bar and/or padlock.
No rivets should be used in their construction.

DOOR AND WINDOW FURNITURE

More than 85% of burglaries are committed by unskilled
opportunist thieves. The correct door and window furniture fitted to buildings will dramatically reduce the risk
from this type of theft and will also increase personal
safety and security. The following *minimum* standards
should apply to all properties, including domestic housing (see UL Standards and NIJ-STD 0306/0318).

Front Entrances

Doors should be of solid wood, with a 1.8–2.0 in (45–
50 mm) minimum thickness.

Three equally spaced hinges are required, fixed with a
minimum of nine 2 in (5 cm) no. 8 brass wood screws.
Hinge materials are of steel, brass, or bronze, with a
closed pin.

Glazing is restricted to the top quarter of the door and
is limited in size to 77.5 in^2 (500 cm^2), with one
dimension not exceeding 2.4 in (60 mm).

A mortise lever deadlock must always be fitted, never a
rim lock on its own. The lock should be a minimum
standard five-lever lock with more than 1000 key
differs.

Bolts should be provided at the top and bottom corners.

A sealed, sheet steel letter catcher should be provided
behind letter boxes. Letter boxes should also have
strong, flap springs.

Door chains or cross-door wire rope devices should
always be fitted for the passing of credentials without full opening of the door.

Door viewers of a minimum 180° angle of vision
should be fitted.

A porch light, controlled from within the building,
should be provided.

Frames should always include a lock bolt striking plate
and metal box staple.

On outward-opening doors or for increased security,
dog bolts should always be fitted.

All locks should include a keyhole cover.

The following should not be used:

rim locks, double throw mechanism deadlocks, locks
with internal or external knob sets

single, glazed, paneled, patio or louver doors

pivots, rising hinges, straps, screwed hinges, plastic or
aluminum hinges, or hinges with open pins

Double Door Entrances

High-quality, clawed bolt, mortise deadlocks should be
used in conjunction with bolts or bars.

Rear or Side Doors

The first four and last three items of "Front Entrances"
apply as minimum standards. There is a great temptation
to use doors with a small glazed panel at the rear. This
should be avoided, as rear doors are often visible from
the road or adjoining properties and glass panels are
easily attacked. If glazing is required for aesthetic
reasons, it should be fitted in a large, embedded (not
externally beaded) panel and should be laminated. To
increase security to all doors, ensure that

Phillips screws are used and clouted to prevent removal
if necessary

the lock's bolt staple does not occupy more than one-
third of the door frame width

two locks are fitted, at least one of which is the mortise
lever deadlock, at one-third door height intervals.
The uppermost lock should be an automatic deadlock latch if possible.

staples are surrounded by sheet steel reinforcing plates

metal grills are fitted over glazed sections, glazed sections are double glazed with polycarbonate, and glazed sections are blocked up or fitted with laminated glass

Windows

All openable windows should be fitted with one of the many patent dog bolt or locking/catch mechanisms currently on the market. All these devices are intended to challenge the determined burglar and to deter the opportunist. The opportunist would only consider forcing a loose catch or casement, or, at best, smashing a pane to get to the window catch. Dog bolts, mortise bolts, and similar devices deter him from doing this by presenting a second line of defense. The professional burglar will be equipped to deal with both the glass and the locking device. If, however, the device is of a screw bolt or particularly unique key arrangement that cannot be opened with the blade of a screwdriver or knife, the burglar will be slowed, increasing the likelihood of discovery. Locking devices should be fitted to all windows and skylights that are openable.

Where additional security is required, but laminated glass is too expensive, decorative security grills or bars should be fitted to the inside of the ground-floor openable windows. Grills should be narrow edged and close grid [4 × 4 in (100 × 100 mm)] maximum. They should be hinged and padlocked.

The most useful type of window lock is the type that will allow ventilation during the day, without lock removal. This will prevent the occurrence of daylight robbery (literally!) while you are in another part of the house.

KEY CONTROL AND DOOR MANAGEMENT

If you do not consider long-term key control and door management, all that you have read in this chapter will be useless. The person who has the key, or access to it, is the master of the lock.

Locksmiths will advise you on detailed plans, but, in general,

issue as few keys as possible

issue keys *only* to permanent staff

keep a list of numbered keys and their respective key holders

carry out a simple, regular, effective, locking routine every day, with periodic senior management checks

do not use key hooks, key boards, etc.; keep keys only in a key safe

never let the staff loan keys to others; make the loaning of keys a dismissive offense

Remember that keys should never be left in locks. As a final plea, remember that good locks and good access control systems can reduce security manpower, but a good, reliable guard, employed to monitor an entrance, should always be considered seriously.

7
Access Control Systems

INTRODUCTION

Access control systems (ACSs) manage entrances and exits. They *improve* security by restricting entry to only those who can prove their right to enter, and by monitoring and recording the actions of those who exercise that right. Access control systems rely entirely on the goodwill of the users of the systems for continued security. If the users do not trust the system, abuse it, or find it difficult to use, it will fail.

The choice of a suitable access system for a particular situation is not easy. Every application is different and the systems must be matched to that particular application. When selecting systems, it is much more important to establish how satisfactorily they would work on your particular building, than to establish that one system has more functions than another. The choice between systems will depend primarily on

the building, its use, and the security level required. A research laboratory will differ greatly in its requirements from a dockyard entrance.

the number of entrances and exits to be controlled

the maximum permissible processing time per person and the number of users

the type of user

whether the system is external and exposed to the general public, or internal and a second line of defense

The basic systems used are

high security—limited entrances and exits, limited numbers of users who have already been screened in some way (e.g., research area, military buildings, etc.)

medium security—multiple entrances and exits with large numbers of users (e.g., single-tenant office buildings, factories, etc.)

low-security systems for the prevention of vandalism and petty theft of residences

low-security systems for the prevention of theft from hotels and multiple-tenanted offices, buildings, etc.

ACSs enhance privacy. They do not *prevent* determined people from entering a building, as this is the role of physical security devices, such as locks, barriers, etc.

PRINCIPLES OF OPERATION AND SYSTEM PARTS

Access control systems are systems that *manage* a number of entry and exit points from a building. They improve security, particularly during daylight hours, by dictating who enters, where, how often, and to what areas. Access systems will not prevent the "inside job" and are easily defeated by collusion.

The fundamental differences between ACSs and normal locks are that

ACSs can identify the "key" from many thousands of other keys, sometimes allowing entry and sometimes refusing entry, without alterations to the locking device

ACSs can record where the entry was made and by whom

ACSs allow very easy time programming, which offers features of the most sophisticated time delay lock system

ACSs monitor the status of the door to indicate whether it is open, shut, locked, unlocked, etc.

ACSs are readily linked to other systems to give backup protection, particularly at times of crisis (e.g., a CCTV may be installed to identify users before permitting final entry)

ACSs incorporate some of the following component parts:

- the key device (card, handprint, etc.)
- the key or card reader
- the processor or programmer
- the command center
- the locking mechanism
- the closing mechanism and barrier
- the signal wiring

The various categories of the system all act under the same principle—that recognition of a binary code, generated electronically, activates a checking procedure within the system. If, after checking, the code is verified, a second signal activates a locking device, allowing entry.

The choice of a particular access system is a matter of trying to match the product to the environment in which it will operate, the level of security required, and the needs of the users. This is, of course, true for any product, but for ACSs it is perhaps even more important than usual that each of these areas be considered in depth. Many ACSs offering splendid capabilities and of great potential have failed in the field due to relatively minor items that were overlooked.

Most ACSs fall into one of the following categories (although the "edges" of each category are sometimes blurred by the manufacturer):

- alphanumeric push-button system
- alphanumeric push-button plus speech and/or visual identification system
- card entry or token entry system
- card entry plus personal identification number (PIN) system
- personal feature identification system

The Key Device

All key devices try to be unique by replacing the conventional metal key. To do this, a variety of key devices have been invented, some of which are described below. Key devices are easily divided into three broad categories.

Cards That Allow Any Person in Possession of the Card to Enter.   Included in this class are

> *magnetic strip cards*—plastic, laminated cards (like credit cards) that have a magnetic strip along one edge onto which a code is printed. The code contains a personal number to identify the user and record her use of the system, and a lock-activating code. The card is inserted into a reader slot. Of all the cards listed here, the magnetic strip card is the most commonly used, but is by no means necessarily the best choice in many applications.
>
> *passive radio frequency (RF) cards*—plastic laminated cards into which a series of small, metallic components have been embedded. These components form what is known as a *tuned circuit*, which will respond in a particular manner when absorbing energy that is generated and transmitted at a radio frequency. If the reading device (which transmits the radio frequency energy) receives a reflected signal from the card, which it recognizes, it will allow entry. These devices usually work on "reflected" RF frequency recognition. No keyhole is required. These cards are

also often called passive proximity cards as they will work at distances up to 3 ft (1 m) from the reader. The readers/cards should be kept clear of metal objects or the signal can be corrupted.

> *proximity cards*—a magnetic strip card that does not require a magnetic card slot. The card has its own magnetic field that initiates an electrical signal within the card reader when the card is passed near the reader.
>
> *active proximity cards*—these cards are similar to passive RF and magnetic strip cards in that no card slot is required. The card is simply passed near the card reader. The primary difference with the active proximity card is that it has within it a very small lithium battery and microchip-tuned circuit. This allows it to send a signal to the reader. Types include those that constantly transmit (short battery life), those that transmit only when switched on near the reader by the user, and those whose readers switch the card on and await a return signal (these have battery lives of 2 to 5 years) (Figure 7–1).
>
> *radio transmission tokens*—small, rechargeable radio transmitters. Each transmitter is recharged daily by the user. The system will allow access to any person carrying the transmitter when she comes within the vicinity of a radio receiver mounted near the door concerned. The control unit identifies the level of security clearance allowed to the token holder.
>
> *induction cards*—cards that contain an electrical circuit with coded information. As the card is pushed into the key reader on the door, it cuts a magnetic field in a particular manner and an electrical current is induced onto the card. This current triggers a device within the card that then transmits the encoded information to the key reader for it to decipher and verify or reject.

Older systems, still commonly encountered, have cards that include electric logic circuits, punched-hole

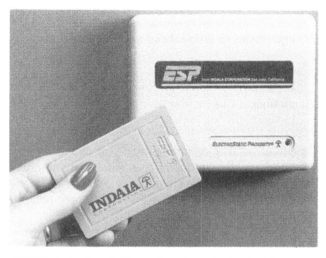

FIGURE 7–1   Proximity card reader (electrostatic)

pattern cards, fibrous patterns, etc. Although most of these proved easy to forge and have thus fallen from grace, a new form of punched-hole card is now emerging and its advantages are the cheapness of card manufacture and ease of recording.

Personal Feature Identification (PFI) or Biometric Systems.

Unlike card systems, PFI devices will only allow entry to the *person* actually identified by the key reader/processor. These devices usually involve the use of an alphanumeric PIN code plus handprint, fingerprint, voice identification, etc.

Although not as commonly used as cards, PFI devices are beginning to make an impact on the market. To date, the main drawback with all these devices has been the high levels of processing power required and their low processing speed. As with card entry systems, there are few unclassified standards and recommendations to give guidance on their use. Until quite recently, PFI was looked upon as purely a high-security device because of the uniqueness of the key, the cost:risk ratio, and the limit on the number of users imposed by slow processing. The era of cheap microprocessor power is rapidly removing these drawbacks and the potential benefits of nontransferable security key systems that uniquely identify the person are enormous.

In the recent past, the largest causes of complaint by PFI system users were they were slow to use (long comparison and retrieval times) and they rejected an authorized person too frequently or allowed incorrect admittance. Most systems are now nearly as fast as conventional *high*-security card systems, but are still not as fast with rush-hour scenarios at particular doors and readers.

The second cause for complaint was mitigated successfully by dividing the problem into two types of rejection:

Type I errors—rejection of a legitimate user who has full clearance
Type II errors—admissions of the wrong person who has no clearance.

Tolerance of normal variations in PFI characteristics for an individual (such as minor cuts on fingers on fingerprint recognition systems by cross-referencing to other fingerprints, etc.) has reduced Type I errors to 0.1% (i.e., 1 in 1000 or once or twice per year per person on *well-used systems*). Type II errors have been reduced as more storage capacity and processing power have become available, making more detailed PFI possible and "forgery" harder. Errors are now thought to be less than 0.01%, or 1 in 10,000, for each individual *on well-used systems*.

The main areas of PFI are

a fingerprinting mechanical reader—these devices store the pattern created by a fingerprint as a dot matrix or code. When an authorized person places her finger on the reading device, an electronic search is made for a match. The system refuses entry (in error) to 0.1–0.005% of the authorized users. A single finger is normally chosen for enrollment into the system; this stored record takes approximately 2 minutes to enter the print. Comparison of the presented finger with the stored pattern and fingerprint verification takes 2–5 seconds. Fingerprint systems are currently the most developed, most widely used PFI systems in the world.

hand profile or hand geometry—just as fingerprints are unique, so is the *geometry* or *shape* of the hand and/or fingers. The user places his hand on a flat pad, which electrically and mechanically maps the shape and compares it with a previously stored record. Increased security is added by requiring the user to enter a personal identity code, which must also be used to match and verify the record. The code, and sometimes the entire map of the hand, can be stored on a card for direct reading without the use of the hand pad device. Three-dimensional (3-D) geometry systems are better than 2-D geometry systems.

signature verification—this system is not used any longer other than in its very high security, dynamic speed and pressure format due to unreliability, forgery, etc. Even bank checks are not accepted based on signature only! Dynamic speed system prices are reducing rapidly. These dynamic systems take approximately 2 minutes to enroll into and 2–10 seconds to verify. They are probably the third most used system after the fingerprint and hand profile systems.

speech verification—the authorized users of the speech verification system record a series of short sentences or words on their own without other users being present. These sentences are then valid "pass" sentences for a set period of time, after which they must be renewed. The processor stores the pattern and modulation of speech for each sentence. The user is then given a unique alphanumeric code. The code is used to activate a processor that will then expect that person's pass phrase to be spoken into a recorder. The recording is compared with the prerecorded phrase. The error rate is 0.5–0.01%.

eye retina pattern system—this very high security system (see Figure 7-2) is similar to the fingerprint system and is becoming the PFI market leader. To use the system, the eye is presented to a viewing slot. A person focuses her eye on a target. An infrared light source tracks across the eye, mapping and recording the vascular pattern at the *rear of the retina*. Enrollment takes 3–4 minutes. Scanning for each entry takes 1 minute and verification usually takes 5 seconds.

The merits of all of these different systems are discussed later in this chapter.

FIGURE 7–2   Retina pattern reader

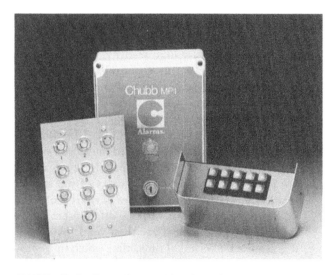

FIGURE 7–3   Numeric combination electric lock access system

Alphanumeric Codes. Alphanumeric codes are basically PIN codes and must be punched onto a touch-sensitive alphanumeric keypad. The pad is a form of combination lock with centralized control and memory capabilities to reject a code or attribute levels of security clearance to any person at any particular place (see Figure 7–3).

PIN codes are used to enhance the security of other cards, magnetic strips, etc., and are similar to those used in the cash card machines of all the major banks.

Key or Card Readers

Key or card readers are obviously matched to the keys used.

The Processor

The processor is a computer with prewritten software for controlling the system stored. The computer will carry out code comparisons; generate rejection or acceptance codes; store information about the user, the number of entries or exits made, and the time and date of use; program no-go times; etc. In addition, the processor operator can program the system to reschedule security clearances and invalidate ("void") lost cards or cards of recently dismissed personnel (see Figure 7–4). Many modern systems delegate all access decision making to the card reader. This action is called *distributed intelligence* and in this form, the processor only logs and monitors data.

The Command Center

The command center is the name usually given to the processor, keyboard, display screen, and printer.

Signal Wiring

Signal wiring varies from system to system. On simple magnetic card entry systems the wiring might be normal, single-core, PVC-insulated cables. On sophisticated systems, coaxial or data transmission cables may be used.

Locking Systems

Despite the electronic nature of security access systems, the most critical area of the system is still the item that actually prevents entry—the lock. The most common locking devices are

- mortise deadlocks that are key *or* access card controlled (common in communal dwellings). The lock bolt rests in an electronically released strike plate or staple. Users enter with a key or card. Other users are permitted entry to a controlled space upon voice identification or visual identification, using CCTV, by activation of a lock release button within the secure space.
- rim locks (not recommended)
- card entry, manual exit only. Locks are the mortise deadlocks, but without the remote release function. Hand bolts are often favored.
- espagnolette bolt and solenoid. The key device deenergizes a solenoid within the door, which releases vertical and/or horizontal bolts.
- on commercial systems, a second lock (mortise deadlock) at one-third door height to provide an emer-

FIGURE 7–4 Decision-making card processor

gency and off-hours barrier to potential intruders or people who have stolen cards.

electromagnetic locks that hold the door secure by the power of magnetism alone

Remember that

the lock and striking plate must always be fully compatible, with full mortise penetration

emergency exit facilities *must* be discussed with fire officers. Paddle handles and break glass, thumb-turn devices are the most common.

electric strike plates should always be heavy duty with continuous operation–rated coils

strike plates should always have a slight play (forward and backward) with the mortise held in the striker or an overload of the coils could result due to lock friction

Closing Mechanisms

Closing mechanisms should be carefully considered. The system will only be as secure as the door. Unless an effective, but operative, door-closing device is used (that will not require Mr. Universe to operate it), the system will either not function at all (because the door will not close automatically behind the authorized user) or it will be abused by the door being propped open. The application determines the most appropriate door-closing device. The following mechanisms should be considered:

normally hinged door with separate mechanical closing device (lowest level of provision)

internally spring-loaded or flush, floor-mounted door-closing device (better)

automatic, electronically operated sliding doors, with "magic eye" closing sensor

Remember that

all hinged, hydraulic door-closing devices should be three-stage and super-smooth

a doorstop or threshold should be provided to prevent doors from *closing* beyond 90°

door-closing devices should have no dead points at either end of their arc

Barriers

single-leaf door—the most common, but often least secure, access-controllable barrier. It is commonly a hollow core construction and hence is totally unsuitable. The *lowest*-level security door for access control would be a one-quarter-glazed (laminated glass), solid wood door that is connected to an alarm system. Solid metal or sheet steel–faced doors are preferable.

air lock—two doors in close proximity open to view from the secure area. The second door can only be opened after the first has been locked behind the person using the system (either by that person or by the people within the secure area). Sliding doors are normally used to save space in the lobby area. This is a very secure system.

turnstile—only really effective if full height, providing a revolving door–type barrier. If of open construction, the turnstile will prevent access but not arson, vandalism, etc. Full-height turnstiles are one of the few ways of preventing two people from entering with one card.

double doors — a security problem, as one half usually has to be fixed and the other half left to swing. The fixed door must be bolted, but must open in the event of fire. This can be achieved by the use of Redlam bolts, but is not recommended. Access-controlled entrances should be separate from final fire exits for easy operation of both sets of doors.

USES AND APPLICATION

The type of access control system chosen depends on its use. Too many manufacturers claim their systems can be used for all situations, with only minor modifications to system parts. Invariably, this is not the case. A telephone and CCTV communal access system is totally different from a commercial computerized system, not only in parts, sophistication, etc., but in basic concept. "Model" units should be chosen with care. If the system is of even moderate value, careful scrutiny of the equipment, demonstrations in its use, and testimonials are essential. None of the devices on the market have existed long enough to become the "Rolls Royce" of access control systems and those that start to approach it are soon overtaken by other trends. The following sections give specific guidance on the types of system commonly found in use for particular applications.

RESIDENTIAL SYSTEMS

System Types

- telephone entry system — conventional key
- telephone entry plus CCTV — unconventional key
- card access versions of the preceding system types

Application

- high-rise or communal dwellings
- elderly persons' housing
- sheltered housing schemes
- university residences
- nursing homes

Objectives

to reduce vandalism to communal areas
to reduce petty or opportunist theft
to reduce nuisance (e.g., children, gate-crashers, dogs)
to improve the sense of security to occupants
to provide a simple, but effective, low-security and privacy-enhancing system

Economic Justification

In addition to improved security, access systems provide other benefits. Refurbishment of a single elevator and redecoration of a communal entrance area damaged by vandalism, can cost more than the entire system's capital cost. The hidden benefits are enormous — tenants regain pride in their dwellings, begin to feel in control of their surroundings, carry out their own repairs, clean communal areas, etc.

System Planning

The eventual success of the system rests with the users. The designer must involve all tenant organizations, maintenance bodies, and other client departments in the design process. The *users,* not the "authority," must be reassured and asked for their opinions of the systems proposed. Their criticisms should be judged fairly. Without this interaction, very embarrassing mistakes can be made. The designer might not know that some tenants are blind, use wheelchairs, cannot operate spring doors, etc. The opinions and advice of the users should be gathered, whenever possible, in confidence during a one-to-one conversation. For refurbishment work, this should take the form of visits to apartments during the evenings or a small exhibition of the proposals. On new works, similar groups of people using existing systems can be asked for their opinions or simply observed in order to spot potential problems.

"Know your customer" is the maxim. Failure to do this has resulted in installing push-button panels that are too high for the wheelchair user, numbering too small for the elderly to see, codes too complex for children, doors with closing devices that are too stiff to operate, and "open door alarms" that respond too quickly!

Once installed, the users must be given instructions on how to operate the system properly and what they must not do!

System Management

Caretakers. The reason why many communal access systems are installed is due to the demise of the responsible, accountable caretaker. Even the most dedicated caretaker cannot be ever-vigilant, due to shared duties, illness, etc. When a system is to be installed in a building that is still fortunate enough to employ a caretaker, the following should apply:

No caretaker's entry or master button should be made available.

A separate assistance telephone should be provided within the secure area and kept within a locked, steel cabinet accessible by a key given to occupants only. This telephone should be used in case of system failure, alarm malfunction, etc.

The caretaker should be given a complete instruction manual that clearly points out how the system should be operated; the potential abuses (e.g., door propping, key passing, etc.), the emergency maintenance telephone number, the cleaning solvents to use and not to use, etc.

The system should always increase security and assist the caretaker in her duties; it should never be an imposition.

Tenants.

Tenants should be instructed in the same manner as the caretaker (if one exists).

Whenever possible, only adult tenants should be given keys to the access control system. Distribution of keys to children increases the likelihood of loss or misuse. Many systems fail because "other people's children" gain access to the communal area by learning the code or by obtaining a key. They only have to create a nuisance once or twice before confidence in the system is threatened. It is essential that this type of pitfall be explained to the tenants and that tenants are made to realize that the eventual success of the system rests with them.

Deliveries.
Free access to nontenants should never be permitted, either by key or tradesman's pass code. To gain access for delivery, the tradesman, mailman, milkman, etc., should always be required to ring the caretaker or a tenant who can identify him on sight. The following measures should be taken:

postal deliveries—a mailbox should be provided at the entrance so that the mailman does not need to enter the building. This can take the form of a single, secure, lockable cabinet that has an external cover and lock-and-key lid to discourage vandalism. Mail could be distributed by the caretaker, responsible tenants, etc., or collected individually. Individual pigeonhole boxes, behind the lid, with rear key access are also useful.

fire brigade—special provisions are not normally required. In case of fire, tenants usually will leave the block and assemble at a predetermined location, thus allowing the fire brigade to enter. If necessary, as with normal residences, the brigade will break down the entry door. The only proviso to this is that where flush-fitting, solid steel doors are used, as is sometimes the case in Europe, the fire brigade should be notified. The access system, lock type, door construction, etc., should all be shown in the

drawings submitted for the fire officer's or building code approval.

ambulance drivers and police—these individuals should be treated the same as the fire brigade and no special provisions should be made

refuse collection—no special provision made. Refuse should be collected from external, locked compounds.

Nuisance Calls.
It should be pointed out to users that nuisance calls cannot be totally eliminated and they are far less stressful than damage caused by vandalism, mugging within the premises, loss of use of elevators, etc. The following general measures can be adopted:

volume reduction on apartments' "call" bell
disable apartments' call bell (timed or manual)
voice cutoff facility for the tenant
visual verification before acceptance of message (prevents verbal abuse)
deterrent indicator lamp on call panels (e.g., "police called")

Maintenance and Responsibility.
The system installer should be made responsible for the entire system. Whenever possible, the scope of the installation should include all doors, hinges, locks, mailboxes, door-closing devices, etc., as well as the actual access control system. In this way, all these items can be covered by the maintenance agreement and failure of any part of the system cannot be blamed on a third party. The agreement should be for a period of at least 2 years and should include

all parts and labor
unlimited repair calls
no hidden costs, such as recommissioning
additional free instructional literature (in quantities at least double the number of tenants)
monthly checks, minor maintenance every 3 months, and annual overhaul on all parts; the opportunity for the client to define those parts that need replacing in addition to those listed in the installer's maintenance agreement (e.g., locks, handles, doors)
repair call response to any tenant or other responsible person listed

The maintenance agreement should be included in the bidder's final price and should be itemized to enable the specifier to ensure that these items are included in the budget. Maintenance agreements should always be negotiated prior to contract award. Minutes of the negotiations and the final agreement should be signed and kept on file. The installer's maintenance engineer should live within 90 miles of the protected premises to give good service. System parts frequently misbehave!

Manufacture and Installation

Whenever possible, the specified company should manufacture, install, test, commission, and maintain the entire system.

In practice it is often found that companies buy items from various manufacturers (often foreign) to construct a system (e.g., a German keypad, an Italian camera, a Japanese telephone). When this occurs, the standard of equipment should be checked for compliance with safety standards and information on approvals from local authorities, licensing bodies, etc., should be sought.

When items of foreign manufacture are installed, the company should be asked to confirm, in its proposals, that it will always keep at least two of these items in stock so that they are available for immediate use. Too often, delivery delay is used as an excuse for poor maintenance response.

All wiring for the systems should be protected in conduits, ducts, etc. Conduits and cable trunking should be recessed, whenever possible. When surface conduits or trunking are used, these should be steel with nonremovable covers. PVC trunking run on the surfaces during refurbishment work should be resisted; it is too easily damaged.

Entrance Design

In new buildings, careful thought should be given to the location and orientation of the entrance. In existing buildings, this is not always possible, although it is often possible and beneficial to change the main entrance position. When entrances are in less than ideal positions, additional thought should be given to considering the subjects discussed in this section. In general,

entrances should be planned in detail, with the architect and responsible fire officer, with direct reference to access control

entrances should be planned only after consultation with eventual users, tenants, etc. A visit to the site is essential to determine and plan routes and movements. Look for routes from parking lots, children's shortcuts, shops, etc.

the entrance should be flush with the external wall, with no reveals or side walls; canopied, but only to give temporary shelter; visible to passing police if possible, visible to other dwellings if not

in existing buildings, all escape routes should be routed to the final exit, if possible. All other entrances should be blocked up and made to match the internal and external appearance. Old, locked doors cause panic during a fire and encourage vandalism.

the external route to the entrance should be clear, with hard surfaces and no steps. The route should not contain large bushes, shrubs, trees, etc., that could cast shadows or conceal pranksters or vandals. This is particularly important with CCTV systems.

the route should be well marked and lit with vandal-resistant lighting structures

the building name should be cast in concrete, brick, or a built-in stainless steel engraved panel

the access control system call panel should be lit by a built-in light mounted on the wall directly above the unit. The light source should have an actual lamp life of at least 4000 hours and should be time/switch controlled and separately supplied. The light should be vandal-resistant and fully recessed with a polycarbonate diffuser.

immediately outside the entrance, there should be no feature that would attract local youths (e.g., shelters, benches, etc.). No garages, stores, etc., should be placed within 60–100 ft (18–30 m) of the entrance, as this encourages propping open doors instead of using keys and, even when a "door open" alarm is fitted, some individuals will always try to overcome the time delay.

external lobbies should not be provided

if two separate entrance doors are required, one should be designated as the final exit for escape and routes should lead to it. External arrangements for the second entrance should always be at least as comprehensive as for the main entrance. Second entrances and exits are invariably back entrances and hence should be designed with even greater thought to potential vandalism.

Entrance Foyers

Total security is impossible and there will always be some occasions when undesirable persons gain entry. With this in mind, the foyer should be designed to the following guidelines:

It should contain no recesses, porches, etc.

It should contain no doors to communal stores.

The doors to the power supply rooms should not be labeled and all doors should be substantial and provided with good locks (mortise and deadlocks).

The foyer should be either totally visible from the outside to passing pedestrians and police patrols or obscured by means of tinted glass so as to not invite entry. The foyer should be well lit. Fittings should be positioned at high levels and recessed into ceilings or walls. Low-level lighting, upward wall-washers, pelmet-lighting, etc., should be avoided as they can be pulled off.

In new or refurbished entrances to apartments, ceiling

tiles, plasterboard, etc., should not be used, as they are too easily damaged. The ambience can be improved by factory-painted, sectional steel grids of various patterns that have lights fitted behind them.

Loose furniture should be avoided. Fixed furniture should be minimal (e.g., a chair by the elevator for the elderly or mothers with children). Furniture should be of good quality, appear clean, and be vandal-resistant. Planters should be kept to the perimeters and be fixed. Loose planters can rapidly disappear!

All control apparatus boxes, connections, etc., for the access system should be built into a concrete or brick wall, be within a cavity in a double-skin brick wall, or be in a concrete enclosure. Access to the panels, boxes, etc., should be from the rear and only by means of a securely locked steel door that is flush with the structure (to resist jimmying). Whenever possible this cupboard should be within a ground-floor caretaker's apartment. No parts, wiring conduits, etc., should be visible from the foyer and the door to the enclosure should not be marked or named.

Doors and Frames

Doors were described generally in an earlier section. However, in particular, one must consider that

doors should be single-leaf, outward-opening, with a no-key escape mechanism in case of fire

frames should be bolt- (not strap-) fixed, with a minimum of six bolts per joint

doors should be 32 in (81 cm) minimum opening (larger for wheelchair use). On single doors, leaf should not exceed 40 in (102 cm).

doors should not open to greater than 93°

at least four hinges should be fitted per door

doorstops should be fitted

no hold-open device should be fitted (e.g., hook and eye or "dead spot" on the closing device)

door handles will be required on the outside; these should be push/pull plates, not loop or "D" types. Handles should be bolted and welded with no internal nuts, etc.

door-closing devices should be concealed, single-action, pneumatic (three-stage), or strong spring type

wiring should be carried through the door frame and hinges (not external flexible loop)

with aluminum doors, ensure the door face is flush to the jamb and not set back from it

aluminum doors should have narrow stiles and special-shaped strike release plates

on double doors, sequential closing devices will be required

Door Glazing and Foyer Glazing.

a glaze-free zone of at least 2 ft (0.6 m) should be maintained around a lock, internal lock device, electrical connection point, etc.

panes should have one dimension under 6 in (15 cm)

glazing should be laminated if possible

no openable glass should be provided in the foyer or the door

no glazing should be allowed in the bottom 3 ft (0.9 m) of the door. If children use the premises regularly, small horizontal or vertical slits should be considered for accident prevention.

door glazing should be *built into* the door laminate and not beaded

foyer glazing should be internally beaded only, to a minimum depth of 1 in (2.5 cm) and fixed with countersunk screws with permanent capping in wood or metal

System Selection and Design Appraisal

Once the building entrance and its environment have been planned and initial consultations with clients, users, etc., have been carried out, the system selection should begin. The chosen system should be capable of meeting the exact needs of the building and its occupants under all situations. If a system that matches the needs cannot be found, there are two alternatives:

do not use an access system; resort to improved locks, a caretaker, etc.

modify the best system found to suit your specific needs

If the second course is taken, the modifications should be carefully thought through and the standard of the end product rigorously tested. If this is not going to happen, it is better to choose the first alternative.

Call Panels and Locks

Although the following guidance applies mainly to alphanumeric panels, the principal areas for concern and consideration are universal to all panel types. The call panel is the most important part of the system. It will receive the most use and abuse, be exposed to the harshest environment, and carry out the most important function. For these reasons, the selection of call panels should only be made when

the product can be examined in detail in its finished form at the offices of the potential purchaser

the product can be seen in operation under similar conditions

testimonials can be obtained from previous users

selected at random from a list provided by the manufacturers

The biggest danger facing a call panel is failure caused by vandalism. To reduce the risk of failure caused by vandalism, ensure that the panel is constructed and installed according to the following groups of guidelines.

When building entrances are within the boundaries of a secure, private development or private road with primary access restricted by foot or car (e.g., guard post, doorman, etc.) then less emphasis can be placed on vandalism and more on aesthetic appearance (see Figure 7-5).

Materials.

The whole of the call panel should be constructed of stainless steel of a substantial gauge that will not deform with hammer blows. The fascia should be of matte, plain finish, not stippled, chromed, polished, etc. This will make cleaning easy and allow for easy removal of paint, glue, etc. All material used on the fascia panel, including any call-button seals, vision panels, etc., should be able to withstand chemical solutions, flames from lighters, strong cleaning solutions, etc., without deteriorating (not plastic). All bolts, screws, hidden fixings, etc., should be

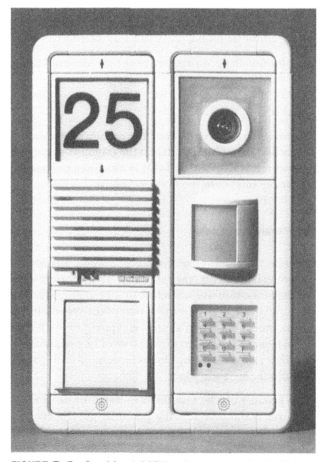

FIGURE 7–5 Combined CCTV and access control unit

of hardened, nonferrous metal (which will not corrode) and of sufficient strength to resist attacks by jimmying, etc. No dissimilar metals that will result in electrolytic action or corrosion should be used. Aluminum should not be used.

Design. The entire panel should be constructed in such a way that it withstands

- solvents
- Superglue
- hammer blows
- insertion of sharp objects
- water from any direction

The above criteria apply to both the secure (internal) and insecure (external) parts of the panel and its contents. Remember that attacks can come from within the building!

The entire panel should have an IP or NEMA classification suited for the environment in which it is to operate.

Call buttons should be solid, with no straight, exposed edges that could be wrenched (half spheres or similar are a good design). When pressed, it should not be possible to obtain any leverage between the button and the fascia panel. The button should be housed in an internal box construction. When buttons project from the panel, they should be shrouded to protect them from blows. The shrouding should have sloping sides to deflect blows from any angle and should be an integrally formed part of the fascia, not a screwed-in or brazed-on part.

Buttons should be indirectly connected to the electrical circuit they operate, so that blows, excessive pressure, etc., will not affect the internal circuitry or life of the switch contacts. Whenever possible, buttons should be returned to their original position by hydraulic (air) pressure, not a spring.

If touch-sensitive, electronic buttons are used, they must be able to withstand the same attacks that might be expected on mechanical buttons.

The external fascia should have no visible fixings; all fixings should be from the rear. The fascia should have cast-in threaded lugs that permit the fascia to be bolted back to the overall enclosure and the surrounding masonry. Fascias should never be surrounded by lightweight blocks, wood (hard or soft), metal trims, etc. These are easily attacked and, after failure, will expose the edges of the fascia. The fascia should be inserted into the structure from the inside (secure side) and should be larger than the external masonry or concrete structural opening by at least 1 in (2.5 cm). By adhering to this configuration, it is possible to create a recessed light pocket in the structure above the fascia (with a polycarbonate-faced opening) that can accommodate a night-light.

Any openings in the fascia for loudspeakers or microphones should be as small as possible. These openings should be of greater structural thickness than the fascia to prevent distortion caused by the use of bars, levers, etc. Perforated openings or slots with covering cowls are best.

Arrangement. The panel should be positioned close to the door and be at a height that is convenient for wheel-chair users and small children. Panels should have either an individual button for each apartment, or buttons numbered 0–9.

The electrical supply, 24V or 12V DC, should come from an isolating transformer that is remotely located. All wiring should be protected within conduits that are steel, seamless, and heavy duty. The remote source should include rechargeable batteries of sealed nickel cadmium, a trickle charger, etc., that are capable of sustaining the entire system for a minimum period of 48 hours. The electrical supply should be via a suitable fused connection unit (unswitched).

All electronics used to process calls should be of standardized modular design of an easily replaced, snap-in connection type to facilitate replacement of any failed circuitry.

It should not be possible to make contact with any electronic or electrical component, microphone port, loudspeaker, etc., through the fascia, including water being squirted through microphone apertures, steel wires threaded through loudspeaker grills, etc.

Facilities. The panel and its associated processing electronics (sometimes referred to as the *encoder*) should offer at least some of the following features:

buzzer or lamp to denote "door open" to blind and deaf persons, respectively

door-propped-open alarm with adjustable time delay. The alarm should first activate a neon light on all the occupants' panels, to afford them the opportunity of investigating and closing the door. If there is no response to this, an alarm should sound after a set time and for a set duration.

an antitamper/antivandalism alarm that alerts the building superintendent or resident who will take the appropriate action

instructions on the use of panel and buttons encoded in braille on the stainless steel panel for blind persons

the ability to enter the number required, followed by a "call" button to reduce wrong number calls and a timed, cutoff feature should a single number be called more than a predetermined number of times per minute, to reduce nuisances to occupants.

Instructions. Instructions on how the call panel should be operated should be short, clear, and concise. The instructions should be *engraved* into the thickness of the stainless steel fascia panel (not separate panel), not painted or highlighted in paint. No abbreviations should be used in the operational instructions.

Whenever possible, personal code lists should be avoided. If this is not possible, only the numbers or codes of the individual apartments should appear on the panel. Names should never be used.

Key and Lock. The various types of systems used to allow tenant entry are

mechanical key and lock—insertion of the key will always throw the mortise bolt of the lock. Normally, the mortise would be remotely thrown by the tenants, who identify visitors by voice or CCTV systems limited to individual apartments.

a key and mortise lock with a pivoting striking plate—the mortise bolt is fixed and the insertion of a key deenergizes an electrical circuit, which normally prevents the striking plate from turning 90°, and releases the mortise from the staple box

proximity card systems—these have no keyhole, simply a vision panel or presentation area near which the card is passed. The card "key" releases either a mortise bolt or striking plate.

Other card systems are available, but most are very easily vandalized. The one exception is the open-slot "drop through" card reader system, which almost approaches the proximity system in resistance to vandalism.

The most common method of access is normal mortise lock and key. If this is chosen,

the main entrance key must be completely different from the dwelling key. (No mastering system should be used.)

the lock should be regularly maintained and serviced

the lock should be changed annually, all old keys returned, and new keys signed for and released

the lock should be frame-mounted (integral, not ring- or surface-fixed) and should be a mortise deadlock latch. The internal release should be a brightly colored lever to facilitate easy escape. The lever should be short and very strong, to resist vandalism. The lock should, whenever possible, be a special key design (e.g., Abloy, Kaba) to prevent easy copying by tenants and proliferation of keys.

the lock, the mortise, and the area surrounding the striking plate release should be reinforced using stainless steel–covered plates that are permanently fixed (dome-head bolt or welding, etc.)

A key that deenergizes the lock striking plate is seldom used, as the wear on electric switching components is greatly increased if both tenants and visitors are admitted by the same process.

Proximity card access for tenants can offer several major advantages over normal key systems:

The card is unmarked and appears similar to a credit card, hence its use is not obvious nor does it display the address of the property.

Lost cards are not easily copied.

No keyholes exist, hence there is no obvious point of weakness in the system; Superglue attacks, broken keys, etc., are not a problem.

The card reader is not open to attack. The reader may be a black metal square protected by a polycarbonate sheet. Some systems require no window at all; the card is simply presented in front of a plain part of the normal, call panel fascia.

Cards are easily changed at short notice, without changing the lock.

Speed of entry is faster than with key locks.

When selecting proximity card systems, look for systems that

have rigid cards guaranteed for at least 5 years

can be reprogrammed easily to change access regimes

have card readers that are not affected by radio transmissions (e.g., CB radio) or paint sprayed over the reader area

use cards that are not affected by temperature, bending, magnetic fields, scratching of surface, etc.

Residents' Panels. Each resident-occupied area will require a panel that allows at least verbal communication with the main entrance call panel. The following guidelines apply:

Panels should be constructed from metal, polycarbonate, or high-impact-resistant PVC.

Panels should be surface mounted and connected to the signal wiring via a jack and socket to allow for easy repositioning or extension.

The position of the unit should suit the individual users. It is no good putting all units in hallways or sitting rooms only to later find that half of the tenants are bedridden or live in their kitchens!

It should be possible for the user to throw a switch to disconnect the unit at will for privacy and to prevent disturbance.

The panel should include volume control, a buzzer or bell, and a flashing indicator.

It should not be possible to open the main entrance door from the panel before positive identification of the caller has been made (by either voice or voice/visual).

Communication should always be occupant-initiated. It should not be possible for the person at the main entrance to speak first.

Communication should be two-way, where one person speaks at a time. One-way "caller only" systems do

not allow the caller to be questioned, which is particularly important on voice only systems.

The means of communication on the panel should be by built-in microphone or loudspeaker if possible, not a telephone handset, as this is easily damaged.

It should not be possible for the resident to permanently deenergize the main entrance door lock release by jamming the release button in the release position, thereby allowing unrestricted access.

Other Controlled Exits and Entries

In many instances, the building considered will have at least one alternative exit. This exit should be uncontrolled. If only a simple latch lock is provided, tenants may be tempted to prop this second entrance open and use it in preference to the main entrance. Either of the following procedures should be adopted or the access control security will fail

Make the alternative exit "exit only" (i.e., no key entry to any person) and include a "door open" alarm system to warn tenants. This door should be separately identified from the main door either by a different buzzer tone or indicator lamp.

Make the other exit a second access control point with its own call panel (i.e., a repeat of the main entrance).

The second option is sometimes necessary for large buildings. If a second panel is provided, the following should apply:

The panel should be identical to that used at the main entrance in order to avoid confusion.

The overall signal processor should be designed *either* to treat the main panel as the priority call panel (storing calls made on the second panel until the system is free) *or* to be able to cope with at least two simultaneous callers, one at each panel. If the first option is adopted, the storing feature should initiate an "in use, please wait" call signal or a "call sent, please wait" indicator on the second panel.

The duration of caller/occupant use of the systems should be adjustable to suit the circumstances. Situations can arise when the retired colonel in #33 is engaged in a lengthy political debate with the milkman, while you jump up and down at the alternative entrance panel!

Closed Circuit Television (CCTV)

CCTV systems offer many advantages over voice only telephone entry systems and are usually not prohibitively expensive. CCTV systems can be found in a variety of forms:

systems that use an integrated call panel camera and residents' panel with a television monitor

systems as above, but with a camera that is mounted remotely from the call panel

systems as above, but use the residents' normal television sets as monitors by incorporating manual or automatic changeover switches

The last system is usually found to be unacceptable for a variety of reasons and should be avoided. The choice between the first two systems depends, to some extent, on the shape of the entrance area. An externally mounted camera offers the following advantages:

When suitably positioned and configured, it can cover a wider field of view.

It can be replaced and maintained while the voice part of the system is still in operation.

It is highly visible and can have a deterrent effect.

Its disadvantages are that

it offers an inviting target to vandals

it must be positioned so that it will not be affected by stone throwing, grappling, etc.

it will need an expensive camera and lens configuration if facial recognition is to be achieved.

On the whole, integrated CCTV camera units are the best for residential access control systems. The camera and lens are completely protected. The lens simultaneously gives facial identification and displays background activity. This is useful if entry is obtained under duress or while the door is open to admit a legitimate user. The camera should have a lens that is specifically selected for the entrance design and should not be of standard supply. The camera lens should not be placed directly in the call panel fascia, but should be protected by a sheet of polycarbonate (just bigger than the lens) and fixed in a frame on the back of the fascia.

The system should be designed to allow one-way, tenant-to-caller, visual identification only.

The TV monitor should only energize and form a picture after positive voice identification has been made. This will prevent visual abuse to the tenant.

The system should provide a self-call facility to allow the tenant to view the entrance without having to wait for a call on the entrance panel. This can be very beneficial, as it allows neighborhood watch patrols to work without fear of attack. It also allows the police to catch vandals in the act more easily.

The camera should be able to operate satisfactorily at low-light (<10 lux) levels and additional, vandal-resistant night-lighting should be provided. The camera housing should include a heater and defroster.

Tenant monitors should include

monitors in the same case as the speaker, recessed flush into a wall cabinet. Surface-mounted, separate, or screen-in-the-telephone-handset types should be avoided. Although very stylish in appearance, they are less robust and, in an effort to miniaturize all the components, picture quality is often sacrificed.

contrast adjustment and brightness control. The quality of the picture on the monitor screen will vary with the light in which the camera operates and eyesight varies greatly from person to person.

When selecting a CCTV system, always insist on a demonstration in conditions similar to those in which the system will operate. Set the conditions yourself. Any system will operate satisfactorily given exceptional camera lighting, short wiring runs, and a subject who is fixed to the spot directly in front of the camera! Will it work at dusk, with no artificial lighting, in a canopied entrance, with a coaxial cable length of 328 ft (100 m), and a visitor standing sideways 10 ft (3 m) from the lens? Can you recognize the visitor now?

COMMERCIAL SYSTEMS

System Types

Commercial systems include

- telephone entry and conventional key
- telephone entry, CCTV, and conventional key
- card access
- card access, telephone, and/or CCTV
- combination locks

Code and Information Processing. Card access control systems can be divided into three fundamental types—stand-alone, off-line; on-line; and distributed processing.

In stand-alone, off-line card access control systems, all functions, electronics, etc., necessary for controlling an individual door lie with the card reader. Readers are programmed, but do not have any interconnecting language highway. The processing power within the reader is limited. With the on-line system, the processing power is much greater and allows for many management functions. Readers in these systems are networked to send information to the remote, single, central processor.

Distributed processing uses standard, stand-alone, off-line readers, but provides the ability for subsequent connection to a full, remote processor.

Stand-alone, off-line systems are more commonly used in buildings with few card readers and/or few cardholders. They are the cheapest of the three systems and have only basic lock control functions. They are perfectly adequate for many smaller or lower-risk installations.

On-line systems are capable of almost infinite expansion and offer not only access control capabilities, but movement monitoring, payroll clocking, energy management, etc. These systems are very expensive and are really only suited for larger buildings with many doors and people to control (50 doors for 500 people is a common minimum cutoff for this system).

Distributed processing using intelligent door controllers supports the middle ground for most commercial installations. These systems allow users to gather information at a door, control a door, etc., without very expensive central controller or data wiring. Hand-held recorders can be used to interrogate the reader or record data for display and printing on a normal computer used for other purposes. As the system expands, groups of readers or all readers can be gradually linked to a dedicated access controller/processor or can automatically download their information to disk on a nondedicated machine. Should a communication link be broken, full capabilities remain and information is passed once the line is reinstated. Most of these distributed processing readers can handle many thousands of pieces of card data. When interlinked and controlled by 80386, 80486, or 68000 series microprocessors, as many as 1500 door readers can be simultaneously controlled.

Preliminary Considerations

Before considering any commercial access control system and trying to prepare a brief, it is important to remember

ACSs are expensive

ACSs are only intended to control the movement of people during working hours. Off-hours, conventional locking may still be required.

it is the *lock*, not the card or reader, that provides the system strength. The lock should not be forgotten or neglected in the face of the electronic wizardry that will be presented by the manufacturers. There is little point in using an ACS when intruders can use their credit cards to access the building!

on sophisticated systems, it is not the central controller that determines the security enforcement, but the operator. The machine cannot think; it can only help the operator identify potential system weaknesses or possible breaches of security. A *person* will need to act — the machine will not.

The primary differences between commercial and domestic or residential ACSs are that

only internal doors are normally considered as viable applications, although perimeter gates, etc., may be included in the system. Main entrances are almost always manned (at least during working hours).

the risk of vandalism is greatly reduced

data-gathering and processing functions are often required

greater effort is made to actually identify the card holder

Objectives

The primary objectives are to

allow access to particular parts of a complex or building to those individuals authorized to enter during working hours

provide a selective, changeable, easy-to-use, easy-to-maintain system

give more security and provide more information than would be possible using a conventional lock-and-key system

Application

- computer suites
- payroll rooms and bullion areas
- process or laboratory areas
- lobbies
- elevators

System Planning

Consultation.   The reason for implementing the system must be established and the system's role clearly defined. At the outset, clients must be made aware of the benefits to their organizations and of the demands the system will make. It must be made clear that no system is a panacea for all security problems — locks will still be required and manpower will not necessarily be reduced.

The Brief.   Once the ACS's role has been defined, the next step must be to determine what the system will be expected to do and how it is expected to do it. Once this is established, manufacturers may be approached and the selection procedure started. The following are some of the data the manufacturers will need:

How many doors? Entrance, exit, or both?

How many users and cardholders?

What is the peak flow rate through each door?

What type of material and area of building are to be protected?

Are there any special restrictions on the system type? For example, would radio systems interfere with the organization's normal work or machinery?

What types of doors are being considered?

Will the system need to be hands-free operation?

What type of data (if any) will the system be expected

to gather (e.g., time of use, cardholder numbers, time control, etc.)?

Will individual card removal or authorization be required?

Will multilevel security be required?

The data-gathering functions of ACSs are often overemphasized by the manufacturers and the client should be advised on the actual usefulness of most of these data. All too often, systems including sophisticated, data-gathering features are purchased because they appear to be state-of-the-art, without considering who will analyze the reports and act on them. Data gathering alone does not improve security.

Computer Suites and Similar Restricted Areas

Risks. Before considering which ACS to use, it is very useful to step back and consider the reason for its use, how it will need to operate, and what can be done to minimize the need for an ACS. The risks faced by computer suites are

- sabotage
- espionage
- theft of information, including how the system functions
- tapping into computer signals

Security. The selected ACS must reduce these risks when the building and computer suite are occupied. The system will not be capable of dealing with attacks made by skilled intruders when the building is unoccupied. To counter this, whatever system is finally chosen, the following items should form an internal part of the door control:

a door contact and/or volumetric detection device

a security lock with restricted, registered key circulation

direct alarm dialing to the police or other security force.

It should not be possible for both main building and computer room ACSs to be inoperative at the same time.

System Use. Unlike the residential ACS, the commercial system does not have to rely on the goodwill and cooperation of a wide cross section of people. Obviously, these will help *any* system function more effectively, but a commercial system is used by fewer people who are regularly reminded by management of the need for proper use of the system. One person should be made solely responsible for every aspect of the system and she should ensure that it is always working and used correctly.

Unlike domestic systems, nuisance calls, deliveries, etc., should not be a problem as an area of this type

should always be totally enclosed within the core of a larger building or complex.

Maintenance. See "System Management" earlier in this chapter.

Manufacture and Installation. See "Manufacture and Installation" earlier in this chapter for general advice. For computer installations, look for systems that are already successfully in use. Computer hardware manufacturers often have specialists who provide advice on systems. Sometimes standards are stipulated by the computer manufacturers' insurance companies!

Location and Entrance Design. The following should be considered during the planning stages in order to reduce the demands on the automatic ACS:

The secured area should be within the core of a larger building or complex.

The whole building should be divided into security zones with greater security clearance requirements increasing progressively toward the heart of the building.

Automatic access control should only be used in those areas of the building that cannot be physically secured by other means (e.g., conventional locks, manned desks, etc.).

Materials, hardware, signal cable routes, etc., which might be potential targets, should be grouped together and located within secure areas. Access to these areas should be carefully restricted to the minimum number of personnel.

When items of equipment, such as air-conditioning units, power distribution units, etc., require servicing, they should be serviced outside of the access-controlled area. Service engineers should be in-house staff with the necessary clearance and should be accompanied on all occasions. On no account should service engineers be given access card codes, which will allow unlimited, unaccompanied access to all areas under the control of the access system.

The most secure areas (usually the central processor, tapes, and drives) should be contained in a single room. The room should have access-controlled doors during working hours and should be protected by an alarm outside of working hours. The room should, if possible, be at the end of a short corridor that leads only to that area and is not a passageway to any other area. The entrance to this corridor should have a door and a sign prohibiting entry.

Doors into this area should be installed in pairs to form an air lock in which both sets of doors are under the control of the system, or should lead to a lobby to which admittance is severely restricted, by either key or access card. On no account should members of the public or unauthorized staff be allowed within the vicinity of the access controlled doors or be able to view them.

Entrance Foyer and Doors. Quite often, a computer floor will require an access door with a ramp. The ramp should, wherever possible, have a level section on the unsecured side of the door, large enough to rest a stationary trolley and with enough room beside it to be able to walk to the ACS panel. The doors to the restricted area should be substantial with good door closing devices. Doors should offer resistance to attack commensurate with the level of security provided by the rest of the enclosure.

The entrance foyer should be overlooked from the secured area by means of a laminated window hatch with lock. This will allow for daytime delivery of items without allowing entry (thus reducing the number of people in possession of cards). It will also allow the use of panic alarm buttons and access door override if an attack on the system is made or unauthorized persons in possession of a key try to gain admittance.

Doors and barriers should be designed with security, means of escape, and pattern of usage in mind. Revolving, enclosed turnstiles are very secure, but difficult to get a trolley through. Single-leaf doors are more easily secured than swinging doors. Simple measures, such as making "personnel only" doors access controlled and keeping delivery doors locked from the inside, will enhance overall security while simplifying the system.

System Selection. The choice of a particular system will depend on the

number of doors to be controlled
number of system users and cards to be issued
relationship and interlocking requirements of the doors
need for data recording
level of security required
frequency of system code or key changing

The important thing to remember is that it is extremely unlikely that the building or area you are considering is in any way unique. For this reason, there are sources of reliable advice, information, references, etc. If the situation appears to be unique, you will need specialized assistance from a security consultant! Before making any start on system selection, eliminate those systems that are unsuitable for your application by making inquiries of the hardware installer and manufacturer, their insurers, and specialized client departments. Banks, government agencies, etc., have special needs and will often specify or provide a list of preferred systems.

COMPUTER SUITES AND SIMILAR RESTRICTED AREAS: SYSTEM TYPES

Radio Transmitter Systems

The radio transmitter system comes within the category of a hands-free system. The system consists of a radio transmitter that is carried in the pocket or on the belt of the authorized user. The lock striking plate on the door responds to the transmitted signal, thereby unlocking the door. Essentially, the system is intended for buildings with many doors in continual use, all of which require constant security surveillance.

Transmitters. The transmitter is the system's "key" device and is issued to every authorized user. It consists of a plastic case measuring 4 × 2 × 0.5 in (92 × 50 × 15 mm). Inside the case, a rechargeable nickel cadmium cell provides power to a small radio transmitter. The transmitter emits a radio frequency signal that may be retuned by the system manager, along with the door receiver units. The power of the transmission is very low and frequencies do not normally cause interference with client equipment. The transmitters are recharged daily in a central recharging unit. During use, they transmit continuously and will usually operate for at least 15 hours without needing to be recharged. The user of the transmitter must be within an adjustable range of reception from the door for the locking mechanism to be activated (usually 3–9 ft, or 0.9–3 m).

Each transmitter can be given a unique code of transmission. This allows multilevel access control, user movement tracking, and use recording to take place. Carriers of the transmitters have no knowledge of the coded frequency they use (it can be changed at will) or access to other areas in the system. Transmitters are not removed from the premises secured by the system, but are stored every night in the recharging unit. Most systems incorporate a limited battery life, so that if removed from site, the batteries will have expired by the time the building is unoccupied, making reentry impossible. Some systems include the ability to activate an alarm at the final exits of the building to prevent unauthorized removal.

The transmitter also has a second function. Normally, doors to the restricted access area would be hooked up to the alarm during unoccupied hours. Using the radio transmitter system, the alarm devices covering these doors may be left energized at all times. The transmitter deenergizes the alarm's transmit function and prevents alarms from ringing while authorized staff carrying transmitters are in the area. If intruders, unauthorized staff, or staff carrying transmitters who are not given that level of security clearance try to enter the area during working hours, the alarm will ring. Out of working hours the system can be programmed to remain active even if an authorized transmitter is present. This reduces the risk of an intruder gaining access by removing a transmitter from the recharging unit.

Door Mechanism. The door mechanism is usually a standard, fully recessed, electric release striking plate and fixed-bolt mortise lock. Each door is provided with a receiver (aerial loop) concealed within the door, wall, or floor.

The received signal is transmitted to the control unit, which signals a relay to deenergize the lock release and allow the door to open. Most systems include an internal manual override button for emergency exit use if the transmitter and receiver fail while the power to lock is still on. Systems usually include a delayed "door open" alarm with an adjustable timer that either can be a self-contained local unit to warn the occupants that the door has not shut properly or can be wired back to the central system controller. This feature will also prevent general direct-to-police alarms from ringing when door contacts are fitted and all persons leave the area temporarily (lunch hour, etc.) with the door propped open. An alarm of this type is always useful on computer suite areas for the simple reason that, despite good air conditioning, staff quite often leave doors propped open during summer in the mistaken belief that this will introduce cooler air into the space. Once the staff have been educated in security and air conditioning, this problem can be reduced.

Control Unit. The control unit is located within a secure area of the building (normally a reception or guard room) and it controls the operation of the electric release locks and their response to the transmitter signals that allow or refuse access, depending on the level of security afforded to a particular transmitter. The control unit houses the power supply standby backup battery, electronic comparison devices, which compare the transmitter signal code with authorized codes. Control units normally include the following features:

- monitor lamps to ascertain the condition of the system
- local alarm and remote repeat (general alarm)
- on/off/test key-in-lock switch
- antitamper alarm
- cancel alarm key-in-lock switch
- system override or manual system switch
- mains-fail/mains-on/battery-on/battery-charging lamps. When additional standby battery capacity (i.e., more than 12 hours) is required, a separate battery and trickle charger is required, which needs to be housed near the control unit.

Monitor Panel. The monitor panel contains indicator lamps that show the status of the system, doors, and other areas. The system can also be switched or tested from the monitor panel, which resembles a standard fire alarm panel. The monitor panel us usually located outside of the secured area and in a position where guards or patrols can regularly check the status of the system. The indicator lamps on most systems will allow the checking of

whether the door has latched shut
unauthorized entry
intruder loitering at a door entry point (alarm)
doors locked to all transmitters for a set period of time unless overridden by the fire alarm system
emergency exit door status
normal access or exit point being used by an authorized user

Charging and Storage Unit. The charging and storage unit is simply a lockable, recharging station for the hand-held transmitters. Transmitter users are given a key that allows them to "park" their transmitter into the station for overnight recharge and lock it in position. If the transmitter is not locked in, or if any attempt is made to change or remove any other transmitter to which a key holder is not entitled, an alarm will sound. Charger units must be surface mounted and their size depends purely on the number of transmitters to be stored and charged.

Advantages. The radio transmitter system offers the following advantages:

fast in actual operation of door
continuous day and night detection of intruders passing through a restricted space or doorway
reduces risk of pass-back of transmitter. Once released and given to second person trying to gain admission, the first person is detected by the intruder detection system.
prevents following of authorized persons through the door. This can be important, as saboteurs and industrial spies make it their business to look as though they belong in the building and outrank the person holding the door.
hands-free; very important if transport of materials, paper, etc., is a frequent event
gives selective access and multilevel security
there are no keys that can be used directly on the system doors or removed from the building
no visible keyhole or card hole to invite tampering
codes can be changed easily
no over-the-shoulder viewing of numerical codes, as used on combination locks

Disadvantages.

high initial expenditure on the charger, monitor, controller, etc.
by fairly simple means can be removed from the site without causing alarm, be substituted with a dummy unit, and be recharged off-site for use by unauthorized persons

because of the complexity of the system using an active device (i.e., continuous transmitting device), the risk of parts failure is greater than with some static systems (e.g., magnetic card); hence, maintenance commitment is greater and is reflected in servicing cost. If it is not, corners are being cut.

greater interwiring of parts than self-contained, stand-alone ACS. Hence, potentially less secure and more expensive.

the coded signal from the transmitter can be remotely recorded from an unsecured area for later use. The door unit controller cannot normally distinguish the authorized transmitted sound from one produced by a cassette tape and speaker.

problems with range and interference, which cause the door to open or stick

Summary. Generally this is a very good system that offers some unique features, but it is only really suited to larger establishments with very heavy door traffic, not too many visitors, and a high security requirement. If used, the radio transmitter system demands good, vigilant security staff who can check system operation and recognize patterns of use of particular staff members in order to be able to detect any out-of-the-ordinary use.

Electromechanical and Mechanical Combination Locks

Push-button, alphanumeric combination locks are primarily intended for the smaller installation where only one or two lightly used doors are involved. They are occasionally used to augment security on magnetic card-in-slot systems, by providing a common pass code to be entered before the use of the card. The code can be used to identify the individual.

Like all systems, an off-hours security lock must be used to secure the door during unoccupied hours. The systems are seldom linked to data-gathering systems when used in their pure form (see Figure 7–6).

Advantages.

self-contained, small, and compact

different codes can be used on different doors, increasing the difficulty of entry into a particular zone

low maintenance and expertise commitment

cheap, easily replaced, and commonly available

no keys, cards, etc.; nothing to carry or remove from site

electrical types activate alarms if wrong combination is used or multiple guesses made. Both mechanical and electrical systems offer time delay locking if the wrong code is selected (entry only).

FIGURE 7–6   Electromechanical combination lock

Disadvantages.

codes easily spotted

button can be secret spray marked to leave a trace of the code for the intruder

code must be changed frequently to maintain security

limited combinations available; guessing at easily remembered codes frequent (e.g., all the sevens)

codes often written down by users in order to aid memory, particularly if frequently changed to increase security! The former defeats the latter.

obvious point of attack, vandalism, etc. Often abused by impatient staff.

slow to operate

not hands-free; some units require two hands, one to hold the open-door button and one to push the door!

mechanical types set into door are low-security rated and can easily be forced

no simultaneous alarm/access control as with the radio system

Selection and Use.

Units should be recessed into structural openings and screened from view.

Always choose a unit that is well engineered and

robust with adequately sized buttons. Cheap, small, stylish units often collapse after 6 months of use.

Always change the combination before installing the lock; never leave the combination unchanged for a period of longer than 6 months.

Never use on doors that lead off passageways, public areas, etc.

Avoid rim-fixed mechanical types as they can easily be forced in daylight by the swift application of a concealed jimmy.

Never use as the sole means of access control for areas containing sensitive, secret, or classified processes or documents that could attract the attention of professional thiefs.

Summary. This is a low-security, cheap, access control system that is suitable for single laboratories, dangerous areas, computer tape storerooms within secure computer suites, etc.

Cards and Biometric Keys

For a general description of the generic types of cards available see "Principles of Operation and System Parts." The following variations on a common principle exist:

card in slot only, stand-alone
card in slot only, centrally controlled and monitored
card in slot plus combination lock, stand-alone or centrally controlled and monitored
card in slot plus conventional key that enables the lock to be centrally controlled and monitored
card in slot plus conventional key, but with keypad PIN
proximity cards

All of the preceding centrally controlled systems can be linked to data-gathering and processing stations.

The card is the key to the system. The system is only as secure as the card used. The most commonly used cards are magnetic cards. It is unnecessarily costly to provide cards with excessive security against copying or forging in relation to the goods or information protected. Remember, they are *keys*. They can be stolen from the user and returned before they are missed. Cards can be provided to criminals via unauthorized copies taken during manufacture or site coding. The security of any card access system is increased immeasurably by

not allowing card removal from site
collection of internal door cards at a front entrance
the use of PINs on card readers
embossing the user's photo onto the card

Magnetic Spot Cards. Used for low to medium security with mass circulation of cards, cards of this type consist of a three-ply laminate. The outer two layers are of PVC, the inner is usually of a barium/Butyl rubber/ferrite mixture. All three layers are fused together and the edges sealed so that the card appears to be solid PVC. The center layer is capable of long-term magnetization with little loss of field strength.

Information on the user, acceptable codes, etc., are stored (or encoded) on the card by passing it through a control unit. The control unit produces a very strong magnetic field by means of an array of electromagnets. The electromagnets may be arranged and rearranged to produce thousands of different magnetic field patterns. In addition, each magnetic field produced by an electromagnet may be focused to give a particular magnetic field shape. Because of the ease of iris focusing, the magnetic fields are circular in coverage and usually the fields are of the same diameter. To encode the information, a card is passed through the magnetic field created by the programmer. Areas of the card that cut lines of the magnetic flux have the barium ferrite layer turned into tiny polarized magnets. Areas not passing under (not "cutting") lines of magnetic flux remain unaffected. The result is a card that now has a central layer made up of a particular array of circular magnetized areas. The magnetized areas represent a 1 and the nonmagnetized areas represent a 0 when presented to a reading device. Information is thus presented to the reader as a series of 1s and 0s, forming a simple binary code. When the information contained on the card is of a personal (and not purely numeric) nature, the binary code must be cross-referenced to an alphanumeric code before readable English can be deciphered.

Advantages.

very commonly used
cheap
cards are durable and relatively flexible
magnetization does not deteriorate
cards can be erased and recoded in the control unit, hence reuse of the same card several times is possible
recoding is simple and rapid

Disadvantages.

The commonness of the card makes forgery relatively easy for the skilled operator; blanks are easily obtained.

Information can be copied from the card and reproduced on blanks if the card is stolen.

It is sometimes claimed that the strong magnetic fields created within the central layer occasionally and inadvertently scramble information contained on credit cards or other security cards of the magnetic strip type if they are stored together. (This has not been proved.)

Summary. This is a cheap, reliable card with slightly improved security over the magnetic strip card. It is more durable than a magnetic strip card, but is not to be used to protect high-security areas.

Magnetic Strip Cards. Used for low-security, mass circulation, magnetic strip (or stripe) cards are the most common system access card on the market. They were also the first card developed. A magnetic material is simply fused onto the surface of a PVC card of substantial thickness (0.03 in). The card is encoded in a similar manner to the magnetic spot card, but bars of magnetized and unmagnetized material on the strip form the binary code, rather than dots. The width of each magnetized bar can be altered. Magnetic strips are widely used on all types of credit cards, for stock control and sales recording, and are the standard against which all other systems are compared for security and price.

These cards are manufactured in two standards: either standard (300 oersted) or HiCo (high coercivity) (4000 oersted). The magnetic strip card can be damaged by stray magnetic fields. The higher the oersted number, the better the card's resistance to magnetic field corruption. HiCo cards are virtually incorruptible. Magnetic cards are also manufactured in long-life versions. Standard PVC cards normally last 1 to 2 years in service. Long-life cards are manufactured from a laminate of polyester and Mylar and usually last over 4 years.

Advantages.

cheapest card on the market
readily available in many formats to contain varying amounts of information
very reliable; error rate in readers is extremely low
all the bugs are removed; development is complete
can store a great deal of information

Disadvantages.

less secure than more recently developed cards, as relatively simple equipment required to lift recorded information from card for transfer to blanks
more susceptible to damage than laminated card (e.g., scratching, cracking, etc.)
data can be erased or scrambled by magnetic fields of relatively low strength
not usually reusable (not really a disadvantage as cards are very cheap)

Summary. This is the cheapest card available. It is perfectly adequate for most low-security ACSs, particularly if a large number of people will be issued cards.

Magnetic Watermarked Cards. Used for medium-security, restricted card circulation, this card is very similar to the magnetic strip card in principle. It uses a bar code of magnetized material to produce a binary code. The essential difference, however, is that with magnetic watermarked cards, the magnetic strip is encoded with its information *before* the card is constructed. The information is not placed onto a blank card with a surface-fused magnetic strip. The preencoded magnetic strip is placed between two layers of PVC and the whole thing is heated until the result is not a laminate or a sandwich, but a homogeneous mass with the magnetized layer dispersed throughout both PVC layers. This results in the card having a magnetic fingerprint or watermark of the type found in stationery, the difference being that this watermark is not visible to the naked eye.

Advantages.

more durable than magnetic strip cards
slightly more secure than magnetic strip cards and much more secure than magnetic dot cards
encoded information very difficult to read and practically impossible to transfer
uncommon
will not scratch, not easily scrambled or damaged by magnetic fields

Disadvantages.

cards more expensive than magnetic strip or dot cards
cards ready-encoded in manufacture

Summary. This card is more secure than previous card types. It is more appropriate for medium-security-risk installations with a limited number of users and restricted card circulation.

Metallic Disks. Used for low security and medium circulation, this type of card reverses the normally accepted practice. Metal disks or oddly shaped pieces are embedded within the body of a PVC card. The disks are given a particular pattern and are of a precise size. They are not magnetized and they carry no information. The reader is provided with an electromagnet. As the card is presented to the reader, the magnet is energized, creating a field across the slot. As the card travels into the reader, the flux of the magnetic field is altered in a particular manner, depending on whether PVC or metallic disks are present. The alteration is read by the reader as a binary code.

Advantages.

very simple
quite reliable
cheap capital cost
information is not actually contained on the card in any readable form or code

Disadvantages.

cards easily copied and forged
code is permanent; cards cannot be reused
not in common use

Summary. These cards are now virtually obsolete, as they have been superseded by magnetic strip and dot cards.

Magnetic Field Effect Cards (Wiegand Effect). Used for high-security, restricted card circulation, field effect cards use the same arrangement as metallic disks, in that the magnetic field is produced by an electromagnet within the card reader. The card itself has short lengths of a specially heat-treated, twisted wire embedded within it. The wire has an outer sheath that will hold a magnetic charge and can store it, and an inner twisted core that will magnetize but cannot store the magnetism. Each length of wire is placed at a particular point within the card and acts as an electronic signal generator when placed in a magnetic field. The generated signal and the sequence of generation is dictated by the strength of the field passing through the card and the position of the wires as they pass through the field. In simple terms, the signal is generated when a field of a particular strength energizes the inner core, but does not have sufficient energy to permanently magnetize the outer skin of wire. As the wire passes into the magnetic field, the inner core absorbs the magnetism.

As it does so, it tries to untwist and return to its original shape. At a certain point, a rapid "snap" action happens on the inner wire. As it moves through the magnetic field, the card generates an electromotive force and current, and transmits it toward the outer skin, which it tries to magnetize. The change from absorbing to transmitting energy generates an electrical current of minute size and hence a signal. These signals are read in the card reader and deciphered by the central controller.

The wires in the card are normally in two regular rows giving a binary 0 signal row and a binary 1 signal row. To make the positions of these binary code embedded wires difficult to find, either identical, nonmagnetic wires or wires generating false key signals are embedded, at random, with the real binary signal generator wires. This makes these cards extremely difficult to copy.

Advantages.

very high security
unique, although tried and tested
very difficult to reproduce (much more so than metallic disk)
no code to be erased, altered, or damaged, hence very robust
the most vandal-resistant card
not affected at all by EMI and RFI fields

Disadvantages.

permanent, not reusable
costly
still possible to forge, given luck and some skill
single card supplier

Summary. This is an excellent system for small, high-security, stand-alone units for pure access control lock activity.

Holographic Identity Cards. Used for high-security, medium circulation, these cards have several advantages over conventional cards of the magnetic type, but the cost of the cards and their readers currently restricts their more generalized use. They are high-security versions of optical photocell read bar code credit cards being used in some bank cash-dispensing machines. Essentially, the card is a two-part, two-function device. The card is constructed from two layers of rigid PVC. The base layer is engraved with information; the top layer acts as a lens and protective cover. The first part of the card identifies the authorized user. This is commonly done by engraving a work number and name, personal details, level of security clearance, code, etc., into the card. In addition to this, most cards have a passport photograph of the authorized user. This part of the card will identify the person and her rights within that area of the building if she is challenged by security staff. Some cards also carry the person's address or company name (logo, etc.). This should be avoided, as a lost card immediately exposes and identifies the authorized person or company, who is then at risk. The manufacturer's name and address is more acceptable on these cards, as lost cards could be returned without risk to the company or individual. Box numbers for return are also acceptable.

The second part of the card contains the hologram. This consists of a holographic generator that transmits and reflects very thin beams of laser light in a precisely defined pattern to create a pattern or picture that can be read by the card reader. The laser continuously scans the card as it is passed through the reader. Information is conveyed in the form of patterns of light (or *holograms*) and transformed into a coded signal.

Advantages.

security almost unrivaled
doubles as an identity pass
forgery almost impossible

Disadvantages.

usually used in conjunction with push-button combination lock as an activation device
larger than most other units
expensive

Summary. This is a very good system, but expense usually limits its use to protection of cash, bullion, or classified information.

Infrared Shadow Cards. Used for very high security, very restricted circulation, infrared shadow cards are PVC cards within which special bar codes are placed. The card appears opaque to ordinary light, but when subjected to infrared light, the bar code casts a shadow behind the card, as thin PVC will transmit infrared light waves. The arrangement of the code is precise and the shadow must be recognized exactly in order for the lock to be released. Very sensitive electronic sensors are located and arranged in a grid within the card reader to collect the shadow as reduced energy level, infrared light. To improve security further, elements of the bar code are placed in PVC layers of varying transmittability, giving the shadow differing tones or contrast, and the sensors are tuned to recognize these different, multiple energy levels.

Advantages.

probably the most secure access card commercially available

forgery almost impossible

Disadvantages.

relatively expensive

can only carry limited, encoded information (although identity codes are almost limitless)

cannot carry nonencoded information (company logos, etc.)

fairly easily damaged by contact with a variety of materials and substances that will affect the transmittability of PVC

Summary.   These cards are normally only considered if recommended by the client's expert security organization or consultant, but they are rapidly gaining favor with many of the larger security organizations.

Proximity Cards.
These cards are low security, mass circulation. All of the foregoing cards require a card-in-slot reader. The major advantage of the proximity card is that it does not. Proximity cards are usually passive radio frequency (tuned circuit) magnetized implants or electrostatic circuits, as described in "The Key Device" in this chapter.

Advantages.

relatively cheep (per user!)

no card reader accessible to user (only a protected sensor), hence reduced likelihood of accidental or malicious damage

faster in operation than many systems, particularly those using push-button panel and communication with a normal card. Some systems are hands-free, similar to radio transmitter systems.

cards usually carry very little information to identify building, user, etc., which improves security

cards very robust and not affected by magnetic field scrambling

can be used in interior or exterior application as nothing to weatherproof

range adjustable 0.8–8 in (20–200 mm) on passive types; up to 3 ft (1 m) on active types

Disadvantages.

not as secure as most normal card reader systems (particularly tuned circuit radio frequency cards)

information on cards is usually limited purely to a personal access authority number, which allows entry and identifies the card

radio frequency cards can be affected by other radio frequency sources (e.g., motors, lights, etc.)

individual cards slightly more expensive

Summary.   This is a very useful alternative to magnetic strip or dot cards where damage to card readers might be a problem. It is also useful in situations where disabled persons are being considered (e.g., blind persons or those with disabilities that make accurate positioning, location, or operation of push-button panels difficult).

Hand Geometry Systems.
There are only two really good manufacturers of hand geometry systems worldwide. Although this limits choices, they have been developing their systems for at least 15 years to make them very reliable and simpler to operate than they first appear. Hand geometry systems map and measure the geometry of the hand by the use of sensors, CCTV scanning, and digitized storage to disk or optical disk. The range of measurements made for comparison can include finger lengths, finger widths, joint placements, palm shape, relative finger length, etc.

Advantages.

less prone to hand damage (e.g., cuts, abrasions, etc.) rejections than fingerprints

simple to use

Disadvantage.

bulk of reader

Summary.   Overall, this system is a direct competitor of fingerprint systems. It has been under market refinement for longer than fingerprint technology and is often preferred to fingerprint systems by users.

Fingerprint Systems.
Fingerprint systems are now reasonably common on single-door or small, multiple-door, high-security, limited access installations, such as research labs, vaults, computer file rooms, etc. There are at least ten reputable manufacturers and installers worldwide who specialize and develop this technology.

Most systems operate on the recognition of single points. This is done by identifying and matching the points created by the ends and junction points on the skin ridges to prerecorded prints on file. Some systems provide additional security by mapping and linking these points with straight lines, to give a matrix or grid mapping for a match or by looking for characteristic patterns such as spirals in the skin pattern of the fingerprint.

Advantages.

multiple manufacturers; competition
simple to use
may be single or multiple print (for same person)

Disadvantages.

disliked among individuals who have to have their
fingerprints stored
damage to fingerprint (e.g., cuts, sores, etc.) easily
upsets such systems; backup prints needed
damage to prints due to materials on hands (grease,
paint, makeup, lacquer)
slower to use than card systems

Retina Eye Pattern.   There is only one manufacturer of
this technology with any serious development program—Eyedentify, Inc., USA. This limits choice! Comparison must be made with other biometric systems or
high-security cards.

The system is not yet widely used, although a significant number of installations perform well. Particular
attention should be paid to any problems associated with
daily operation of the system by those wearing contact
lenses and those who have light-sensitive eyes.

Voice Recognition.   Developed by ECCO Industries,
USA, this system records the spectrum of audio frequencies generated by the voice box of the user as she repeats
a password or phrase. This template (or pattern) is stored
for comparison. Three different templates are stored in
a person's file. The file is accessed by a PIN code and the
person using the system then repeats the password. The
password is compared with the stored phrase. If a good
comparison is found, then verification is given, access is
permitted, and the last entry is used to update the person's master voice template. This updating allows for
minor variations, throat infections, etc.

Advantage.

cheaper than many other biometric systems

Disadvantage.

less well proven or used than fingerprint or hand
geometry systems

Biometric Systems: Generally.   The fundamental advantage of biometric systems is their ability to accurately
identify a person with a right to enter and not a card or
key with any carrier. The biggest disadvantages are cost
per door and operation speed. Costs are going down
every year, but it is possible that the promised cost
breakthrough may be cut away by the introduction of
multiple use "smart cards." Operational speeds are determined by ease of use and speed of retrieval. Operational
use times are at least two to four times that of the slowest
card (but still only seconds) and do not really vary greatly

between systems. Retrieval speed depends, to some extent, on the computer storage media and speed (e.g., optical disk versus floppy disk, etc.) and the use of PINs.
PINs are used by users to select their master print file for
comparison with the current print. This speeds retrieval
and comparison, but reduces overall security.

Smart Card.   Smart cards are credit cards with built-in
microprocessor processing and information storage
facilities. They are able to store huge amounts of personal data on the card. These details can only be given
to those requesting information by the card user entering pass codes, etc., into a reader (in response to questions) or by authorized retrieval by the police or similar
groups. These cards are currently under development by
the major bank groups (particularly in Great Britain) for
primary use in cash/credit, transfer, point-of-sale
facilities/utilities, direct withdrawal, etc. It is likely that
every single adult in any first world country will soon
carry some form of smart card. Their introduction will
have a dramatic impact on many fields that could use the
data storage feature [e.g., access control cards (high
security), drivers' licenses, medical history, insurance,
etc.]. The card becomes a multilevel, multiple use personal, secure organizer! In the United States, information can be obtained on systems and their capabilities
from the Smart Card Industry Association (SCIA). Bio-
metric-activated unique *traceable* contents, encryption
security, and radio proximity features mean that with
falling production costs smart cards could replace all
current access card technologies within the next 10 years.

Card Readers

The card reader eventually installed will depend entirely
on the type of card chosen and the level of security
required.

**Magnetic Strip, Magnetic Key, and Magnetic Dot Card-
in-Slot Readers (See Figures 7–7 and 7–8).**   The
smallest units are magnetic "Dip" key readers that are 2
× 2 × 3 in (50 × 50 × 75 mm) or magnetic strip or
swipe pass through readers that are 1 × 1 × 3 in (25 ×
25 × 75 mm). Typical card only readers measure 4 × 3.5
× 6 in (100 × 90 × 125 mm). If numeric or
alphanumeric push-button panels are added to either of
the above, the size increases to approximately 5 × 8 ×
6 in (125 × 200 × 150 mm), and far fewer units of this
type are flush fitting. Units incorporating push panels
should always include a cowl to prevent over-the-
shoulder viewing of the number selected. "Dip" card
readers, in which the whole card is inserted into the
reader, are often less reliable than swipe readers, particularly in external uses.

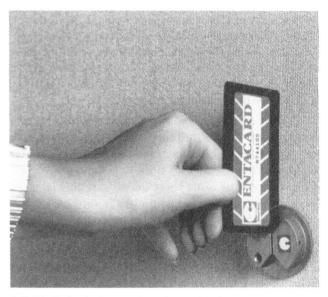

FIGURE 7-7 Swipe card reader and swipe card

FIGURE 7-8 "Dip" card reader (left) and swipe card reader with PIN (right)

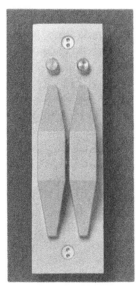

FIGURE 7-9 Infrared shadow pass through card reader

Holographic, Field Effect, and Infrared Shadow Pass Through Readers (See Figure 7-9). These cards are normally read by sliding one edge of the card between two parallel plates while holding the opposite side of the card. For this reason, they are mostly surface mounted. Sizes range from 2.75 × 1 × 2 in (70 × 30 × 50 mm) for field effect, to 6 × 4 × 12 in (150 × 100 × 300 mm) for holographic types.

Both the field effect and holographic card readers are manufactured in horizontal card entry or vertical card entry forms. Generally, horizontal card entry readers are more reliable, as card flexing is minimized by more extensive internal card guides. This reduces the incidences of card misreads (a hardware failure). When card readers are used externally, anticondensation heaters are often required within the units.

Proximity Readers. Proximity card readers (see Figure 7-1) require large sensors, although this is offset by the fact that they may be totally hidden behind a layer of smoked glass. Typical size is 18 × 10 × 2 in (450 × 250 × 50 mm).

On the whole, there is little to choose between any of the above cards or readers for speed or ease of use. None is completely resistant to deliberate vandalism. Proximity readers are possibly more resistant to accidental damage or abuse, as they are usually more robust, recessed flush, and therefore hidden.

The best card reader units will inform the controller of

card number (maximum of, say, 20 characters)
personal number (at least 5 characters minimum)
door activated, door number, door status (e.g., propped, not latched, etc.)

In some establishments, the provision of a panic button might be required within the secure area to take the card reader out of action, thus preventing an unauthorized person in possession of a stolen card from entering the secure area.

Central Control and Processing Units

The central control and processing unit is required on all non–stand-alone, single-door systems. Basically it is a small computer with memory and automatic comparison programs, and other features that can be added by programming with particular software. The cost and complexity of the control station is a function of the number of doors controlled, the data-gathering technology, and the level of card control required. When central processing and control is required, it is often essential to let the client's accountants look into the economics of leasing the system versus buying. An important element in the

FIGURE 7-10 System components and central processor for a large access control system

calculations will be the need for regular maintenance and attendance during the first year of use.

The parts of a typical ACS are shown in Figure 7-10. It should be noted that on smaller systems, controller, processor, programmer, and printer can be contained within a single desktop unit and not as a series of individual units. This cuts down the space requirement and makes installation simple, as no interconnecting cables are required. For larger buildings, group access controllers, which verify card authority, and power door locks can save extensive cabling to the central control and processing unit.

It is not uncommon for card reader design to change rapidly and for central processing software to lag behind in development. When purchasing, it is *essential* that the purchaser ensure the readers, cards, processor, and software are of the same revision, are compatible, and have been operating successfully elsewhere for at least 1 year.

System Uses. All systems offer at least some of the following features. Those that are of real benefit to the security of the building are prefixed by an asterisk (*). Given the necessary central control capacity, the following might be carried out:

*pure access control of doors
*simultaneous access control and alarm inhibits (radio transmitter system)
personnel-tracking routes through building
personnel time and motion studies (e.g., time spent in particular areas)
*monitoring and recording of guard patrols
control of use of photocopiers, withdrawal of materials
*closing routine check of door locks and alarm systems
*fault recording
door operations and planned maintenance of locks and card readers

Central Control Unit Functions (See Figure 7-10). Whenever possible, the door control device and central control unit should be matched to give as many of the following functions as possible:

adjustable door unlocking time (2-30 seconds), which can be varied for every door on the system

door-not-closed alarm—alarm will sound on the panel if the door unlock time is exceeded. In high-security areas this feature should, when possible, differentiate between a door that has been propped open, a door that has attempted to latch but failed, and a lock that has had an item placed in the mechanism to simulate the mortise bolt. These features can be added using a combination of door hinge contacts, mortise bolt end contacts, etc.

antitamper alarm—antitamper alarms should be fitted to door card readers on medium- to high-security-risk systems and on all central control units

voided card use alarm (muted)—this alarm shows and records any attempts to use a card that has been lost and voided from the system

forged card alarms—some systems will allow the control unit to distinguish between an object blocking the card reader and a forged card

code recording comparison and validation—on most systems, this feature is carried out entirely by the central processor and control unit. On larger systems it can be beneficial to distribute some of these functions, namely comparison and validation, to the actual card reader. This is often referred to as *distributed intelligence.* Systems of this type offer the advantages of reduced data loading onto the central unit (hence a smaller unit is required and there is less likelihood of a jam of information and data processing during rush hour) and longer continuity of security (at a reduced level) during power failure, rather than the total lockout that occurs with some systems.

remote lock activation (either momentarily or indefinite)—this feature allows the controller to seal the building to potential intruders during an emergency and disallows all card use while still allowing emergency exit

remote lock deactivation—this feature is opposite of the preceding feature. All doors left free swinging during an emergency (e.g., fire during working hours).

card voiding—this feature is an extremely important function. Staff members leave or are replaced, lose cards, get sick, etc. Under any of these conditions, cards that were once carried must be withdrawn from circulation. This is difficult, as people often take cards home. In order to maintain security, the system must be reprogrammed to refuse entry to that card. The speed and ease with which the user can do this will determine how well the practice is carried

**FIGURE 7–11 Graphics
display of controlled access
point status**

out in the long term and, hence, the security of the building.

anti–pass back — systems differ in their approach to trying to remove the possibility of staff members passing their cards under the door to someone who has forgotten or lost her card, or worse still, to an intruder. Simple systems employ a nonconsecutive use feature where the same card may not be used consecutively in the lock. This can cause problems with rooms with several entrances and exits, as a person may leave through an uncontrolled exit and try to return via the same entrance. The simplest way of overcoming this is obviously to make all doors controlled and to make exit use indistinguishable from entry use. Better systems employ the "no-further-entry-until-exit" principle to restrict card pass back. Some systems improve upon this by including a time delay before further entry, which increases the need to loiter and the risk of detection.

last card memorization and sequence analysis — some systems allow the use of individual cards (or all cards) to be analyzed for potential abuse. Using this method, the controller scans the data record looking for cards that have made dual entry within very short

time periods (anti–pass back), excessive use of cards through particular doors, etc. This can also be expanded to include personnel tracking and attendance features, but on most systems, neither of these is warranted. The personnel-tracking feature will not tell you the person's location, only the card's location (another person might be using it), unless the reader is a personnel identification type (hand geometry, etc.). The attendance feature is only really useful when linked to payroll, time, and flextime recording. During an incident (e.g., fire, explosion, etc.), few security staff are likely to wait for the printer to give forth! Roll call is only really justified when the installation is extremely large or the potential risk or consequence great (e.g., petrochemical works, nuclear reactors). Even at these locations it is essential that a remote station be found to house the controller and printer securely.

duress — all systems should include a duress function whenever possible. On medium- or high-security systems, it is essential. This feature can take the form of an additional code number added to the push-button code, card entry followed by immediate partial withdrawal and reentry, etc. In order to reduce the risk to the card carrier, the alarm should only be raised at the central controller and decisions on whether to deny access, seal the building, etc., should be made and carried out by the responsible person, not automatically.

Central Control Unit Features.   When selecting control units, it is important to remember that the time they will really prove their worth most is when the system is under attack. Often the attack will be sudden, hence the main criteria for the selection of the unit should be that it is extremely easy to use, it can be read easily and quickly, and the presented information makes sense and has real value.

Too often, in the past, the screen or printer has informed a security guard of complete gobbledygook! Therefore, when selecting a control unit, look out for

VDUs that give a graphic display of the building plans, showing door alarms activated, card readers being abused, etc. (see Figure 7–11)

printed messages on the screen or printer that are in English, not codes. Assess whether messages can cope with full descriptions (e.g., "The card belonging to Mr. Jones has been left in card reader 3, room 5, first floor") or whether there is a restriction on the message length.

Ensure that programmers allow programming to be carried out easily. If possible, the process should be in English, not code, and should include on-line help screens (not user handbook references). A good system might allow a dialogue along the lines of

Program: What card do you wish voided?
Operator: Card 213 to be voided.
Program: Card 213 not recognized as previously issued. Do you require list of cards, if so at what security level?

Obviously this is only possible on larger systems, but the general point applies to even the simplest system.

Use of the control unit should be restricted by either key or card to authorized persons only. Multilevel access should be possible by means of personal entry codes, which might allow the security guard only to monitor the screen, activate alarms, etc., a junior manager to change card use and void cards, and a senior manager to analyze records and list personnel details, times of use, etc. Three or four levels of access is common on better systems.

On large systems incorporating CCTV surveillance, it should be possible to use the VDU and select any controlled door to view the activity at that door while having details and recorded pictures of the card user superimposed on the screen. This will enable visual identification to be matched to the card record. If possible, choose a VDU unit with a screen that will tilt and rotate. Somebody will be watching it for long periods at a time (particularly if CCTV is incorporated) and if glare, height,

and position irritate the user, she will become less vigilant.

Units should store data on miniature floppy or optical disks, not tapes. Tapes use more space and if speedy retrieval is required, tape drive equipment must be sophisticated and therefore expensive.

Control units should carry out "intelligent" actions. For instance, if a personal number is mispunched on a call panel once, it should simply record the fact. If it happens twice, it may give the control user a warning alarm and bring up details of the affected door, its location, etc. On the third occasion, it should refuse entry automatically, in case the fourth guess is correct. If, by the fifth try, no action has been carried out by the control unit operator, the door should automatically ring the general intruder alarm.

Standby Power

Intruders, realizing the premises are covered by an access control system, might attempt to disable the system by cutting the power supply. If they have more time, they might wait for a power cut! On all systems, standby battery power should be provided. In buildings where general standby facilities are provided by a central battery bank, this might suffice. In buildings with continual power supplies, battery backup may not be required. In buildings with standby power generation, a battery unit will still be required because the warm-up time of the generator might exceed the time needed for entry through a door (3–4 seconds) or the permissible downtime of the controller.

When selecting standby batteries ensure that the manufacturer explains the capacity to your satisfaction. It is no good knowing that the battery will give "8 ampere-hours," if the current demand of the locks and other system parts is not known. The standby capacity should be explained in the form of "x ampere-hours, provided by trickle charger and sealed nickel cadmium cells capable of maintaining all system functions, lock operation, etc., for a continuous period of y hours with a maximum 1000 total door operations, able to recharge to full capacity in 8 hours."

HOTELS, CLUBS, AND MOTELS

Access control to hotels, etc., is a fast-expanding market. Hotels are still prime targets for thieves, either professional or opportunist. The average hotel will require five or more access levels for security:

- management—all areas including cash rooms, etc.
- emergency use—master key
- maids or cleaners—floor key

- housekeepers—ancillary and floor key
- maintenance staff—master or plant area keys

Traditionally, these levels are provided by key mastering and suiting. This has disadvantages because

keys of whatever kind give unlimited access for the life span of the lock in the door. This means thieves can pick a time that suits them to defeat the lock.

locks are changed infrequently due to labor costs. Keys, however, are frequently lost. Hotels have tried large key tags, tags that ring alarms if taken through certain areas, etc., neither of which greatly improves security, as keys are often lost or stolen inside the hotel.

to cut key-mastering costs, maids, etc., often have keys

that give access to many more rooms than they intend to clean that day. The turnover of cleaning staff in most hotels is alarming and it is not unknown for the cleaning staff to gain employment simply to copy or remove keys.

keys are easily copied and most are readily cut from

commonly available blanks. If they were not, the hotel would soon have half its rooms locked permanently due to lost keys.

Access systems for hotels usually take one of two forms—centrally controlled and monitored, and stand-alone "key maker" systems.

Centrally Controlled and Monitored Systems (System 1)

Centrally controlled and monitored systems are usually comprised of

door lock and card reader with electrically operated (mains), low-voltage lock release
communication cable (usually a coaxial cable) that links all the room card readers in the hotel
the central controller and key maker units

These systems are either one- or two-way interactive. In the one-way system, a card is manufactured in the key maker. The card may be magnetic strip or, more often, a punched-hole card. A random number generator within the controller gives the card a number, which is then translated to the card. The control operator introduces this card by giving it a room number. The controller then memorizes the new card number, couples it to the room number, and erases any previous numbered card code given to that room. When the card number is appointed to a room, the controller sends a signal to the card reader telling it to recognize only that card (or one of the other six levels of master card). The lock will then

always accept that card until a new card is produced that supersedes it.

With a two-way, interactive system, the lock and card reader can report to the controller unit to notify the controller that a false or forged card has been tried, that forced entry is being attempted, or that a door is propped open or has failed to latch shut following an exit. The two-way system is usually far too expensive for all but the biggest and best establishments and often even for these applications it is overcomplicated for the staff being asked to carry out security duties.

Stand-Alone Key Maker Systems (System 2)

With stand-alone key maker systems, the automatic communication between card reader and controller is removed. A key maker unit makes a card as described in the preceding section. A hand-held programmer is then used by management security staff to program the lock to accept the card issued. Most stand-alone units are battery-powered units and hence there is an additional maintenance cost.

Advantages.
System 1.

reprogramming locks and reader is automatic
when guests pay bills, their room number may be entered, canceling all cards issued to them or their family groups
locks do not have to be replaced, as total reissue of cards and reprogramming of locks can take place daily, if required
with two-way systems, management can tell whether maids, etc., are in rooms or attacks on locks are taking place and deactivate lock operation from the controller

System 2.

simple to operate and cheaper than system 1 (capital cost only)
no wiring between rooms
locks, readers, etc., may be fitted later far more easily than for system 1

Disadvantages.
System 1.

capital cost
cabling required
available management capabilities (e.g., knowing when rooms are being cleaned, etc., can jam the system and obscure its primary security function)

System 2.

with some systems, locks must be programmed manually each time the card is changed (although the person programming the card has no knowledge of the code). Better systems automatically remember the code on insertion of a new card.

no alarm functions

When considering the cost of an installation of this type, or preparing a brief, due allowance must be made for

capital cost of equipment

card blank costs

maintenance requirements (in-house and manufacturer) and cost

how many times the locks will operate on a single battery (stand-alone) and replacement costs (labor and materials)

When selecting systems, look for

systems that give adequate mastering levels to suit client's operational needs

cards that are not easily reproduced

units that include full mortise-deadlocking latches with internal dead bolt thumb turns (these are more secure than a half mortise dead bolt latch with "do not disturb" or "privacy" button)

systems that sound a local door buzzer on the unit if the door fails to latch (when card is removed most locks close automatically) or system that will not allow card removal from the unit until latch catches; have instructions telling user to "shut door again" or "call management"

units with an additional capability to revert to standard key-locking doors in case of system failure

Avoid systems that have conventional knob-set locks, as these are extremely unsecure. Ensure that the lock does not form part of the actual handset and that, even with handset removed, access to the lock is not available (otherwise picking is still possible and security reduced).

Costs for centrally-controlled, two- or one-way systems vary enormously depending on the number of rooms, physical disposition, etc., and a large element of the cost will be cost of actual installation (particularly on later installation). More conventional and commercial card ACSs can be used in hotels, etc., either for all doors or for some areas, goods, etc.

8
Security Lighting Systems

INTRODUCTION

When concentrating on the specifics of security lighting it is easily forgotten that no matter what form it takes, all security lighting is trying to achieve one goal—to improve vision. If people could see effectively in the dark, unaided by electrical means, the need for security lighting would drop to almost nothing and a high proportion of all nighttime crimes would be reduced.

To criminals, darkness has several advantages: it conceals their activities, thus providing them with more time and less risk of detection; it reduces the risk of surprise discovery, as approaching patrol guards and lights can be seen; it improves their ability to escape without recognition if discovered, and it increases the risk of ambush to security personnel and hence has a psychologically deterring effect on their diligence (in the interests of self-preservation).

Conversely, good security lighting will reduce the area of concealment, increase the likelihood of recognition, reduce the risk of personal attack, and generally make the task of defense easier. Security lighting designs can range from the simple, internal schemes found in many modern office blocks to the high-security external schemes used on strategically important facilities. The former type of scheme is well within the scope of the informed architect or security manager if common sense and logic are applied to the problem. The latter schemes are not within their range of experience or abilities and should only be carried out by designers who specialize in security lighting (often within the client's own organization). In between these extremes, there are many types of schemes that also require the advice of the specialized lighting engineer. This is particularly so with area and "facade" lighting, where the client believes that aesthetics are often as important as security!

Whatever the scheme, sound assessment of alternatives or proposals can only be made by a person with some knowledge of the subject. The following sections provide a basis in the fundamental principles of security lighting and give as much raw data as possible. The data sheets at the end of the section give details of lamps, scheme comparison, etc., and should be used as reference material.

Beware!

It is widely believed that for better viewing, it is enough to throw more light on a particular area; this is generally true. However, in security lighting, things are not quite that simple. The ability to see effectively depends on the level of illuminance not only immediately above or on the object, but also on the surrounding area. The ability to see also depends on the length of time the eyes have to adapt to changing lighting conditions, the reflectance of viewed objects, etc.

While security lighting might require careful design and calculation, simple common sense should never be forgotten or ignored:

Smaller objects are harder to see than large objects under lower lighting levels. What size objects do you hope to detect? A vehicle? A crawling man?

The smaller the object, the higher the illuminance and contrast that may be needed in order to see the object.

Moving objects are easier to spot in low light levels than stationary objects.

A man dressed in black, standing against a black wall that is well lit will still be difficult to detect. A man dressed in black, standing against a white wall will not be so difficult to detect.

Illuminate the attacker, not the defender.

Minimize places to hide.

You can think of many more. Test design proposals against them.

CATEGORIES OF SECURITY LIGHTING

Security lighting is normally categorized as *external* [i.e., street and approach lighting, flood and facade lighting (see Figure 8–1), area lighting, and perimeter or boundary lighting] or *internal* [i.e., minimim level, small area lighting (shops) and quick-response, large area floodlighting (factories)].

Deciding into which category your area falls will greatly assist in determining suitable lamps, illuminance levels,

FIGURE 8–1   Floodlight lighting for security purposes using SOX lamps

etc., when selecting from manufacturers' data or using reference books. It will also be valuable in clarifying the objectives of the security lighting in a particular area.

Remember that standards other than security standards may apply when security lighting is being considered. While it might be perfectly workable for security lighting to provide illumination to only, say, 0.1 footcandles, health and safety standards may require ten times this level as a minimum. Simply classifying a system as *security lighting* will not help you in court if a guard is hit by a truck or falls downstairs!

Classification of Risk and Definition of Terms

As with all security design, the design of security lighting is very much linked to the category of risk (see Table 8–1) for which the system is designed, which, in turn, is usually a measure of the value of the goods within a certain area or importance of that particular area to the client.

Alphanumeric classifications are always misleading, but frequently used instead of design descriptions, specifications, and briefs. Hence, it is important to know what each alphanumeric classification represents.

No single design scheme will fit into only one category, unless it is very simple. Hence, all areas of a large- and small-scale plan of the building and grounds under con-

sideration should be studied and given some thought before deciding, with the client and security advisers, on the actual risk level to be adopted in a particular area.

IES Classifications

The Illuminating Engineering Society of North America (IES), in its lighting handbook, gives general classifications for security lighting:

> *surveillance lighting*—lighting to detect and *observe* intruders
> *protective lighting*—lighting to discourage or deter attempts at gaining entrance, acts of vandalism, etc.
> *safety lighting*—lighting for safety, to permit *safe* movement of guards and other authorized personnel

These are, once again, *not* definitions of a need, risk, or solution, but simply an attempt at a shorthand description. Always, always try to define each area of design by writing and/or drawing

the risk envisioned
the *level* of illuminance thought appropriate (footcandles or lux)
the luminaire chosen
the lamp selected
the position of the luminaires

Table 8-1   List of Generally Accepted Classifications of Risk and Corresponding Class of Security Lighting Scheme Design

Grade of Security	Class	Typically
1. Deterrence of vandals and casual opportunist criminals with little or no skill	D Basic level	Low risk. Good weatherproof or robust fitting suitably wired, but with no special switching or wiring methods.
2. Protection of low values against a deliberate criminal, or protection of high values against vandalism/pilfering, etc.	C Intermediate level	Moderate risk. A security lighting system using robust light fittings. Lights would be switched from a central point or photo or time controller not accessible to normal visitors. System can have dual function, e.g., publicity lighting window display.
3. Protection of high or moderate values against a deliberate criminal with moderate level of skill	B Advanced level	High risk. Vandal-resistant fitting must be used, and very securely mounted. Wiring to be in solid drawn steel conduit with clutch head screws/fixings (to be permanently fixed if possible or concealed). Photoelectric control as a basic switching option. Locked, separate distribution point for supplies with "signed out" keys to lock. Key luminaires on system to be battery standby, or totally self-contained and have quick ignition and fast restrike times.
4. Protection of high values against organized crime; high level of skill	A Maximum level	Extreme risk. All measures of Class B plus: Fittings recessed in brick work, concrete, etc., or buried. Wiring buried or concealed. Duplicate overlapping system, e.g., alternate lamps on separate circuits. Full standby power generation for supply with maximum interruption time of 5 seconds before fully reinstated. Link to alarm system to indicate tampering with wiring (balanced relays). Inclusion of CCTV to monitor areas.
5. Defense if terrorist target	T	No standard approach. The advice of various bodies involved in protection against terrorism MUST be sought and no work of any kind should take place until these bodies have submitted their recommendations.

As we shall see, a description like "Grade 1, Class D, Level 1, No DB Security Lighting" may look impressive on a design document, but it is really a poor substitute for a full design description.

Functions and Objectives

We shall later see that security lighting design schemes are often given functional classifications—deterrent, revealment, prevention, or detection. Revealment is, by far, the greatest functional class in security lighting designs. However, with all these classes, one should always ask, "What are my objectives?" Security lighting is provided because *seeing* is believing. If you could identify friend from foe and animal from human by sense of smell or by hearing (like many animals can), you would not need lighting!

What are the levels of "seeing" an object?

Level 1—Detection; I know it is there.
Level 2—Recognition; I know it is a man.
Level 3—Identification; He is not a member of the staff.

Guards must be given sufficient light of the right type, direction, continuity, uniformity, etc., to move from Level

1 to Level 3 in a very short period of time as an intruder moves across a very small area. If they cannot, your lighting design has failed. Consideration of the IES's recommended lighting levels shows that at a minimum, at any time, the difference between simple detection and good recognition ("high level of detection" in IES definition) is when illuminance jumps from 0.5 lux to 5 lux!

ILLUMINANCE LEVELS

The provision of a certain level of illuminance will not of itself provide adequate security lighting. However, in purely practical terms, one must start somewhere and that point should be with standard equipment. For this reason, the quantity of light flux *falling onto* an area (or the *illuminance*) is chosen as the basis for design. The units of illuminance are lux (or lumen/m²) in System Internationale (SI) units and footcandles (lumen/ft²) in imperial units. Illuminance may be viewed as the density of light on a subject with the lumen being the flow of light radiating outward from the lamp. Illuminance may be regarded as being diametrically opposite in effect to fog. The greater the amount of water vapor, as fog, within a particular area (or volume), the greater the obscuration;

FIGURE 8-2 Safety and surveillance at low cost using 18W SOX lamps

the greater the flow of light, in lumens, onto an area, the greater the illuminance. Low-cost security lighting is illustrated in Figure 8-2.

Accepting that illuminance levels simply provide a relative level scale on which to base one's subjective appreciation of what is being provided, Table 8-2 shows generally accepted illuminances for various tasks in security lighting:

$$1 \text{ lux} = 1 \text{ lumen/meter}^2$$

$$10 \text{ lux} = 1 \text{ footcandle}$$

Under health and safety legislations, an employer or site owner has a duty to ensure sufficient light is available at all times for employees to carry out their work *safely.* A "reasonable" level should be provided. This may be just a fraction of a footcandle for some tasks, to many footcandles for large vehicles maneuvering close to buildings after dark. Your insurers will normally advise "reasonable" levels of illuminance, but they are normally simply referring back to standards and codes of practice that have been used in successful defenses for negligence claims in courts. Foremost of these standards is the *IES Applications Guide* published by the IES. Table 8-3 shows IES criteria for illuminance of their categories of lighting.

In the case of security lighting, the illuminance level will usually result from a combination of the security luminaire output and the district brightness (DB) of adja-

cent, normally lit areas (e.g., streetlights). When designing, great care should be taken in evaluating the DB contribution and in higher risk areas, it should be ignored, as it might not be on when it is needed most! In general, it is best to design a security lighting system solely on the security luminaire contribution to provide a minimum acceptable, reasonable level of illuminance and to view the DB as a bonus.

QUALITY AND OTHER CONSIDERATIONS

Luminance, Uniformity, Appearance, Contrast, and Cover

Luminance. Once the required illuminance level has been established, the next step in the design process is to ensure the effectiveness or quality of the light within a particular area. In very simple terms, the amount of light falling onto an area is its *illuminance level;* the amount reflected from it is its *luminance.* (We quite often refer to things being *luminous* in the dark.) The difference between the two amounts is the reflectance and absorption of the surface onto which light is falling (i.e., luminance = surface illuminance × % surface reflectance). In security lighting, we are not really concerned with the illuminance of areas, objects, etc. What we are really concerned with is the luminance of the person we wish to detect!

Table 8-2 Recommended Illuminance Level for Various Tasks

Level of Illuminance (lux) E_h	Task (DB = district brightness)	Comparison or Example
<0.05 or <0.005 footcandles	None. Base level of illuminance at which eye will perceive presence of an object by photopic vision (or mainly by use of rod receptors). No perception of color.	Dark room, no artificial light, moonless sky, i.e., starlit night in open country.
0.2 or 0.02 footcandles	None. Base level for vision proper without excessive adaptation time.	Moonlit night in open country with no other sources; minimum for safe movement.
0.5 or 0.05 footcandles	Minimum recommended level for any area considered under any risk at all.	Midway between poor streetlights on a small residential back street.
1.0 or 0.1 footcandles	Minimum recommended level for areas with no DB to allow positive detection of person.	Very poor residential street lighting level. Recommended level for safe nighttime pedestrian lighting or internal emergency lighting.
1.5 or 0.15 footcandles	Minimum possible level in immediate vicinity of area of focus of normal CCTV (Newvicon).	As above.
5 or 0.5 footcandles	Recommended average level to be achieved throughout schemes to allow positive orientation. Colors just perceivable (Class C lighting). Safe facial identification level (if 0.8 lux cylindrical).	Normal good residential street lighting level, or internal level, typically used in large industrial storage areas; 0.6 footcandles is recommended level for medium activity general pedestrian/ parking areas.
10 or 1.0 footcandles	Recommended on all doors, windows, etc., that are particularly vulnerable or of high risk. Class C with medium DB. Class B with low DB.	Well-lit larger residential street lighting. Often requires minimim level for external security lighting.
<15 or <1.5 footcandles	Class C moderate risk with high DB. Class B medium risk with DB. Class A extreme risk with low DB.	Busy "B" road lighting.
20 or 2.0 footcandles	Recommended for perimeter with street lighting nearby and for facial recognition at distance. Class B high risk with high DB. Class A extreme risk with medium DB.	Main road lighting. Good attractive facade floodlighting. Vehicle storage/parking lots.
30 or 3.0 footcandles	Class A extreme risk into high DB.	Strategic buildings; oil, gas, water, military.
50 or 5.0 footcandles	Recommended for checkpoints.	Entrances, gate houses, etc.
100 or 10 footcandles	Recommended for *continuously* active CCTV work.	75–100 lux is normal level recommended for recreation or training in sports halls.
200 or 20 footcandles	Checking documentation continuously. On desk in gate house or reception.	Lighting level found in most ancillary areas of office blocks.
400–500 or 50 footcandles	Normal reading level in offices.	Standard office illumination level.
1000 or 100 footcandles	Drawing office or detailed working level.	>1000 lux color television requirement.
5000 or 500 footcandles	Overcast daylight.	
14,000 + or 1400 + footcandles	Bright sunny day.	

E_h = measured at the horizontal plane (the most common form). Good security and CCTV lighting calculations will consider the vertical illuminance, E_v (CCTV works), and the cylindrical illuminance, E_c (facial recognition to guards).

A burglar in dark matte clothing will have far less luminance than one in bright white or yellow clothing with a highly reflective surface under identical illumination levels.

The burglar will not normally oblige us with the latter type of clothing, so we must somehow improve our odds of seeing him under a certain illuminance level. This can be achieved in many ways. The first is to increase illuminance levels in areas that have matte surfaces with highly absorbent texture so that more light will be reflected onto the person and then from him to our eyes, cameras, etc. The second is to improve his luminosity—his apparent brightness to our eye *at any instant*. This is achieved by reducing the amount of adaptation the eye must undergo

Table 8–3   IES Recommended Security Lighting Levels

Category	Task	Illuminance*	
Surveillance	Detection and observation of intruders in large, open areas using standard or glare lighting systems:	*Lux*	*Footcandles*
	Average value anytime	2	0.2
	Absolute minimum anytime	0.5	0.05
	Confined areas as above		
	Average value anytime	5.	0.5
	Absolute minimum anytime	1	0.1
	Pedestrian and vehicular entrances		
	Average value anytime	10	1.0
	Absolute minimum anytime	2.5	0.25
	CCTV surveillance	As required by camera, see Chapter 5; 10 lux usual minimum or average.	
Protection	To discourage or positively deter potential intruders by increasing risk of detection to a high level	Average value 5.0 lux with 10 lux in sensitive or particularly vulnerable areas.	
Safety	To reduce probability of accident, collision, mugging, traffic accident, etc.	As required by IES code for particular industry, production area, building type, etc. *Safety always takes precedence over security level.*	

*All levels are on vertical or *observed* plane 3.3 ft (1 m) above ground level. Uniformity should not be lower than 0.75 if possible, or a minimum of 0.5 where not.

in order to adjust to new lighting levels when a person is moving from one area to another. Two examples of how luminosity depends on the eye adapting are

> going from a very bright room to a dark garden and bumping into a tree not seen because the eye has not had time to undergo a full adaptation
> becoming "blinded" by light when moving back from the dark garden into the extremely bright room

To improve the apparent luminance of a person, we must try to grade the security lighting so that situations with large differences in illuminance levels do not occur over short periods of time or within small areas. This is particularly important to security personnel doing their rounds after dark.

Uniformity.   The correct gradation of lighting levels is normally referred to in lighting design as *uniformity*. Good uniformity of lighting across the lit area results in very little falloff in illuminance levels between lamp positions.

Luminaires must be placed to produce acceptable uniformity of lighting levels between sources and the lit area. This uniformity is the ratio of the maximum illumination level to the minimum illumination level in the lit area. Rules of thumb are

> average uniformity is 3:1   (both measurements
> maximum uniformity is 8:1   taken on horizontal
>   plane)
> discomfort contrast is 10:1
> luminaire and lamps should be of equal size and mounting height

fewer, wider luminance pattern luminaires with good overlap and uniformity are better than many smaller, narrower, luminance pattern luminaires

Appearance.   Another extremely important criteria of good security lighting is that of the *appearance* of the intruder in the protected area. The appearance is a combination of the contrast between the luminance of the intruder and the background illuminance level, the difference in color between background and intruder, and the direction and size of the shadow cast by the intruder. Of these three factors, the most important is the contrast.

Contrast.   The concept of night vision contrast is not as easy to explain today as it was in the days of black-and-white television! In those days, it was self-evident that a good level of contrast between the various types of white, through gray, to black gave good picture definition at the low luminance levels of which the black-and-white television sets were capable.

To understand how and why contrast is so important, we need to know a little of the makeup of the human eye. The eye gathers available light and focuses it on the retina at its rear. The retina is comprised of photo receptors called *cones* and *rods*. Connections to nerves from these cones and rods convert the photons of light energy into chemical messages to the brain.

Rods are the predominant receptors at low lighting levels, as they are far more sensitive. Rods give people a limited ability to see in the dark. The cones are the predominant receptors in higher illumination conditions. In between these extremes different proportions of rods and cones come into play.

Rods cannot stimulate color vision in the brain. Subsequently, we only see gradations of gray in an unlit or poorly lit area. The stronger these gradations or *contrast,* the more clearly we see the object. So, we must try to enhance the contrast in a scene at any of the available lighting levels. The simplest ways of achieving this are to

increase the background reflectance of the area intruders must cross (as they will not oblige us with reflective clothing!). This will make intruders appear grayer in contrast to the background.

turn intruders into larger intruders by ensuring they cast a good, long shadow. This requires correct side displacement of luminaires.

Cover. In all security lighting systems (interior and exterior), the extent of an area without "cover" for the intruder is critical to successful detection. If a sprinter can run 100 yd (91 m) in less than 10 seconds, then an average person can cover 10 yd (9 m) in less than 3 seconds. This is not a long time to recognize a person or distinguish a person from a large animal, and certainly not very long to differentiate between friend and foe.

In general, clear zones should be 10 yd (9 m) or more if possible. If clear zones of this size are not possible, then the contrast must be improved between intruder and background. If the exterior district brightness is good, this could mean actually reducing perimeter security lighting levels below normal design values to silhouette intruders against the background.

Other Lighting Techniques

Ensuring that adaptation, luminosity, and uniformity of illuminance are considered in the security lighting system design, will normally only result in a coarse scheme. By thinking about the nature of the problem a little further, a far more refined and efficient design can result at no extra cost (in fact, you usually save money).

Think about what you are trying to do, to what, and for whom! For instance, floodlighting an area used as a parking lot will be of little use if the glare resulting from incorrect positioning of lights blinds the guard watching the lot.

The following paragraphs explain techniques that may be used to enhance the performance of any security lighting scheme.

Glare Control and Generation

Glare is defined as a *condition of vision,* in which there is discomfort, a reduction in the ability to see significant objects, or both, due to unsuitable distribution, range of luminance, or extreme contrasts in space or time. There

are several key factors, in this somewhat long definition, that need an explanation in order for you to fully understand the significance of glare in security lighting.

Glare is a condition produced in the eye that gives *discomfort or disablement;* it is not a property of a light source. Glare is the result of the way in which the illumination from the source interacts with its surroundings. Glare will change with lamp position relative to an object, the position of observation, the reflectivity of the object surface, and many other factors.

Glare can have two effects; it can cause actual discomfort, eyestrain, etc., and it can reduce one's ability to recognize objects (disability glare).

Glare is usually the result of the combination of two conditions:

• contrast
• saturation

Contrast. If the amount of light *reflected* within the field of view, from any object, varies by more than a ratio of 10:1, we usually say that the contrast is too great and will produce discomfort. The eyes will normally try to encompass more than they can actually see if the level of illuminance at the edges of the field of view is much less than that in the center. This will cause divergent vision. However, the eyes like to *converge* onto an object and focus; *divergence* causes strain to the muscles controlling the eyes.

Saturation. Saturation glare is the result of the eyes' unwillingness to cope with extremely high levels of reflected light. This will seldom have any significance in security lighting as illuminance levels are inherently low.

While discomfort glare is unpleasant, disability glare is far more important in terms of security, as it may mean the difference between seeing and not seeing an intruder. Disability glare results when the eye is trying to focus on a normal target and it is disturbed by a bright light at the very edge of the scene. To reduce the potential effects of discomfort by divergent vision, the eye tries to mask the bright object by becoming less sensitive all around the edge of the scene. This veiling can mean that the guard, while more comfortable, has involuntarily become "blind" to intruders near the bright source at the edge of her field of vision. The more normal the angle of the light source to the eye, the more acute the glare and veiling.

Glare as a Positive Force

Glare is commonly used in everyday language in a negative manner. In security lighting schemes, it can be either a negative or a positive factor. Generated correctly and used sensibly, glare can enhance many security lighting designs. All security lighting schemes should be designed to illuminate the intruder or potential intruder and not to illuminate the building security guard.

By using glare as a positive force, we can kill two birds with one stone. We not only illuminate intruders, we also dazzle them, making it difficult for them to see beyond a certain point or even disorientating them when they are caught. As glare depends, to a large extent, on the contrast conditions and the direction of observation, in order to *generate glare* we must ensure either that lights are directed into the normal field of view or that a high level of contrast exists within a small area, thus preventing quick adaptation.

Using glare as a positive force, we are trying to ensure production or disability glare only. Criminals will tolerate some discomfort glare as their incentives are higher than those of the security guard. It should be noted that these positive glare methods should never be used when there is any risk of disabling glare to people outside the area to be protected. Beware! Light can travel long distances at night in open country! Never use this method near adjacent roads. Figures 8–3 and 8–4 show various ways in which the positive glare effect can be used to enhance revealment lighting on perimeter areas.

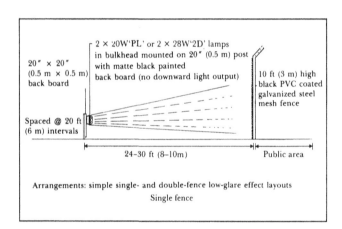

Figure 8–3   Perimeter glare lighting to reveal

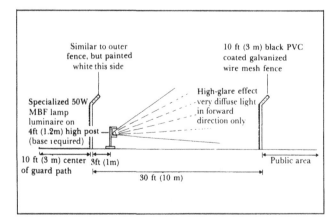

Figure 8–4   Glare lighting for double–fence mount bulkhead

Figure 8–5   Glare as a negative force—disabling the view from a gate house

Glare as a Negative Force (See Figure 8–5)

The judgment of whether glare is a negative or positive force in security lighting depends, in more ways than one, on your point of view. To the security guard in her box being dazzled by approaching headlights, glare would be seen as a negative force. To the terrorist driving that car at the barrier, it seems a positive force. In most security lighting schemes, one will inevitably come across some area where glare is a problem to the designer. In most cases, the problem is easily solved and the benefits to the design can be enormous. Problems with glare are invariably caused by

luminaires that are not properly sited or aimed
lamps that are not properly shielded to diffuse light
surfaces with high reflectivities too close to a lamp or luminaire

Glare Control

Although much thought has been given to and research carried out on the subject of glare within artificially lit buildings, little practical evaluation of glare as an outdoor phenomenon has taken place.

Many designers make the mistake of using the *glare control mark* as a recommendation when choosing outdoor lighting system types. This index of the *glare production* of the luminaire is only of any real use for internal lamps in very regular patterns and no faith should be placed on it out of doors. Rather than place their faith in manufacturers' "numbers," designers should

ensure that no lamp is positioned so that it shines directly into the eyes of observers, passersby, and, most particularly, motorists
choose lamps with some form of direction control in the form of a louver, diffuser, etc.

never choose luminaires that have one, fixed, position and always allow at least 15° adjustment in two planes if at all possible

never backlight movable objects. If the objects are moved, glare could result.

try and ensure that post-type lamps produce intensities lower than 1500 candelas when the product L × A$^{0.25}$ is used (L being the luminance and A being the area over which this occurs)

never put lamps close to highly reflective surfaces

position luminaires at regular intervals

try to place lamps of the same type and intensity in the same area

try to use a few, wide-angle beam luminaires to get good uniformity rather than smaller pools of light from many smaller sources

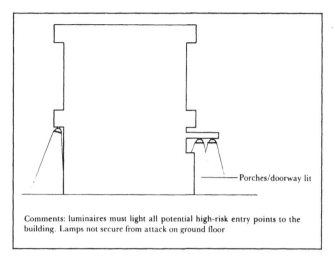

Comments: luminaires must light all potential high-risk entry points to the building. Lamps not secure from attack on ground floor

FIGURE 8–6   Partial facade deterrent lighting

FORMS OF SECURITY LIGHTING

The types of security lighting required for a particular building depend on many factors. However, no matter the type of building, the security lighting scheme will contain some, if not all, of the following elements by which we try and classify its function.

- deterrent or protective
- revealment or surveillance
- prevention
- detection

Deterrent or Protective Lighting

A well-lit building perimeter or open area will always cause a criminal to think very long and hard about the risks of being seen at work and caught red-handed. Even a modest amount of lighting, for instance a porch light over a domestic front door, will have an enormous deterrent effect. (See Figure 8–6.) Lighting to deter has one weakness—the deterrent is actually *the thought of being seen.* If the building is a remote, unmanned store in open country, this form of lighting will have no real value.

Lighting to deter is usually the cheapest form of security lighting, as it simply provides light on very high risk, small areas, such as doors, gates, etc., in order to shorten the time criminals can work unseen. The intention is to not actually reveal their activities, but to put doubt in their minds so that they think about the risks involved in attacking the premises. Deterrent lighting is commonly used on city stores, supermarkets, domestic houses, etc. Vandal-resistant luminaires with small lamps illuminate doorways, passageways, parking spaces, etc. The lamps are usually time-switched or photoelectrically controlled and require only standard wiring and conduits.

Revealment or Surveillance Lighting

The philosophy on which revealment lighting is based is that the deterrent effect has failed and, despite some psychological dissuasion, the criminal has decided the potential rewards outweigh the risks. Because of this, revealment lighting is normally only considered on buildings or areas with a security classification of Level 2 or above and is normally termed *Class C* (moderate risk). The pertinent point in the security classification is vulnerability to the *deliberate* criminal. This form of security implies that some thought has gone into the plan of attack. Thus, a good deal of thought should go into the design scheme to reveal the criminal's presence. There are four types of revealment lighting:

- perimeters or boundaries
- open areas, grounds, etc.
- building facades
- building interiors

Whichever type is considered, it is important to remember that revealment, in terms of security lighting, only means providing illuminance levels of 0.1–1.5 footcandles (1.0–15 lux) in appropriate places. It does *not* mean daylighting the subject. As with deterrent lighting, revealment lighting depends on the alertness and judgment of security personnel for its effectiveness. A fast-moving object seen under revealment lighting could initially appear to be a man when, in fact, it is the factory cat casting a large shadow! However, with revealment lighting, security personnel will eventually be able to make the distinction and act appropriately. It is essential that revealment lighting be designed to aid security personnel, not to hinder them. Glare must be controlled in line of sight so that guards can recognize the intruder's appearance easily. The lighting should *reveal the criminal,* but not spotlight the security guard. For instance, a secu-

FIGURE 8–7 Partial facade lighting to deter for safety reasons

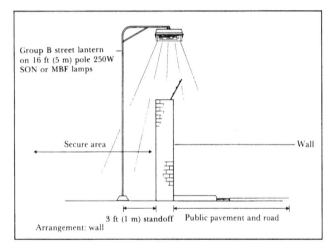

FIGURE 8–8   Perimeter revealment lighting (surveillance)

rity guard within a checkpoint house should not be more highly illuminated than the area she is surveying, lest she be revealed and her movements become obvious to the criminals.

The following paragraphs give broad details of the aims of each type of lighting and their effects on building design. Figures 8–7 through 8–12 show typical applications and arrangements of fittings.

Perimeter or Boundary Lighting to Reveal

The aim of perimeter security lighting is to enable security personnel to detect the movement of intruders as they are about to climb a boundary fence, straddle a wall, or drive through gates; or to highlight persons loitering

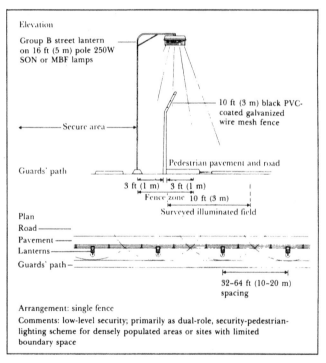

FIGURE 8–9   Perimeter revealment lighting (surveillance)

outside a boundary and report them to passing police patrols.

If the security personnel who patrol a building are housed either in a highly illuminated internal room or external gate house, their effectiveness is reduced. As mentioned earlier, this will be because their eyes will take time to adapt to the new, lower level of illuminance and their movements might be apparent to intruders, who could thus avoid or attack them.

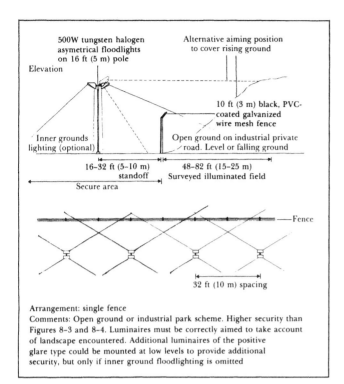

FIGURE 8–10 Perimeter revealment lighting (surveillance)

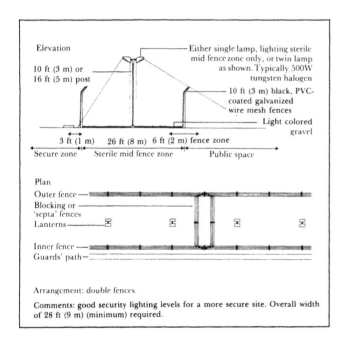

FIGURE 8–11 Perimeter revealment lighting (surveillance)

In order to prevent this, consideration should be given to the design of the area from which security personnel operate. The following provisions should ensure that guards are not put at any unnecessary, avoidable disadvantage:

Ensure that the room faces directly onto the main entry gate in the boundary (i.e., with windows primarily looking along the gate and fence).

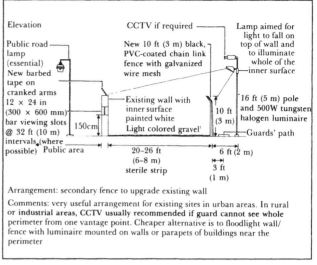

FIGURE 8–12 Perimeter revealment lighting (surveillance)

Ensure that two exits from this room are available and convenient for "doing the rounds." One exit can be used for checking the gate normally viewed and for checking personnel entering or leaving the grounds by the gate. The other exit should not be directly visible from the front gate or road and should allow security guards to leave the hut without being observed. This will ensure that nobody is tempted to climb over the front gate after seeing guards walk away to do their rounds in the full blaze of the gate security lighting.

Use tinted glass or tilted, restricted-height-and-pane-size windows to prevent people from seeing in. Use white-painted glazing bars and only allow glazing on facades at 90° to one another.

Use directional task lighting with dimmer circuits to allow persons inside the room to examine documents, etc.

Ensure interior surfaces are subdued with matte colors.

Gates, Barriers, and Entrances. In addition to the design of security huts, gates and entrances should receive special consideration in the design of a security lighting system. Most thefts of any importance are of items that require transportation from the premises. Criminals are very ergonomically minded and will always prefer to drive a vehicle onto the premises to load stolen goods, rather than to carry individual items to the boundary (where they might be seen by a passing police patrol). So, good security lighting at the gate will help security personnel see unauthorized vehicles enter or leave, identify registration plates, and assist police patrols. It will also help with the "inside job." Security personnel are loath to check company or private vehicles, or personnel they recognize, because this behavior is thought antisocial, is time-consuming, and is difficult under nighttime conditions. Good lighting at a barrier or gate will allow security

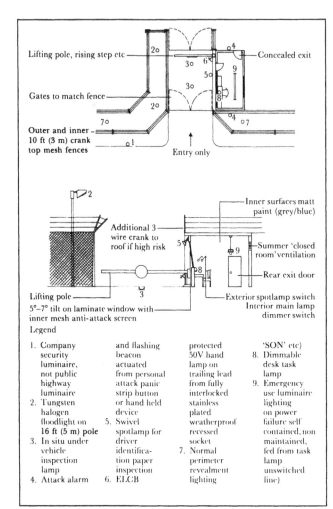

FIGURE 8–13 Gate house and barrier areas

personnel to see under, above, or inside vehicles to look for stolen goods; more importantly, it will allow the driver to be recognized (company vans are often stolen for the purpose of stealing from the company).

Depending on the type of premises protected, a gate installation could contain the following elements (see Figure 8–13):

post- or roof-mounted, high-level general floodlighting mid-height, externally and internally switched, swivel-mounted spot lamps at 1.5 ft and 6 ft (0.5 m and 1.8 m) for inspection of vehicles, drivers' documents, etc.
a ground-leakage, current-protected power point; flexible, retractable, trailing lead; and a low-voltage, high-efficacy hand-lamp for searching into vehicles

Note: It should be remembered that all equipment that is likely to be hand-held, such as a trailing lead, hand-lamps, and swiveling spotlights, should always be of the safety, double-insulated type or be supplied at very low voltage. Similarly, all socket outlets that might be used to provide power to such lamps should be of the current-operated, ground-leakage, breaker-protected type with a trip to the "off" position on fault levels of greater than

10mA. This will ensure some degree of safety from the potentially lethal conditions in which this equipment could be used. Designs should comply with the *NEC code.*

Walls or Fences? See Figures 8–8 through 8–12. When considering revealment lighting, it is important to remember certain factors that will affect the choice of whether revealment lighting should be used:

Is the building protected by a high fence? If so, lighting will reveal persons cutting their way in or out, picking locks from the outside, climbing, etc. In this condition, revealment lighting has reached its maximum potential, because security personnel may detect and deter intruders and dissuade or help apprehend them in relative safety *before the building defense has been breached.*
Is the building protected by a wall? In this case, the potential is reduced, as security guards can only spot potential intruders if they patrol externally, and hence put themselves at risk. If they do not patrol externally, more reliance is placed on police patrols. By the time revealment lighting has played its role, the building defense will have been breached and less time will be available for security guards to do their jobs.
If the building is protected by a wall, can an open area be lit behind the wall? If so, this will improve detection by removing a potential area of concealment. This area need not be a barren strip (as is often suggested), but could contain flower beds, ground cover plants, etc., which would still provide little cover for the intruder.
If the wall is near buildings, could it be floodlit from these buildings? If so, this will provide direct illumination of intruders as they climb over the wall. Floodlighting from a distance has the very substantial benefit of being almost vandalproof. If buildings are immediately adjacent to the wall or fence, could downward floodlights or streetlights be used? (This type of lighting may need the permission of the local authority or responsible highway department.)
Are any objects, materials, temporary buildings, etc., likely to be stacked or built between the actual light source and the wall, fence, or line of sight of the security guard? If so, then obviously revealment lighting becomes totally ineffective and should not be considered in that area. (Area revealment lighting might be better suited.)

Open Field Sites and Rural Sites. In terms of revealment lighting, open field sites and rural sites are the easiest conditions for which to design a security lighting system. This is because the surrounding lighting levels are likely to be very low (i.e., a low district brightness prevails) and hence the contrast will be high. Because of the prevailing gloom of the surrounding area, a level as low as 0.8 to 1.0 footcandles (8–10 lux) on the fence face and interior zone might well provide quite good revealment lighting. However, with a level this low, positive glare is unlikely to reach its full potential effect, as the number and efficacy of luminaires would probably be low. By directing fittings outward, toward intruders, glare could be used beneficially. Care should be taken not to cause disturbance to others at some distance, in houses or on the roads. Glare is less important here than general contrast revealment and so schemes that utilize low-cost, small-output luminaires mounted at regular intervals on posts some distance from an internal fence or directly mounted to shine upward on an internal wall face are probably the most cost-effective. Miniature low-pressure sodium (mini-SOX) lamps are well suited to this purpose, as are miniature fluorescent bulkhead fittings. Low district brightness will mean that post-top luminaire designs would only be required in areas or buildings with a high-risk classification (see Data Sheets 6, 7, and 8 in Appendix 5).

Towns, City Center, or Industrial Estate Sites, Roads, and Pavement. On these high-risk buildings, glare can be effectively used *only* if the boundary illumination is quite high, while the internal and external illuminances are low.

The district brightness here will almost always be higher than in open field sites. Hence, in order to achieve the same contrast, the illuminance levels will need to be higher. Vandalism is also likely to be more of a problem. Generally, space is at more of a premium and hence roads and pavement are likely to be immediately adjacent to the boundary wall or fence. The incidence of crime is likely to be higher.

These conditions warrant different solutions from the open field site. Glare effects can only be used on fences not facing roads or on walls. High illuminance levels require more efficacy from lamps, if operating, installation, and maintenance costs are not to become prohibitive, due to large numbers of luminaires. For these reasons, post-top or high-level, roof-mounted luminaires with high efficacy and good directional properties are favored.

Vandalism can be a real problem in city areas. A lamp that fails could be due to natural aging, vandalism, or removal. If lamps are not suitably protected within luminaires with very good, robust, vandal-resistant properties, maintenance personnel will fight a losing battle against damage and the system efficiency will be substantially reduced. Failed lamps will soon become the norm and security personnel will not notice luminaires from which lamps may have been deliberately removed.

Underlighting. Seeing is believing, but does it work? Many architects and designers are faced with clients or

public organizations that resist the cost of security/road lighting on the grounds that it cannot be proved successful in reducing crime. "Perhaps the criminals just happened to go elsewhere." While there are many 1970s studies into urban crime reduction, a 1988 study conducted by the Centre for Criminology and Police Studies in Great Britain has shown marked reductions in all forms of crime after a significant period by simple enhancement of street lighting levels to reduce street crime in a deprived inner city area. The report concluded that "good lighting cannot only be regarded as effective crime prevention strategy, it is desirable in its own right. It improves the appearance of an area, encourages people to use the streets, enhances public safety and road safety and can and does reduce crime."

Open Area Lighting to Reveal

See Figures 8–14 and 8–15. Essentially, open area lighting can be given the general term *floodlighting*. This term gives an impression of the types of fittings and lamps that might be successfully used in such a design scheme. The objective with open area lighting is always to reveal the criminal passing through an area or working within it. However, the type of applicable lighting design depends largely on the type of area and the risk classification.

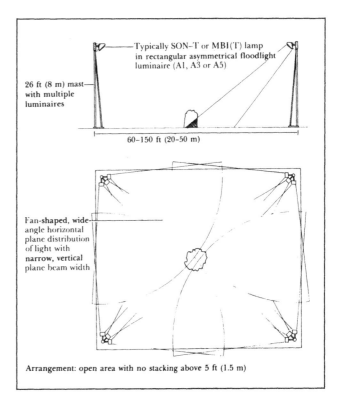

FIGURE 8–14 Area lighting to reveal

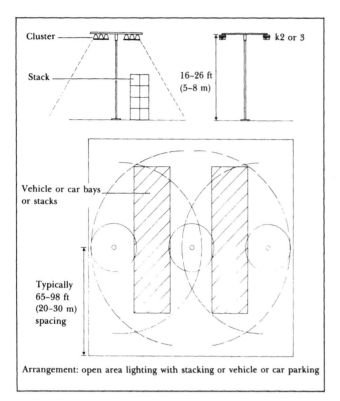

Cluster

Stack

16–26 ft
(5–8 m)

k2 or 3

Vehicle or car bays
or stacks

Typically
65–98 ft
(20–30 m)
spacing

Arrangement: open area lighting with stacking or vehicle or car parking

FIGURE 8–15   Multiple lamp cluster or outreach arm pole

It is essential to remember that for perimeter lighting to be effective, a low brightness is required. Hence, area lighting should not normally be installed within 65–98 ft (20–30 m) of perimeter lighting or the former could reduce the effect of the latter!

If an area is large, open, and normally empty at night (for instance, an office parking lot), a scheme utilizing very few, directional luminaires, mounted high in each corner, might give the required illumination. However, the same scheme might cause shadows in an area where goods are stacked, which would make it ineffective and unacceptable because of the higher risks involved. Obviously, the solution is either use more lamps and luminaires to remove shadows and reposition them within the area to be lit or use high-mast lighting (see Figure 8–15). This obviously implies higher costs.

When considering area lighting to reveal, the cost plays a more important role than in, perhaps, any other type of lighting.

High operating and maintenance costs require that careful consideration be given not only to the type of design scheme, but also to the lamps, luminaires, and posts and fixtures used. Cost comparisons should always be carried out, even quite early in the design process.

In some situations, the security and building function can be combined (e.g., the same pole may carry normal, nightshift, loading bay lighting and a reduced number of the same fittings on the same pole used later in the evening for security pupuses).

Costs/Benefits. Obviously, before a client will consider the use of area revealment security lighting (or any type of security lighting) she will want to know not only its cost, but its benefits. The client will probably ask, "If I do not adopt this scheme, what risks do I run?" As in security lighting, there is no true payback period; the benefit depends largely on the accuracy, fullness, and truthfulness of the client's brief. If the answer to any of the following questions is "yes," the benefit factor value and the worth of the scheme will be high:

Will any *likely* damage caused by the activities of a criminal on the building facades result in high inconvenience or loss? For example, arson (e.g., small fire in perimeter rooms), actual facade fabric damage (e.g., stained-glass windows or rare artifacts that require large expenditures on restoration or replacement), or water damage, loss of air conditioning, and indirect damage to goods in perimeter rooms.

Could materials stored within the open area be easily stolen or damaged? Are the articles of high value? Remember, a single, badly damaged car, or cars with the stereo cassette removed, could cost more than a security luminaire or complete scheme.

Is the area of high amenity value and would lighting provide a particular sense of security? For example, lighting of forecourts, parking areas, etc., near old persons' homes, single-parent housing developments, etc.

The design and appraisal of the true worth of an open area revealment scheme should be based on a brief that includes the following information:

building use
building contents or storage area contents
high-value areas
final exit position
surrounding occupancies and building types
assessment of vandalism risks in the area
surrounding street lighting provisions

Building Facade Lighting to Reveal

Building facade lighting can take several forms, all of which have the benefit that they usually provide security plus an enhancement of the nighttime appearance of the building. These forms are

total facade lighting — floodlighting (see Figures 8–16 through 8–19)
partial facade lighting — lighting of doorways, etc., from external sources or lighting of ground-floor walls only (see Figure 8–20)
internal facade feature lighting (see Figure 8–21)

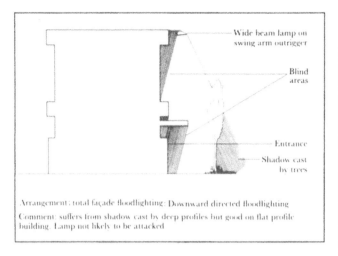

Wide beam lamp on swing arm outrigger

Blind areas

Entrance

Shadow cast by trees

Arrangement: total façade floodlighting: Downward directed floodlighting

Comment: suffers from shadow cast by deep profiles but good on flat profile building. Lamp not likely to be attacked

FIGURE 8–16 Building facade lighting to reveal

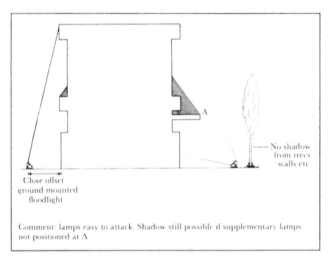

A

No shadow from trees walls etc

Close offset ground mounted floodlight

Comment: lamps easy to attack. Shadow still possible if supplementary lamps not positioned at A

FIGURE 8–17 Upward-directed floodlighting

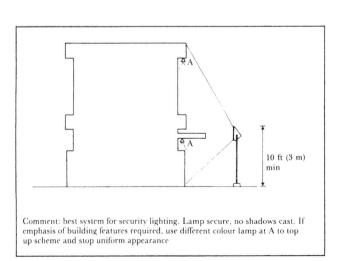

A

A

10 ft (3 m) min

Comment: best system for security lighting. Lamp secure, no shadows cast. If emphasis of building features required, use different colour lamp at A to top up scheme and stop uniform appearance

FIGURE 8–18 Remote floodlighting

FIGURE 8–19 Remote, post-mounted facade lighting

FIGURE 8–20 Partial facade lighting

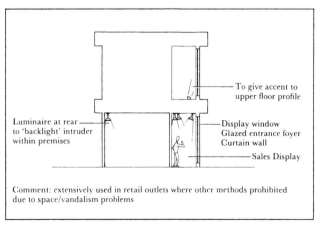

To give accent to upper floor profile

Luminaire at rear to 'backlight' intruder within premises

Display window
Glazed entrance foyer
Curtain wall

Sales Display

Comment: extensively used in retail outlets where other methods prohibited due to space/vandalism problems

FIGURE 8–21 Internal, facade, and feature lighting

It should be remembered that the vast majority of criminals gain entry on the ground floor and hence it is usually not necessary, from a security stance, to light the entire height of a facade. (Although in some cases the advertising this provides might actually pay for the scheme!)

Total Facade Lighting. Total facade lighting usually requires the use of several, high-efficacy luminaires mounted externally. Depending on the type of building and its form and position, one or more facades may be floodlit. Total facade floodlighting is only really cost-effective (in security terms) when there is a low ambient illuminance. If the surrounding illuminance levels are high, either because of the presence of good street lighting or the use of open area lighting, facade lighting on this scale will have little additional deterrent or revealment effect. However, if ambient levels are low (e.g., factories with unlit, open forecourts; public buildings with unlit gardens; etc.), the deterrent and revealment factors are considerable.

When considering total facade lighting, the plan outline and side elevation/section of the facade are of prime importance. It is fairly obvious that a building facade that has little detail; has flush-fitting windows, doors, etc.; and is perpendicular, will present few problems when designing for total facade lighting. However, a typical, neoclassical, public building with deeply recessed windows, large mullioned details, porches, doorways, pillars, etc., is a different proposition. Here, the design must be correct or the resulting shadows will reduce the security effectiveness considerably (although the aesthetic appeal might be quite high).

When considering the building facade, the options are usually

downward-directed floodlighting (roof-mounted; see Figure 8–16)
upward-directed floodlighting (ground or basement; see Figure 8–17)
remote floodlighting (on poles; see Figure 8–18)

Downward-directed floodlights

will not illuminate deep, recessed porches or windows effectively
will not be subject to vandalism or accidental damage
will not obstruct passage around the perimeter of the building
will not cause blinds to be drawn by occupants at night
will illuminate not only the facade of the building, but also a zone of ground in front of it
will require less frequent lens cleaning than upward-directed floodlights
will require no buried, specially protected cables (i.e., armored or ducted)

This "whole facade" type of lighting is usually eminently suited to modern, modularly designed factories in which elevations are simply straight, plain, sides and where vandalism and obstruction could be a big problem. The overall effect is bland and not aesthetically pleasing, but in a factory-type application these are not prime considerations.

Upward-directed floodlighting is usually considered on older buildings where the dual roles of appearance and security are important. It is normally only used when the building plot is so restricted that floodlights cannot be placed at any reasonable distance from the building. This is quite often the case with modern office blocks, older town halls, and other public buildings in city areas. Ideally, this type of lighting should never be used, as much better overall effects will be obtained in terms of security and aesthetics from a floodlight placed some distance from the building.

With this lighting form, it should always be remembered that the overall appearance of the building will never approach its normal daytime appearance. This is because under daylight conditions illumination of the surface is usually a mixture of downward-directed light and spectral ground or surface reflection. Consequently, the shadows cast at night by floodlighting will be almost the opposite of those during the daytime.

In general terms, the points to consider in the design of the upward-directed facade lighting scheme are as follows:

Will the luminaire be easily vandalized in the chosen position?
What level of attack should it withstand (e.g., stamping, hammer)?
Could the lamps be protected in special pits or recesses? Can these be drained?
Could the lamps cause glare conditions for passersby or motorists?
Will the angle of incidence of the luminaire beam on the wall result in low modeling due to the spectral surface?
Will shadows cast by porches, etc., be a risk? If so, could top-up lighting be warranted?
Could the building be more effectively lit at less cost using partial facade and/or internal facade lighting?

Unless the building is protected by a fence or wall, this type of lighting has a poor security rating. It is too easily open to attack, masking, etc., to inspire real confidence. In city areas, where police patrols are frequent, the system can still be considered as having some security value, but in open fields or buildings situated on their own grounds, this value is very low indeed.

Remote floodlighting of a building is, whenever and wherever possible, the best solution to facade security lighting. Economically, it is generally cheaper than other schemes, because fewer fixtures with greater efficacy and longer lamp lives can be used. Hence, wiring costs are reduced (providing cable runs are not of excessive length).

The ground rules for the security and positioning of floodlighting of this type are as follows:

Ensure that luminaires are located far enough away or protected from potential vandalism, are suitably weatherproofed, and are vandal-resistant.

Ensure that the luminaires are projected onto the building from the normal direction of view of security personnel. (This will sometimes coincide with the normal direction of view of the public.)

For rectangular buildings on open fields or spacious sites, aim the luminaires from a point just off the projected diagonals of the building plan (see Data Sheet 2 in Appendix 5).

Decide on the illuminance level required. (This depends on building material and finish, and the district brightness.)

Decide on the lamp type and intensity to be used, and the position and distance away from the building. This decision should be based on rough economic considerations (e.g., number of lamps, number of circuits, circuit lengths, etc.).

Generally, economic considerations result in

Scheme A — low buildings; mid-intensity lamps in luminaires with narrow beams placed close to the building and aimed steeply

Scheme B — higher buildings; high-intensity lamps within widespread beam luminaires aimed from and at shallower angles

Scheme C — tall buildings with particular, shadow problem areas; spotlighting

Typically, Scheme A would use luminaire type A2, tungsten halogen 1kW + symmetrical distribution type; Scheme B would use luminaire type B2, with MBF or SON lamps at 250W + ; and Scheme C would use projector luminaire type B3-1, medium angle, downward, long-distance floodlighting with SON lamps at 250W + or B3-2, narrow angle floodlighting. For shadow elimination from a long, offset distance, use short-life (1000-hour) tungsten filament projector lamps at 300W + [type B1 GLS (L)].

Ensure that chosen luminaires have good quality ocular reflectors with a parabolic cross section and polycarbonate front glasses.

Ensure that the lamps used are aimed correctly once they have been installed and then locked in position.

Using the quick calculation methods shown on Data Sheet 5 (see Appendix 5), check the value of illuminance actually achieved against that required. Allow a sufficient maintenance factor and coefficient of utilization in accordance with the manufacturer's data.

Check that lamp projection will not be a hazard to others.

In terms of aesthetic considerations,

offset, diagonal lighting will enhance the building's features

lighting placed in front of surrounding structures or trees will bring these items out in silhouette — an improved dramatic effect

if color appearance is important, mix the lamp types used to tone down either the yellow SON/SOX, or white tungsten halogen or fluorescent (MBIF) lamps (see Data Sheet 2 in Appendix 5)

Building Interior Lighting to Reveal

Interior lighting to reveal can be divided into two classifications:

premises that will be internally lit, but intruders will only be seen by passersby or police patrols (e.g., shops with large window displays)

premises that employ security personnel to carry out internal patrols (e.g., large office blocks, banks, factories)

Shops. The first class is the simplest in technique, but most expensive to run. This is because, if security relies on revealment to persons on the street, the scheme will only be successful if

the field of view is clear, which is usually only the case in modern shops with large display windows and racks at right angles to the windows

stock is very strongly backlit so that those persons removing it are in high silhouette

stock is generally lit to a higher level of illuminance than necessary in most security schemes [i.e., 3–5 footcandles (30–50 lumens)]; reduced-level, general illumination

If the field of view is clear, the decision has to be made either for backlighting or reduced-level lighting. The choice will, of course, rest on the particular circumstances. The layout and content of some premises may actively prohibit the use of backlighting (e.g., racks across the line of view). If, however, backlighting can be used, it is generally less expensive than reduced-level, general lighting. If the cost of stock occupying the floor space is modest (e.g., grocery stuffs, retail clothing, hardware), the use of one or more rows of fluorescent light fixtures left on all night (or suitably timed) will provide adequate security. If the stock is of a more expensive type, then it should not be left on display, but locked away securely. When this is not possible because of the work involved in handling the goods, reduced-level lighting is the only alternative. The aim is still to reveal the presence of intruders to the passing public or police, so luminaires should be arranged to illuminate stock without creating glare.

A typical arrangement might consist of one or more spotlights or fluorescent lamps (SL; low-voltage, single-ended, tungsten halogen; and 2-D are suitable lamps) being directed from above onto the highest-value goods, which are placed toward the front of the shop, but at some distance from any breakable window. Another scheme is to retain certain areas of normal lighting and leave certain rows of fluorescent fixtures on all night. Obviously, if lamps normally used in display and general lighting (i.e., GLS and warm, white fluorescent lights) are left on all night, costs will be high due to the low efficacy of filament and warmer (low color, temperature) fluorescent lamps. Although discharge lamps such as SOX, SON, and mercury have a much higher efficacy, at present they are not looked upon (in many display area situations) as an option, because they are not of a daylight color appearance and generally give poorer color rendering. This prejudice against discharge lamps is being gradually overcome by

the introduction of lamps of the discharge type that have better color-rendering properties due to the addition of advanced phosphorous coatings

the realization that people do not generally window-shop at night with the intention to buy

the realization that, in many situations, perfect color rendering is not a preeminent requirement

Of the discharge lamps presently on the market, the SON lamp represents the most acceptable choice in terms of economy, but has poor color rendering. If this is important, the second choice is SON-DELUX (deluxe version of SON in terms of its color-rendering property). MBI lamps of the diffusing, coated variety are a good choice, although larger (longer) fluorescent tubes with high efficacy (e.g., "polylux" 4000° color temp white) are just as good and do not require special fittings or control gear.

Generally, when reduced-level lighting is recommended, PAR (encapsulated blown-glass, tungsten, mirror-reflector) lamps and GLS filament lamps should be avoided, as they have a low efficacy and short life. Lamps of this type used on security applications alone would require changing every 3–6 months; if used during the day as well, they would seldom last longer than 3 months. Compare this with 1–2-year replacements for discharge and fluorescent lamps under the same conditions.

A good compromise design commonly used in larger shops is the use of one or two recessed, ceiling-mounted, 55W SOX (SLI) lamps centrally positioned within the shop. These lamps give a monochromatic, yellow color illumination and only require changing approximately every 2 years. To offset the uniformity of color, window displays of small goods, of low value, can be lit to very

high daylight or colored illuminance levels by the use of fluorescent tubes in good quality luminaires. This will give "sparkle" to the window and a "warm glow" to the interior. It must be emphasized, however, that the window display luminaires should be aimed inward and should have good diffusers or louvers with high-cut angles or passersby will not be able to penetrate the glare of the window area to view the interior.

As the practice of using these special SOX, yellow spectrum–colored discharge lamps for security only purposes is becoming more widespread, the awareness that failure of the lamps to ignite is usually a result of interference and disconnection by burglars makes their use even more beneficial.

In the past, the main argument against the use of discharge lamps in interior security was the long *restrike time* (i.e., the time taken to reignite the light and achieve full lumen output after power is reinstated; see Data Sheet 3 in Appendix 5). However, this argument was based on the assumption that lights would only be switched on during the night if the presence of an intruder were suspected or, if power failure was caused by the intruder by switching off the supply, the long restrike time would endanger the security staff or policemen, as the criminal would be warned of their presence. Also, when lighting was reignited it would give insufficient revealment. These arguments are countered by the following:

Lamps are cheap to run, so they can be left on all night. Special lamps mean failures are suspected immediately.

Special electronic ignition circuits are now available that will give 1-minute restrike times to SON lamps (formerly 5 minutes) or instantaneous restrike to hot SOX lamps.

If special luminaires are used, separate circuits are required, which can be made secure from interference by locking isolators, etc., and switching on and off can be arranged by the use of a high-level, disguised, upward-facing photoelectric cell. This cell will ignite the security lamp when the normal house-lighting level falls away when switched off by staff leaving the premises, and is switched off when they return and illuminate the premises in the morning. This removes the need for any accessible light switch.

Offices, Factories, and Banks. Here, the prime consideration is the safety of security personnel. When patrolling inside the building, security personnel should have the benefit of adequate lighting to reveal. If this is not provided, most personnel will be too aware of the dangers they face and perhaps will perform their duty less diligently.

The arguments used for the application of special, security-lighting luminaires using discharge lamps apply just as much to factories, offices, and banks as they do to shops. The biggest difference, however, is the security and positioning of the lamp-switching device. Switching by photocell overcomes these problems. If, however, lamps are *only* required during security rounds, they can be pre-time-switched from a central point or switched locally by means of lockable, secret key, grid switches (if instantaneous response lamps are used). In the latter case, the switch should be outside the area to be inspected and should be two-way switched from two well-lit positions. This allows

anyone entering or leaving the area to be seen

anyone switching the secret key to be seen

security personnel to be guided to a place of relative safety. If the lights within the area are deliberately switched out, the security guards can see where the alternative switch is, and can go to it and reignite the lamps.

In factories, the advantages of discharge lighting for security are that

factory spaces tend to be higher than shop spaces, hence larger lamps can be used to give the same evenness of distribution. This reduces *capital* costs and increases efficiency. Wide-angle reflectors may be used.

many factories already use discharge luminaires and hence the addition of security lighting might only mean the minor modification of switching and wiring of certain fittings

the height of the fixtures above the floor reduces the the likelihood of malicious damage or preplanned interference with the lamp or gear

Areas of Special Attention. In certain parts of a factory or bank, goods or materials will be stored that are a prime target for thieves. These areas could be output warehouses with rack upon rack of television sets, kitchen appliances, or precious metal stocks. When considering these areas, the following points should be noted:

These areas are likely to be of interest to professional criminals, not opportunists.

Seventy percent of robberies from such areas are the result of some inside information, hence protection of equipment must be a prime consideration.

In all storage areas, goods are continually on the move. The security lighting should be flexible enough to provide easily adapted and easily repositioned lights that at the same time cannot be moved by unauthorized persons. Luminaires that are suspended from a structural roof grid by a locking bracket will allow this.

There are many forms of lighting such spaces, but basically, lamps used should, in this case, have instantaneous ignition (i.e., tungsten halogen floods or spots). The luminaire should have a polycarbonate

diffuser to prevent the lamp from being easily broken. The luminaire should be mounted high and out of reach. The luminaire suspension should allow for position change (i.e., a swiveling base or sliding base that can be bolted in the desired position.

All lamp wiring should be at a high level (protected in conduits or steel wire, armored wiring) and local switching should not allow unauthorized de-energization.

Lamps should be wired in small groups only, so a circuit failure (malicious or accidental) does not result in blackout conditions.

A means of increasing the illuminance in the area to normal daytime levels should be provided in a position of safety.

IMPORTANT: The author has noticed in several letters and papers on the subject of internal security lighting, the suggestion that emergency lighting systems could be used in a dual role. This is NOT advisable and would be severely criticized by any fire officer. Use of the emergency system increases the general wear and tear on the components, reduces the effectiveness of batteries (central or integral) by constant discharge, causes lowering of lamp lumen output, etc. If a security adviser ever recommends this, get a new security adviser!

Prevention

Security lighting can only be said to prevent crime when used in conjunction with surveillance devices, which would have little deterrent effect were the lights not in operation. A classic example of this would be CCTV systems using simple cameras that would not function at night without very high artificial illumination. Provision of security lighting could either make the camera visible to the intruder, leading him to think he is being or will be observed, or, if lumen levels are high enough, actually allow this to take place.

Detection

Detection in security lighting has only become common since the advent of cheap, solid state electronics. Because of its relative novelty, there are no standards covering any of the complete systems. For this reason, great care should be taken in assessing the usefulness of a particular piece of equipment before recommending its use to a client.

Basically, detection security lighting consists of linking lamps that have very short strike and warm-up times (i.e., fluorescent, tungsten halogen) to some device that will bring the lamps on if the presence of an intruder is sensed. All of these detection systems rely on the effect that sudden powerful illumination of an intruder can cause shock, surprise, disorientation, and panic. The aims of such systems are threefold:

to detect the intruder
to illuminate the intruder
to alert security personnel

Unlike the previous deterrent and revealment methods, detection systems are dormant for the majority of time and are designed to act after deterrents have failed and before revealment (at the building) becomes necessary. The obvious benefit of such a system is the reduction in lamp running time. Many devices are being marketed as alternatives to deterrent and revealment lighting based on these criteria. However, no device has yet been marketed that has the reliability level or reassurance quotient of the more usual methods of security lighting. Detection systems should be viewed as useful additions to the general spectrum of available security lighting systems and not a single solution. A common error in the application of these systems is to use types with long-range projector lamps and very sensitive microwave, ultrasonic, or passive infrared sensors to replace large areas of revealment lighting in front of buildings. However, because of poor equipment, poor design, or poor positioning, many devices either fail to detect or frequently give false alarms. Because of this, criminals have learned to "test the water" by throwing large, light objects into the monitored area to activate the suspected system. If the system is activated, they might give up or try some other means of attack. If the system does not react, thieves will be encouraged by the cover of darkness. Systems that "cry wolf" too often are obviously useless. A better use of these systems is in providing positive glare as a surprise element across the facade of a build-

ing. By relying on shorter-range detection, system reliability increases. Using perimeter lighting as normal, a building facade remains in darkness with a cross-facade pattern of detection devices linked to high-level illuminance, and alarms. This reduces the requirements for area floodlighting and building facade revealment lighting.

Sensing Devices.   Most detection lighting systems rely on sensing devices that are constructed to UL Standards and most are commonly linked to alarms. Table 8–4 gives the device type, reliability rating, etc., for the more common types of sensors used.

When deciding whether or not to recommend detection as a form of security lighting, the following should be assessed:

the reaction time—the speed of response
the false alarm rate—failure of component parts
the tolerance of the system. Will it light lamps if animals, snow, etc., enter its detection zone? This is particularly important with microwave and ultrasonic detectors.
the limits of the system. Over what range will it detect? The detection range limits its usefulness. If a space is 98 ft (30 m) wide, but a device only detects within a 49 ft (15 m) wide zone, this results in 49 ft (15 m) of area of blindness in which the criminal can act undetected.
the switching capacity. How many lamps, of what rating, can be switched from a single unit? Is a master control unit needed or recommended?
Does the system operate on standard voltages?
Are the switching and sensing devices enclosed in robust, weatherproof, and resistant enclosures? Are they included in the price or are they extra? Are they installed by others?

Table 8–4   Intruder Sensors for Use in Ignition of Security Lighting Systems

Device	Description	Reliability Rating	Uses
Infrared ray (active)	Projected light from transmitter focused on photocell. When beam broken, current in cell ceases and alarm activates.	B (Internal use)	Across small passages, door openings, etc.
Differential pressure sensors (liquid)	Pressurized liquid-filled tubes buried in ground. Detect pressure vibration in ground and transmit to fluidic devices as shock waves.	A In correct environment	Used on military installations as a perimeter "zone" detector.
Passive infrared	Senses bodily heat radiated from an intruder.	B	Small passages and rooms.
Ultrasonic movement detector	Sound waves transmitted at high frequency— 20–40kHz. Interruption of waves causes variation of pitch due to Doppler effect.	A	External and internal spaces of medium to large size.
Microwave (or radar) detector	Extremely high frequency—10.7GHz. Action as ultrasonic.	A In correct environment	As above.

A = Good   B = Fair

How many sensing units and luminaires will be required to ensure overlaps occur and no blind spots exist?

Is the detection and sensing element equally responsive throughout its entire quoted range or is the quoted range a maximum? Does the signal strength, etc., improve at close range?

Can the sensors be easily and readily adapted or coupled to sensors manufactured by others?

How much of the system could be supplied and installed by a standard electrical contractor? Specialized installation and lack of alternative supplies or manufacturers will increase the costs. Installation by a specialist does, however, have the benefit of increased reliability.

Will spares of this equipment be available in 5 years? Is the company fairly well established?

How much of the system and its components are covered by UL standards?

Where can a local installation be seen in operation and tested? Who can be contacted to give an unbiased appraisal? Has the system been government agency or defense tested?

Decisions on detection-type systems should never be made based on published propaganda. Actual installations will provide far more useful (and truthful) information.

A far more reliable type of detection lighting is the linking of tried and tested internal use detection devices made by reputable security specialists for security lighting systems. By means of relays and contactors, a signal present in, say, a magnetic door alarm or pressure pad can be used to give an audible alarm and a police alarm signal, and, in addition, instantly illuminate a desired area. This type of system has the advantage of simultaneous alarm and lighting, unlike many security lighting detection systems that are, essentially, sophisticated light switches. What is often forgotten by manufacturers and promoters, is that security personnel might be on the far side of a large building and not even realize that the security lights have come on!

SCHEME PLANNING

The following paragraphs try to explain the logical sequence of events that should make up the successful design and installation of medium to large systems. It is not exhaustive and, although on paper, the order is, I hope, logical, many projects in practice are far from this ideal. In many instances, security systems and particularly security lighting are an afterthought or detail in which the architect initially shows little interest. Little wonder that the resulting design may upset the architect's aesthetic concept of the building by use of ill-considered lighting. The time to provide constructive criticism is during the design, not once the scheme is complete. If even cursory attention is paid to the following topics, the resultant design should be more acceptable to all the parties concerned.

The Brief

The Building.

    use
    contents
    consideration of working hours (e.g., whether security
        guards will be on station)
    areas with hazardous or valuable materials
    grade of security required
    internal or external lighting
    detection or revealment, etc.

Cost Restraints.

    Is there a separate budget for security work or is it part
        of mechanical and electrical services?
    Will work be carried out by specialists or will electrical
        contractors be used?
    Will security lighting be measured and specified? Or
        will an estimated sum be reserved for spending late
        in the contract?
    What is the budget? Is it enough?

Form of Contract.   Will the contract allow for the use of certain nominated suppliers, alternative materials, lowest tenderer, or certain nominated specialized contractors?

Statutory Requirements or Guidelines.   Do any statutory levels of illumination need to be maintained? If the client is the government, state, bank, armed forces, local authority, police, or metropolitan borough, do they publish or have any guidelines, technical notes, or suggested minimum standards for security lighting?

Liaison.

    Who are the client's end users of the system?
    Can they be approached to help formulate the brief?
    Who will be responsible for operating the system once
        it is completed?
    Who will maintain it?
    Will the brief require any refinements or additions
        during the contract? If so, who authorizes additions
        to the architect or supervising officers?
    Will a supervisor of works and/or a resident engineer
        be appointed whose duties include supervision of
        the security lighting system installation?

Preferences.

    Does the client have any existing security provisions?
        With whom? What types of contracts?

Would they or could they carry out the works?
Should they be invited to tender?

Defenses and Barriers: Liaison.

Who provides? When?
Fence or wall? If a fence, will it include bastions or
septa?
Can you see through the fence at shallow viewing
angles?
Can luminaires be mounted on outriggers, saving pole
costs?
Can wiring be buried at the same time the fence bottom
is dug in, secured, or the wall foundations dug?

Assessment of Site Conditions

Global Conditions. Obtain maps of the area and surrounding streets. Get information from the town or city
council on any likely modifications to the existing area
(e.g., new road plans, new housing, etc.).

Surveys.

local conditions — clean air, pollution, presence of corrosive fumes, salt-laden air, etc., that could affect choice
of materials. High winds a problem?
local terrain — rise and fall of land in relation to building and others in vicinity. Elevations showing the
surrounding building and road levels should be
drawn. Will cable trenches be dug easily if required?
approaches — determine likely means of approach to
building from existing roads; map any obstructions
flora and fauna — map any large trees, etc., in area to be
lit that might affect siting of lamps or projection of
light
surrounding buildings — map types of buildings around
the site, levels, problems associated with light spill,
etc. Note types of construction of buildings and surfaces in order to assess reflectivities.
surrounding lighting — assess the level of nighttime ambient lighting around the site (direct brightness) and
adjacent buildings (street lighting levels). Find out if
and when lighting is switched off (particularly street
lighting).

Statutory Bodies

Crime Prevention Officer. Inform crime prevention
officers of intentions. Arrange to meet them to discuss
local crime and vandalism conditions.

Highway Department. If any area of the site is to be
serviced by local authorities, security lighting might also
include safety lighting (e.g., street lamps to give minimum local authority lighting levels). Check on type of
street lighting, plans to modify it, type of switching, etc.
Ask whether lamp standards could serve dual street and
security lighting roles (particularly on perimeters).

Fire Officer. Determine whether any fire switches to
energize or deenergize particular areas of lighting will
be required. This is important for facade revealment
lighting.

Council. Planning permission applications will need to
give some indication of nighttime appearance. A building that looks like Alcatraz after completion, situated in
a mixed residential and commercial area, will almost certainly incur the wrath of tenant associations and could
cause the client postcontract problems.

Site Planning

Obtain a site plan or a drawing of the proposed site plan.
Indicate the orientation of the building, facade types,
main entrances, approach routes, areas of storage, parking, likely security routes, trees, shrubs, and other hiding
places.

Building Planning

Obtain building drawings that show

building plans and elevations at each level
types of facades, recesses, reveals, doorways, upstands,
overhangs, mullions, and windows
areas within the building and their uses; areas requiring internal security lighting; positions of final exits,
main entrances; position of any central security room

Outline Design

Take each, in turn. Consider briefly and document the

philosophy to be used (deterrent, etc.)
means of achieving this philosophy (system types)
levels of illuminance to be adopted
types of equipment to be used (luminaires, post heights,
etc.)
outline wiring routes and potential clashes
builder's work required
approximate number of lamps required

Discussion

Armed with the above, discuss the proposals with the
design team and client. If possible, get a firm commitment and agreement, particularly on the philosophy to be
adopted. The major cause of postcontract argument and

complaint is that other parties "thought" they knew what they were getting and this was not supplied.

Detailed Design

Calculations.   Calculations should cover

circuitry—routes, loadings, voltage drop, protection from overload, shock, mechanical damage, derating factors, ring mains, spurs, individual circuits, fuse ratings (particularly important on discharge lighting), cable types, final termination

routing—positions of subcircuit origin, center of load, use of feeder pillars, trenching, builder's methods of installation and protection

lumen calculation, etc.—*use the appropriate calculation* to ensure correct, resultant illuminance. To carry out such calculations the following details will be required:

- lamp lumen output
- maintenance factors
- derating factors
- utilization factors
- distances to subject
- reflectivity, glare, etc.
- beam angles and divergence
- color-rendering properties
- contrast

Remember, the value of illuminance (in lux) resulting from the calculation must be applicable to the use of the system. For example,

E_h (horizontal illuminance)—general internal security lighting

E_c (cylindrical illuminance)—facial recognition for patrolling guards

E_v (vertical illuminance)—CCTV system illuminance level

Each of these results from a different formula.

Component Choice.   Discussions on

cable types

protection of cables—in earthenware ducts, armored, in conduit, etc.

luminaires—degree of vandal resistance, weather-tightness, corrosion protection

lamps—life/capital, cost/lumen, output/mounting position, ease of replacement

switching—manual, central, time-switched, photoelectric, monitored circuitry

post types—aluminum, steel, concrete

Drawings.   Drawings must include as much information as possible, as the person installing the system is unlikely to have read your specification or the company's quotation.

suitable scale

necessary details (e.g., post foundations, cable entries, puddle flanges)

luminaire schedule

who does what

positioning and aiming accuracy

Specification

If the design of the system is being left to a specialized designer/contractor, it is extremely important that you are sure you know what the contractor is proposing. Do not accept a quotation with a vague description of the work, sketch, and a price. The quotation should be the last page of a comprehensive package of design calculations, specifications, and drawings that cover all aspects of the work listed in the preceding sections. Do not be afraid to ask to see the calculations!

If this design is done in-house and you intend to appoint a specialized contractor for the installation, your specification should be at least as good as that which you would expect from a specialized design/contract company.

The specification of a security lighting installation should inform the contractor of the aims of the design and tell them the equipment, circuitry, etc., that they shall use and the method of installation, standards of workmanship, and materials required to achieve this.

Particular importance should be given to

luminaire position

luminaire type and particular catalog specification

the confidentiality of the bid, drawings, and installed works (cables)

the importance of setting up and commissioning the systems (particularly important in floodlighting) and the required attendance during this period

the particular importance that will be placed on the witnessing of *all* tests

Selection of Bidders

The architect or engineer should make formal prebid inquiries that request, at the minimum, the following information:

company size and personnel (permanent staff, working only)

company turnover and profitability

past experience in this type of work with stated examples and contacts, which can be visited if necessary normal insurance coverage

whether the company has its own facilities for design, working, and particularly for preparing record/as installed drawings

Comparison of Returned Bids

Bidders should be scrutinized to ensure

specified manufacturers have been used and not cheaper alternatives

lamps, luminaires, and posts quoted are those specified

numbers of lamps and luminaires are on schedule of rates and quantities

any exclusions are noted (e.g., builder's work by others, post holes by others, duct laying by others)

commissioning period bid on is correct and the monies are sufficient for adequate attendance

bid is not based on a one-visit concept and the contractor can meet programs of other trades, coordinate her work with theirs, etc.

Contract Progress

Checks should be made at regular intervals by the supervising officer or his representative to ensure that

cables to be buried are of correct size, do not have any subterranean jointing, and are properly laid and protected in accordance with specifications

correct luminaires and lamps, as specified, are actually being installed

routes of buried cables are *exactly* in accordance with design and a true record drawing is being made

any modifications to designed cable routes do not affect other services that might be buried

route changes are noted and do not come within any secure area

Testing and Commissioning

All testing for compliance with regulations should be fully witnessed.

Checks on positioning and aiming of luminaires should be carried out.

If detection systems or automatic switching devices are used, these should be exhaustively tested for correct operation, before handover, by a responsible person. Their correct operation should be verified in writing, to the architect or supervising officer.

Full measurements of illuminance and illumination levels on all areas should be taken.

Maintenance

No security lighting system should be installed before a planned maintenance agreement schedule has been drawn up, and a person or company appointed to carry out the work involved. The maintenance agreement should include

daily and weekly visual checks on all lamps and luminaires (usually by guards with instructions from installers)

weekly checks on correct operation of system and component parts (usually by guards with instructions from installers)

6-month testing of circuits and cleaning of luminaires

repair, painting, etc., of *slightly* damaged or corroded parts

replacements of failed items or those beyond repair

regular, complete lamp replacement program in accordance with lamp manufacturer's lamp mortality rate

emergency lamp replacement arrangements (24-hour repair service with 12-hour response) with countercharging for failure to meet obligations

Maintenance and Switching

Maintenance costs can be high and maintenance times long when luminaires are needed. Maintenance should be reduced by

using poles or posts of aluminum for corrosion resistance, lightness, and ease of treatment. Steel corrodes, needs galvanizing, painting, or both, and is heavy.

using poles that are flange base plate mounted, not root mounted (i.e., buried directly in the ground). This allows poles to be removed or taken down for maintenance more easily and reduces hidden corrosion problems of root poles. Ensure concrete footing is the correct size to withstand turnover forces of maximum likely wind speeds and so that poles will also withstand these forces. Some older security lighting systems do not look pretty after even quite moderate storm force winds. Have the footing stand 6 ft (2 m) clear of the ground, so that the base plates never lie in standing water or vegetation. Cover the base plate and flange with bitumen to reduce corrosion.

on high poles, for example, over 26 ft (8 m), use hinged-at-base, raise-and-lower poles to allow easy changing of lamps without special equipment or trucks. Remember to ensure the direction in which the pole is to be lowered and that the surroundings will always allow this! Do not put your security fence in the way!

wiring the system via photoelectric cells that are frequent, but not mounted on top of luminaires (where they are difficult to access). Alternatively, have each luminaire or group of luminaires controlled by lamp-top photoelectric cells that have internal column wiring and a base plate switch inside the door that allows override of the photocell to the lamp should the photocell fail.

LAMPS AND LUMINAIRES

Introduction

The choice of the correct lamp and luminaire for a particular security lighting task is probably the most difficult part of any design scheme. In order to decide on the most suitable lamp and luminaire, designers should initially think in general terms and ask themselves the following questions:

What is the purpose — spotlighting, area floodlighting, facade floodlighting, general illumination?

What degree of control over the spread of light from the luminaire is required? Can the distribution be general bat-wing, directed forward, etc?

What degree of physical protection from vandalism is required?

What level of weatherproofing, corrosion resistance, etc., is required?

Armed with the answers to these questions, the designer will turn to catalogs of manufactured luminaires and attempt to select a suitable type. There are literally thousands of luminaires manufactured in the United States alone, so the task will be difficult. Luckily, many manufacturers, realizing the need for good, specialized luminaires for security lighting, have begun to make special ranges. These luminaires are, however, invariably more expensive than their similar all-purpose "brothers." Data Sheets 1–8 in Appendix 5 show the various types of lamps and luminaires commonly found in the security lighting field.

In the selection of any particular lamp for a scheme, the major areas for consideration are

total lumen output per watt per lamp (i.e., the efficacy or efficiency of the conversion of electrical energy into useful light)

the life of the lamp (i.e., the number of hours it will run at a high percentage of its initial lumen output). The longer the life, the lower the replacement maintenance costs.

the restrike time (i.e., the time it takes to reestablish full lumen output after a power failure)

the distribution of light from the lamp when it is enclosed in a given luminaire

When scrutinizing proposals look to see whether these criteria have been properly considered. Consult Tables 8–5 through 8–9 for help. In addition to the markings described in Tables 8–5, 8–7, and 8–9, all details that are necessary to ensure proper installation, use, and maintenance should be given either on the luminaire, on a built-in ballast, or in the manufacturer's instructions provided with the luminaire.

Table 8–5 Classification of Luminaires According to the Type of Protection Provided Against Electric Shock (from British Standard 4533)

Class	Type of Protection	Symbol Used to Mark Luminaires
0*	A luminaire in which protection against electric shock relies upon basic insulation; this implies that there are no means for the connection of accessible conductive parts, if any, to the protective conductor in the fixed wiring of the installation, reliance in the event of a failure of the basic insulation being placed on the environment.	No symbol
i	A luminaire in which protection against electric shock does not rely on basic insulation only, but which includes an additional safety precaution in such a way that means are provided for the connection of accessible conductive parts to the protective (grounding) conductor in the fixed wiring of the installation in such a way that the accessible conductive parts cannot become live in the event of a failure of the basic insulation.	No symbol
ii	A luminaire in which protection against electric shock does not rely on basic insulation only, but in which additional safety precautions such as double insulation or reinforced insulation are provided, there being no provision for protective grounding or reliance upon installation conditions.	▣
iii	A luminaire in which protection against electric shock relies upon supply at safety extra low voltage (SELV) or in which voltages higher than SELV are not generated. The SELV is defined as a voltage which does not exceed 50V AC rms between conductors or between any conductor and ground in a circuit which is isolated from the supply mains by means such as a safety isolating transformer or converter with separate windings.	◈

*Class 0 luminaires are not permitted in the United Kingdom.

Table 8–6 The Degree of Protection Against the Ingress of Solid Bodies (First Characteristic Numeral), Moisture (Second Characteristic Numeral), and Mechanical Impact (Third Characteristic Numeral) in the Ingress Protection (IP) System of Luminaire Classification

First Characteristic Numeral	Short Description	Brief Details of Objects That Will Be "Excluded" from the Luminaire
0	Nonprotected.	No special protection.
1	Protected against solid objects greater than 2 in (50 mm).	A large surface of the body, such as a hand (but no protection against deliberate access). Solid objects exceeding 2 in (50 mm) in diameter.
2	Protected against solid objects greater than 0.5 in (12 mm).	Fingers or similar objects not exceeding 3 in (80 mm) in length. Solid objects exceeding 0.5 in (12 mm) in diameter.
3	Protected against solid objects greater than 0.1 in (2.5 mm).	Tools, wires, etc., of diameter or thickness greater than 0.1 in (2.5 mm). Solid objects exceeding 0.1 in (2.5 mm) in diameter.
4	Protected against solid objects greater than 0.04 in (1.0 mm).	Wires or strips of thickness greater than 0.04 in (1.0 mm). Solid objects exceeding 0.04 in (1.0 mm) in diameter.
5	Dust-protected.	Ingress of dust is not totally prevented but dust does not enter in sufficient quantity to interfere with satisfactory operation of the equipment.
6	Dusttight.	No ingress of dust.

Second Characteristic Numeral	Short Description	Details of the Type of Protection Provided by the Luminaire
0	Nonprotected.	No special protection.
1	Protected against dripping water.	Dripping water (vertically falling drops) shall have no harmful effect.
2	Protected against dripping water when tilted up to 15°.	Vertically dripping water shall have no harmful effect when the luminaire is tilted at any angle up to 15° from its normal position.
3	Protected against spraying water.	Water falling as a spray at any angle up to 60° from the vertical shall have no harmful effect.
4	Protected against splashing water.	Water splashed against the enclosure from any direction shall have no harmful effect.
5	Protected against water jets.	Water projected by a nozzle against the enclosure from any direction shall have no harmful effect.
6	Protected against heavy seas.	Water from heavy seas or water projected in powerful jets shall not enter the luminaire in harmful quantities.
7	Protected against the effects of immersion.	Ingress of water in a harmful quantity shall not be possible when the luminaire is immersed in water under defined conditions of pressure and time.
8	Protected against submersion.	The equipment is suitable for continuous submersion in water under conditions which shall be specified by the manufacturer.

Third Characteristic Numeral	Short Description	Details of the Type of Protection Provided by the Luminaire
0	No resistance to impact.	Exposed lamp.
1	150 g weight from 15 cm.	Plastic enclosures to resist.
2	250 g weight from 15 cm.	Toughened plastics for enclosure.
3	250 g weight from 20 cm.	Toughened plastics for enclosure.
5	500 g weight from 40 cm.	Toughened plastics for enclosure and polycarbonates or GRP.
7	1.5 kg weight from 40 cm.	Metal cases of aluminum, steel, or polycarbonate.
9	5 kg weight from 40 cm.	Metal cases of aluminum, steel, or polycarbonate.

Table 8-7   Ingress Protection (IP) Numbers Corresponding to Some Commonly Used Descriptions of Luminaire Types and the Symbols That May Be Used to Mark a Luminaire in Addition to the IP Number

Commonly Used Description of Luminaire Type	IP Number*	Symbol That May Be Used in Addition to the IP Classification Number	
Ordinary	IP20†	No symbol	
Dripproof	IPX1	▪	(one drop)
Rainproof	IPX3	⬛	(one drop in square)
Splashproof	IPX4	△	(one drop in triangle)
Jetproof	IPX5	△△	(two triangles with one drop in each)
Watertight (immersible)	IPX7	▪▪	(two drops)
Pressure-watertight (submersible)	IPX8	▪▪ m	(two drops followed by an indication of the maximum depth of submersion in meters)
Proof against 1 mm diameter probe	IP4X	No symbol	
Dustproof	IP5X	◆	(a mesh without frame)
Dusttight	IP6X	◆	(a mesh with frame)

*Where X is used in an IP number in this code, it indicates a missing characteristic numeral. However, on any luminaire, both appropriate characteristic numerals should be marked.
†Marking of IP 20 on ordinary luminaires is not required. In this context an ordinary luminaire is one without special protection against dirt or moisture.

Table 8-8   The Degree of Protection Against Ingress or Damage in the NEMA System: Common Classes

NEMA Type	Description
1	General purpose
2	Dripproof, indoors
3	Dusttight, raintight, sleet- and ice-resistant outdoors
3R	Rainproof, sleet- and ice-resistant outdoors
3S	Dusttight, raintight, sleetproof, and iceproof indoors and outdoors
4	Watertight and dusttight
4X	Watertight, dusttight, corrosion-resistant outdoors and indoors
9	Hazardous locations, potentially explosive atmospheres
12	Industrial, dusttight, and driptight
13	Oiltight and dusttight

Note: The words *tight* and *proof* are not interchangeable in enclosure terms. *Tight* means total exclusion of matter; no particles will enter. *Proof* means while particles below a certain size may enter, these would normally have no adverse effect on internal working or product life in use.

Table 8-9   Information That Should Be Marked on Luminaires

1. Mark of origin.
2. Rated voltage(s) in volts. (Luminaires for tungsten filament lamps are only marked if the rated voltage is different from 250V.)
3. Rated maximum ambient temperature if other than 77°F (25°C) (t^a . . . °C).
4. Symbol of class ii or class iii luminaire, where applicable (see Table 8-5).
5. Ingress protection (IP) number, where applicable (see Tables 8-6 and 8-7).
6. Maker's model number or type reference.
7. Rated wattage of the lamp(s) in watts. (Where the lamp wattage alone is insufficient, the number of lamps and the type shall also be given. Luminaires for tungsten filament lamps should be marked with the maximum rated wattage and number of lamps.)
8. Symbol for luminaires with built-in ballast or transformers suitable for direct mounting on normally flammable surfaces, if applicable (see Table 8-8).
9. Information concerning special lamps, if applicable.
10. Symbol for luminaires using lamps of similar shape to "cool beam" lamps when the use of a "cool beam" lamp might impair safety, if applicable (▧).
11. Clearly marked terminations to identify which termination should be connected to the live side of the supply when necessary for safety or to ensure satisfactory operation. Ground terminators should be clearly indicated (⏚).
12. Symbols for the minimum distance from lighted objects, for spotlights and the like, where applicable.

Luminaire Housings and Lamp Protection

Security lighting is only of benefit if it always provides light when you need it. If a lamp fails regularly or unexpectedly, the designer has failed in her objective. Lamps can fail for a number of reasons:

  failure of electrical supply
  end of life
  accidental damage
  deliberate attack

WIRING TECHNIQUES AND MATERIALS

The security lighting installation on any premises is only as secure as its wiring. Wiring can often be the weak link in the chain. Wiring usually radiates from some central point to several different loads, some of which are security or essential loads, and others that are nonessential. Wiring also crosses boundaries from secure to nonsecure areas, and often changes direction and environments, even in quite small installations. Because of these factors, great care must be given to the sizing, rating, fusing, routing, and protection of cables.

It may be obvious, but, if so, it is well worth repeating—the overall level of security accepted as a minimum, determines the type of protection given to cables. If the security classification is low, wiring can be carried out in any manner that conforms with normal practices, recommendations, and regulations. Cables will simply need to be correctly sized and conductors afforded some protection against low-risk, mechanical damage. In such cases, PVC/PVC flat cables laid under floorboards may suffice in, for example, domestic or small, light industrial workshops. However, should the security classification require protection from terrorists or professional criminals, then measures such as monitored, grounded, conduits; steel wire armor (SWA) cables with antitamper cores, etc., might be required to sense tampering, cutting, or other malicious damage, and to provide mechanical protection against accidental damage.

Once the security classification has been determined, the wiring scheme may begin to take shape. All schemes should start with basic questions and aims clearly defined. The following paragraphs lay down a suggested order in which a scheme might take shape and gives general advice on each aspect.

Supply Point and Nature

  What is the voltage and frequency of the power supply available on the premises?
  If the supply is not to proper standards, will variations in supply voltage, frequency, etc., significantly affect the luminaire's performance? Will the lamp strike on 90% nominal voltage? The lamp manufacturer should be consulted if there is any doubt.
  What is the type of supply to the premises (i.e., overhead/underground)? Is it likely to fail? If the building has a high-security classification, failures of statutory authority supply caused by accident, malicious damage, weather, etc., should be taken into account and the consequences assessed. If the consequences of failure are too horrendous, could supplies be underground rather than overhead, as these are generally more secure? The various electric companies grade their area networks and will tell you the chances of failure. Could new supplies from a different source be brought onto the site? If none of these are possible or cost too much, but the security provision is high, a standby electrical generator must be considered.
  If even a short break in supply and hence lamp failure for up to 5 minutes is not allowed, interruptible standby supplies should be considered.
  Where is the point of the power supply? If the power supply to the site is in an insecure area, it must be made secure. If the main distribution point is in a position removed from the center of loading or the center distribution, a submain to a more central security lighting board might be required. This submain might need protection and the distribution board certainly will.

Wiring and Distribution

The cable chosen must be suitable for the job. This obviously means it must be sized correctly and it must ensure continuance of supply. Generally, the more complex the nature of the cable, the less vulnerable to changes of natural environment, less susceptible to damage, etc. Table 8–10 lists cable types that are suitable for general use.

In general terms, PVC/SWA/PVC cables are commonly thought of as superior for external conditions and mineral-insulated, metal-sheathed (MICC or MIMS) cables are preferred for interior situations. It is worth remembering that in either internal or external wiring, the relative costs of cable from, say, small core PVC/PVC to PVC/SWA/PVC are small compared to the installation labor cost. So, pick a good quality cable that is appropriate and do not compromise.

Specially Sheathed Cables and Cable Cores

In areas where cables are exposed to harmful environments, such as extremes of temperature, corrosive ground water, fumes, etc., special sheathing will be required. Ensure that the designer has taken this into account.

Table 8-10 Wiring Types

PVC/SWA/PVC (Polyvinyl chloride/steel wire armored/polyvinyl chloride cable)

Advantages	*Disadvantages*
1. Armored against accidental damage.	1. Armoring is ferrous and will corrode if subsheath is damaged.
2. A cheap, readily available, easily worked cable.	2. Larger diameter for small current carrying capacity.
3. No specialized training needed to terminate ends.	
4. Can be laid directly into duct or supported from catenary wire.	

MIMS (Mineral-insulated, metal-sheathed cable), MICC, etc.

Advantages	*Disadvantages*
1. Smaller diameter than SWA.	1. Not armored and susceptible to "pinhole" damage, ingress of moisture at terminal, etc.
2. Squashing, etc., does not cause failure.	2. Expensive to terminate.
3. Insulation not affected by high temperatures.	3. Time-consuming to install.
4. Fire-resistant.	4. In external or high security interior would need to be sheathed with a cable-protecting duct.
5. Not easily tampered with (mineral insulation easily absorbs moisture when exposed, causes short circuits in live cables).	
6. Very pliable.	
7. High current-carrying capacity.	

PVC-sheathed, single core cables would usually be drawn into conduits. Two types of cables exist—solid core and stranded core. In general, stranded cores are stronger and more pliable than solid cores, and they are superior in buried or concealed installations that might require rewiring at some future date.

Application: External Lighting

Conduits should be polyethylene or similar plastic (buried), galvanized steel, nonferrous or aluminum—steel-galvanized unions, connections, etc., should not be used or electrolytic corrosion will result.

Where conduits, ducts, etc., are used, drain holes should be used and boxes containing connectors should be packed with hydrophobic insulating material to prevent tracking caused by condensation or moisture penetration.

The ground connection to boxes should be run separately from the conductor armoring. It is not good practice to rely on the fortuitous connection of the male gland ring only to achieve a good ground.

Normal PVC sheathing will harden and age if exposed to sunlight; rubber sheathing will rapidly deteriorate under the same circumstances.

Annual maintenance should take place. Bituminous paints, lead oxide, etc., should be used to reduce the effects of corrosion on boxes.

Suspended cables should always be mechanically protected (either by conduit or SWA) and a catenary should always provide the necessary support.

There are many very tough, rigid, or flexible-sided plastic and polymeric conduits available that, if carefully selected, will provide protection equivalent to steel con-duits and be far more suitable for use, particularly if cables are buried.

As intruders normally do not possess wiring diagrams for the security lighting, they are unsure of the effect of simply cutting through an unknown cable. Total darkness might result, alarm bells may ring, etc. One school of thought says that this being the case, cables should be run on the surface or clipped to the fence. The latter is impractical due to ground maintenance needs (removing weeds and cutting shrubs). Fences mainly expose the PVC cables to daylight, which deteriorates the sheath and makes it possible for professionals to tamper with a cable. Maintenance is simple, as is testing, jointing, etc.

The opposing school of thought thinks the risk of fence mounting is too high and prefers cables to be buried at a depth of 2 ft (60 cm).

Often, economics and common sense prevail. If the ground is soft, bury the cable so that it is protected and difficult to find and tamper with. If the ground is rocky, put the cable on the fence or the trenching cost might be 40 times that of a simple cable antitamper monitor.

Distribution and Supply

The following list gives the most commonly encountered methods of supply and distribution found in security lighting designs.

- ring
- radials from a secure box in a secure duct
- dual supply points
- standby supply
- standby self-contained battery source
- monitored lighting circuit
- coaxial protection

Ring and radial circuits are usually arranged using three-phase neutral and ground cables. Every fourth luminaire is connected to the same phase conductor as the first luminaire on the run in the sequence [e.g., 1R(red), 2Y(yellow), 3B(blue), 4R, 5Y, 6B, 7R, etc.]. This achieves good security in that if a fuse should blow on one phase, only every third luminaire is de-energized and 80% lighting levels between, say, fittings 2Y and 4R might still be possible.

When assessing the method selected by a particular designer, the basic rule to remember is that it is far easier to switch off the lighting by pulling out the fuse or using a switch, than to cut the wiring. Most criminals, if attacking security lighting, will, in order of preference

attack the fitting (air rifle, stones, etc.)
dismantle the fitting
darken the fitting using material, etc.
pull the fuse or switch
tamper with the wiring

The order of preference is determined by ease of attack, lack of expertise, lack of danger, etc. If the fitting and accessories have been correctly chosen and positioned, then the first three approaches become time-consuming, will attract attention, and increase likelihood of detection. This leaves the last two options. As most criminals who attack any commercial building will have some prior knowledge of the basic layout and contents, they would probably also know the positions of mains electrical distribution cupboards supplying lighting, alarms, etc. It is, therefore, essential that these have sturdy doors and frames and good locks. In addition, individual switch fuses and isolators should be able to be padlocked in the "on" or "off" position. (This will also provide a safety measure during maintenance work on remote fittings to prevent accidental re-energization.)

Only professional criminals will attempt to tamper with the live wiring of security lighting. The main reason is that time is required to find and identify cables, and there is danger in bypassing or cutting them, etc. The end result of this work might be the de-energization of one or two lamps if they are wired radially or by the spurred tee method. These results could be far more easily achieved by the first three methods of attack.

Most professional criminals will not extinguish all the lamps on a particular circuit or development, as this inevitably draws attention to their activities. A far more common ploy is to attack certain lamps, say 12 hours before the crime, make the damage appear as vandalism, and rely on the fact that unless the company concerned operates a highly efficient maintenance scheme, those lamps will remain damaged for the next few days!

Fusing, Switching, and Cable Sizing

The system designer must be a competent electrical engineer who will take full account of the following:

Cable derating factors should be used for bunching and particularly for trenching, etc.

Due to third harmonic currents, full-sized neutrals must always be used with discharge lighting (i.e., SON, MBIF, etc.).

If multicore cables are used, manufacturer's advice should be obtained on suitability and rating.

If MICC cables are used between lamps and sources, or lamps and control gear, then heavy-duty 660/1000V types should be used.

No diversity should be applied as all lamps will usually switch-start and run together.

Cables should be generously sized for current consumption and voltage drop as additional wiring can be expensive should more luminaires be required. However, if there is some spare capacity, a circuit may be extended.

Where the exact luminaire type has not been determined (when trying to size cables), a factor of 1.8 × the kW sum of the lamps should be used to determine the kVA and hence current of the circuit.

Pure impedance loads (i.e., tungsten filament lamps) have running current equal to their starting current, hence no special fuses or switches are required.

Discharge lighting lamps required special consideration of fusing and switching.

Power Factor and Switching. When a discharge lamp has a voltage across its terminals, the only impedance between the live and neutral conductors is that provided by the gas between the electrodes. The electrodes themselves are designed to discharge electrons readily to the filler gas, promoting ionization of that gas. As they are designed to promote electron emission (i.e., current), the current that passes between the lamp's electrodes would continue to rise when a voltage is applied. This dramatic rise in current must be checked if the lamp is to have any life or it will burn out the electrodes. Ohm's Law does not apply during starting of the lamp, as the discharge takes place in a gas and not a metal, and a negative slope volt-amp curve prevails.

An inductive element is used to "choke" this rise in current to a maximum level. This inductive element has a power factor of 0.5 lagging, which would cause the kVA of the lamp to consume a large current for a given power, hence requiring large supply cables.

If the power factor is improved by means of a capacitor, the current consumption is reduced.

In addition to this capacitor, the discharge lamp requires a high potential difference (or gradient) to promote the actual ionization arc in the gas. This is generally a voltage of 400–700V. This is generated by a thyristor placed across the electrodes, which provides a high voltage with a high frequency. In high-pressure sodium lamps, a step-up transformer is also used to give a 3000V pulse.

The presence of these capacitors and thyristors limits the current surge on start (with the choke in circuit to approximately 20 times the running current). Once the lamp has ignited, this current consumption falls to one and one-half times the warm running current until the lamp has fully warmed up (usually 20–30 minutes). The results of this are that

> switches must cope with the current discharge of capacitors when closing and high-voltage surge on breaking. Hence, if power factor capacitors are used, quick action make-and-break switches should be used. If they are not present, slow make-and-break switches should be used.
>
> switches for discharge lighting control should have a rating of at least two times the normal warm running current

Many high-mast lamps have low-level chokes and capacitors; any interconnecting wiring must be capable of taking at least one and one-half times the normal running current. Fuses and circuit breakers must not break during the initial current surge and hence should be of the motor-rated type with a rating of at least three to four times the running current. Voltage transient surging will worsen, the lower the impedance of the live conductor to the point of application. A surge arrester might be required on large installations of lamps close to main substations.

The current used to size the cable should be property assessed. In discharge luminaires this can mean checking summations of lamp starting/running and capacitor current against one another. This is not a job for the unskilled.

COSTS

No prospective client or architect, bound by the financial constraints on most projects, is likely to consider a security lighting scheme unless it can be shown that costs have been very carefully considered.

All too often, security lighting (and general external lighting) is an easy target for cutting costs. The target is an easy one because, in many instances, those with little experience in security matters can see no evident return on the capital expenditure involved, nor the necessity for certain items of equipment. Toilets, chairs, and heating equipment are obvious necessities, not "that lamp on the corner of the building." There are only two ways to counter such attacks on the necessity for security lighting and they should always be used together. The first is an

explanation of the aims of the system, its principle of operation, and the protection it affords (consequential loss, vandalism, etc.). The second is a cost analysis that is complete and easily understood. All too often, the aims of the system are given without the costs and pertinent questions such as, "How much are the lamps?" "How much will it cost me to run?" find the architect or engineer with no answer. This immediately undermines the client's confidence.

Always present the most energy efficient and cost-effective design that is appropriate for the particular level of security required by the site. In order to ensure that this is the case, the following rules should be checked against the design before actual costs are determined in order to assess whether better designs might exist:

Rule 1 – Use the most efficient suitable light source available.

Rule 2 – Use the light output from the lamp effectively.

Rule 3 – Maintain equipment in good working order.

Rule 4 – Use energy-saving devices and systems that are effective and reliable (i.e. detection systems).

Table 8–11 shows the elements of a typical cost/design analysis that should accompany all proposed schemes. To use element 3.0 of Table 8–11, the designer of the system should carry out a true cost-comparison study.

Table 8–11   Cost Summary/Design Analysis

Where schemes are large, the analysis should include:

1.0	Projected professional fees for design.
2.0	Budgeted cost of each option:
2.1	Capital expenditure: lamps and control gear. wiring. luminaires, number required. bulk purchase discount. posts/bases/brackets, etc. builder's work. labor.
2.2	Maintenance: planned cleaning. planned lamp replacement. luminaire and lamp life/replacement. access and attendance costs by others.
2.3	Running costs: lamp life and replacement cost (see Table 8–13). efficacy lumens/watt or footcandles/watt. total load (control gear). power factor/kVA used. electricity tariffs, night rates, etc. daily check costs.
3.0	Budgeted cost of scheme chosen by discounted cash flow (DCF) and for cost/1000 lumens provided.
4.0	Cost of scheme as percentage of value of stock, etc., protected.
5.0	Reduction in insurance premiums possible.

Table 8–12   Approximate Layouts for Security Lighting Luminaires to Give 0.2 Footcandles (2 Lux) Minimum Illumination Levels*

1.0	Perimeter or wall facade mounted luminaires (For fences see Data Sheets 8 and 9 in Appendix 5.)

1.1		SON-T and SON-E 70W bracket mounted:				

	Height (H)		*Spacing (S)*		*Throw (T)*	
	Feet	*Meters*	*Feet*	*Meters*	*Feet*	*Meters*
	9.84	3	39.3	12	18.6	6
	11.48	3.5	45.9	14	22.9	7
	13.12	4	52.5	16	26.2	8
	14.76	4.5	59.0	18	29.5	9
	16.40	5	65.6	20	32.8	10
	18.0	5.5	72.1	22	36.0	11
	19.6	6	78.7	24	39.3	12

1.2	8.2	2.5	32.8	10	27.8	8.5
	9.84	3	41.0	12.5	34.4	10.5
	11.48	3.5	49.2	15	41.0	12.5
	13.12	4	57.4	17.5	45.9	14.0
	14.76	4.5	59.0	18	49.2	15.0

1.3	9.84	3	30.3	24.5	18.0	5.5
	13.12	4	80.2	27.5	19.6	6
	16.40	5	98.4	30	21.3	6.5
	19.6	6	101.7	31	22.1	6.75

2.0   Area lighting (See Data Sheets 10, 11, and 12 in Appendix 5 for further information.)
500W linear tungsten halogen on 26.2 ft (8 m) pole, spaced 78.7 ft (24 m), aimed 40° down:

	Height (H)		*Spacing (S)*		*Throw (T)*	
	Feet	*Meters*	*Feet*	*Meters*	*Feet*	*Meters*
	26.2	8	39.3	12	45.9	14

Giving illumination levels of 2.1 footcandles (21.0 lux) at periphery, 7.0 footcandles (70 lux) center.

70W SON-D on 26.2 ft (8 m) pole, spaced 98.4 ft (30 m), aimed 14° down:

	26.2	8	49.2	15	49.2	15

Giving illumination levels of 0.5 footcandles (5 lux) at periphery, 1.5 footcandles (15 lux) center.

2.1   Economic mounting heights:

Lumens per lamp	*Height (H)*	
	Feet	*Meters*
0–4000	14.76	4.5
4000–6000	19.6	6
6000–10,000	26.2	8
10,000–25,000	29.5	9
25,000–40,000	32.8	10
40,000 +	49.2	15

2.2   Economic spacing and maximum throw

a) For shovel reflector floods and wide-angle floods (MBF, SON):

H	Smax to periphery from last luminaire	Smax between luminaires	Tmax
T	1H	3H	3H

b) For asymmetric linear tungsten halogen floods and medium-angle floods (TH and MBF/SON):

H	Smax to periphery from last luminaire	Smax between luminaires	Tmax
T	⅔H	1.5H	4H

c) For large areas (parking lots, etc.) using large SON or MBF lamps in medium-angle or narrow-angle flood at heights recommended by 2.1:

H	Smax to periphery from last luminaire	Smax between luminaires	Tmax	Dmax
T	H	3H	3H	6H

*Always consult luminaire manufacturers for free scheme design based on particular luminaires.

D = Distance between continuous rows in line of throw with an intermediate construction over H, e.g., stacked goods.

H = Mounting height.

S = Spacing between luminaires on same facade.

T = Throw of even illumination at *x* ft (m) from base of luminaire post or wall.

Source: Courtesy of The Electricity Council of Great Britain.

Table 8–13 Typical Running Costs Comparison per Annum Using 55W SOX 100%

Lamp	Running Cost (%)	Lumens	% Cost/1000 Lumens	Lamp Life (hours)
18W Mini-SOX	33.5	1756	19.0	5000
2 × 8W fluorescent	33.5	800	41.8	2000
1 × 16W 2D	29.4	1050	23.0	5000 +
35W SOX	93.5	4390	21.3	10,000
55W SOX	100.0	7200	13.8	10,000 +
70W SON	113.6	5500	20.6	10,000 +
75W GLS (ID)	100.2	940	106.5	1000
100W GLS	127.4	1220	104.4	1000
150W TH	206.2	2800	73.6	4000
200W TH	245.6	3200	76.7	4000
300W TH	402.7	5000	80.5	4000
500W TH	668.4	2000	334.2	2000

General: SOX, SON, and Mini-SOX lamps are generally the most economical lamps to run. GLS, tungsten halogen, and MBFU are the least. Tungsten halogen still used where color appearance/restrike time important. All SOX and SON lamps should be "instant" restrike type. Note: short life/low lumen output of GLS/tungsten halogen lamps.

Annual hours run: 4000 hours assumed.

The method commonly used—discounted cash flow (DCF)—is relatively simple, but outside the scope of this book. Although it is a little tedious to carry out by hand, many building service computer software packages include this method. It is important that realistic estimates are made for annual increases in fuel maintenance and capital expenditure on lamp replacement, as these can seriously affect final calculations.

Table 8–12 gives details of various schemes that have proved effective and economical. In general, most security lighting schemes require major reappraisal or rewiring and relamping every 10 years. For this reason, all cost comparisons should assume a 10-year life.

The DCF calculation should include

capital cost of the entire installation (a price estimate)
running costs per year (based on known "hours run" and tariffs; see Table 8–13)
annual labor costs for maintenance, lamp replacement, and testing (estimated)

annual capital cost for replacement lamps (from known lamp life)
annual percentage increase in electricity cost (often estimated at 5%)
annual percentage increase in labor cost (from projected labor cost indices)
annual percentage increase in material costs (from past evidence)

Having carried out the cost comparison using DCF, it is often wise to carry out a simple check to show where the monies on each scheme would be spent. This will highlight the benefits of the design that might not be readily appreciable on either the draft design drawings or in the DCF. One of the simplest methods to use is the cost-per-1000-lumens comparison. For this, a table should be constructed along the lines of Table 8–14 (page 300) using data sheet information and the results presented to the client as a histogram.

Table 8–14 Cost-per-1000-Lumens Assessment/Comparison: Typical Calculation Sheet

Schemes		A	B	C	Comments
System description	As proposed by specialist				
1. Lamp description no. off	Data Sheet 4				
2. Luminaire class	Data Sheets 1 and 2				
3. Luminaire total circuit watts running	Manufacturers				
4. Lamp life (hours)	Manufacturers (at full output)				
5. Hours run per annum	Client: usually in excess 4000 hours				
6. Lamp replacement per annum	$5 \div 4$				
7. Lamp's lumen output total	Manufacturer for size chosen				
8. Capital cost luminaire	Inclusive of gear/lens				
9. Capital cost lamps	At date of design				
10. Capital cost poles	If applicable				
11. Running cost per annum	$3 \times 5 \times$ tariff				
12. Initial cost per 1000 lumens per annum	$\dfrac{8 + 9 + 10}{(7 \div 1000)}$				
13. Running cost per 1000 lumens per annum	$\dfrac{11}{1000}$				
14. Lamp costs per 1000 lumens per annum	$\dfrac{9 + [(5 \div 4) \times 9]}{(7 \div 1000)}$				
15. Total cost per 1000 lumens per annum	$(12 + 13 + 14)$				

Appendix 1
Suggested Readings: Security Books and Periodicals in the United States and the United Kingdom

The following books, periodicals, etc., provide very valuable additional information on security systems, suppliers, etc.

ASISNET, Security News Database Information, Inc., 1725 K St. NW, Washington, DC 20006, 202-429-8730. Provides latest security news updates.

The Association of Burglary Insurance Surveyors, Security Surveyor (monthly periodical published by Paramount Publishing, Paramount House, Hertfordshire, England).

Bernard, R.L. *Intrusion Detection Systems 2E.* Stoneham, MA: Butterworths, 1988.

Cadell, Vivian. *Burglar Alarm Systems.* London, England: Newnes Technical Books, 1984.

Fennelly, L.J. (Ed.). *Handbook of Loss Prevention and Crime Prevention 2E.* Stoneham, MA: Butterworths, 1989.

Hamilton, Peter (Ed.). *Handbook of Security* (2 vols.). Norwell, MA: Kluwer Harper Handbooks, 1988.

Lyons, Stanley L. *Exterior Lighting for Industry and Security.* London, England: Applied Sciences Publishers, 1990.

Lyons, Stanley L. *Security of Premises: A Manual for Managers.* Stoneham, MA: Butterworths, 1988.

National Institute of Justice. National Criminal Justice Reference Service (NCJRS) Bulletin Board. Customer Service: 800-851-3420; 301-251-5500. To call the bulletin board: 301-738-8895. Use modem rates from 300 to 2400 baud and modem settings at 8-N-1. Free.

The National Locksmith. National Publishing Co., 1533 Burgundy Pkwy., Streamwood, IL 60107, 312-837-2044. Fax: 312-837-1210. Editor: Marc Goldberg. Circulation: 20,000. Monthly. Subscription: $28. Adv. A technical magazine for locksmiths: product reviews, codes, servicing information, computerized locksmithing.

Poyner, Barry. *Design Against Crime: Beyond Defensible Space.* Stoneham: MA: Butterworths, 1983.

Sanger, John. *Basic Alarm Electronics.* Stoneham, MA: Butterworths, 1988.

Sanger, John. *More Kinks and Hints.* Stoneham, MA: Butterworths, 1986.

Schnabolk, Charles. *Physical Security.* Stoneham, MA: Butterworths, 1983.

Schum, John L. *Electronic Locking Devices.* Stoneham, MA: Butterworths, 1988.

Security and Fire Protection Yearbook and Buyer's Guide (annual periodical published by Paramount Publishing, Paramount House, Hertfordshire, England).

Security Industry Buyer's Guide (annual periodical published by C&P Telephone of Virginia, Bethesda, MD).

Security Journal (quarterly periodical published by Butterworths, Stoneham, MA, and ASIS).

Security Letter (semimonthly periodical published by Security Letter, Inc., of New York).

Security Letter Source Book (annual periodical published by Butterworths, Stoneham, MA).

Security Management (monthly periodical published by ASIS, Washington, DC).

Sennewald, C.A. *Effective Security Management.* Stoneham, MA: Butterworths, 1985.

Trimmer, H.W. *Understanding and Servicing Alarm Systems 2E.* Stoneham, MA: Butterworths, 1981.

Underwood, Graham. *The Security of Buildings.* Stoneham, MA: Butterworths, 1987.

Walker, Phillip. *Electronic Security Systems 2E.* Stoneham, MA: Butterworths, 1988.

Weber, Thad L. *Alarm Systems and Theft Prevention 2E.* Stoneham, MA: Butterworths, 1985.

Appendix 2
Sources of Further Information

STANDARDS

Underwriters Laboratories (UL)

UL 365   Burglar Alarm Units—Police Connect
UL 437   Key Locks
UL 464   Audible Signal Appliances
UL 603   Power Supplies for Use with Burglar Alarm Systems
UL 606   Linings and Screens for Use with Burglar Alarm Systems
UL 608   Burglary Resistant Vault Doors and Modular Panels
UL 609   Burglar Alarm Systems—Local
UL 611   Burglar Alarm System Units—Central Station
UL 634   Standard for Connectors and Switches for Use with Burglar Alarm Systems
UL 636   Holdup Alarm Units and Systems
UL 639   Intrusion Detection Units
UL 681   Installation and Classification of Mercantile and Bank Burglar Alarm Systems
UL 687   Burglar Resistant Safes
UL 752   Bullet-Resisting Equipment
UL 768   Combination Locks
UL 771   Night Depositories
UL 827   Central Station for Watchman, Fire Alarm and Supervisory Services
UL 887   Delayed Action Time Locks
UL 904   Vehicle Alarm Systems and Units
UL 972   Burglary Resistant Glazing Materials
UL 983   Surveillance Cameras
UL 1023   Household Burglar Alarm System Units
UL 1034   Burglary Resistant Electronic Locking Mechanism
UL 1037   Antitheft Alarms and Devices
UL 1076   Alarm System Units—Proprietary Burglar
UL 1610   Central Station Burglar Alarm Units
UL 1635   Digital Burglar Alarm Communicator System Units
UL 1638   Visual Signaling Appliances

UL 1641   Installation and Classification of Residential Burglar Alarm Systems
UL 140   Relocking Devices for Safes and Vaults
UL 291   Automated Teller Systems
UL 294   Access Control Units

American Society for Testing Materials (ASTM) (Committee F12)

F12.10   Security Systems and Services
F12.30   Anti-Counterfeiting Systems
F12.40   Detection and Surveillance Systems and Services
F12.50   Locking Devices
F12.60   Controlled Access, Security Search and Screening
F12.80   Bank and Mercantile Vault Construction
F12.91   Definitions and Nomenclature
F761.82   Residential Chain Link Fences

National Institute of Justice (NIJ)

NIJ-STD-0301.00   Magnetic Switches for Burglar Alarm Systems
NIJ-STD-0302.00   Mechanically Actuated Switches for Burglar Alarm Systems
NIJ-STD-0303.00   Mercury Switches for Burglar Alarm Systems
NIJ-STD-0304.00   Passive Night Vision Devices
NIJ-STD-0305.00   Active Night Vision Devices
NIJ-STD-0306.00   Physical Security of Door Assemblies and Components
NIJ-STD-0324.00   Telephone Diallers (Digital)
NIJ-STD-0318.00   Physical Security of Sliding Glass Door Units
NIJ-STD-0316.00   Physical Security of Window Door Units
NIJ-STD-0322.00   Telephone Diallers with Taped Voice Messages

Security Industry Association (SIA)

Derived Channel (Data Exchange) Channel Standard (1988)

Digital Communication Standard (1989)

SIA Ad-Hoc Low Voltage Wire Committee Position Paper (1989) on "Interpretation of 1987 National Electrical Codes"

Article 725—Remote Control Signaling Circuits

Article 760—Fire Protective Signaling Circuits

Article 800—Communication Circuits

British Standards Institute

BS 4737

Part 1, 1986 and 5804, 1987 under revision—specification for installed systems with local audible and/or remote signaling

Part 2, 1986 and 5805, 1987 current—specification for installed systems for deliberate operation

Part 3, 1988—General Requirements and Specifications for System Components (also IEC 839-2-2):

3.1   Continuous wiring
3.2   Foil on glass
3.3   Protective switches
3.4   Radiowave doppler detectors (amended 1982), under revision
3.5   Ultrasonic movement detectors (amended 1982), under revision
3.6   Acoustic detectors (amended 1980), current
3.7   Passive infrared detectors (amended 1980), under revision
3.8   Volumetric capacitive detectors (amended 1980), current
3.9   Pressure mats (amended 1980), current
3.10 Vibration detectors (amended 1980), current
3.11 Rigid printed circuit wiring (amended 1980), current
3.12 Beam interruption devices, 1986, current
3.13 Capacitive proximity detectors (amended 1980), current
3.14 Deliberately operated devices, 1986, current
3.30 Specification for PVC cables for interconnection wiring, 1986

Part 4—Codes of Practice:

4.1   Code of Practice for Planning and Installation, 1987
4.2   Code of Practice for Maintenance and Records, 1986
4.3   Code of Practice for Exterior Alarm Systems, 1988

Part 5—General:

5.2   Recommendations for Symbols and Diagrams, 1988

BS 5979 (1987), Code of Practice for Remote Centres for Intruder Alarm Systems

BS 4166 (1978 Revision), Specification for Automatic Intruder Alarm Terminating Equipment in Police Stations

BS 6329, Specification for Modems for Connection to PSTN

BS 6707 (1986), Specification for Intruder Alarm System for Consumer Installation

BS 6789, Part 3, Apparatus for Automatic Calling and Answering

BS 6799 (1986), Code of Practice for Wire Fire Intruder Alarm System

BS 6800 (1986), Specification for Home and Personal Security Devices

BS 7042 (1988), Specification for High-Security Intruder Alarm System in Buildings

BS 8220, Guide for Security of Buildings Against Crime

Part 1: Dwellings (1986)
Part 2: Offices and Shops (1987)
Part 3: Warehouses (1990)

British Standards for Architectural/Physical Security Materials

Doors

BS 4787, Internal and External Wood Door Sets, Door Leaves, and Frames

BS 1285, Specification for Wood Surrounds for Steel Windows and Doors

BS 1245, Specification for Metal Door Frames (steel)

Windows

BS 952, Glasses for Glazing (types)

BS 5050, Glazing

BS 5051 (1988), Glazing; Bullet Resistant Glazing

BS 5367 (57) (1985), Code of Practice for Installation of Security Glazing

BS 5516, Code of Practice for Patent Glazing

BS 5544, Specification for Antibandit Glazing

Locks

BS 3621 (1980), Specification for Resistant Locks

BS 5872, Locks and Latches for Doors in Buildings

BS 2088, Performance Tests for Locks

Loss Prevention Council (LPC)

British testing organization tests to levels above British Standard and certifies like UL.

LPS 1167, Specification for Ultrasonic Doppler Intruder Detectors (1989)

LPS 1168, Specification for Microwave Doppler Intruder Detector (1989)

LPS 1169, Specification for Passive Infrared Detection Services (1989)

International Standards

International Electrotechnical Commission (IEC)

IEC 839, Alarm Systems (1988)
IEC 839-1-1, General Requirements
IEC 839-1-2, Power Units

IEC 839-2-2, Requirements for Detectors
IEC 839-2-3, Requirements for Infrared Beam Interruption

Comité Européen des Assurances (CEA)
    German standards

Standards Association of Australia (SAA)
    AS2201.3, Intruder Alarm Systems (1985)

Internationally Agreed Quality Standards for Quality Assurance Management, Provision, and Installation

BS 5750 (Britain)
Euronorm (EN) 29000 (Europe and USA)
International Standards Organisation (ISO) 9000 series (worldwide)
ISO 9000 (Japan)

Appendix 3
Security Associations—United States and United Kingdom, Short Form List

For full information, refer to annual guides, such as the *Security Letter Source Book*.

Academy of Security Educators and Trainers (ASET), Chesapeake Beach, MD 20732

Airport Security Council (ASC), New York, NY 11003

American Bankers Association (ABA), Washington, DC 20036

American Federation of Police, North Miami, FL 33161

American Hardware Association, Schaumburg, IL 60173-4796

American Safe Deposit Association, The (TASDA), Greenwood, IN 46142

American Security Educators, Downey, CA 90240

American Society for Industrial Security (ASIS), Arlington, VA 22209, or Washington, DC 20006

American Society for Testing Materials (ASTM), Philadelphia, PA 19103

Associated Locksmiths of America, Dallas, TX 75204

Association for Loss Prevention & Security (ALPS), Portland, OR 97215

Carnahan Conferences on Security, Lexington, KY 40506-0046

Central Station Electrical Protection Association (CSEPA), Washington, DC 20036

Committee of National Security Companies, Inc. (CONSCO), Memphis, TN 38112

Communications Security Association (CSA), Frederick, MD 21701

Computer Security Institute, Northboro, MA 01532

Defense Institute of Security Assistance Management, OH 45433-5000

Defense Investigation Service: Industrial Security Branch, Washington, DC 20324

Door and Hardware Institute, McLean, VA 22102

Electronic Industries Association (EIA), Washington, DC 20006

Information Systems Security Association (ISSA), Newport Beach, CA 92658

Insurance Services Office (ISO), New York, NY 10038

International Association for Hospital Security (IAHS), Lombard, IL 60148

International Association for Shopping Center Security (IASCS), Atlanta, GA 30340

International Association of Campus Law Enforcement Administrators (IACLEA), Hartford, CT 06105

International Association of Chiefs of Police (IACP), Arlington, VA 22201

International Association of Professional Security Consultants (IAPSC), Daytona, FL 32725

International Association of Security Services (IASS), Northfield, IL 60093

International Biometric Association, Washington, DC 20036

International Fence Industry Association (IFIA), Stone Mountain, GA 30083

International Security Management Association (ISMA), Boston, MA 02110

International Society of Crime Prevention Practitioners (ISCPP), Alexandria, VA 22308

Jewelers Security Alliance of USA, New York, NY 10017

[The] Justice Department: Law Enforcement Assistance Agency (LEAA), Washington, D.C.

Meradcom, Fort Belvoir, VA

National Association of Chiefs of Police, Arlington, FL 33161

National Association of School Security Directors, West Palm Beach, FL 33416

National Association of Security Personnel (NASP), Ocala, FL 32670

National Bureau of Standards (now National Institute of Technology and Standards), Washington, DC 20234

National Burglar and Fire Alarm Association (NBFAA), Washington, DC 20036

National Cargo Security Council (NCSC), Washington, DC 20006

National Classification Management Society, Rockville, MD 20852

National Council of Investigation and Security Services (NCISS), Severna Park, MD 21146

National Retail Merchants Association (NRMA), New York, NY 10001

National Security Industry Association, Washington, DC 20036

Office of Science and Technology Policy of National Security Council, Washington, DC 20036

Sandia Laboratories, Albuquerque, NM

Security Equipment Industry Association (SEIA), Santa Monica, CA 90405

Security Industry Association (SIA), Santa Monica, CA 90405

Underwriters Laboratories, North Brook, IL 60062

US Department of the Army, The Pentagon, Washington, DC 20310 (Civilian Personnel Policy and Security Policy)

Advice and assistance may also be obtained from *state* organizations such as the

- State Alarm Association
- State Burglar and Fire Alarm Association
- State Licensing Authorities

GREAT BRITAIN—ORGANIZATIONS THAT PROVIDE INFORMATION ON SECURITY SYSTEMS AND PHYSICAL SECURITY

For full details, refer to annual guides such as the *Security and Fire Protection Yearbook (UK)*.

Association for the Prevention of Theft in Shops (APTS)

Association of British Security Officers (ABSO)

Association of Burglary Insurance Surveyors (ABIS)

Association of Chief Police Officers

British Computer Society (Security)

British Institute for Security Laws (BISL)

British Security Industry Association (BSIA)

Electrical Contractors' Association (Security Section) (ECA)

European Economic Community Association of Airport and Seaport Police (AASP)

Fire Protection Association (FPA)

International Maritime Bureau

International Professional Security Association (IPSA)

Loss Prevention Council (LPC)

Master Locksmiths' Association (MLA)

National Approved Council for Security Systems (NACOSS)

National Association of Security Services (NASS)

National Museums Security Advisor, London

National Supervisory Council for Intruder Alarms (NSCIA/LPC)

For security advice relating to disabled or handicapped people, contact

the Centre on Environment for the Handicapped, London

the Home Office

local police and crime prevention officers

Appendix 4
Metric Equivalents

LINEAR MEASURE

1 centimeter .... 0.3937 inch
1 inch .... 2.54 centimeters
1 decimeter .... 3.937 in .... 0.328 foot
1 foot .... 3.048 decimeters
1 meter .... 39.37 inches .... 1.0936 yds.
1 yard .... 0.9144 meter
1 dekameter .... 1.9884 rods
1 rod .... 0.5029 dekameter
1 kilometer .... 0.62137 mile
1 mile .... 1.6094 kilometers

SQUARE MEASURE

1 sq. centimeter .... 0.1550 sq. inch
1 sq. inch .... 6.452 sq. centimeters
1 sq. decimeter .... 0.1076 sq. foot
1 sq. foot .... 9.2903 sq. decimeters
1 sq. meter .... 1.196 sq. yards
1 sq. yard .... 0.8361 sq. meter
1 hectare .... 2.471 acres
1 acre .... 0.4047 hectare
1 sq. kilometer .... 0.386 sq. mile
1 sq. mile .... 2.59 sq. kilometers

MEASURE OF VOLUME

1 cu. centimeter .... 0.061 cu. inch
1 cu. inch .... 16.39 cu. centimeters
1 cu. decimeter .... 0.0353 cu. foot
1 cu. foot .... 28.317 cu. decimeters
1 cu. yard .... 0.7646 cu. meter
1 cu. meter .... 0.2759 cord
1 cord .... 3.625 steres
1 liter .... 0.908 qt. dry .... 1.0567 qts. liq.
1 quart dry .... 1.101 liters
1 quart liquid .... 0.9463 liter
1 dekaliter .... 2.6417 gals. .... 1.135 pks.
1 gallon .... 0.3785 dekaliter
1 peck .... 0.881 dekaliter
1 hektoliter .... 2.8378 bushels
1 bushel .... 0.3524 hektoliter

WEIGHTS

1 gram .... 0.03527 ounce
1 ounce .... 28.35 grams
1 kilogram .... 2.2046 pounds
1 pound .... 0.4536 kilogram
1 metric ton .... 0.98421 English ton
1 English ton .... 1.016 metric tons

Appendix 5
Data Sheets

DATA SHEET 1   Commonly used security luminaires for exterior use (IP54 or better)

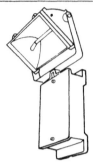

Large area floodlight A3 or A5
Usually building or wall mounted
above 15m
Uses linear double-ended lamps:
400, 600 or 1000W SON(T) or
1000W, 2000W MBF
Dimensions: 750mm wide
            650mm long
            1000mm deep
Distribution of light Z (wide
symmetrical fan in horizontal
plane, asymmetric in vertical
plane)
Weight: in excess of 30kg

Building façade floodlight B2
Offset wide angle building or wall
floodlight or short pole mounting
Uses lamps suitable for vertical
mounting sizes: 500W, 1000W
MBF; MBI lamps to 2000W, SON
250W, 400W, 1000W
Dimensions: 650mm wide
            400mm long
            400mm deep
Distribution of light: Y
symmetrical fan in horizontal and
vertical planes medium angle

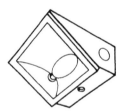

General purpose floodlight A1
For area lighting using:
150W, 250W, 400W SON,75W
90W SOX, 250W 500W MBF
Dimensions: 600mm wide
            600mm long
            300mm deep
Distribution of light Y or Z
(medium or wide fan) wall or pole
mounted
Weight: 22kg

General purpose floodlight A2
For area lighting using 300W,
500W, 750W or 1000W double-
ended linear tungsten halogen
lamps
Dimensions: 250mm–350mm wide
            250mm long
            150mm deep
Distribution of light Y or Y/Z
(medium fan)
Weight: 4kg

DATA SHEET 1   (continued)

Small area floodlight or perimeter
fence floodlight A3-1
Usually building, or 8m pole
mounted
Uses: 200W, 300W, 500W double-
ended linear tungsten halogen
lamps open wire guard or
polycarbonate cover
Dimensions: 190mm wide
190mm long
120mm deep
Distribution of light Y/Z
Weight: 0.75kg

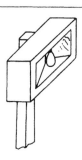

Perimeter lighting luminaire for
maximum glare generation
Used in 1.2m post at, say, 10m
intervals facing perimeter fence
and generating a disabling glare
Uses: 35W SON, 50W MBF lamps
Must be highly vandal resistant
Dimensions: 300mm wide
200mm long
350mm deep
Distribution of light: direct forward
no backward component. Medium
fan
Weight: 8.0kg

Perimeter fence or internal road
lighting K2-3
General street lighting style
luminaire for 5m–6m mounting
height on column or wall bracket
Includes photoelectric cell switch
Uses: 35W, 55W or 90W SOX
lamp
Dimensions: 160mm wide
450mm long
205mm deep
Distribution of light: ovoid
12×30m approx
Weight: 4.0kg

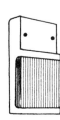

General purpose security bulkhead
luminaire
For wall-mounting Most efficient
type MINI SOX lamp. Spaced
18m apart at 4m above ground
Uses: 2×18W fluorescent, 18W
SOX or SOX or GLS lamps

Dimensions: 165mm wide
90mm long
290mm deep
Weight: 2.0kg

DATA SHEET 2 Recommended illuminance values in lux for building floodlighting*

Approx reflection	Typical materials	Condition	Low district brightness	Medium district brightness	High district brightness
0.8	White brick	Clean: Fairly clean: Fairly dirty:	15 20 45	25 35 75	40 60 120
0.6	Portland stone	Clean: Fairly clean: Fairly dirty:	20 35 65	35 55 110	60 90 180
0.4	Concrete	Clean: Fairly clean: Fairly dirty:	30 45 90	50 75 150	80 120 240
0.35	Middle stone	Clean: Fairly clean: Fairly dirty:	35 50 100	55 90 180	90 140 280
0.3	Dark stone	Clean: Fairly clean: Fairly dirty:	40 55 130	60 90 220	100 150 360
0.25	Yellow brick †	Clean: Fairly clean: Fairly dirty:	45 65 130	75 110 220	120 180 360
0.2	Red or Blue brick †	Clean: Fairly clean: Fairly dirty:	55 80 160	90 140 280	150 230 450

* Based on reflectance for white light. When using coloured sources on areas of similar colour, such as sodium lamps for yellow brick, plan for 70–50 per cent of recommended illuminance.
† Type of pointing and lightness of mortar have a major effect on the reflection factor for brickwork. The area of mortar can be as much as a seventh of the whole. The effect of light on strong coloured surfaces depends more on colour than illuminance.

(Courtesy of The Electricity Council)

Suggested floodlighting sources for facade materials and grounds

Material	Colour	Suggested light source
Concrete Portland stone	White:	Tungsten and TDH, SON, MBF, MBI
Brick	Red: Yellow: Blue grey:	Tungsten, THD, SON, red fluorescent Tungsten, THD, SON, SOX gold fluorescent THD, MB, MBF, MBI
Red sandstone and ragstone	Red-brown:	Tungsten, TDH, SON, gold fluorescent
Grass and foliage	Green:	MB, MBF, MBI, green fluorescent

Floodlighting should not attempt to duplicate the daytime appearance of the building or object since the main direction of light is usually reversed. The best decorative floodlighting installations are those which exploit these differences rather than try to minimise them.

DATA SHEET 3 Common lamp types for security lighting applications

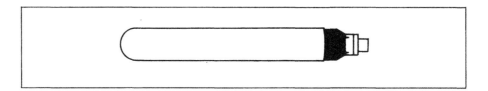

SOX (SLI) low-pressure sodium lamps

Wattage range:	10–180; Lumen Range 1000 to 33,000
Efficacy:	67–145 Lm/W; SOX E 165 Lm/W
Life:	Circa 8000 hours Linear Lamp (18,000 "U tube")
Restrike	Igniter circuits instantaneous while lamp is hot
Color:	Yellow (only)

These lamps produce the most light for a given input and are very economical in use. They are not recommended where the best visibility is required at extreme range. 18W good for exterior bulkheads.

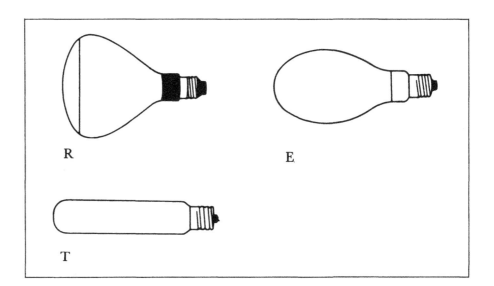

SON high-pressure sodium lamps (HPS)

Wattage range:	35–1000; Lumen Range 2000 to 126,000
Efficacy:	55–126 Lm/W
Life:	24,000 hours
Restrike	Igniter circuits usually within 1 minute Others 5 minutes or longer to full output
Color:	Pink/orange

A general purpose lamp with economical running costs. Best lamp for large area or floodlighting schemes. Lamps are either reflector (R), elliptical (E), or tubular (T).
SON "Delux" and "White" SON give enhanced color rendering.

DATA SHEET 3 (continued)

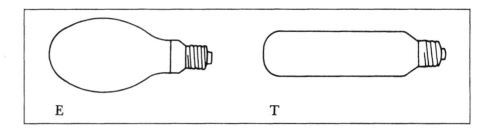

MBI metal halide lamps (clear glass); MBIF (Elipse E coated glass)

Wattage range:	75–3500; Lumen Range 5000 to 300,000
Efficacy:	50–85 Lm/W
Life:	7500–20,000 hours depending on operating conditions
Restrike	Igniter circuits usually within 1 minute Others 4–5 minutes to full output
Color:	Blue/white

For use where color rendering is of prime importance, e.g., color CCTV. Lamps are either elliptical (E) or tubular (T).

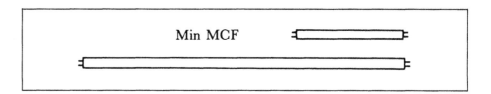

MCF/U tubular fluorescent lamps (miniature/U tube, etc.)

Wattage range:	8–125; Lumen Range 1200 to 16,000
Efficacy:	20–75 Lm/W (high efficacy colors)
Life:	1.2 m and longer 9000 hours; others 5000 hours
Restrike:	Instantaneous

A general purpose lamp for short-range use. Very cold weather can significantly reduce light output. Small external bulkheads or general internal. First cost greater than tungsten halogen but more economical.

DATA SHEET 3 (continued)

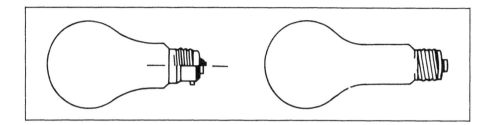

GLS lamps (general service lamp: tungsten "bulb")

Wattage range:	8–2000
Efficacy:	8–23 Lm/W
Life:	1000 hours (some ratings are available in 2000-hour or longer versions)
Restrike	Instantaneous
Color:	Any

Smaller wattages useful for infill purposes but short life and low efficacy make them uneconomical for large-scale use. Superseded by SL/PL/2D and L.Volt TH. Do not use.

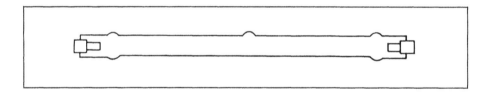

TH tungsten halogen lamps

Wattage range:	200–2000; Lumen Range 3400 to 36,000
Efficacy:	16–22 Lm/W
Life:	2000 hours
Restrike	Instantaneous
Color:	White

A floodlighting lamp for short- to medium-range distances. Its high energy consumption makes it uneconomical for use on a large scale.

Mainly perimeter lighting where restrike required.

Extra low voltage, single-ended types (6V/50W) are now replacing all PAR 38/GLS lamp applications (hand-held or projected spot-beams of 200 lux, 2 m diameter at distance of 3 m).

DATA SHEET 3 (continued)

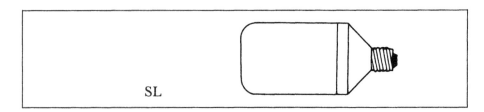

SL

2D or SL miniature compact fluorescent lamps

Wattage range:	8, 13, 18, 25
Efficacy:	41–50 Lm/W
Life:	10,000 hours plus
Restrike	Instantaneous
Color:	White

Excellent interior or exterior bulkhead luminaire lamp for patrolled areas. Replaces GLS. See also PL range.

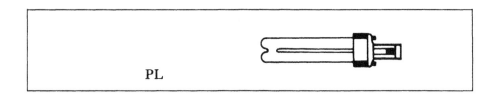

PL

PL miniature fluorescent lamps (2L, 4L, etc.)

Wattage range:	7, 8, 11, up to 60
Efficacy:	59–78 Lm/W
Life:	10,000 hours plus
Restrike	Instantaneous
Color:	White

More compact, more efficient version of SL and 2D.

E   E

MBF color-corrected mercury discharge lamps

Wattage range:	40–1000; Lumen Range 910 to 301,000
Efficacy:	30–55 Lm/W
Life:	16,000–24,000 hours
Restrike	Five minutes or longer to full output
Color:	Blue/white

A cheaper lamp than MBI but now largely superseded for security purposes by SON or MBIF.

DATA SHEET 4 Circuit efficacy in lumens/watt (L/W) and lighting design (LDL) of principal lamp types and ratings

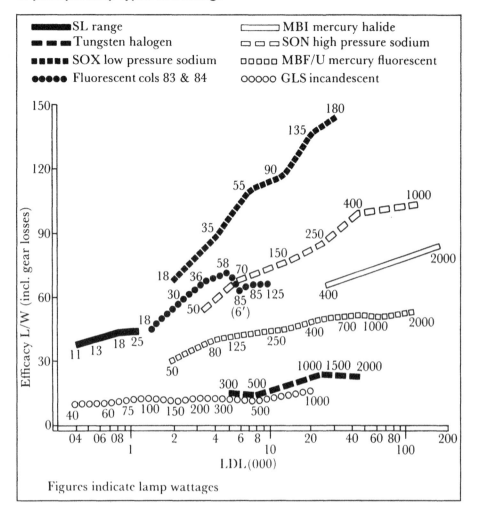

Figures indicate lamp wattages

DATA SHEET 5   Exterior area lighting calculations

To establish the number of luminaires necessary to provide a given luminance the following approximate method may be used:

Number of luminaires =

$$\frac{\text{Area (m}^2\text{)} \times \text{illuminance required (lux)}}{\text{Beam factor of luminaires} \times \text{lamp output (lumens)} \times \text{maintenance factor}}$$

(Beam factor × waste light factor = utilization factor commonly used in interior design.)

The beam factor and lamp output should be obtained from the relevant manufacturers. The maintenance factor will depend upon the location of the installation and the frequency of cleaning and guidance should be sought from the manufacturer.

As a first approximation, assuming a combined beam factor and maintenance factor of 0.25 will suffice.

The equation can be rewritten to find the average illuminance from a given number of luminaires as follows:

Average illuminance (lux) =

$$\frac{\text{Number of luminaires} \times \text{beam factor} \times \text{maintenance factor} \times \text{lamp output}}{\text{Area (m}^2\text{)}}$$

DATA SHEET 5 (continued)

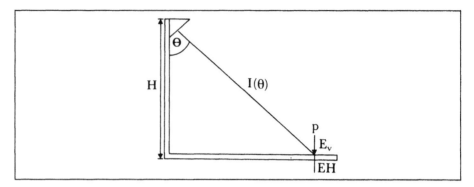

Where it is necessary to find the illuminance at a given point, a more complicated method is used:

To find the horizontal illuminance (EH) at the point P

$$EH = \frac{I(\Theta) \, Cos^3(\Theta)}{H^2}$$

where: H = mounting height (m)

I (Θ) = the intensity from the luminaire at the angle (Θ) (candelas)

(Θ) = angle between the downward vertical and a line drawn from the luminaire to the point of measurement.

Again an allowance for maintenance should be made and also, where extreme distances are involved, an allowance for atmosphere absorption.

Where the illuminance on a vertical plane (Ev) facing the luminaire is required:

$$Ev = \frac{I(\Theta) \, Cos^2(\Theta) \, sin(\Theta)}{H^2}$$

The following table shows the illuminances achieved for every 1000 candelas from the luminaires, for mounting heights of 5 m, 8 m, and 10 m, horizontal intervals up to 60 m from the mounting position.

Unmaintained illuminance per 1,000 candelas (lux) Distance (m)	Ev Lux	EH Lux
Mounting height = 5m		
10	7.15	3.58
20	2.28	0.57
30	1.07	0.18
40	0.61	0.08
50	0.39	0.04
60	0.27	0.02
Mounting height = 8m		
10	4.76	3.81
20	2.0	0.8
30	1.0	0.27
40	0.59	0.12
50	0.38	0.06
60	0.27	0.04
Mounting height = 10m		
10	3.54	3.54
20	1.79	0.89
30	0.95	0.32
40	0.57	0.14
50	0.38	0.08
60	0.27	0.04

More detailed calculations should be based on manufacturer's data and information of the type given in Data Sheets, 8, 9, and 10.

(Courtesy of The Electricity Council)

DATA SHEET 6 Criteria for perimeter lighting

While perimeter lighting is only one of a number of security lighting techniques it has been the subject of some comprehensive research and as a result it is possible to specify the lighting conditions with some precision.

Where a guard must survey the area outside the perimeter fence it is important that the fence itself should not be a distraction. The ratio of fence to field illuminance should be as follows:

Maximum fence/field illuminance ratios	
Fence and field conditions	Maximum allowable illuminance ratio fence/field
Black plastic wire mesh fence and a grass field	10:1
Green plastic wire mesh fence and a grass field	2:1
Galvanised wire mesh fence and a grass field	1:1
Note: the same ratios apply to a tarmac search area but for a concrete search area these maxima may be doubled.	

Full-scale trials also reveal that detection performance of the guard was enhanced if he moved continually, as this helped him to resist the temptation to focus on one section of the fence in preference to the whole field.

Other criteria that have been established are those necessary to reveal the intruder either while he is still beyond the fence or "alternatively" while he is between the fence and the line of luminaires. These may be summarized as follows:

Criteria for security lighting of fences		
Purpose	Criterion	Position
To reveal intruders outside fence	(a) Vertical illuminance of 1.0 lux minimum	Illuminance 0.3m above the ground on a vertical plain parallel to the fence at the distance where first detection is required.
	(b) Vertical illuminance uniformity ratio not to exceed 5:1	Vertical illuminance as above, along the line where first detection is required.
	(c) Vertical illuminance ratio for fence and field not to exceed 1:1 for a black fence and grass field (see previous table for other conditions)	Vertical illuminance to be measured on the inside of fence at eye-level and as above in field.
To reveal intruders between fence and line of luminaires	(a) Minimum horizontal illuminance of 3 Lux on average	Horizontal illuminance measured at ground level.
	(b) Horizontal illuminance uniformity ratio not to exceed 5:1	Horizontal illuminance measured at ground level.

(Courtesy of The Electricity Council)

DATA SHEET 7   Comparison of security lighting equipment for perimeter fence lighting schemes (see Data Sheet 8 for comments)

Lamp/luminaires: scheme	Mounting height (h)	Spacing (s)	Distance back from fence	Average illuminance on fence face 2m above ground	Average illuminance in fence zone 4m wide	Glare effect to intruder	Evenness of illuminance in sterile strip	Visibility of intruders approaching fence
	(m)	(m)	(m)	(lux)	(lux)	(See notes for explanation)		
A 500W tungsten halogen lamps in asymmetric floodlights aimed at 15m from base.	3	10	10	18*	15*	3	5	3
A1 As above aimed at 15m from base	2	10	10	50	40*	2.5	3	3
As above aimed at 6m from base	3	10	10	40	16*	2	3	4
A2* 2 × 500W tungsten halogen lamps each in asymmetric floodlights diverging 90° and aimed at 15m from base	3	20 (between clusters)	10	25*	10*	2	4	3.5
B* 2 × 20W fluorescent tubes in diffusing luminaire mounted horizontally and aimed straight ahead	1.2	6	10	3*	2*	4	3	2

Lamp luminaires: scheme	Mounting height (h)	Spacing (s)	Distance back from fence	Average illuminance on fence face 2m above ground	Average illuminance in fence zone 4m wide	Glare effect to intruder	Evenness of illuminance in sterile strip	Visibility of intruders approaching fence
	(m)	(m)	(m)	(lux)	(lux)	(See notes for explanation)		
B1 2 × 20W fluorescent tubes in open luminaire mounted horizontally and aimed straight ahead	1.2	6	10	9	8	3.5	5	2
C 50W MBF lamps in bulkhead units giving no sideways or backwards light	1.5	6	10	9	8	5	5	2
C1 As above	2.7	6	10	9	8	3	5	4
C2 As above	1.5	18	10	3.5*	6*	4	3	3
E 400W SON/TD lamps in projector	8	20	10	3.5*	6*	4	3	3
E1 250W SON/TD lamps in projector	5	10	10	68	48	5	5	5
E2 150W SON lamp in projector	4	6	10	88	24	4.5	4	2.5

DATA SHEET 7 (continued)

Lamp luminaires: scheme		Mounting height (h)	Spacing (s)	Distance back from fence	Average illuminance on fence face 2m above ground	Average illuminance in fence zone 4m wide	Glare effect to intruder	Evenness of illuminance in sterile strip	Visibility of intruders approaching fence
		(m)	(m)	(m)	(lux)	(lux)	(See notes for explanation)		
E4	70W SON lamp in flat front bulkhead	1.5	6	10	14	14	4	5	2.5
E5	90W SX lamp projector	4	6	10	53	34	NA	NA	NA

Evaluation of systems of security lighting

A number of systems have been studied in practical tests to establish their effectiveness. An arbitrary and subjective scale of assessment was used ranging from 5 (excellent) to 1 (poor), and a panel of observers, ranging in age from twenty-five to sixty-six years, had their results averaged to provide the values given in the above table.

Notes
(1) All the results were obtained using a fence 3m high constructed of heavy-gauge wire mesh covered in black PVC. If galvanised mesh is used, the visibility assessments must be reduced by at least one step.
(2) All the results were obtained under conditions of low district brightness. If the district brightness is significant, the assessments for glare effect on intruder and concealment of defenders by glare from lights must be reduced by at least one step.
(3) Assessments are made on a 5-point scale: 5 – excellent, 4 – superior, 3 – satisfactory, 2 – moderate, 1 – poor. The assessments given are the consensus of experienced observers.
(4) The visibility of an intruder approaching the fences is that when a dark clothed intruder is 30m from the fence. The visibility of an intruder actually within the fence zone was assessed as 5 in every case.
(5) All the systems described used luminaires spaced 10m back from the fence. If the luminaires are only 6m back from the fence, all assessments of visibility of intruders approaching the fence must be reduced by at least one step because the brightness of the fence mesh (even of black PVC covered wire) masks the defenders' ability to see out.
(6) The illuminances are vertical illuminances, measured with the photocell facing towards the lights in each case. Those in the fence zone and outer area had the cell 300mm above ground level.

(Courtesy of The Electricity Council)

DATA SHEET 8   Lighting systems based on Electricity Council recommendations
Lighting system type "A"

Type of luminaire: Open-fronted asymmetric floodlight
Lamp: 500W tungsten halogen (TH)
Rated life of lamp: 2000 hours
*Luminous efficacy: 19 Lm/W

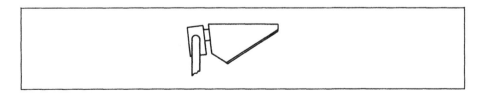

DATA SHEET 8 (continued)

Arrangement of luminaires

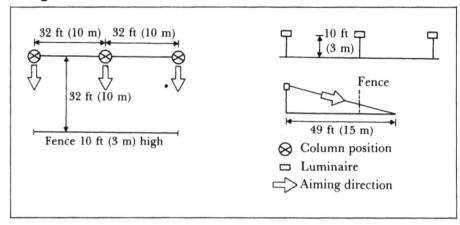

Illuminance readings

Average vertical illuminance on fence 6 ft (2 m) above ground: 18 lux
Average vertical illuminance within strip 6 ft (2 m) wide on both sides of the fence
1 ft (0.3 m) above the ground: 15 lux

Distance from fence (ft)	16	32	49	64	82	98	131	164	196
Distance from fence (m)	5	10	15	20	25	30	40	50	60
Vertical illuminace facing toward line of luminaires at height of 1 ft (0.3 m) above gound (lux)	27	19	15	9	9	7	4	3	2

Advantages and disadvantages of using type "A" system

The glare effect that would be experienced by intruders and the revealment of intruders approaching were satisfactory. The concealment of defenders behind the luminaires and the uniformity of illuminance in the strip between the luminaires and the fence were very good. This lighting system has the advantage that if the lamps are extinguished due to an interruption in supply, full light output is obtained as soon as the supply is resumed. Tungsten halogen lamps have a relatively short life and relatively low luminous efficacy, and their running and maintenance costs are comparatively high.

Lighting system type "B"

Type of luminaire: Weatherproof diffusing (twin lamp)
Lamp: 20W tubular fluorescent, white (600 mm, MCFU)
Rated life of lamp: 7500 hours
*Luminous efficacy: 40 Lm/W (approximately)

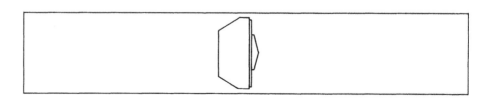

DATA SHEET 8 (continued)

Arrangement of luminaires

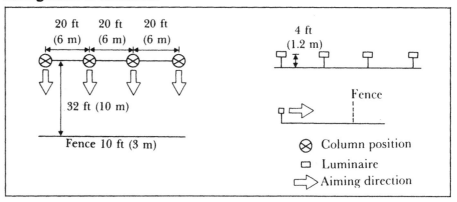

Illuminance readings

Average vertical illuminance on fence 6 ft (2 m) above ground: 3 lux
Average vertical illuminance within strip 6 ft (2 m) wide on both sides of the fence
1 ft (0.3 m) above the ground: 2 lux

Distance from fence (ft)	16	32	49	64	82	98
Distance from fence (m)	5	10	15	20	25	30
Vertical illuminace facing toward line of luminaires at height of 1 ft (0.3 m) above gound (lux)	5	3.5	2	1.5	1.2	0.8

Advantages and disadvantages of using type "B" system

The glare effect that would be experienced by intruders and the concealment of defenders behind the luminaires were very good. The uniformity of illuminance in the strip between the luminaires and the fence was satisfactory but the revealment of intruders approaching the fence was poor. There are no restrike period problems, if there is a temporary interruption in supply. Due to long lamp life and high luminous efficacy, running and maintenance costs are relatively low.

Lighting system type "C"

Type of luminaire: Bulkhead
Lamp: 50W mercury discharge (MBF)
Rated life of lamp: 7500 hours
*Luminous efficacy: 30 Lm/W (approximately)

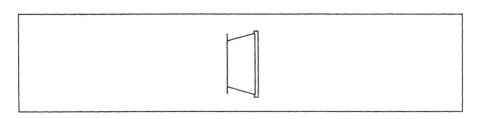

DATA SHEET 8 (continued)

Arrangement of luminaires

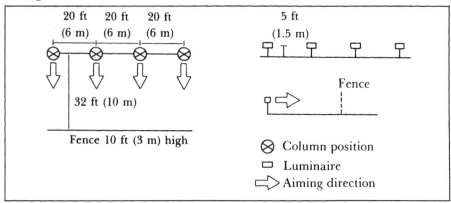

Illuminance readings

Average vertical illuminance on fence 6 ft (2 m) above ground: 9 lux
Average vertical illuminance within strip 6 ft (2 m) wide on both sides of the fence 1 ft (0.3 m) above the ground: 10 lux

Distance from fence (ft)	16	32	49	64	82	98	131
Distance from fence (m)	5	10	15	20	25	30	40
Vertical illuminace facing toward line of luminaires at height of 1 ft (0.3 m) above gound (lux)	4	3	2	1.8	1	0.9	0.6

Advantages and disadvantages of using "C" system

The glare effect that would be experienced by intruders, the concealment of the defenders behind the luminaires, and the uniformity of illuminance in the strip between the luminaires and the fence were very good. The revealment of intruders approaching the fence, however, was poor. A great disadvantage is that, should there be a temporary interruption in supply, a period of about 5 minutes is required before the lamp will restrike. If this is unacceptable, it would be necessary to provide very expensive "no break" equipment. Due to long lamp life and high luminous efficacy, running and maintenance costs are relatively low.

Lighting system type "D"

Type of luminaire: Weatherproof asymmetric floodlight
Lamp: 1500W tungsten halogen (TH)
Rated life of lamp: 2000 hours
*Luminous efficacy: 21 Lm/W (approximately)

DATA SHEET 8 (continued)

Arrangement of luminaires

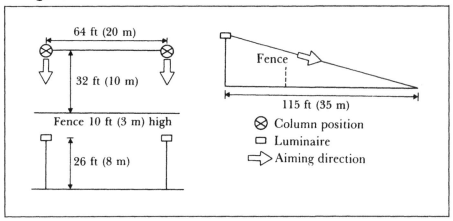

Illuminance readings

Average vertical illuminance on fence 6 ft (2 m) above ground: 44 lux
Average vertical illuminance within strip 6 ft (2 m) wide on both sides of the fence
1 ft (0.3 m) above the ground: 10 lux

Distance from fence (ft)	16	32	49	64	82	98	131	164	196	295
Distance from fence (m)	5	10	15	20	25	30	40	50	60	90
Vertical illuminace facing toward line of luminaires at height of 1 ft (0.3 m) above gound (lux)	18	32	49	48	50	46	34	27	20	10

Advantages and disadvantages of using type "D" system

The glare effect that would be experienced by intruders, the revealment of intruders approaching the fence, and the concealment of defenders behind the luminaires were all very good. The uniformity of illuminance in the strip between the luminaires and the fence was poor. The lighting system has the advantage that if the lamps are extinguished due to an interruption in supply, full light output is obtained as soon as the supply is resumed. Tungsten halogen lamps have a relatively short life and relatively low luminous efficacy, and their running and maintenance costs are comparatively high.

Lighting system type "E"

Type of luminaire: Asymmetric floodlight
Lamp: 400W high-pressure sodium (SON)
Rated life of lamp: 6000 hours
*Luminous efficacy: 90 Lm/W (approximately)

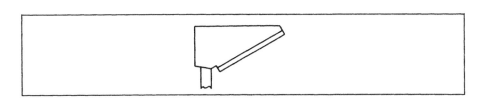

DATA SHEET 8 (continued)

Arrangement of luminaires

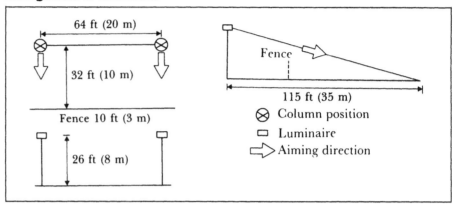

Illuminance readings

Average vertical illuminance on fence 2 m above ground: 67 lux
Average vertical illuminance within strip 2 m wide on both sides of the fence
0.3 m above the ground: 35 lux

Distance from fence (ft)	16	49	82	131	196
Distance from fence (m)	5	15	25	40	60
Vertical illuminace facing toward line of luminaires at height of 1 ft (0.3 m) above gound (lux)	42	32	20	10	8

Advantages and disadvantages of using type "E" system

The revealment of intruders approaching the fence and the uniformity of illuminance in the strip between the luminaires and the fence were very good. The glare effect that would be experienced by intruders and the concealment of defenders behind the luminaires, are however, very poor. If there is a temporary interruption in supply, lamps will restrike in under a minute. If this is unacceptable, it would be necessary to provide very expensive "no-break" equipment. Due to long lamp life and high luminous efficacy, running and maintenance costs are relatively low.

Relative percentage costs (excluding stand-by plant)			
Type of lighting system	Installation costs %	Running costs %	Maintenance costs %
A	100	100	100
B	135	15	15
C	115	15	30
D	115	150	75
E	140	40	35

(Courtesy of The Electric Council)

DATA SHEET 9   Recommended luminaire mounting height for floodlighting schemes

Lamp: type	Normal rating W	Through life: lighting design lumens	Recommended minimum mounting height (h) m
Tungsten: general lighting service GLS	200	2,720	4.5
	300	4,300	6.0
	500	7,700	7.5
	750	12,400	9.0
	1,000	17,300	9.0
	1,500	27,000	12.0
Tungsten halogen TH	300	5,100	6.0
	500	8,500	7.5
	1,000	22,000	9.0
	1,500	33,000	12.0
	2,000	44,000	15.0
High-pressure mercury MBF	125	5,800	6.0
	250	12,500	7.5
	400	21,300	9.0
	700	36,500	12.0
	1,000	58,000	15.0
	2,000	110,000	18.0
Mercury-tungsten MBT, MBTF	160	2,560	4.5
	250	4,850	6.0
	500	12,300	9.0
High-pressure sodium (HPS) SON	150	15,000	12.0
	250	25,500	12.0
	310	32,000	12.0
	360	36,500	12.0
	400	45,000	15.0
	600	62,000	15.0
	1,000	110,000	15.0

(Courtesy of The Electricity Council)

DATA SHEET 10   Typical utilization factor for area floodlights

Type of asymmetric flood-lights for use with horizontally mounted lamp, and typical lamp range	Ratio of area depth to mounting height		
	2:1	3:1	4:1
Open tungsten halogen 300–1000 W T/H	0.5–0.6	0.5–0.6	0.5–0.6
Enclosed project, narrow beam, specular reflector 300–2000 W T/H 240–400 W SON, MBF, MBIF	–	0.5	0.5
Enclosed projector, wide beam, diffused reflector 300–2000 W T/H 240–400 W SON, MBF, MBI	0.3–0.4	0.3–0.4	–

(Courtesy of The Electricity Council and The Chartered Institution of Building Services Engineers)

DATA SHEET 11   Spacing of floodlights relative to mounting height (h)

Type of luminaire	Maximum spacing	Limit of throw	Maximum end spacing
Wide angle	3 h	3 h	h
Medium angle	1½ h	4 h	⅔ h

(Courtesy of The Electric Council and The Chartered Institution of Building Services Engineers)

Index